AF334958

Fundamentals of Neurodegeneration and Protein Misfolding Disorders

BIOLOGICAL AND MEDICAL PHYSICS, BIOMEDICAL ENGINEERING

More information about this series at http://www.springer.com/series/3740

Martin Beckerman

Fundamentals of Neurodegeneration and Protein Misfolding Disorders

 Springer

Martin Beckerman
Asheville, NC, USA

ISSN 1618-7210   ISSN 2197-5647   (electronic)
Biological and Medical Physics, Biomedical Engineering
ISBN 978-3-319-22116-8   ISBN 978-3-319-22117-5   (eBook)
DOI 10.1007/978-3-319-22117-5

Library of Congress Control Number: 2015950467

Springer Cham Heidelberg New York Dordrecht London

Printed on acid-free paper

Springer International Publishing AG Switzerland is part of Springer Science+Business Media
(www.springer.com)

Preface

The dawn of the twentieth century was a remarkable time. During that brief period, Max Planck and Albert Einstein launched modern physics. J.J. Thomson and Ernest Rutherford, who rest near Isaac Newton in Westminster Abbey, had discovered the electron and the atomic nucleus, and Niels Bohr and Erwin Schrödinger, the Braggs, father and son, and many others too numerous to mention soon followed with their pioneering work. This creative explosion was not limited to physics. It was also a time for great discoveries in genetics and neuroscience. During that time, Thomas Hunt Morgan affirmed the centrality of the gene and Santiago Ramón y Cajal established the primacy of the neuron.

This age of discovery did not end there. Instead, it laid the foundation for another set of major events—the emergence of modern molecular biology brought on by the discovery of the alpha helix and beta sheet, the DNA double helix, the three-dimensional structure of proteins, and the emergence of the fundamental dogma of biology: DNA → RNA → protein. These discoveries were made possible in large measure by the Braggs and their colleagues who, in the 1930s, 1940s, and 1950s, transformed X-ray crystallography into a tool for exploring molecules of biological significance. The events, times, and places surrounding the discoveries by Pauling, Corey and Branson, Crick and Watson, and Perutz and Kendrew are beautifully described by Judson in *The Eighth Day of Creation*.

Today, research into the scientific disciplines is funded by a number of governmental bodies, chief of which in the United States are the National Institutes of Health, the National Science Foundation, and the U.S. Department of Energy through its Office of Science. It was not always so and their existence today is due in large measure to Dr. Vannevar Bush, head of the U.S. Office of Scientific Research and Development during WWII. He had shown that a partnership between government, university, and industry could successfully produce results in the national interest that would otherwise be impossible to achieve. He articulated this observation and recommended continuation of this partnership in a landmark letter to President Franklin D. Roosevelt, dated July 25, 1945. Entitled *Science—The Endless Frontier*, this document clearly articulated the arguments for creation of the aforementioned agencies and their accompanying laboratories.

I entered the field of nuclear physics in 1971 working as a postdoctoral fellow at the Weizmann Institute of Science. My area of interest at that time was the sub-barrier fusion of heavy-ions, the term "heavy-ions" referring to atomic nuclei heavier than helium. The motivation for studies such as these was twofold. First, there was the goal to better understand the great cosmic edifice known as stellar nucleosynthesis in which the elements are synthesized in phases in stars and inter-stellar gasses, so that the Earth and all of us within it are literally created out of stardust. The second goal was to explore ways of reaching the fabled superheavy "island of stability" whose existence was first posited by Glenn Seaborg in the 1960s, and was predicted by nuclear theory to be located near or about nuclei with 194 neutrons and from 114 to 126 protons. The challenge in both problem classes is keeping the energy as low as possible, hence the focus on sub-barrier (i.e., sub-Coulomb barrier) energies. That effort on my part continued through my time in the Laboratory for Nuclear Science at MIT in the late 1970s and early 1980s.

By that time many people who had started out, like me, in nuclear physics and other branches of physics, chemistry, biology, engineering, and their subfields had joined together to create new disciplines with names ranging from cybernetics, bioengineering, and medical physics to biophysics, neuroscience, and complex systems science. These new fields were supported by the aforementioned agencies and their counterparts in the UK, on the Continent, and elsewhere across the globe. Continuing to this day, this support has enabled researchers in these fields to produce a constant stream of scientific insights resulting in practical benefits to the public in accordance with the charges put down by Bush in his letter to President Roosevelt.

There is perhaps no better proof of the value of these endeavors than the creation of the miracle drugs. The notion of a "magic bullet" aimed at a specific disease agent is due to Paul Ehrlich. He had created the first of these, arsphenamine (Salvarsan), a drug for the treatment of syphilis, in 1910. The most famous of the miracle drugs is, of course, penicillin, discovered by Alexander Fleming in 1928. In the years following that discovery it became apparent that penicillin was difficult to mass produce. At that point Vannevar Bush and his Office of Scientific Research and Development stepped in and sponsored the development of a method of mass production. That support enabled the development of the deep-tank fermentation method by Pfizer, at that time a small company located in Brooklyn, resulting in mass production of penicillin and shortly thereafter of streptomycin. That drug had been discovered earlier, in 1943, by Selman Waksman (who coined the term antibi-otic) at Rutgers University in a project sponsored by Merck and Co.

With no claim to originality, I followed this well-trod path into the exciting new fields. My entry point was not so much my earlier experimental work but rather my modeling and simulation background, in particular, my experience with Monte Carlo simulation methods beginning with the Metropolis algorithm, its siblings, and its descendants, both stochastic and deterministic. That method was first introduced in a 1953 paper authored by Nicholas Metropolis and two husband-and-wife teams, Arianna and Marshall Rosenbluth, and Augusta and Edward Teller. The core idea, the Monte Carlo method, had been proposed earlier by Enrico Fermi, and, more formally, by Stan Ulam and John von Neumann in 1947.

That path was, of course, not the only one taken by researchers into the new disciplines. These endeavors are intrinsically multidisciplinary. The researchers brought with them their unique expertise and skills in their field of interest whether it be mechanical, experimental, or computational. Those skill sets functioned synergistically to enable the research groups to successfully attack the selected scientific problem. It is this character that I have attempted to capture in this textbook.

Neurodegenerative disorders, or dementias, are not new and have been the subject of discussion since ancient times. That these disorders may have a physical even treatable origin is, however, a relatively new concept. The groundwork for its emergence was laid in the mid to late nineteenth century with the birth of pathology, in which for the first time the light microscope was used to carry out detailed postmortem examinations of body tissues taken from deceased individuals suffering from various illnesses.

Chapter 1 begins with Alois Alzheimer and his now-legendary presentation given on 4 Nov 1906. His patient, Auguste Deter, a 51-year-old woman had been suffering from what is now known as early-onset Alzheimer's disease. In his talk, Alzheimer described the anomalous extra- and intracellular deposits he had found at her autopsy. Amyloid deposits were first observed by Rudolf Virchow, the founder of modern pathology, in the mid-1800s. Alzheimer's key observation, one that pointed to a physical cause for his patient's illness, was the presence of similar deposits in her brain tissue. The chapter covers the subsequent analyses of these deposits, first using stains that make visible features of interest, and then using electron microscopy and X-ray crystallography. The chapter next discusses similar findings in other forms of neurodegeneration and the overarching presence of misfolded proteins in all of these deposits.

Chapters 2 and 3 cover protein folding basics. Chapter 2 begins with the development of the field from the pioneering chemical denaturation studies of protein folding and unfolding; to the discovery of the alpha helix and beta sheet, to DNA's primacy, and to how a protein folds into its three-dimensional operational (functional) form discovered using X-ray crystallography and nuclear magnetic resonance. The first of these, the chemical denaturation studies, enabled Christian Anfinsen to formulate his thermodynamic hypothesis. The later studies and the realization that proteins are dynamic entities continually sampling an ensemble of conformations form the conceptual basis of the energy landscape picture presented in Chap. 3. Those discussions are accompanied by overviews of the key experimental and computational methods that are the exploratory tools of the discipline.

Chapter 4 begins with an examination of the insoluble amyloid fibrils that have been the subject of so much attention over the years. It emerges that the underlying phenomenon is far more complex and interesting, and challenges our understanding of protein dynamics. A variety of aggregates are generated, ranging from small soluble oligomers up to large insoluble aggregates, amyloidogenic and amorphous. The shift in attention from large amyloids to the smaller soluble oligomers that is occurring today is examined in this chapter along with an initial look at a class of proteins referred to as "intrinsically disordered," which have a prominent role in neurodegeneration.

It is believed that the strong age dependence seen in neurodegenerative disorders arises from a progressive decline over time in protein quality control. Ordinarily, misfolded and incorrectly processed proteins are either repaired or removed rapidly. Failure of the protein quality control system to do so leads to the inclusions seen in the diseases. The protein quality control system is an extensive one. It consists of molecular chaperones, the ubiquitin-proteasome system, the autophagic-lysosomal pathway, and quality control elements connected to the endoplasmic reticulum/ Golgi apparatus where proteins are synthesized, finished, and shipped to their cellular destinations. Beginning with the molecular chaperones, Chaps. 5 and 6 examine each of these systems.

The notion of a prion, a proteinaceous infectious particle, is an astonishing one. Chapter 7 examines the evidence for the existence of prions, proteins which by themselves without the involvement of any nucleic acids can produce and transmit a disease. The biophysical process that makes this possible is called "templated conformational conversion." In this process, a prion seed interacts with and converts a normal non-disease-causing form into one that is structurally more like itself and competent to convert others in the same way. The prion phenomenon is now fairly well established and is forcing a reassessment of many of the assumptions in the field on how neurological diseases progress from one region to another in the brain.

Chapter 8 is devoted to Alzheimer's disease. The amyloid hypothesis of neurodegenerative disease was posited with Alzheimer's disease in mind. Since its original formulation that theory has undergone a revision that redirects its focus to the small soluble oligomers believed to be more toxic than the fibrillary aggregates. A consistent observation is that Alzheimer's disease and, similarly, many of the other neurological disorders arise from an imbalance between production and clearance. Another important observation is the co-occurrence of inflammation in Alzheimer's disease and other neurological disorders. That amyloidogenic deposits and inflammation are somehow interconnected was noted by Virchow over 150 years ago. Chapter 8 examines these interlocking aspects and important risk factors such as the ApoE4 allele.

Proteins are the workhorses of the cell and when they do not attain their proper functional form diseases result. Neurons because of their exceptional metabolic requirements and elongated morphology are especially vulnerable to shortcomings in cellular logistics, the movement of matériel from one locale to another. Chapter 9 examines how these aspects come into play in Parkinson's disease. In this disorder, an impressive number of risk factors have been uncovered that implicate failings in the protein quality control system. A major objective in the field has been to identify the specific toxic effects and understand how they trigger the beginning stages of disease. Chapters 8 and 9 present the major findings on this subject. Both discuss how mutations, aberrant proteolytic processing, errant post-translation modifications, and damaging environmental toxins produce dysfunctional synapses, impair logistics, damage mitochondria, and cause neuronal death.

The completed sequencing of the human genome in 2001 was a monumental accomplishment. It is now emerging that the human genome possesses far more complexity than previously thought. For example, unstable repeat expansions are present in DNA that can increase in length from generation to generation and in doing so cause disease. There are noncoding regions of DNA that under certain circumstances can encode. And the so-called junk DNA may not be junk after all. Chapter 10 introduces the unstable repeat disorders. In some diseases such as Huntington's disease, unstable repeats are situated in protein-coding regions. In others, for example, in fragile X syndrome and several types of mental retardation, expanded repeats are located in non-protein-coding regions. These repeats generate unusual secondary structures that, in turn, cause errors in replication, repair, recombination, and transcription.

Chapter 11 discusses amyotrophic lateral sclerosis (ALS) and frontotemporal lobar degeneration (FTLD). It is only in the last few years that the predominant factors responsible for familial forms of these diseases have been uncovered. Significantly, these genetic factors are involved in multiple stages of RNA processing. Unstable repeats are again encountered in these disorders along with toxic peptides and RNA molecules. Chapter 11 ends with an overview of another critical family of disorders, the tauopathies, which includes not only ALS/FTLD but also chronic traumatic encephalopathy brought on by repeated mild trauma to the head.

In his *Experimental Medicine* of more than 160 years ago, Claude Bernard discussed the transformation of experimental physiology into a science. I have tried to show in this text that his vision has been realized, and furthermore, that these fields are the very embodiment of Karl Popper's concept of a science, which advances through a continual process of positing (theorizing) followed by rigorous testing and refinement of the idea(s). In addition, I have attempted to convey to the reader a feeling of the excitement engendered within the discipline by the new discoveries and the emerging ideas. I hope to have inspired as well as inform, and in this small way contribute to the overall endeavor.

Asheville, NC, USA   Martin Beckerman

Acknowledgments

I wish to thank my colleagues Francesca-Fang Liao and Elias Greenberg for their careful review of the manuscript and for their valuable technical comments and suggestions. I am also grateful to my editors David Packer and HoYing Fan at Springer Science+Business Media for their encouragement and support throughout this project.

Contents

Chapter 1
Introduction

Martin Beckerman

Neurodegenerative disorders are diseases of the nervous system in which neurons deteriorate and die off and neural circuits become disabled. The best known of these are Alzheimer's disease and Parkinson's disease. These diseases affect millions of individuals worldwide. Some of these disorders have been known since ancient times. Description of symptoms typical of Parkinson's disease appear in ancient Indian and Chinese medical texts, and then again in classical Greek and Roman writings. It was described by Galen (138–201) and by Zhang Zihe (1156–1228) in his classic *Ru Men Shi Qin*. Parkinson's symptoms were described neurologically for the first time by the English surgeon James Parkinson (1755–1824). In his 1817 paper entitled "An essay on the shaking palsy", he described its main symptoms— tremulous motion, reduced muscle strength, and bent posture. A far-more complete clinical description of Parkinson's disease including an increased emphasis on slow movement as a major clinical feature was provided by great French neurologist Jean-Martin Charcot (1825–1893) widely regarded as a father of modern neurology. He named the disorder Parkinson's disease, and in his famous 1872 lectures carefully delineated differences between it and other movement disorders.

The "term movement disorder" encompasses a broad spectrum of neurodegenerative diseases that produce unwanted or hyperkinetic movements such as tremors. Included in this grouping are both Parkinson's disease and Huntington's disease. The former is strongly identified with failing dopaminergic neurons located in the *substantia nigra* region of the midbrain while Huntington's disease involves neurons located throughout the brain with some regions impacted more than others. Huntington's disease is named for the American physician George Huntington (1850–1916) who described this disease in 1872. In his paper, he used the term chorea, taken from the Greek *choreia*, which means to dance, and refers to the irregular and spasmodic muscle movements that first signal the presence of the illness. The use of that term was not new, but rather was first applied to Huntington's disease by the Renaissance physician and founder of the field of toxicology, Paracelsus (1493–1541).

© Springer International Publishing Switzerland 2015
M. Beckerman, *Fundamentals of Neurodegeneration and Protein Misfolding Disorders*,
Biological and Medical Physics, Biomedical Engineering,
DOI 10.1007/978-3-319-22117-5_1

Huntington's disease has an interesting if not outright disturbing history in the United States and Europe. It was the subject of a largely discredited 1916 paper by Charles Davenport. In his paper, he connected the entry of this disease in this country to three families that migrated in the seventeenth century from the village of Bures in Suffolk County, England to Connecticut. He then linked members of these families to women charged with practicing witchcraft. This theme was later amplified in a notorious 1932 paper by Percy Vessie that asserted that people suffering from Huntington's disease were prone to criminality and other forms of social misconduct. These arguments were subsequently amplified and used as a justification for forced sterilization and restricted immigration as part of the eugenics movement that flourished for 40 years or more in the United States and Europe.

One of the main characteristics of Parkinson's disease, Huntington's disease, and other forms of neurological disorder is their increase with age. That there is a connection between aging and cognitive decline is not a new concept. It can be traced back to Pythagoras who noted in the seventh century B.C. that mental decline takes place during the later stages on life. According to Solon (c.630 B.C. to c.560 B.C.), the Athenian law giver and statesman, a person's mental condition has a bearing on the legality of late-life alterations to a will. Galen referred to late-life mental decline as "morosis", and thought of old age as a disease condition in itself. Galen was an accomplished anatomist, and performed numerous dissections of monkeys and pigs, human dissections being prohibited. His tract became the standard medical text for the next 1400 years. It was eventually updated and supplanted by Paracelsus and Andreas Vesalius (1514–1564), who was able to use human cadavers in his anatomical investigations.

The emergence of neurological disorders as diseases having underlying physical causes took considerable time. For a long time women suffering from movement disorders were regarded as witches, while individuals undergoing late-life mental decline were often treated as criminals and locked up in prisons or asylums (Fig. 1.1). The term "cognitive" refers to tasks carried out by the brain involving knowing,

Fig. 1.1 Treatment of dementia through the years. (**a**) *Left-hand panel*: Woodcut depicting a victim of a witch trial. (**b**) *Right-hand panel*: Philippe Pinel, a champion of the humane treatment of the mentally ill, shown at the infamous Bicêtre asylum (from Berchtold *Neurobiol. Aging* 19: 173 © 1998 Reprinted by permission from Elsevier)

thinking, remembering, organizing, and judging. However, rather than applying the term "cognitive decline" to late-life neurological disorders the far-more pejorative designation "senile dementia" was assigned. This was an exceptionally broad term due in large measure to the inability of anatomical investigations to discriminate and pinpoint physical correlates of the symptoms being exhibited. Senile dementia was taken to encompasses forms of madness such as the ones depicted by Shakespeare in *Hamlet* and *King Lear*. At other times it included illnesses such as syphilis and schizophrenia. The term "hardening of the arteries" was applied to the syndrome, while individuals showing similar signs earlier than age 65 were regarded as suffering from an early-onset dementia, or *dementia praecox*.

Substantial progress in cataloging and discriminating between the various types of brain disorders began with the work of early pathologists such as the British physician William Cullen (1710–1790). He introduced the term "neurosis" meaning a nervous disorder and medical condition, and included senile dementia under that heading. Further steps were taken during the nineteenth century. These included the advent of more humanitarian conditions for the treatment of the mentally ill and improvements in clinical diagnosis. The notion of age-related dementia was refined by Samuel Wilks (1824–1911), who noted in 1864 that anatomical changes (atrophy) in the brain accompanied senile dementia. The decline in brain vasculature was noted and these observations produced the leading theory of senile dementia at the end of the nineteenth century—the decline in blood flow to and in the brain results in a loss of brain oxygenation and this was the cause of senile dementia.

Alzheimer's disease (AD) is the leading cause of senile dementia. It is estimated that 13 % of the people over the age of 65 suffer from AD. It is considerably worse for people over the age of 85—for them the likelihood of exhibiting Alzheimer-like symptoms increases to 45 %. The statistics are indeed grim. Worldwide, 5 million new cases of AD are diagnosed every year (or one new case every 7 s). Given the attention that this illness now receives it is perhaps surprising how little attention it got in 1906 when the German pathologist Alois Alzheimer (1864–1915) carried out his groundbreaking work. At that time the (juvenile) illness he had diagnosed was regarded as a new illness, a form of presenile dementia having only the most tenuous connection to the more common senile dementia seen in elderly individuals. That situation gradually changed with the realization that most of the cases of senile dementia had the same histopathological signs as the condition described by Alzheimer, namely, extracellular senile plaques and intracellular neurofibrillary tangles.

1.1   Alzheimer Discovers Amyloid Deposits in the Brain

In order to make progress past the anatomical level of observation, and distinguish between different kinds of neurological disorders, greater resolution of brain tissue was needed. That was provided by the light microscope. Invented by the Dutch lens-maker Anton van Leeuwenhoek (1632–1723) and improved upon by the English scientist and inventor Robert Hooke (1635–1703). This device enables the user to observe cells, and their organelles, and other macromolecular substructures. Contrast is important, and advances in the field have occurred whenever new and improved dyes have been discovered that make the macromolecular structures of interest stand out against the cellular background. The amount of detail made visible by a light microscope is limited by the wavelength of the probe being used. When used with the right dyes the light microscope can resolve structures down to about 0.2 μm, or 200 nm (half the 400 nm nominal wavelength of visible light).

The light microscope is the primary tool of the pathologist. This field has its modern beginnings in the pioneering studies of Rudolf Virchow (1821–1902) in the mid-nineteenth century. Virchow is regarded as the father of modern pathology. In 1854, he applied the term "amyloid" to describe unusual deposits found in the liver, the spleen, and other organs and tissues of people who had succumbed to illnesses ranging from tuberculosis to syphilis. These deposits were initially thought to be fatty or waxy in nature; the term amyloid had been coined to a few years earlier, in 1838, by the botanist Mathias Schleiden in his description of starch and cellulose. The correct identity that these deposits were proteinaceous was made a few years after Virchow's initial discoveries.

The first steps in discriminating between the different kinds of neurological disorders were taken by Charcot who combined careful clinical observations with anatomical studies and tissue pathology. He was the first to describe amyotrophic lateral sclerosis (ALS). In his lectures delivered at the Hôpital de la Salpêtrière in Paris, he distinguished between upper and lower motor neurons and began the process of separating the different motor diseases from one another. In the United States, this disorder is referred to as Lou Gehrig's disease after the famous baseball player who died from it in 1941, but on the Continent and elsewhere it is known as Charcot's disease.

The next and arguably the most famous in a series of landmark events was the 1906 presentation by Alzheimer (Fig. 1.2). In it, he described his postmortem analysis of brain tissue taken from Auguste Deter, a 51-year-old patient (in 1901) suffering from early-onset dementia. Using a light microscope, and utilizing the newly developed Golgi silver staining techniques (see Appendix 1), Alzheimer found that the tissue was permeated with extracellular deposits that he called "miliary foci" and are now referred to as extracellular plaques. He also observed that the cells themselves contained dense bundles of fibrils. He then made the conceptual leap that these deposits were in some manner causally connected to the patient's dementia.

1.2   Amyloids Are Discovered Elsewhere in the Brain and in Different Tissues of the Body

Alzheimer's findings were not unique to the dementia he was exploring, or the story would end there. Instead, his discoveries were preceded by those of Andrew Pick (1851–1924) who reported in 1892 that deposits were present in brain tissue of deceased patients suffering from senile dementia. These aggregates are now known

Fig. 1.2 Alzheimer's research group at the Royal Psychiatric Clinic of the University of Munich in 1910. Lewy is standing at the far right and Alzheimer is third from the right in the *upper row* (from Goedert *Nat. Rev. Neurol.* 9: 13 © 2013 Reprinted by permission from Macmillan Publishers Ltd)

as Pick bodies, and these deposits along with an extreme atrophy in the frontal and temporal regions of the brain characterize the dementia. Clinically the disorder observed by Pick differs from Alzheimer's disease. In the former case, social and personality deficits predominate in the early stages whereas in the latter instance memory loss occurs first. A few years after Alzheimer's report, in 1912, Frederic Lewy (1885–1950) described the presence of intracellular deposits (now known as Lewy bodies) in the brains of patients with Parkinson's disease. Thus, it was becoming clear that deposits of some sort of protein are associated with a large number of different diseases of the brain.

The question then naturally arose as to whether the brain deposits were the same or similar to those seen elsewhere in the body by Virchow and his successors. That question was answered in the affirmative by the development of a test specific for amyloid in 1927. In that year, Paul Divry (1889–1967) and Marcel Florkin (1900–1979) published results of their studies of various body amyloid tissues stained with a dye called Congo red and viewed using polarized light. When that is done, the amyloid tissue exhibits an apple-green birefringence. In a second 1927 paper, Divry showed that the same apple-green birefringence can be seen for extracellular plaques and again in 1934 for the intracellular tangles. The dye, Congo red, used by Divry originated in the textile industry where it had been invented in 1883 and was given that mysterious-sounding name Congo red as a marketing ploy. Shortly thereafter it was tried as a potential histological stain for tissues. Although it is no longer used to stain textiles, it is still widely used to detect the presence of amyloid in the manner established by Divry, that is, in conjunction with polarized light and the revealing of apple-green birefringence patterns (Fig. 1.3).

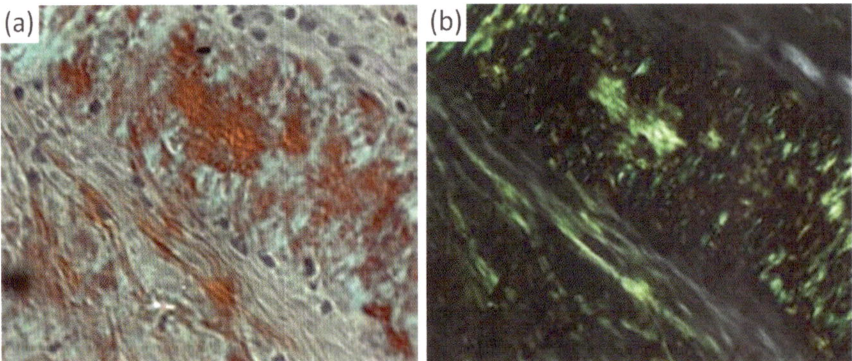

Fig. 1.3 Detection of amyloid tissue using Congo red dye (**a**) Congo red stained section viewed under a light microscope. (**b**) Same section viewed using an optical microscope and a pair of crossed polarizers showing apple-green birefringence (from Howie [from the cover] *Lab Invest* 88: 232 © 2007 Reprinted by permission from Macmillan Publishers Ltd)

1.3 Amyloid Structure Is Revealed Using the Electron Microscope

Light microscopes are not able to reveal the underlying structure of amyloids. They simply do not have sufficient resolution, but an electron microscope does—about three orders of magnitude greater resolution than a light microscope. The existence of birefringence patterns for these deposits makes it likely that their underlying components are arranged in an orderly fashion rather than in an amorphous manner as most people thought. In 1959, Cohen and Calkins used an electron microscope (Appendix 2) to examine the submicroscopic structure of amyloids taken from different organs of the body (spleen, liver, and kidney). They found that amyloids do indeed possess an orderly substructure. The fibrils are straight and rigid—from 75 to 120 Å in width and vary in length from 1000 to 16,000 Å (Fig. 1.4). The fibrils, in turn, are composed of protofibrils; these filamentous subunits range from 25 to 35 Å in diameter. Two or more of these units, twined about one another, form the fibril. Finally and significantly, they observed a similar structural arrangement for all the amyloids they examined.

To summarize, by 1960 it had become apparent that amyloids deposits were associated in some as yet to be determined manner with a diverse collection of illnesses of the body and mind. These aggregates were present in extracellular spaces and within cells of various tissues and organs. Amyloids are proteinaceous aggregates characterized by: (1) birefringence patterns when stained with Congo red and viewed under polarizing light microscope conditions, and (2) composed of long-twisted fibrillar structures when examined using an electron microscope.

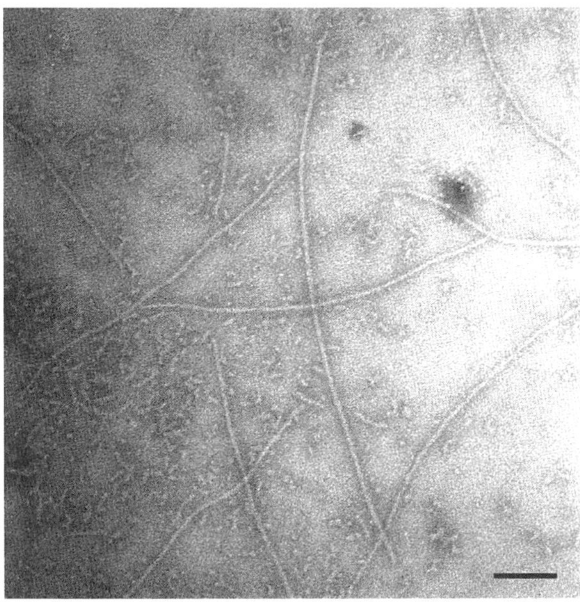

Fig. 1.4 Electron micrograph of Aβ_{1-42} amyloid fibrils formed in water from Aβ (1-42), and stained with 0.1 % phosphotungstic, acid showing long straight fibrils of 70–80 Å diameter within a fibrillar aggregate background. Bar = 1000 Å (from Serpell *Biochim. Biophys. Acta.* 1502: 16 © 2000 Reprinted by permission from Elsevier)

As is the case with any major findings in science, answering one set of questions leads to the next set of perhaps even better questions. In the case of the amyloids, the first question that arises follows from the striking similarity of the structures observed in different parts of the body. The question is: are all amyloid deposits generated by the same protein or are amyloid deposits generated by different proteins perhaps specific to the tissues or disease states to which they are associated?

1.4   Amyloids Are Produced by Many Different Proteins

To answer this type of question, the primary amino acid sequences of the amyloidogenic proteins needed to be determined. This was accomplished for the first time in 1971 and then over the followed couple of years by several research groups starting with efforts headed by George Glenner and Earl Benditt. These investigators found that the proteins were not the same, but rather each one amyloid protein was unique to the disease associated with their presence. The primary structure, that is, the amino acid sequences, of the amyloidogenic proteins found in different tissues and organs had little resemblance to one another; they normally fold into a diverse set of 3D structures, and unlike the amyloids are soluble. The diverse set of amyloidogenic proteins, their precursors, and the disorders associated with them are summarized in Table 1.1.

Table 1.1 The systemic amyloidoses

Amyloid protein	Disease	Precursor protein	Syndrome
AL/AH	Immunoglobulin light/heavy chain amyloidosis	Immunoglobulin light/heavy chains, N-terminal LC variable domain fragments	Produced by elevated clonal population of B cells; multiple organs affected; kidney, and heart failure
AA	AA amyloidosis, Familial Mediterranean fever	Serum amyloid A (SAA) protein	Acute phase protein; concentration elevated in inflammation/injury; Affects liver, spleen, and kidneys
ATTR	Senile systemic amyloidosis, Familial amyloid polyneuropathy	C-terminal fragments of wild-type transthyretin, mutant transthyretin	Leading cause of polyneuropathy and cardiomyopathy
$A\beta_2M$	Hemodialysis-related amyloidosis	β_2-Microglobin	Concentration elevated in kidney disease; accumulates in bones and joints
AApoAI	ApoAI amyloidosis	N-terminal fragment of apolipoprotein AI	Polyneuropathy and cardiomyopathy
AApoAII	ApoAII amyloidosis	N-terminal fragment of apolipoprotein AII	Polyneuropathy and cardiomyopathy
AApoAIV	ApoAIV amyloidosis	N-terminal fragment of apolipoprotein AIV	Polyneuropathy and cardiomyopathy
AGel	Familial amyloidosis of Finnish type	Fragments of mutant gelsolin	Buildup in the eye, progressive polyneuropathy
AFib	Hereditary fibrinogen amyloidosis	Mutant fibrinogen A α-chain	Renal failure; visceral, vascular, cardiac, and neurological involvement
ALys	Lysozyme amyloidosis	Lysozyme mutants	Buildup in the viscera, especially liver, kidneys, and gut; renal disease/liver rupture
ACys	Hereditary cystatin C amyloid angiopathy	Cystatin mutants	Buildup in tissues and cerebral arteries; cerebral hemorrhage

The amyloid diseases caused by the misfolded proteins are listed in *column 2*. The genetically encoded precursor proteins and the specific disease-causing fragments are described in *column 3* along with a short précis of the syndrome in *column 4*

The deposits of waxy materials puzzled over by Virchow in the 1850s occurred in the spleen, liver, and kidneys. They were most likely due to what is today called AA amyloidosis involving buildup of amyloid A (AA) protein over time as a consequence of long-lasting infections and chronic inflammation produced, for example, by tuberculosis and rheumatoid arthritis. Amyloid A protein deposits are derived from an overproduction of serum amyloid A (SAA), an acute phase protein serving as a precursor for the AA protein. These buildups occur most often in the heart but can occur in other tissues and organs as well as observed by Virchow.

In systemic amyloid diseases such as AA amyloidosis discussed in the previous paragraph, the amyloids collect in multiple organs and tissues of the body. In another disease of this type, immunoglobulin light chain amyloidosis (AL), B-cells secrete light chain peptides that circulate through the body and collect in the kidneys and also in the heart and liver leading to organ failure and death. Light chain amyloidosis is associated with B-cell myelomas. In contrast to the abovementioned systemic amyloidoses, there are a number of diseases in which amyloid deposition is specific to a single organ or tissues of the body. An example of this class of amyloids diseases is Type 2 diabetes in which aggregates of islet amyloid polypeptide (IAPP, amylin) form in the pancreas. These islet amyloid deposits of IAPP are believed to be a major contributor to β-cell failure.

1.5 Atomic Level Discovery of the Cross-Beta-Pleated Sheet Conformation

As indicated in Table 1.1 the amyloidogenic proteins are anomalous deposits built from nonamyloidogenic precursors or from fragments thereof. In order to probe further what has occurred to produce the anomalous deposits one has to examine the atomic level structure of the proteins. This requires using more powerful techniques than light or electron microscopy. The paramount experimental techniques for exploring protein structure at atomic level of detail are X-ray crystallography and nuclear magnetic resonance (NMR). These tools, and especially X-ray crystallography, were used to explore the composition of the deposits in the later 1960s and produced a breakthrough set of findings. Specifically, two sets of results were reported—one set by Eanes and Glenner in 1968 and the other by Bonar, Cohen, and Skinner in 1969. In brief, these groups reported that the amyloidogenic proteins have folded into three-dimensional forms quite different from their normal ones. In both studies, the amino acid residues are folded back and forth to form a series of β-pleated sheets. The protein chains in this configuration were then stacked in a perpendicular orientation one on top of the other along the fibril axis. A modern rendering of this type of structure is presented in Fig. 1.5.

Transthyretin (TTR) is a carrier protein that transports thyroxine (T4) and retinol (vitamin A). In its amyloidogenic ATTR form, it is associated with three diseases in the elderly—senile systemic amyloidosis (SSA), familial amyloid polyneuropathy (FAP), and familial amyloid cardiopathy. This protein became the subject of a number

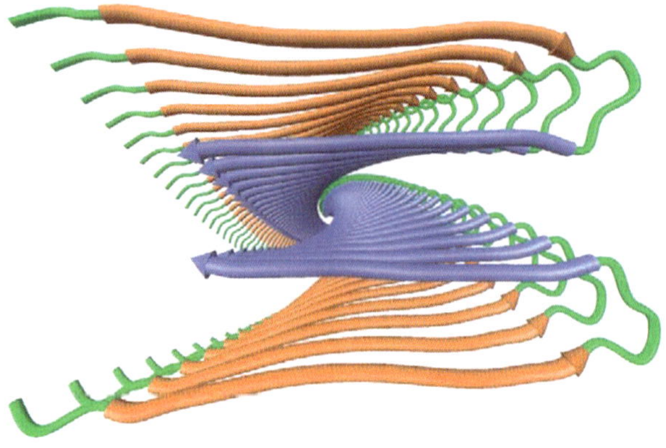

Fig. 1.5 Minimalist model of the cross-beta structure of an Alzheimer's disease Aβ fibril as determined by means of NMR and X-ray crystallography. The view is down the long axis of the fibril (from Tycko *Curr. Opin. Struct. Biol.* 14: 96 © 2004 Reprinted by permission from Elsevier)

of pioneering studies of amyloid formation. One of these was that while TTR was normally soluble, if the pH conditions in their environment were lowered, the proteins partially unfolded and then refolded into an alternate 3D conformation. In that shape, the TTR proteins became aggregation-prone, insoluble, and were capable of forming amyloid fibrils. This type of result demonstrating that the biochemical and biophysical environment has as influence on how proteins fold will be examined in greater detail in the next two chapters. Suffice it to say that at the time the results solidified the main conclusion from both the biophysics and biochemistry—normally soluble proteins have folded into a conformation different from their functionally expected one resulting in the formation of insoluble (and detergent-resistant) amyloids.

Thus, an amyloid is a material satisfying three criteria. It:

- Exhibits *apple-green birefringence* when viewed under polarized light using an optical microscope;
- Possesses a *fibrillar structure* when seen with an electron microscope; and
- Is organized into structures composed of *β-pleated sheet* when examined by means of X-ray crystallography.

The linkage between these levels is a tight one—it was shown by the Glenner group in 1972 that the atomic level, beta-sheet arrangement is indeed responsible for the properties revealed by Congo red staining.

1.6   The Brain Protein Aggregates Are Characterized Biochemically and Genetically

It is interesting to note that Alzheimer's disease was initially regarded as not being particularly important; and due to its early onset it was not regarded as the same illness as senile dementia. It took an editorial in the *Archives of Neurology* in 1976

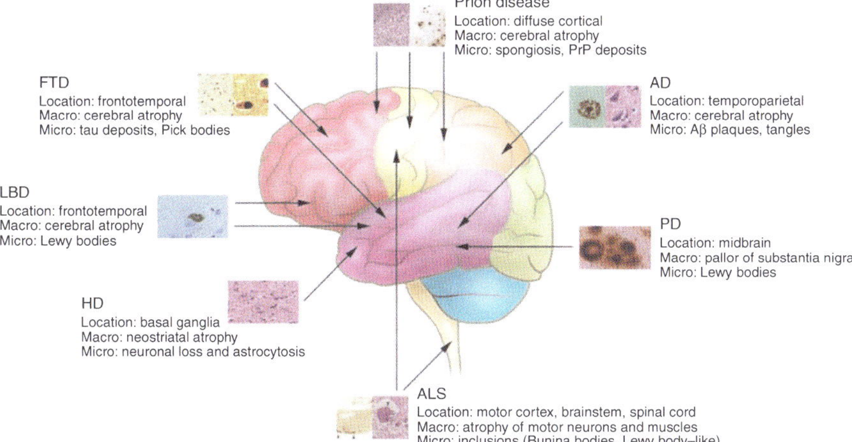

Fig. 1.6 Neurodegenerative disorders, their characteristics, and regions primarily affected. *FTD* frontotemporal dementia, *LBD* Lewy body disease, *HD* Huntington's disease, *ALS* amyotrophic lateral sclerosis, *PD* Parkinson's disease, *AD* Alzheimer's disease (from Bertram *J. Clin. Invest.* 115: 1449 © 2005 American Society for Clinical Investigation)

by Robert Katzman to establish in the minds of researchers that these were the same disorder and to sound the alarm regarding its rising prevalence. The Alzheimer's Association was established in the US a few years later in 1979.

That Alzheimer's disease is not the only form of senile dementia has been realized still more recently with the emergence of frontotemporal lobar dementia (FTLD) as the second most prominent form of dementia in the young after Alzheimer's disease. In Alzheimer's disease, cognitive functions decline, beginning with learning and episodic memory (current and past experiences) and progressing to thinking, loss of language, and further decline in memory. In FTLD/FTD the first, clinically relevant signs involve changes in behavior and personality. As depicted in Fig. 1.6, each class of disease is associated with a specific region of the brain, each giving rise to distinct sets of clinical symptoms.

One of the first sets of goals in the field was to characterize the extra- and intracellular deposits of proteins associated with the neurodegenerative diseases. As indicated in Fig. 1.6 extracellular and especially intracellular deposits are present not just in Alzheimer's disease and Parkinson's disease but in all the others as well. First, there was a need to identify the composition of the extracellular plaques and neurofibrillary tangles of Alzheimer's disease, the Lewy bodies seen in Parkinson's disease, and the others. Secondly, there was a need to tie the constituents whenever possible to genetic mutations in familial cases of the disease, or to other abnormalities, thereby providing evidence that these deposits are causally connected to the neurodegeneration. These efforts required the concerted application of an ensemble of newly developed biochemical and genetics techniques.

1.6.1 Modern Genetics Comes of Age

The modern field of genetics has its beginnings in the early 1900s with the efforts by the British geneticists William Bateson (1861–1926) and Reginald Punnett (1875–1967). Punnett introduced the public to the new field with his book *Mendelism* in 1905; Bateson coined the term "genetics" (taken from the Greek word for "to give birth") in 1906, and the two of them launched the *Journal of Genetics* in 1910. Gregor Mendel (1822–1864) is well known today as the father of genetics. His laws of inheritance were neglected prior to their independent rediscovery in 1900 by Hugo de Vries and Carl Correns, and shortly thereafter Bateson and Punnett put forth the idea of gene linkage as an explanation of departures they observed from Mendel's laws. With these activities the modern field of genetics was born. (The term gene linkage refers to the propensity of genes that are located in close proximity to one another on a chromosome to be passed on together during meiosis.)

These events were accompanied by the recognition that chromosomes were the seat of genes. The association of genes with chromosomes was first posited by Walter Sutton in 1903 and that identification was firmly solidified in 1915 by the publication of the studies by Thomas Morgan and his coworkers. The final step was working out the number of human chromosomes. That number was initially established in the early 1920s as 48, but was eventually and conclusively reduced to the correct number, 46, in 1956–1958. It is now well established that there are 22 paired somatic chromosomes; the 23rd, or sex chromosomes, consists of an XX pair for females and a XY pair for males. Each human chromosome has two arms—a short arm (denoted by the symbol "p") and a long arm (indicated by the symbol "q"). These two arms are joined at the centromere.

1.6.2 Connections Are Established

The extracellular plaques seen in AD are composed of fibrils of the amyloid-beta protein, a proteolytically cleaved short peptide of around 40–42 amino acids generated from the genetically encoded amyloid precursor protein (APP). The first step in this identification came in 1984 when Glenner and Wong succeeded in determining the amino acid sequence for the principal protein found in the cerebrovascular amyloidosis associated with AD. They named this peptide the "amyloid-beta (Aβ) protein". This same protein was isolated from extracellular plaque core by Masters in 1985.

Down syndrome is the leading cause of mental retardation, with a frequency of occurrence of about 1 per 700 births. It is named after the British physician John Langdon Down (1828–1896), who in 1866 published an essay in which he described the condition seen by him in the children's asylum that he operated as superintendent. The association of Down syndrome with a genetic abnormality took nearly a century to establish. That this was indeed the case had been put forth in the 1930s but it was not until 1959 that Jérôme Lejeune (1926–1994) and Patricia Jacobs working independently identified its cause as the presence of an extra copy of chromosome 21, that is, there were three and not two copies of chromosome 21 (trisomy 21).

By the 1980s, it had become clear that persons with Down syndrome had a greatly increased likelihood of developing Alzheimer's disease later in life. Glenner and Wong had posited that the gene encoding the Aβ protein might be located on chromosome 21, and in both conditions there are insoluble deposits of plaques and tangles. These clues firmly pointed to a connection between abnormalities on chromosome 21 and Alzheimer's disease. Those clues led to major breakthroughs in the 1987–1991 time period. During that time (1) the APP gene was isolated, (2) linkages between early-onset familial Alzheimer's disease (FAD) and genetic markers on chromosome 21 located near the APP gene were found, and a mutation in the APP gene leading to FAD was discovered.

The neurofibrillary tangles (Fig. 1.7), the second major pathological sign of AD, consist primarily of the microtubule-associated protein tau, and encoded by the MAPT gene at 17q21.1. The first step in uncovering that connection was taken by Kidd in 1963 with his examination of the composition of the tangles using a light microscope. He found that the tangles consisted of paired helical filaments (PHFs). These structures, in many cases, filled the cytoplasm to such an extent that the cells appeared swollen. The discovery that PHFs were built from tau was made by Goedert and coworkers in 1986 using chemical and immunological approaches. Finally, in 1988, Kirchner and his colleagues showed that the PHFs were amyloidogenic, i.e., they exhibited the characteristic cross-β X-ray diffraction pattern.

Lewy bodies are the defining pathological feature of Parkinson's disease and dementia with Lewy bodies (DLB). An important first clue as to its composition was the discovery in 1990 that the disease could be inherited in a (Mendelian) autosomal dominant fashion. This was followed by the finding in 1997 that a missense mutation in the gene coding for the protein α-synuclein was present in familial Parkinson's disease. That led to the use of antibodies for α-synuclein resulting in the discovery that this protein was indeed the major constituent of the Lewy bodies and Lewy neurites. These events were preceded by the finding that the gene responsible for the familial cases of PD being studied mapped to chromosome 4q21–23, and were followed by the characterization in 2000 that α-synuclein, like Aβ and tau, produced a cross-β X-ray diffraction pattern.

1.7 Huntington's Disease Is an Unstable Repeat Disorder

The search for the gene responsible for Huntington's disease (HD) spanned more than a decade. It began with the recognition in the 1950s that HD was prevalent among people living near and around Lake Maracaibo in Venezuela. A team of physicians and geneticists that included Jim Guesella and Nancy Wexler descended upon this region in 1979. Using for the first time restricted fragment length polymorphism and genetic linkage analysis these researchers isolated the gene to chromosome 4 in 1983. A team of 58 scientists was then formed that undertook to find the elusive

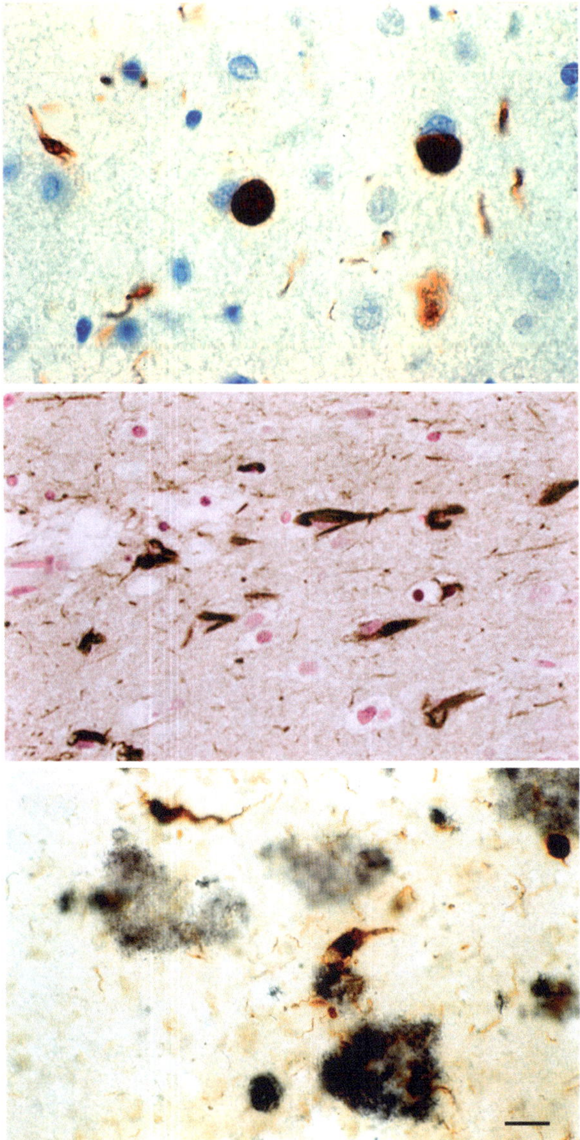

Fig. 1.7 Deposits of tau. *Upper panel*: Pick bodies and abnormal inclusions composed of tau (*brown*) in FTD and sporadic Pick's disease. *Middle panel*: Neurofibrillary tangles and neuropil threads composed of tau (*brown*) in chronic traumatic encephalopathy. *Lower panel*: Neuritic plaques made of Aβ (*blue*) and neurofibrillary tangles and neuropil threads composed of tau (*brown*) in Alzheimer's disease (from Spillantini *Lancet Neurol.* 12: 609 © 2013 Reprinted by permission from Elsevier)

gene. That gene hunting effort took 10 years and in 1993 the team reported the discovery of the gene responsible for the disease, now named huntingtin (htt).

The huntingtin gene, which occupies chromosome locus 4p16.3, harbors a particular kind of defect. In HD sufferers, there is an excessively long stretch of CAG (cytosine, adenine, guanine) repeats in the Htt gene. These sequences are unstable. They increase in length from generation to generation, and with it the corresponding age of onset of the disorder decreases and the severity of the illness grows. HD is one of several disorders that exhibit this type of genomic instability, each one encoding an excessively long stretch of glutamines (Gln, Q) amino acids in a specific protein. Another of these polyQ disorders is spinal and bulbar muscular atrophy (SBMA). That disease was first described by Kennedy and coworkers in 1968 but it was not until 1991 that La Spada and colleagues isolated the cause of the disease as the presence of expanded CAG repeats in the gene encoding the androgen receptor.

PolyQ expansions are not the only ones that have been discovered and linked to neurological disorders. Later in the same year, 1991, a pair of papers appeared reporting that Fragile X syndrome arises as a consequence of an unstable CGG repeat in the fragile X mental retardation 1 (FMR1) gene. In this case, the repeat expansion occurs in the nonprotein-coding 5′UTR (untranslated region) of that gene. Other neurological disorders have been traced to unstable repeat expansions in nonprotein-coding regions, as well. That collection of illnesses includes myotonic (muscular) dystrophy and Friedrich's ataxia among several others.

1.8   The Amyotrophic Lateral Sclerosis—Frontotemporal Lobar Degeneration Clade

Amyotrophic lateral sclerosis (ALS) is an adult onset disease typically seen in individuals 50–60 years of age. It is the most common disease of motor neurons with a worldwide incidence of 2 per 100,000 per year. This disease affects motor neurons in the motor cortex, brain stem, and spinal cord. Motor neurons may be divided into two classes. Upper motor neurons have their cell bodies in the motor cortex and connect to the lower motor neurons situated in the brain stem and spinal cord. They connect to the muscles controlling face, jaw, and tongue movement and swallowing actions (brainstem neurons, corticobulbar tract) and those regulating torso functions and lower limb movement (spinal cord neurons, corticospinal tract). There are more than five different kinds of motor neurons disease. ALS affects both upper and lower motor neurons. The term "amyotrophic" denotes the atrophy of muscle fibers resulting in a loss of signaling capability. The second part of the name, "lateral sclerosis", refers to the hardening (sclerosis) of the lateral columns of the spinal cord observed at autopsy.

The first step in uncovering the proteins responsible for ALS came in 1993 with the finding that missense mutations in Cu/Zn superoxide dismutase 1 (SOD1) were present in about 15–20 % of the cases of familial ALS (FALS). The situation remained that way until the 2006–2009 time period when mutations in two other proteins—transactivating response region (TAR) DNA-binding protein 43 (TDP-43) and fused

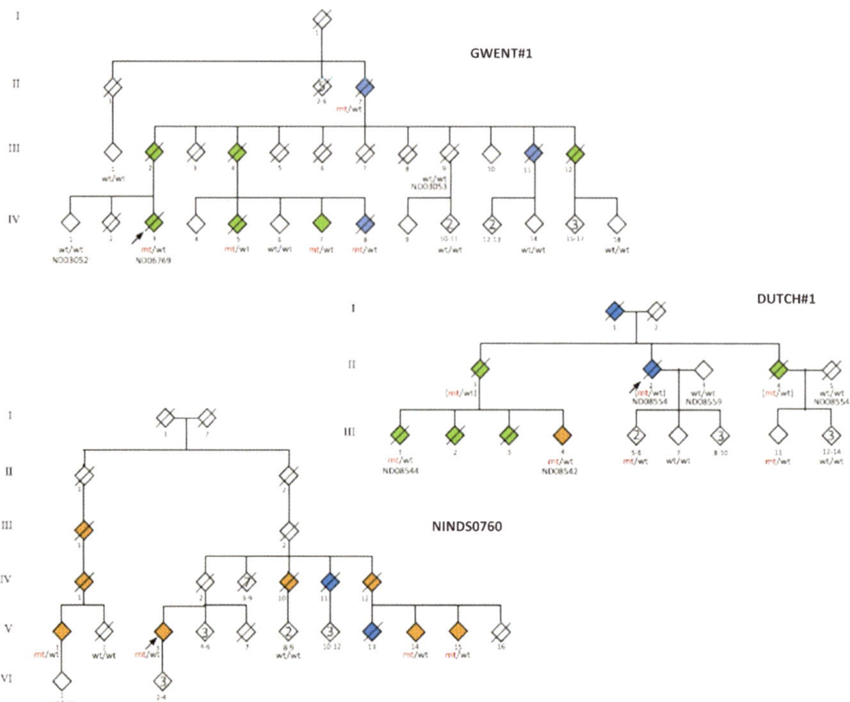

Fig. 1.8 Family trees showing the results of genetic linkage analyses of the 9p21 gene locus demonstrating the common occurrence of ALS and/or FTD in families of Dutch, UK, and Finnish descent. *Solid symbols* denote infected subjects and *slashes* indicate they are deceased. Mutant alleles are designated by the abbreviation mt and wild type by wt. Probands are identified by *arrows*. *Blue diamonds* denote ALS, *orange* FTD, and *green* both ALS and FTD in the same individual (adapted from Renton *Neuron* 72: 257 © 2011 Reprinted by permission from Elsevier)

in sarcoma (FUS) were discovered to be present in a substantial fraction of the intracellular inclusions observed in not only ALS but also in frontotemporal lobar degeneration (FTLD). Mutations in these two proteins could directly account for another 10 % of the instances of FALS cases and some of the nonfamilial, or sporadic, cases, too. Most interestingly, these same two proteins were found to have a causal role in FTLD thereby providing a possible explanation of the observation that in many instances people suffering from ALS later developed frontotemporal dementia.

The linking together of ALS and FTLD was greatly strengthened by discoveries in 2011. In that year, two groups reported that an unstable GGGGCC hexanucleotide repeat expansion in the noncoding region of the chromosome 9 open reading frame 72 (C9ORF72) gene could cause both ALS and FTLD. That defect is the single most prominent cause of ALS uncovered to date. It can account for much (~40 %) of the missing fraction of FALS cases and for about 10 % of the FTLD instances. Figure 1.8 presents the results of a genetic linkage analysis of the C9ORF72 gene locus in several European families whose members suffer from

ALS and/or FTLD. It illustrates the joint occurrence of the two illnesses, one affecting neurons in the motor cortex and the other cells in the frontotemporal cortex in these cohorts.

1.9 Transmissible Spongiform Encephalopathies Are Prion Diseases

The transmissible spongiform encephalopathies are far-less well known to the general public than are Alzheimer's disease or Parkinson's and Huntington's diseases or amyotrophic lateral sclerosis. In this grouping, are three diseases affecting humans—Creutzfeldt-Jakob disease (CJD), Gerstmann-Sträussler syndrome (GSS), and kuru. Other TSEs target mammals; the best known of these are scrapie, a disease of sheep and goats, and bovine spongiform encephalopathy (BSE), otherwise known as mad cow disease (Table 1.2). These diseases are not common. The human forms affect roughly one person per million worldwide. Yet studies of these diseases have played an important role in furthering the science of neurodegeneration and protein misfolding.

The transmissible spongiform encephalopathies are not a new disease, and have been known for at least 200 years. Scrapie is a disease of sheep and goats, and veterinarians in England, France, and Germany had examined infected animals in the hope of isolating the infectious agent. That the brain was the site of infection was recognized early on, and attempts were made to transmit the disease through inoculation of brain and transfusion of blood. These efforts were made difficult by the exceptionally long-incubation period of scrapie (and shared by other prion diseases). Finally, in 1936, it was shown to be transmissible from sheep to goats leading to the hypothesis that a virus was the causative agent.

Table 1.2 Transmissible spongiform encephalopathies (prion diseases)

Animal and human prion disease	Host
Scrapie	Sheep, goats
Bovine spongiform encephalopathy (BSE)	Cattle
Chronic wasting disease (CWD)	Mule deer, white-tailed deer, Rocky Mountain elk
Transmissible mink encephalopathy (TME)	Mink
Feline spongiform encephalopathy (FSE)	Cats
Exotic ungulate encephalopathy (EUE)	Nyala, greater kudu
Kuru; Creutzfeldt-Jakob diseases (CJD, sCJD, fCJD, vCJD, iCJD); Gerstmann-Sträussler-Scheinker syndrome (GSS); Fatal familial insomnia (FFI)	Humans

CJD Abbreviations: *s* sporadic, *f* familial, *v* variant (acquired from eating beef tainted with BSE), *i* iatrogenic (acquired through a medical procedure)

Interest in these diseases was further stimulated in the 1950s when a large outbreak of an obscure disease called kuru occurred among the Fore people of Papua-New Guinea. This outbreak attracted the attention of two epidemiologists, Carleton Gajdusek and Vincent Zigas, who traveled there in order to study the disease, and identify its origins and means of transmission. They discovered that this fatal disease was being transmitted from one individual to another though cannibalistic feasts involving brain tissue. This novel means of transmission took on added significance a few years later when William Hadlow noted the similarity between scrapie and kuru. Those observations lead to a landmark experiment in which kuru-infected brain extracts taken from humans were shown to induce the disease in chimpanzees. Similarities of these diseases to another even more obscure disease, Creutzfeld-Jakob disease, were noted and from these studies there arose a new concept—that of a transmissible dementia.

These finding left open the question as to just what was the agent responsible. Over the next 20 years efforts to isolate the infections agent were carried out leading to the gradual realization that neither a virus nor a viroid nor any other nucleic-acid-containing agent was responsible for these diseases. Instead, the notion was put forth by Tikvah Alper in 1966/67 and in more detail by J.S. Griffith in 1967 that a bare protein perhaps aberrant in form might be responsible. This hypothesis was formalized by Stanley Prusiner in a landmark 1982 paper where he coined the term prion, and argued that the causative agents in all these diseases were misfolded forms of that protein.

This hypothesis has implications that extend far beyond that of the rare prion diseases. If extended to proteins such as $A\beta$, tau, and α-synuclein, it may provide an explanation for how protein misfolding disorders might propagate not only from one person to another but also from one region of the brain to another. For that reason the prion hypothesis has been investigated extensively in the laboratory using a variety of test subjects. Its main tenets are, as follows:

- A misfolded form of the prion protein is the sole infectious agent responsible for animal and human prion diseases;
- This agent behaves like a living microorganism—it propagates from one host to another, and replicates itself, and conveys toxicity, in both locales;
- It replicates itself through a templating process whereby the misfolded form seeds the conversion of the normal form into the disease-causing ones; and
- It exhibits strain effects due to differences in conformation.

1.10 The Genetic Underpinnings of the Neurodegenerative Diseases Are Revealed

The two sets of goals posed earlier have been mostly achieved. First, the deposits have been characterized biochemically and their key components have been identified. Depending on the specific illness, the deposits are primarily composed of

Table 1.3 Neurodegenerative diseases

Neurodegenerative disease	Misfolded protein(s)
AD-PD	
Alzheimer's disease	Aβ, tau, α-synuclein
Alpha-synucleinopathies: Parkinson's disease Lewy body disease Multiple system atrophy	α-Synuclein
FTD-ALS	
Amyotrophic lateral sclerosis (ALS)	TDP-43, FUS
FTLD-tau (tauopathies): Pick's disease Corticobasal degeneration Progressive supranuclear palsy	tau
FTLD-TDP	TDP-43
FTLD-FUS	FUS
FTDP-17	tau
Chronic traumatic encephalopathy	TDP-43, tau
ALS	SOD1
PolyQ disorders	
Huntington's disease	Huntingtin (htt)
Spinobulbar muscular atrophy	Androgen receptor
Spinocerebellar ataxias	Ataxins
Prion diseases	PrP^{Sc}

Abbreviations: *FTLD* frontotemporal lobar degeneration, *FTDP* frontotemporal dementia and parkinsonism linked to chromosome 17, *TDP-43* transactivating response region (TAR) DNA-binding protein 43, *FUS* fused in sarcoma, *SOD1* Cu/Zn superoxide dismutase 1, *PrP^{Sc}* prion protein, scrapie form

misfolded Aβ peptide, or tau, or α-synuclein, or TDP43. Secondly, there is evidence linking aberrant forms of these proteins to the specific illnesses. These observations enable the placement of the various neurological disorders into specific (but imperfect and not necessarily exclusive) clades. This organization is presented in Table 1.3. It further enables the illnesses to be characterized as being a *tauopathy*, or an *α-synucleinopathy*, or a *TDP-43 proteinopathy*.

An important milestone leading up to these pioneering achievements was the development of restriction fragment length polymorphism (RFLP) analysis technique by Botstein in 1980. This approach utilizes DNA restriction enzymes, proteins that recognize specific DNA sequences and catalyze their cleavage. This technique greatly increased the number of markers for human DNA making possible the construction of detailed gene linkage maps such as the one shown in Fig. 1.8. Other major achievements were the development by Mullis in 1983 of the polymerase chain reaction (PCR) technique, which amplifies small amounts of DNA, and creation by Southern in 1975 of the Southern blot, a method for tagging specific DNA sequences.

It is customary to distinguish between familial and sporadic forms of neurodegenerative diseases. The former refers to instances where there is a well-established family history of the disease and the genes responsible follow Mendelian forms of inheritance. In these situations, the diseases can be further divided according to the inheritance characteristics of the disorder—either autosomal dominant or autosomal recessive (autosomal—any of the 22 nonsex chromosome pairs).

A landmark in the field was the identification of mutations in APP Alzheimer's disease gene discussed earlier in the chapter. However, this gene was not the only one found to convey an increased risk of developing AD. Instead, its discovery was followed by that of three other Alzheimer disease genes. Mutations in the PSEN1 and PSEN2 genes that encode protein involved in APP processing were discovered to impart an increased risk of developing AD as did mutations in the gene that encodes apolipoprotein E (implicated in Aβ clearance). The former two genes lead to early-onset autosomal dominant AD, while mutations in the (non-Mendelian) ApoE gene are strongly associated with late-onset familial AD.

The finding that the processing and clearance of the Aβ peptide was centrally involved in familial forms of AD led to the positing of the *amyloid cascade hypothesis* of AD by Selkoe, Hardy and Higgins in 1991–1992. *The main tenet of the hypothesis is that the deposit of Aβ peptidein extracellular plaques arising from an imbalance between Aβ production and clearance leads to Alzheimer's disease.* This hypothesis had broad support from numerous findings in other major neurodegenerative diseases. However, there were also problems with this picture arising principally from the absence of correlations between these deposits and cognitive decline. Those observations were accompanied by a growing number of studies pointing to the possible involvement of the small, soluble *oligomeric* species of the same misfolded protein as the main culprits in the diseases. These combined discoveries led to a shift in emphasis and a revision of the hypothesis. The diseases are, indeed, *protein misfolding disorders* (*PMDs*), but the primary disease-causing (toxic) species may be the oligomeric forms and not the far-larger and insoluble fibrillar amyloid deposits themselves.

1.11   Neurodegenerative Diseases Are Protein Misfolding Disorders

The presenilins and ApoE (especially the ApoE4 allele) are not the only additional causal agents and risk factors uncovered. Some of the most prominent of these additional purveyors of catastrophe are listed in Table 1.4. A number of them act in a Mendelian manner while others operate in a non-Mendelian way. These findings were made possible by an accelerated pace of discovery using genome wide association studies (GWAS) in the years following 2005. These studies were directed at common polymorphisms, while rarer, genetic mutations necessitate the unraveling of polygenic, non-Mendelian forms of AD; these have been searched for using massively parallel next generation methods. ApoE was found some time ago by older methods and remains the most important of the non-Mendelian factors uncovered while a

Table 1.4 Neurodegeneration risk factors

Gene	Locus	Protein
Alzheimer's disease		
PSEN1	14q24.2	Presenilin 1
PSEN2	1q42.13	Presenilin 2
ApoE	19q13.32	Apolipoprotein E
Parkinson's disease		
PARK2	6q26	Parkin
PARK7	1p36.23	DJ-1
PARK8	12q12	Leucine-rich repeat kinase 2 (LRRK2)
PINK1	1p36.12	PTEN-induced putative kinase 1
VSP35	16q12	Vacuolar protein sorting 35 homolog
EIF4G1	3q27.1	Eukaryotic translation initiation factor 4 gamma 1
ALS/FTD		
C9orf72	9p21.2	Chromosome 9 open reading frame 72
PGRN	17q21.32	Progranulin
VCP	9p13.3	Valosin-containing protein
OPTN	10p13	Optineurin
ATXN2	12q24.1	Ataxin 2

All the genes in this table with the exception of ApoE follow Mendelian forms of inheritance. The protein(s) corresponding to the C9orf72 locus has/have not yet been identified

growing number of susceptibility loci have been found recently using the new methods. The last two entries for PD and several for ALS/FTD found using the new methods as are an even larger number of rare mutations not included in the short list.

The entries in Table 1.4 strongly support the concept that the protein misfolding disorders arise from failures in *protein quality control*. This term is a broad one and touches upon almost every component and organelle of the eukaryotic cell. Prominent among these are:

- The machinery responsible for production and clearance of proteins.
- Molecular chaperones that are responsible for refolding of partially folded proteins and prevention of aggregation in the endoplasmic reticulum, mitochondria, and cytosol.
- Components of the ubiquitin-proteasome systems and the autophagic-lysosomal pathway used for disposal of damaged, aggregated, and unwanted proteins.
- Nuclear DNA and RNA processing machinery that carries out replication, transcription, and splicing.
- The mitochondrial network that supply energy (ATP) and maintains calcium homeostasis.
- Transport granules and organelles used to shuttle mRNAs, receptors, neurotrophic factors, and other essential molecules into and out of synapses.
- Cytoskeleton motor proteins and the actin and microtubule rails over which they move that are central to all logistical operations.

To begin, the Alzheimer's presenilins and ApoEs implicate defects in processing and removal in the excessive buildup of Aβ oligomers and fibrils. The list of Parkinson's disease-associated risk factors appearing in Table 1.3 is revealing. These entries include protein involved in ubiquitination, mitochondrial homeostasis, vacuolar transport, and protection against oxidative damage. The protein risk factors uncovered with respect to ALS and FTLD provide further support for the involvement of protein quality control in the disease pathogenesis and expand their scope. TDP-43 and FUS are RNA processing proteins while many of the others have roles in protein quality control dealing with chaperoning and disposal through the ubiquitin-proteasome system (UPS) or through the autophagic-lysosomal pathway. These susceptibility findings provide valuable information on how the diseases develop. How these data all fit together to produce the various PMDs remain a great unsolved challenge in the field today.

Lastly, it must be pointed out that sporadic cases of AD are far-more prevalent than familial ones. They account for more than 95 % of all the cases of Alzheimer's disease! In these instances, there is no family history of the disease and, thus, a specific gene that is responsible cannot be identified. How sporadic disease instances come about is perhaps an even greater mystery.

1.12   Summary

1. Long thought of as disembodied dementias, the physical bases for neurodegenerative diseases are at long last being uncovered. As always in science, the speed at which experimental and conceptual breakthroughs take place determine the overall rate at which progress in a given field is made. The discovery of the physical bases for neurodegeneration was tied intimately to advances in multiple contributing sciences—in microscopy, in biophysics and biochemistry, and in genetics. The combined efforts of researchers in these fields spread over more than 90 years has led to the emerging realization that motor disorders and cognitive dementias are brought in some as yet not understood way from the physical presence of misfolded proteins.

   The first breakthrough was the discovery by Alzheimer and others that anomalous deposits of protein were present in brain tissues of individuals that had succumbed to neurodegenerative disorders. These findings were made possible by the development of new staining methods that enabled researchers to view details in brain tissues that were not previously discernible. These discoveries were followed by examinations of the brain aggregates through electron microscopy and X-ray crystallography. By these means it was established that many of the deposits were amyloidogenic and similar to deposits found elsewhere in the body in many illnesses. The deposits in the brain and body satisfied the following diagnostic conditions:

   • Light microscope/Congo red staining/polarized light: Apple-green birefringence
   • Electron microscope: Rigid, nonbranched fibrils 7.5 to 10 nm in diameter
   • X-ray diffraction: Crossed beta-pleated sheet structures

Table 1.5 Protein misfolding disorders

Disease	Region affected	Symptoms
Alzheimer's disease (AD)	Temporoparietal	Cognitive decline; reduction in learning, memory, reasoning, and understanding spatial relationships
Parkinson's disease (PD)	Midbrain	Movement disorder; tremors and slowed movement
Lewy body disease (LBD)	Frontotemporal	Cognitive decline: Loss of concentration and attention; adverse behavioral changes
Amyotrophic lateral sclerosis (ALS)	Motor cortex, midbrain, spinal cord	Motor neuron disease: Progressive weakness and atrophy of voluntary muscles of face, torso, and limbs
Frontotemporal dementia (FTD)	Frontotemporal	Cognitive decline: Alterations in personality and behavior, apathy, and judgment deficits
Huntington's disease (HD)	Basal ganglia	Movement disorder: irregular, rapid and uncontrolled muscle movements (chorea)
Spinocerebellar ataxias (SCAs)	Cerebellum	Coordination decline; irregular, poor control of gait, hands, speech, and eye movement
Prion disease	Diffuse cortical locations	Movement deficits and cognitive decline

Listed in *columns 2 and 3* are the regions affected and typical (early) clinical symptoms

2. As individuals age, their likelihood of developing brain disorders rapidly increases. Originally grouped together under the general heading of senile dementia, over the past century and a half the dementias have been examined in the clinic, anatomically, and through tissue pathology, and can now be distinguished from one another. Table 1.5 presents the most prevalent of these disorders and ways they are identified clinically. As can be seen in the table each disease affects a specific part of the brain and the corresponding presence or absence of specific symptoms provides clues as to the identity of the disorder.

3. A small group of misfolded proteins are found in the anomalous intracellular deposits present in neurodegeneration; these deposits, or aggregates, are referred to as intracellular bodies or intracellular inclusions. Three proteins are especially prominent; they occur in multiple diseases, and give rise to families of diseases named for them:

- Tau and the tauopathies
- α-synuclein and the α-synucleinopathies
- TDP-43 and the TDP-43 proteinopathies

   Many, but not all, of these proteins form amyloids. For example, proteins subject to CAG repeat instabilities exhibit several morphologies while TDP-43 aggregates are not amyloidogenic at all. The key shared property is: (1) that populations of misfolded proteins are present; (2) they are present in excess of safe numbers, and (3) as a result are prone to form insoluble and detergent-resistant aggregates. (4) Lastly, the identities of the primary toxic species are yet to be determined but the preponderance of the evidence points to the small soluble oligomeric species as the primary culprits.

4. The abovementioned trio of protein aggregators and partners (tau, α-synuclein, and TDP-43) listed in Table 1.3 are not the only mutation and misfolding-prone contributors to neurodegeneration. Advances in genetics have made possible the discovery of a rapidly increasing number of neurodegeneration risk factors. Some of these are inherited in a Mendelian fashion while others are not. Instead, multiple factors combine in some as yet unknown fashion to enhance the disease risk. The partition of diseases into familial (Mendelian) and sporadic (thought to be mainly environmental) instances, the latter being predominant, is undergoing a major revision. Like the three chief misfolders, many of these risk factors are present in multiple diseases. The totality of findings raises the possibility that there exists a unity of mechanism and an as-yet-to-be-discovered pharmacology that will halt if not reverse the progression of these diseases.

5. In conclusion, neurodegenerative diseases arise as a consequence of misfolded proteins and peptide fragments derived from precursor proteins. They are present in large numbers because of failures in protein quality control. The maintenance of protein health requires the participation of most if not all the main organelles and systems of the eukaryotic cell. It not only includes the primary protein quality control machinery—molecular chaperones, the ubiquitin-proteasome systems, the autophagic-lysosomal pathway and by means of endoplasmic reticulum-associated degradation (ERAD)—but also the mitochondrial network, the nuclear machinery responsible for replication, transcription, and splicing, and the cellular systems and subsystems that carry out transport in logistically challenged neurons.

Appendix 1. Ramón and Cajal and the Birth of Neuroscience

In order to render the structure of the neurons and the circuits they form visible, Alzheimer used the silver staining technique pioneered by Camillo Golgi and improved upon by Santiago Ramón and Cajal (1852–1934). These scientists shared the 1906 Nobel Prize in physiology for their pioneering work in neuroanatomy, pathology, and physiology. Up to that time it was unclear how neurons were arranged to form brain networks. The leading theory at that time was known as the reticular theory. According to this theory of brain organization, neurons were physically joined to one another by means of their axons and dendrites and formed a neural continuum. Ramón and Cajal showed in his experimental work that this theory was incorrect. Rather, neurons in a circuit were contiguous not continuous; they were separated from one another by a narrow cleft, later named the synapse by Charles Sherrington. And information traveled in one direction—down axons of one neuron, across the synaptic cleft, and onto dendrites and cell body of a second neuron in a neural circuit. This conceptual model marks the beginning of neuroscience. It was unambiguously confirmed many years later using electron microscopy, which provided sufficient resolution to remove any lingering doubts as to the existence of synaptic gaps physically separating nerve cells.

Appendix 2. The Electron Microscope

The electron microscope is based on the hypothesis first enunciated by the French physicist Louis de Broglie (1892–1987), in his 1923 PhD thesis, that particles such as electrons have wave properties. According to de Broglie (who was awarded the Nobel Prize in physics in 1929 for his work), the wavelength of any moving electron is inversely proportional to its momentum. The faster the electron moves, the shorter its wavelength. In 1931, the German engineers Max Knoll (1897–1969) and Ernst Ruska (1906–1988) made use of this discovery in their invention of the transmission electron microscope (TEM) (Fig. 1.9). Microscopes of this new type employ (1) a beam of fast moving electrons in place of a light beam, and (2) a set of magnetic lenses instead of optical lenses to guide and shape the beam, and to focus onto the chosen imaging media the electrons that have been transmitted through the prepared specimen.

Fig. 1.9 The first commercially produced electron microscope produced by Siemens in 1939 (from Ruska [1986 Nobel lecture] 1987 *Bioscience Reports* 7: 607 © The Nobel Foundation and reprinted with their permission)

The invention of the TEM was followed in 1981 by the invention of a different kind of electron microscope, the scanning tunneling microscope (STM), by Gerd Binnig and Heinrich Rohrer. The STM exploits yet another quantum mechanics discovery—quantum tunneling—to enable researchers to examine surface structure in great detail. Binning and Rohrer shared the 1986 Nobel Prize in physics with Ernst Ruska. The two kinds of electron microscope enable researchers in hospitals and universities to study cellular substructure at 1000-times greater resolution than was possible using a light microscope.

Bibliography

Historical Development of the Field

Berchtold, N. C., & Cotman, C. W. (1998). Evolution in the conceptualization of dementia and Alzheimer's disease: Greco-Roman period to the 1960s. *Neurobiology of Aging, 19*, 173–189.

Cohen, A. S., & Calkins, E. (1959). Electron microscopic observations on a fibrous component in amyloid of diverse origins. *Nature, 183*, 1202–1203.

Divry, P., Florkin, M., & Firket, J. (1927). Sur les propriétés optique de l'amyloide. *Comptes Rendus des Séances de la Société de Biologie et des Filiales, 97*, 1808–1810.

Eanes, E. D., & Glenner, G. G. (1968). X-ray diffraction studies on amyloid filaments. *The Journal of Histochemistry and Cytochemistry, 16*, 673–677.

Glenner, G. G., Page, D. L., & Eanes, E. D. (1972). The relation of the properties of Congo red-stained amyloid fibrils to the β-conformation. *The Journal of Histochemistry and Cytochemistry, 20*, 821–826.

Kirchner, D. A., Abraham, C., & Selkoe, D. J. (1983). X-ray diffraction from intraneuronal paired helical filaments and extraneuronal amyloid fibers in Alzheimer disease indicates cross-β conformation. *Proceedings of the National Academy of Sciences of the United States of America, 83*, 503–507.

Schleiden, M. J. (1838). Einige Bemerkungen über den vegetabilischen Faserstoff und sein Verhältnis zum Stärkemehl. *Annalen der Physik und Chemie, 43*, 391–397.

Virchow, R. (1853). Über eine im Gehirn und Rückenmark des Menschen aufgefundene Substanz mit der chemischen Reaktion der Cellulose. *Virchows Archiv für Pathologische Anatomie, 6*, 135–138.

Wexler, A. (2008). *The woman who walked into the sea: Huntington's and the making of a genetic disease*. New Haven: Yale University Press.

Wilks, S. (1864). Clinical notes on atrophy of the brain. *The British Journal of Psychiatry, 10*, 381–392.

Mutations, Protein Misfolding and Neurodegeneration

DeJesus-Hernandez, M., Mackenzie, I. R., Boeve, B. F., Boxer, A. L., Baker, M., Rutherford, N. J., et al. (2011a). Expanded GGGGCC hexanucleotide repeat in noncoding region of C9ORF72 causes chromosome 9p-linked FTD and ALS. *Neuron, 72*, 245–256.

Glenner, G. G., & Wong, C. W. (1984). Alzheimer's disease: Initial report of the purification and characterization of a novel cerebrovascular amyloid protein. *Biochemical and Biophysical Research Communications, 120*, 885–890.

Goate, A. (1991). Segregation of a missense mutation in the amyloid precursor protein gene with familial Alzheimer's disease. *Nature, 349*, 704–706.

Goedert, M., Wischik, C. M., Crowther, R. A., Walker, J. E., & Klug, A. (1988). Cloning and sequencing of the cDNA encoding a core protein of the paired helical filament of Alzheimer's disease: Identification as the microtubule-associated protein tau. *Proceedings of the National Academy of Sciences of the United States of America, 85*, 4051–4055.

Katzman, R. (1976). The prevalence and malignancy of Alzheimer disease: A major killer. *Archives of Neurology, 33*, 217–218.

Kidd, M. (1963). Paired helical filaments in electron microscopy of Alzheimer's disease. *Nature, 197*, 192–193.

Kwiatkowski, T. J. Jr., Bosco, D. A., LeClerc, A. L., Tamrazian, E., Venderburg, C. R., Russ, C., et al. (2009). Mutations in FUS/TLS gene on chromosome 16 cause familial amyothrophic lateral sclerosis. *Science, 323*, 1205–1208.

Masters, C. L., Simms, G., Weinman, N. A., Multhaup, G., McDonald, B. L., & Beyreuther, K. (1985). Amyloid plaque core protein in Alzheimer disease and Down syndrome. *Proceedings of the National Academy of Sciences of the United States of America, 82*, 4245–4249.

Neumann, M., Sampathu, D. M., Kwong, L. K., Truax, A. C., Micsenyi, M. C., Chou, T. T., et al. (2006a). Ubiquitinated TDP-43 in frontotemporal lobar degeneration and amyotrophic lateral sclerosis. *Science, 314*, 130–133.

Polymeropoulos, M. H., Higgins, J. J., Golbe, L. I., Johnson, W. G., Ide, S. E., Di Iorio, G., et al. (1996). Mapping of a gene for Parkinson's disease to chromosome 4q21–q23. *Science, 274*, 1197–1199.

Renton, A. E., Majounie, E., Waite, A., Simón-Sánchez, J., Rollinson, S., Gibbs, J. R., et al. (2011a). A hexanucleotide repeat expansion in C9ORF72 is the cause of chromosome 9p21-linked ALS-FTD. *Neuron, 72*, 257–268.

Rosen, D. R., Siddique, T., Patterson, D., Figlewicz, D. A., Sapp, P., Hentati, A., et al. (1983). Mutations in Cu/Zn superoxide dismutase gene are associated with familial amyotrophic lateral sclerosis. *Nature, 362*, 59–62.

Spillantini, M. G., Crowther, R. A., Jakes, R., Hasegawa, M., & Goedert, M. (1998). α-Synuclein in filamentous inclusions of Lewy bodies from Parkinson's disease and dementia with Lewy bodies. *Proceedings of the National Academy of Sciences of the United States of America, 95*, 6469–6473.

St. George-Hyslop, P. H., Tanzi, R. E., Polinsky, R. J., Haines, J. L., Nee, L., Watkins, P. C., et al. (1987a). The genetic defect causing familial Alzheimer's disease maps on chromosome 21. *Science, 235*, 885–890.

The Huntington's Disease Collaborative Research Group. (1993). A novel gene containing a trinucleotide repeat that is expanded and unstable on Huntington's disease chromosomes. *Cell, 72*, 971–983.

Vance, C., Rogeli, B., Hortobágyi, T., De Vos, K. J., Nishimura, A. L., Sreedharan, J., et al. (2009). Mutations in FUS, an RNA processing protein, cause familial amyotrophic lateral sclerosis type 6. *Science, 323*, 1208–1211.

Prions

Alper, T., Cramp, W. A., Haig, D. A., & Clarke, M. C. (1967). Does the agent of scrapie replicate without nucleic acid? *Nature, 214*, 764–766.

Castilla, J., Saá, P., Hetz, C., & Soto, C. (2005). In vitro generation of infectious scrapie prions. *Cell, 121*, 195–206.

Deleault, N. R., Harris, B. T., Rees, J. R., & Supattapone, S. (2007). Formation of native prions from minimal components *in vitro. Proceedings of the National Academy of Sciences of the United States of America, 104*, 9741–9746.

Gajdusek, D. C., Gibbs, C. J., Jr., & Alpers, M. (1966). Experimental transmission of a kuru-like syndrome to chimpanzees. *Nature, 209*, 794–796.

Gajdusek, D. C., & Zigas, V. (1957). Degenerative disease of the central nervous system in New Guinea: Epidemic occurrence of "kuru" in the native population. *New England Journal of Medicine, 257*, 974–978.

Griffith, J. S. (1967). Nature of the scrapie agent: Self-replication and scrapie. *Nature, 215*, 1043–1044.

Prusiner, S. B. (1982). Novel proteinaceous infectious particles cause scrapie. *Science, 216*, 136–144.

Saborio, G. P., Permanne, B., & Soto, C. (2001). Sensitive detection of pathological prion protein by cyclic amplification of protein misfolding. *Nature, 411*, 810–813.

Wang, F., Wang, X., Yuan, C. G., & Ma, J. (2010). Generating a prion with bacterially expressed recombinant prion protein. *Science, 327*, 1132–1135.

The Amyloid Hypothesis and Protein Misfolding Disorders

Hardy, J. A., & Higgins, G. A. (1992). Alzheimer's disease: The amyloid cascade hypothesis. *Science, 256*, 184–185.

Selkoe, D. J. (1991). The molecular pathology of Alzheimer's disease. *Neuron, 6*, 487–498.

Selkoe, D. J. (2002). Alzheimer's disease is a synaptic failure. *Science, 298*, 789–791.

Walsh, D. M., Klyubin, I., Fadeeva, J. V., Cullen, W. K., Anwyl, R., Wolfe, M. S., et al. (1992). Naturally secreted oligomers of amyloid β protein potently inhibit hippocampal long-term potentiation in vivo. *Nature, 416*, 535–539.

Genetic Risk Factors and Techniques

Botstein, D., White, R. L., Skolnick, M., & Davis, R. W. (1980). Constriction of a genetic linkage map in man using restriction fragment length polymorphism. *American Journal of Human Genetics, 32*, 314–331.

Gusella, J. F., Tanzi, R. E., Anderson, M. A., Hobbs, W., Gibbons, K., Raschtchian, R., et al. (1984). DNA markers for nervous system diseases. *Science, 225*, 1320–1326.

Lander, E. S., & Schork, N. J. (1994). Genetic dissection of complex traits. *Science, 265*, 2037–2047.

Shendure, J., & Hanlee, J. (2008). Next generation DNA sequencing. *Nature Biotechnology, 26*, 1135–1145.

Chapter 2
Protein Folding: Part I—Basic Principles

By the early 1900s proteins had been studied for more than a hundred years. Chemists had begun to extract proteins such as albumin and wheat gluten from animals and plants at the beginning of the previous century. The Dutch chemist Gerardus Johannes Mulder (1802–1880) had found that each one of these proteins had nearly the same composition. He thought that all were variations on a single primordial substance composed of carbon, nitrogen, hydrogen, and oxygen with variable amounts of sulfur and phosphorus. In 1838, the Swedish chemist Jöns Jacob Berzelius (1779–1848) coined the name "protein" for this universal substance. These events were followed by groundbreaking investigations carried out in the later 1850s and 1860s by August Kekulé (1829–1896) that helped lay the foundations for structural chemistry and protein science. In his studies, Kekulé uncovered the tetravalent character of carbon, proposed its ability to bond to other carbons in long chains, and discovered the ring structure of benzene.

That proteins were enormous molecules was recognized right at the start. The molecular structure deduced by Mulder, $C_{400}H_{620}N_{100}O_{120}P_1S_1$, was far larger than any other molecules studied at that time. In order to better understand its structure and composition, proteins were hydrolyzed with dilute acids or bases, and the breakdown products analyzed. By this means (or similar ones), the 20 amino acid constituents of proteins were discovered one by one over a period of time that stretched from 1819 (leucine) to 1936 (threonine). These efforts culminated in 1902 (by then 16 of the 20 amino acids had been correctly identified). In that year, the central concept that proteins are linear chains of amino acids held together by peptide bonds was put forth simultaneously by Emil Fischer (1852–1919) and Franz Hofmeister (1850–1922), Fisher having coined the term *peptide bond* to describe the linkage.

One of the chief properties that interested early protein scientists was the ability of proteins such as albumins, globulins, and hemoglobin to coagulate when subjected to heating. Harriette Chick (1875–1977) was a well-known British nutritionist, biochemist and one of the first protein scientists. She and Lister Institute director

© Springer International Publishing Switzerland 2015

M. Beckerman, *Fundamentals of Neurodegeneration and Protein Misfolding Disorders*,
Biological and Medical Physics, Biomedical Engineering,
DOI 10.1007/978-3-319-22117-5_2

Charles Martin (1866–1955) published a series of papers from 1910 to 1912 in which they established that there were two distinct steps in the "heat coagulation" (their quotes) of egg albumin and hemoglobin. The first step was that of denaturation and this step was separate from the second precipitation step that depended on pH and salts but not heat. These findings established denaturation as a distinct process that altered the functional properties of the proteins, resulting in their loss of catalytic ability, for example.

Among the people studying the responses of proteins to changes in temperature, pH, and salts was the Chinese protein scientist and founder of modern Chinese biochemistry Hsien Wu (1893–1959). In 1931, he became the first to recognize that protein denaturation was a process in which protein underwent a conformational change. During this same time period—from 1925 to 1936—denaturation of a variety of proteins was examined by several groups. In a 1925 paper, Mortimer Louis Anson (1901–1968) and Alfred Mirsky (1900–1974) reported that the unfolding of hemoglobin was reversible. This result was an exceptionally significant one and was tested with other proteins. It was found to be not unique to hemoglobin but rather other proteins such as trypsin could be reversibly unfolded and then refolded as well. However, egg albumin could not. Its importance was conveyed to a broad audience by Alfred Mirsky and Linus Pauling (1901–1994) in a 1936 paper entitled "On the structure of native, denatured, and coagulated proteins", and by Max Perutz (1914–2002) through a 1940 popular presentation entitled "Unboiling an Egg".

The chapter will begin as it must with the discovery of the alpha helix (Fig. 2.1) and beta sheet by Pauling, Robert Brainard Corey (1897–1971), and Herman Russell Branson (1914–1995), and with the discovery of the 3D protein structures of myoglobin and hemoglobin by John Cowdery Kendrew (1917–1997) and Perutz. It will then resume the discussion of denaturation and reversibility with the most famous of the denaturation/reversibility experiments—those of Christian B. Anfinsen (1916–1995)—leading to his thermodynamic hypothesis.

2.1 Discovery of the Alpha Helix and Beta Sheet

The discovery of the alpha helix and beta sheet by Pauling, Corey, and Branson was a seminal event. In a series of eight articles appearing in the *Proceedings of the National Academy of Sciences* in early 1951, they described the forms that these preeminent structures should take and their stabilization by hydrogen bonds between amide (NH) and carbonyl (CO) groups. In alpha helices, the hydrogen bonds form between a carbonyl group on the ith residue and the amide group on the $(i+4)$th residue lying below it. In beta sheets, the hydrogen bonds form between carbonyl group lying on one strand and the amide group situated immediately adjacent to it on another strand. The beta strands can be oriented in either a parallel or an antiparallel manner.

In the alpha helix, the backbone forms the inner portion of structure while the side chains rotate outward. Alpha helices are about ten residues in length and these

Fig. 2.1 Linus Pauling and Robert Corey in 1951 with their model of the alpha helix, Courtesy of the Archives, California Institute of Technology

structures account for about a third of the amino acid residues in a typical protein. Individual beta strands are typically six amino acid residues in length and they account for a quarter of the residues. Turns allow the chain to reverse direction. They along with loops are usually located on the protein surface.

A number of essential ingredients entered into their search for the two repeating hydrogen-bonded structures. The first component was the need for precisely determined bond strengths and bond angles. That was determined in previous work. The second ingredient was one that was missed in previous attempts to find these regular structures. That was the observation that the N–C peptide bond was partially double in character and therefore quite rigid. That point is illustrated in Fig. 2.2 showing that the six atoms are roughly coplanar leaving the only degrees of freedom represented by the two torsion angles, φ and ψ.

The third component was that of steric constraints, namely, two molecules cannot occupy the same position in space. These constraints limit the angular regions available for peptide bonding. A primary tool for visualizing these constraints and revealing the underlying structure of the polypeptide chain is the Ramachandran plot, a tool developed in 1963 by the Indian physicist Gopalasamudram Narayana Ramachandran (1922–2001). A pair of these plots is presented in Fig. 2.3.

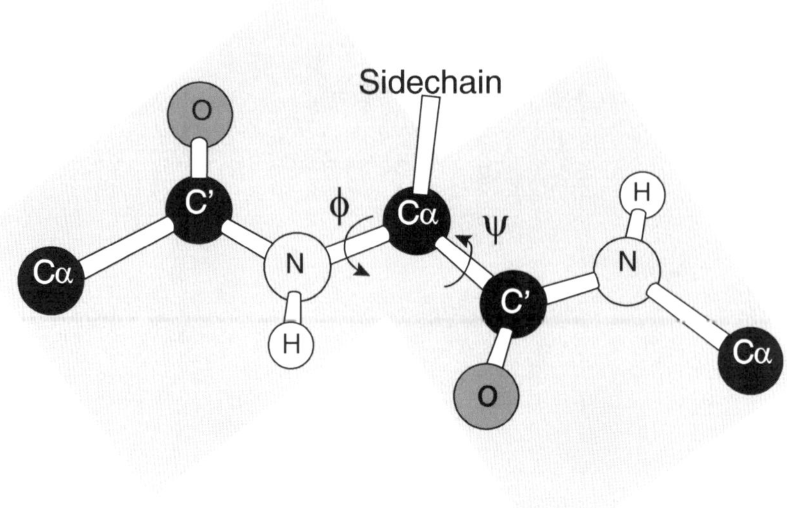

Fig. 2.2 The planar peptide bond and associated torsion (rotation) angles φ and ψ (from Rose *PNAS* 103: 16623 © 2006 National Academy of Sciences, U.S.A. and reprinted with their permission)

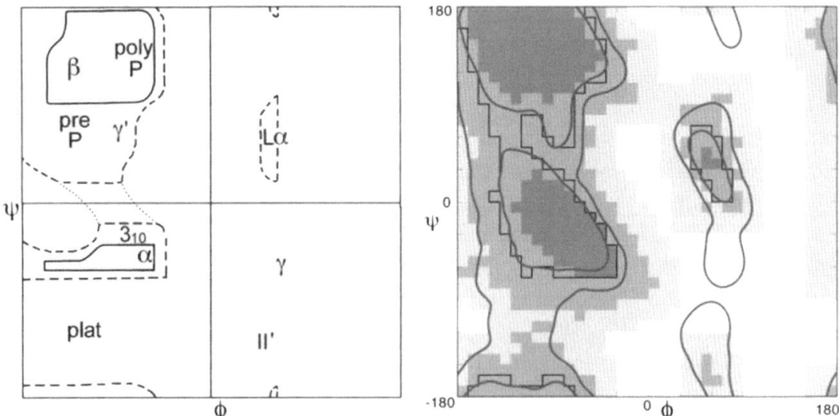

Fig. 2.3 A pair of Ramachandran plots. Not all values of ψ and φ are allowed due to steric constraints. Shown in the *left panel* is an annotated plot of the allowed ψ and φ values as first presented by Ramachandran. Depicted in the *right panel* is a refined and updated Ramachandran plot. In this diagram, the *darkest shading* indicates which areas have high probability of being occupied and by progressively lighter shading those regions that might still be occupied but with lower probability as determined from the data. The large empty regions correspond to (ψ, φ) values that are sterically forbidden. The annotations to the *left panel* highlight the presence of three broad regions populated by alpha helices (α), left-handed helices (Lα) and beta sheet (β), plus several smaller areas (from Lovell *Proteins* 50: 437 © 2003 John Wiley and Sons, and reprinted with their permission)

Lastly, they abandoned the idea that there had to be a whole integer number of residues per turn. Instead they came up with two solutions—a 3.7 residue per turn helix and a 5.1 residue per turn helix. Their result of what must be one of the great triumphs of model building in science was the alpha helix and the beta sheet.

2.2   The Birth of Molecular Biology

2.2.1   X-Ray Diffraction and the Double Helix of DNA

The preeminent method for exploring the shape and internal structure of proteins at atomic level detail is X-ray crystallography. The use of X-rays to explore protein structure had its beginnings over a century ago. X-rays were discovered in 1895 by Wilhelm Roentgen (1845–1923) leading to his award of the first Nobel Prize in physics in 1901. A few years later, in 1912, Max von Laue (1879–1960) (along with Paul Knipping and Walter Friedrich) discovered that crystals could act as a diffraction grating for the scattering of X-rays. This discovery was immediately followed in by those of William Lawrence Bragg (1890–1971) and his father William Henry Bragg (1862–1942) that established the fundamental relationship between X-ray diffraction patterns and atomic positions in a crystal known as Bragg's law (see Appendix 1). These pioneering studies focused on simple crystals such as ZnS (von Laue) and NaCl (the Braggs). Using this new physics technique the Braggs made the important finding that atoms of salt are not organized as a molecule but rather as ions and the monumental discovery (along with Laue) of the wavelike properties of X-rays.

These discoveries led to the award of the Nobel Prize in Physics to Laue in 1914 and to the Bragg's in 1915. WL Bragg was 25 at the time, the youngest-ever recipient of the Nobel Prize in Physics. It is worthwhile to note the role that technology played in making the discoveries by the Bragg's possible. These breakthroughs could not have been made without being preceded by development of (1) the capability to produce high potency (high voltage) electricity, and (2) the ability to generate high vacuum, plus (3) the invention of dry-plate photography.

Moving forward in time, the extension of X-ray diffraction techniques to large or even moderate biomolecules was regarded by many as impractical if not outright impossible. In spite of these misgivings, efforts to do so began in the 1920s and 1930s with the pioneering studies of fibrous materials such as wool by William Astbury (1898–1961). He had studied in the Bragg's laboratory and then moved to Leeds, the center of the British textiles industry. At Leeds he successfully carried out X-ray diffraction studies of fibrous materials including keratin (the material out of which wool is made). He was the first to propose that protein chains could be stabilized by forming hydrogen bonds, and proposed that the fibers formed a helix that uncoiled when the fibers were stretched. In doing so, he laid the foundation for the discovery of the alpha helix by Pauling and Corey. His insights into the physics of wool were immortalized in a limerick by A.L. Patterson and presented at the beginning of a 1938 paper by Astbury entitled "X-ray adventures among the proteins":

> Amino-acids in chains
>> Are the cause, so the x-ray explains,
>> Of the stretching of wool
>> And its strength when you pull,
>> And show why it shrinks when it rains

Astbury also carried out early studies of DNA together with his student Florence Bell. It was already known that chromosomes contain both proteins and nucleic acids. However, at that time DNA was regarded as being too simple a molecule to be the repository of genetic information. Instead, the far more interesting and complex proteins had to be responsible for the storage, with the linear and simpler DNA merely serving in a structural capacity. Astbury and Bell found that purine and pyrimidine bases were stacked along the axis of the molecule, but the full analysis had to wait for Crick and Watson and the X-ray diffraction studies of DNA by Maurice Hugh Frederick Wilkins (1916–2004) and Rosalind Elsie Franklin (1920–1958) in which the existence of two distinct structural forms was first made apparent. The missing ingredients to finding the structure were then supplied by Crick and Watson using the new data. In arriving at their famous solution, they had to get past the prevalent notion at the time of simple crystalline solid arrangements. In their place, they followed Pauling and Corey and introduced helices with nonrational number of repeating units per 360° turn. Finally, they made the great leap forward by proposing the complementary base-paired, double helix that no one before had thought of. Once that structure had emerged it became apparent that is could serve as the information storage. That conclusion appeared, in one of the greatest understatements of science, as a single sentence near the end of their brief two-page paper announcing their discovery of the structure of DNA:

> It has not escaped out notice that the specific pairing we have postulated immediately suggests a possible copying mechanism for the genetic material.

2.2.2 *The First Protein Structures Are Solved by Kendrew and Perutz*

The first clear picture of what a protein looks like in three dimensions was presented by John Kendrew (1917–1997) in 1958 with his publication in *Nature* of the crystal structure of myoglobin. This landmark event was followed in 1960 with the publication of a more refined myoglobin structure by Kendrew and with the publication also in *Nature* by Max Perutz (1914–2002) of the three-dimensional structure of hemoglobin. These efforts had their antecedents in the earlier work of Astbury discussed above and also in a study by John Desmond Bernal (1901–1971) and his student at the time, Dorothy Crowfoot Hodgkin (1910–1994). They obtained the first X-ray diffraction picture of a nonfibrous protein, pepsin, and published their results in 1934 in *Nature*. That effort served as inspiration for Perutz who had joined Bernal at Cambridge in 1936.

The fundamental challenge faced by Perutz and others with solving the three-dimensional structure of a complex biomolecule from X-ray diffraction data is known as the "phase problem". In order to extract electron densities from the patterns of light and dark spots, one needs to know not only the positions and intensities of the light spots but also its phases. This is not contained in the data. For small molecules one can use additional information and guesswork to overcome the phase problem, but this approach becomes progressively more difficult to execute as the number of atoms in a unit cell increases. This necessitates the development of alternative methods that compensate for the missing phase information. Perutz created one of these methods, known as *isomorphous replacement*, to solve the phase problem. In his approach, heavy atoms (having many more electrons than the atoms of the biomolecule) are added to the protein of interest without altering the crystal formation (hence the appellation "isomorphous"). The additional atoms do not shift the locations of the light and dark spots but do alter their intensities. From the patterns with and without the heavy atoms enough information is now present to extract the needed phases.

The solution found by Perutz involved attaching mercury atoms to the hemoglobins in a way that left the locations of the spots alone while changing the intensities. That method was published in a pair of papers that appeared in 1954. The second crucial innovation utilized by Perutz was that of a Patterson map, an inventive way of using just the intensities to produce a map of the relative positions of pairs of atoms in the structure. That crucial intermediate step was pioneered in 1934 by Arthur Lindo Patterson (1902–1966).

When the 3D protein structures were finally solved (Fig. 2.4) the responses throughout the field were surprise (and perhaps dismay) at the complexity and lack of any obvious order to the arrangement of amino acid residues. Pauling's alpha helices were clearly present in the crystal structures but these were arranged in ways that were not even remotely obvious. That response to the structure was best summarized by Kendrew in his 1958 publication in *Nature*:

> Perhaps the most remarkable features of the molecule are its complexity and its lack of symmetry. The arrangement seems to be almost totally lacking in the kind of regularities which one instinctively anticipates, and is more complicated than has been predicted by any theory of protein structure.

2.3   Anfinsen's Postulate and Thermodynamic Hypothesis

2.3.1   *The Divide Between Chemistry and Biology Is Breached*

In 1969, at a joint press conference, two groups reported that they had successfully synthesized an enzyme, ribonuclease A, composed of 124 amino acids. One team situated at Merck Research Laboratory was led by Ralph Franz Hirschmann (1922–2009) and the other located at Rockefeller University was headed by Robert Bruce Merrifield (1921–2006). These researchers solved the previously thought insurmountable problem of how to construct in the laboratory a linear chain of amino

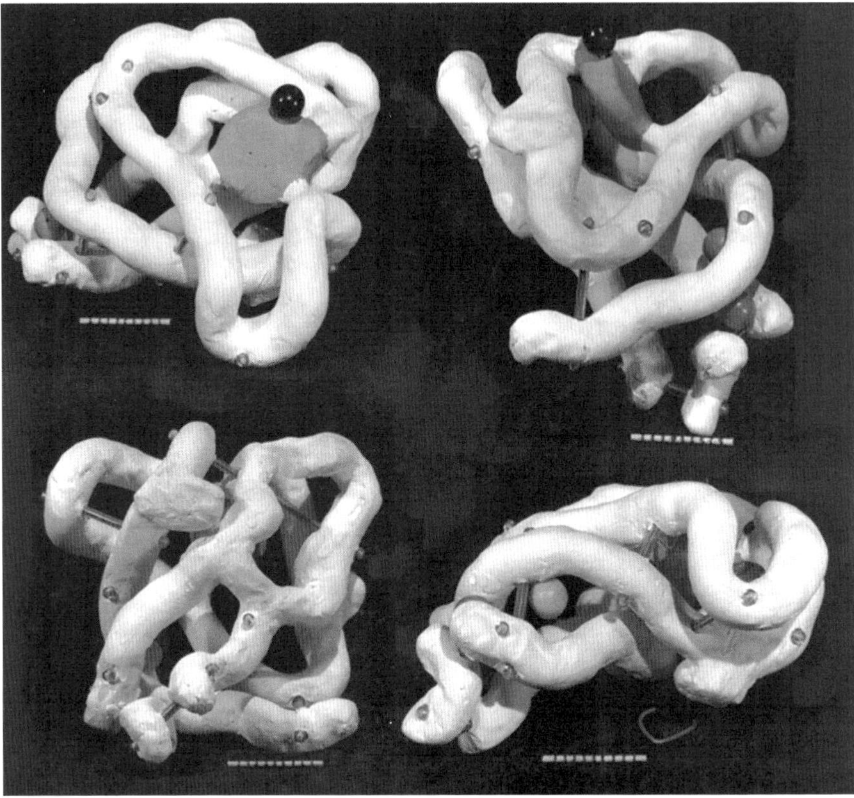

Fig. 2.4 The structure of myoglobin at 6 Å resolution solved by Kendrew in 1958. Presented in this figure is a high res reconstruction of the original Kendrew drawing from his 1958 *Nature* article (from Fersht *Nat. Rev. Mol. Cell Biol.* 9: 650 © 2008 Reprinted by permission from Macmillan Publishers Ltd)

acids in the correct fashion to make a biologically viable protein given that each amino acid has multiple binding sites and only one of these in each instance is the correct one. These efforts were following in 1971 with the publication by Merrifield of a paper reporting that the linear sequence of amino acids joined via peptide bonds can determine by itself the enzyme's three-dimensional (tertiary) structure. (Merrifield was awarded the Nobel Prize in Chemistry in 1984 for his invention of solid phase peptide synthesis, the method he used to synthesize ribonuclease A.)

2.3.2   Anfinsen's Postulate

The first and most important set of clues of how a protein folds into its complex shape comes from the denaturation and refolding experiments. The most famous of these experiments are those of Christian Anfinsen (1916–1995). He showed in his studies of ribonuclease that not only can it be made to refold after unfolding even under extreme conditions, but its enzymatic activity will be restored, as well. He then postulated that *all the information needed to fold a protein is encoded in its primary amino acid sequence*. He further hypothesized that *the process of folding and unfolding obeys the laws of thermodynamics*, and this assertion has come to be known as the *thermodynamic hypothesis*. Given that a cell functions far from thermodynamic equilibrium this statement is in itself a remarkable one. These studies took place in the 1950s and early 1960s and led to his being awarded the 1972 Nobel Prize in Chemistry (along with Stanford Moore and William Howard Stein who elucidated the connection between the chemical structure and catalytic activity of ribonuclease).

Thermodynamic stability is an essential property of a protein. To be useful, a protein's native state must be stable in the thermodynamic sense. Such states, once formed, do not change appreciably in time. The effects of small perturbations and of thermal fluctuations are rapidly damped out and the behavior of the protein is not appreciably altered. These are equilibrium states in the language of thermodynamics. To establish stability in the thermodynamic sense one must show that (1) the native folded conformation is a function of state alone, that is, it does not depend on path or process used to get there, and (2) that the native state is situated at a global minimum in the Gibbs free energy. The reversible unfolding and refolding experiments established that under physiological conditions a protein's folded (native state) was a well-defined thermodynamic equilibrium state corresponding to a minimum in the Gibbs free energy. Anfinsen's framing of folding in the language of thermodynamics followed and incorporated the earlier views of Hsien Wu, Tim Anson, Alfred Mirsky, and Linus Pauling.

Recall from basic biochemistry that the change in Gibbs free energy of a chemical process, in this case that of a protein undergoing folding, is:

$$\Delta G = \Delta H - T\Delta S \tag{2.1}$$

where, ΔH is the change in enthalpy of the system and $T\Delta S$ is the corresponding change in entropy multiplied by the temperature. Further recall that the enthalpic contribution is the sum of two terms, the change in internal energy of the system, ΔE, plus the amount of work done on the surroundings by the system, $p\Delta V$, with ΔV denoting the change in volume and p the pressure (assumed constant). Under the aqueous conditions in which protein folding occurs the $p\Delta V$ contribution to the enthalpy can be neglected and the change in enthalpy is simply the change in internal energy. The entropic term may be understood in terms of the number of configurations available to the protein at a given energy. As noted by Anfinsen as well as Pauling and many others early on, a denatured protein has many configurations available to it while the native state possesses just a few low energy configurations. Thus, the amount of entropy (disorder) is high when the protein is denatured and low when it is folded tightly into its three-dimensional functional form. This change renders the $-T\Delta S$ term strongly positive. The condition for a reaction to occur spontaneously is for

$$\Delta G < 0 \tag{2.2}$$

In protein folding this requirement is achieved by the strong negative character of the enthalpic term, that is, through the decrease in internal energy that more than compensates for the opposing entropic contribution.

While there are a multitude of states corresponding to the denatured state, there are usually only a few similar states corresponding to the native state and its conformation is essentially unique. In order for the native state to be stable there must be an appreciable gap in energy between the native state and nearby nonnative ones. When the differences are appreciable, it is difficult for small perturbations and thermal fluctuations to induce transitions to the nearby higher energy states. Whenever the energy gaps are small, the proteins will only be marginally stable.

Putting this all together along with the discoveries of messenger RNA and codons results in the central operating principles that guide all of biology along with biochemistry, biophysics, and other related disciplines:

- Genetic information is encoded in DNA, which is transcribed into RNA, which is then translated into the protein's primary sequence.
- The primary sequence contains all the information needed to fold the protein into its native-state 3D structure.
- The native-state 3D structure determines the cellular function.

2.4 Native-State 3D Structure

Structure-function relationships are of fundamental interest in science. They are the subject of a famous dialog on bone structure (form) written by Galileo Galilei (1564–1642), which appeared in his "Two New Sciences" in 1638. More recently, they played a central role in the classic treatise "On Growth and Form" published in 1919 by D'Arcy Wentworth Thompson (1860–1948). In proteins, the forms of interest are the native-state 3D structures as these determine the physiological function(s) that the protein is capable of performing.

2.4.1 The Protein Structure Hierarchy

There are four basic layers of protein structure—primary, secondary, tertiary, and quaternary. This particular hierarchy of structural elements was introduced by Kaj Ulrik Linderstrøm-Lang (1896–1959) in the Lane medical lectures delivered at Stanford University in 1951 and published in 1952 by the Stanford University Press. The primary structure is simply the covalent structure of the polypeptide chain including any covalent disulfide bonds that form during folding. Its secondary structure refers to the alpha helices and beta sheets that form through hydrogen bonding and *turns* that connect them. Portions of the polypeptide chain that do not form these regular structures are termed *random coils*. As mentioned earlier, alpha helices are typically about ten residues in length while individual beta strands are usually about six residues in length and form hydrogen bonds with other beta strands to form the beta sheets.

It is customary nowadays to add two additional layers in-between the secondary and tertiary structures. *Supersecondary* structural elements are commonly occurring combinations of secondary structure elements. Examples are the helix-turn-helix and the Greek key (two pairs of antiparallel beta strands). Small supersecondary elements are also referred to as *motifs*. The other additional structural elements are the *domains*. These are the chief functional units in the hierarchy of protein structures, and are situated above the secondary and supersecondary structures and below the tertiary structure.

Domains are compact, conserved, three-dimensional parts of a protein capable of carrying out a specific cellular function independently of other parts of the protein. An example of domain structures is presented in Fig. 2.5 in which the 3D native structure of the nonreceptor tyrosine kinase Src is presented. As can be seen in this figure the protein chain folds into several structurally distinct domains. The SH2 (Src-homology-2) and SH3 domains are recognition modules that direct the protein to the correct binding partners. Catalysis is initiated through tyrosine phosphorylation in the activation loop. SH2 domains bind phospholipids and phosphotyrosine-rich sequences such as pYEEI, while SH3 domains bind proline-rich sequences such as PXXP characteristic of PP II helices.

Domains such as SH2 and SH3 are highly modular and are found in several hundred proteins. In recognition of the importance of domains, there are several open-access databases devoted to archiving their properties. Two of the most prominent are CATH (Class, Architecture, Topology, and Homology) and SCOP (Structural Classification Of Proteins). The tertiary structure of a protein is its 3D native-state structure. These consist of one or more protein domains. A protein possesses a quaternary structure in those instances where two or more polypeptide chains associate and each folded chain operates as a subunit.

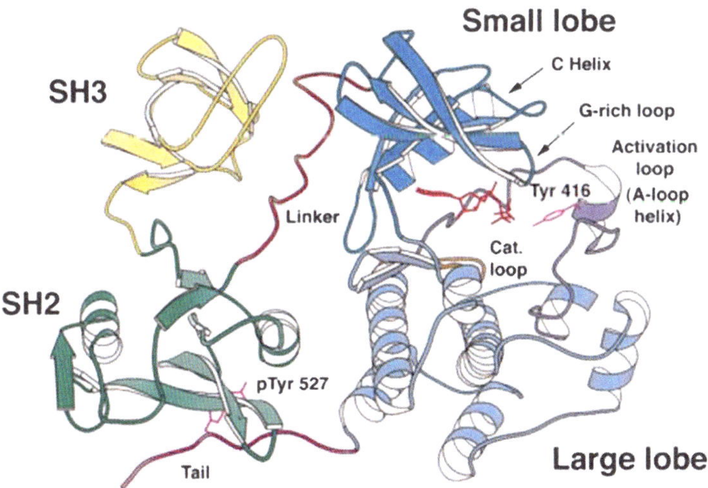

Fig. 2.5 Src domain organization. The amino acid residues in this enzyme are organized into four domains: an N-terminal SH3 domain, a C-terminal SH2 domain, and two catalytic domains connected by a linker sequence (from Xu *Mol. Cell* 3: 629 © 1999 Reprinted by permission from Elsevier)

It is customary to represent amino acids that form well-defined secondary structure elements as ribbons. This type of representation was introduced by Jane Richardson in 1981 as a way of converting long list of atomic coordinates determined by means of X-ray crystallography or NMR into something that can be visualized and understood. Since that time, amino acids that form alpha helices have been depicted as coiled ribbons, and those that form beta strands have been shown as flat arrow-like ribbons.

2.4.2 Semi-Independent Folding and Assembly

The picture of a protein constructed in a modular fashion from functionally distinct domains leads naturally to the notion that these structural units may fold independently into their compact tertiary structures. That idea gives rise to the concept of a foldon. These are independent folding units and thus defined through their folding capabilities whereas domains are defined in terms of their functional properties. Ideally, the two ways of defining these crucial structural elements coincide—the independent structural elements fold independently, as well—while allowing that boundaries may vary slightly from one to the other.

One of the most important sets of finding from H/D exchange experiments by Walter Englander (to be discussed in Chap. 3) pertains to the existence of semi-independent cooperative folding by cytochrome c. This protein consists of 103 amino acid residues organized into five foldon units. As uncovered by Englander these units are assembled in a sequential manner during protein folding. Each foldon functions as a cooperative folding entity and the various stages of native structure coalesce out of the random coil one at a time as illustrated in Fig. 2.6. More generally, the notion of stepwise folding and assembly is a common one, and the steps can occur either in parallel or sequentially. A similar theme is present in the *zipping and assembly model* introduced by Ken Dill in 2007 where small secondary structure elements are formed through hydrophobic zipping (zippering) and these elements then assembly to form the tertiary structure of the protein.

2.5 Hydrogen Bonding, Hydrophobic "Forces", and Steric Constraints Direct Protein Folding

So, how does a protein fold? The answer is that *the folding of a protein from an open largely unstructured conformation into a compact 3D native shape is directed by macromolecular forces*. The three main contributors are *hydrogen bonding, the hydrophobic effect*, and *steric constrains*. Hydrogen bonding is especially important for the formation of secondary structures—the alpha helix and beta sheet. The importance of this driving force was amply illustrated by Mirsky, Pauling, and Corey, and by numerous studies carried out since then. These secondary structures

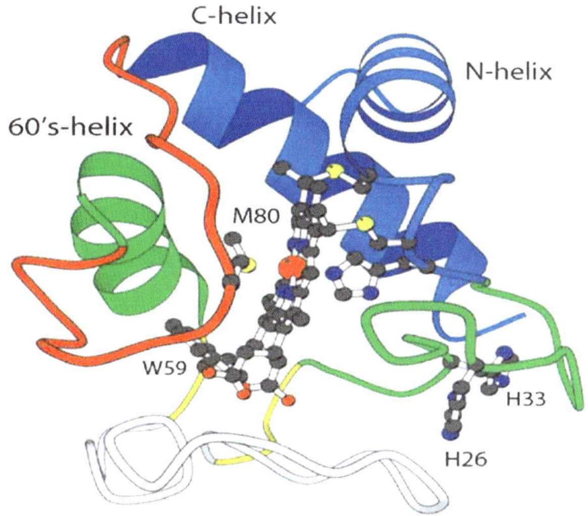

Fig. 2.6 Stepwise assembly of cytochrome c from independent foldons as determined by hydrogen exchange. The various foldons are *color coded* and assemble in a sequential manner starting with the *blue units*, then the *green units*, and next the short *yellow connectors*, then the *red loop*, and lastly the nested *yellow* (*white*) *loop*. The protein is depicted covalently-bonded to a heme group (from Maity *J. Mol. Biol.* 343: 223 © 2004 Reprinted by permission from Elsevier)

account for the vast majority of hydrogen bonds in a protein. In general, hydrogen bond strengths depend on the partner characteristics and microenvironment conditions under which the residues undergoing hydrogen bonding are operating—polar or nonpolar, or charged or hydrophobic, or solvent-exposed. Bond strengths span a broad range from −0.5 to −4.5 kcal/mol., and average about −1.3 to −1.5 kcal/mol.

The hydrophobic effect is the other major contributor. Walter Kauzmann (1916–2009) championed its importance in protein folding in a famous 1959 review article. In that paper, he noted that during protein folding hydrophobic residues are brought into the interior of the structure, that is, they are buried in the interior, while hydrophilic ones are exposed to the solvent. A short time later Charles Tanford (1921–2009) concluded that the stability of the native conformation was entirely due to the hydrophobic effect. This aspect is still under investigation; an emerging consensus is that Tanford's assertion is perhaps too strong. A recent estimate is that the two types of forces contribute to protein folding in roughly a 60–40 split, with 60 % coming from the hydrophobic effect and 40 % from hydrogen bonding. The magnitude of these contributions varies somewhat with protein size. Small proteins cannot bury their hydrophobic residues as completely as can large proteins. As a result the hydrophobic contribution is reduced in small proteins.

Just what is the hydrophobic effect? To answer that question it is worthwhile to consider what happens when oil and water come into contact. That water and oil do not mix has been known for a long time. The phenomenon whereby pouring oil on water stills wave motion was remarked on by Pliny the Elder (AD 23–AD 79).

That same phenomenon was studied scientifically many years later by Benjamin Franklin (1706–1790). A century after Franklin, Lord Rayleigh (1842–1919) exploited the phenomenon in his estimation of molecular sizes. The relevance of hydrophobic behavior for proteins was recognized early on by the X-ray crystallography pioneer John Bernal (1901–1971) and in the Nobel-Prize winning studies of oil films by the surface chemist Irving Langmuir (1881–1957).

In brief, the hydrophobic effect arises from the propensity of water molecules to form hydrogen bonds with other water molecules. Typically each water molecule forms three or four hydrogen bonds with nearby water molecules. The result is a three-dimensional lattice work of water molecules. A protein consisting as it does of 20 different kinds of amino acids has a far more complex set of interactions with water than does oil. Some amino acids are polar, while others are nonpolar; some have a net charge, others do not. Hydrophilic (water-loving) residues are polar, and can reorient themselves and form bonds with water molecules. Water molecules have a far greater affinity for other water molecule than for nonpolar amino acids, and the latter are therefore referred to as being *hydrophobic* (water-hating).

Water is a poor solvent for proteins and this aspect underlies the ability of proteins to spontaneously fold into their various conformations. Water is, in fact, essential. In the absence of water, a protein could not fold into its native state nor could it carry out essential functions such as catalysis. Water surrounds the protein, fills its pockets and grooves on its surface and occupies voids in its interior. In those instances when polar amino acids occupy a protein interior, hydrogen bonds to water molecules alleviate their disruptive influences of the charged groups upon protein stability.

A number of hydrophobicity scales have been published. A consensus list of the most hydrophobic residues is presented in Table 2.1. In terms of their electrostatic properties, these amino acids are, as expected, nonpolar.

The third set of macromolecular determinants of protein structure are the steric constraints on the allowed positions of residues. These constraints were the subject of the previously mentioned investigations carried out around 1960 by Ramachandran. The only protein crystal structure known at the time was that of Kendrew's myoglo-

Table 2.1 Hydrophobic residues shown alphabetically

Hydrophobic amino acids
Alanine (Ala, A)
*Cysteine (Cys, C)
*Isoleucine (Ile, I)
*Leucine (Leu, L)
Methionine (Met, M)
*Phenylalanine (Phe, F)
Tryptophan (Trp, W)
*Valine (Val, V)

The most hydrophobic residues are prefixed by an *asterisk*. Others included in the list are both weaker and more variable in their placement in hydrophobicity scales. Listed in the parentheses are the three- and one-letter codes for these amino acids

bin and he supplemented that data with crystal structures obtained for small peptides. Calculations were done over several months using an electric desktop calculator in lieu of having a computer available to him, and published in 1963. Since then there have been a number of updates and improvements that make use of the newer computer analysis techniques and rapidly expanding number of protein crystal structures available in the Brookhaven protein data bank (PDB). These plots are widely used as a means of validating and refining proposed protein structures.

There are four basic Ramachandran plots. The standard plot (e.g., Fig. 2.3) shows the sterically allowed and forbidden regions for the 18 nonglycine, nonproline amino acids. There are only small variations in Ramachandran plots arising from the different side chains of these 18 amino acids hence these plots are usually combined into a general plot. The allowed (φ, ψ) regions for proline are restricted because of its pyrrolidine ring and it gets its own Ramachandran plot as does glycine. Because of its small size lacking as its does a side chain, glycine can adopt a larger than average range of torsion angles. The fourth type of Ramachandran plot, preproline, is needed because of the restrictions placed by proline on the residues immediately preceding it in a polypeptide chain.

2.6 Levinthal's Paradox

The thermodynamic principle that governs protein folding has two parts. First, protein folding is directed by the macromolecular forces felt by each of the atoms in the protein. Secondly, driven by these forces the protein arrives at a native state that is at a global minimum in the Gibbs free energy. That is, in the language of Anfinsen's postulate, a protein initially in an unfolded denatured state spontaneously folds into its global free energy minimum using only the information coded in its primary amino acid sequence.

At this point an often-posed question is: how does a protein "find" its global minimum in free energy? This type of question was brought into sharp focus by Cyrus Levinthal (1922–1990) in 1968. He pointed out that for a protein consisting of 100 amino acids with perhaps three bond states per amino acid there would be some $3^{100} = 5 \times 10^{47}$ possible configurations of the chain. Assuming that a protein can sample configurations at a rate of 10^{13} per second it would take 10^{27} years to probe each and every one of them. Not only is that not going to happen, but in fact, many of the small globular proteins of roughly 100 amino acid residues in size can fold in a millisecond or less into their native shape.

The Levinthal paradox is largely a rhetorical one. The answer to why it does not happen has several parts. First, a protein does not carry out a global search for Gibbs free energy minima. Instead, as discussed above, its folding is directed by the macromolecular forces along certain pathways. Secondly, there is far more to the fundamental relationships between genes (DNA) and proteins that would appear from a cursory examination. The primary sequences have been carefully sculpted by evolution to enable folding into a stable free energy minimum in physiologically relevant time-frames. Those proteins that do not fold properly do not survive.

In protein folding, physics and evolution work together to produce a number of mechanistic strategies that greatly shorten the folding time. These strategies make efficient use of all the macromolecular forces. They are best illustrated in two-state folding of simple motifs, larger supersecondary structure elements, and small globular proteins. These entities fold in remarkably short time frames in the microsecond to millisecond range. They will be examined in the remainder of this chapter along with the specific strategies exploited with such great success.

2.7 Two-State Folding

The simplest and fastest pathway that a protein can take as it spontaneously folds into its 3D native state from an initially synthesized, unfolded, or denatured state is through two-state kinetics. This folding process is depicted in Fig. 2.7 showing the existence of two stable states and a transition barrier separating one from the other. This mechanism was strongly suggested by Anfinsen's experiments and by earlier ones and is often included as part of his thermodynamic hypothesis. This situation is the protein folding counterpart to classical chemical kinetics in which reactants and products are replaced by the stable denatured and native states, and the only impediment is the activation barrier, the height of which determines the

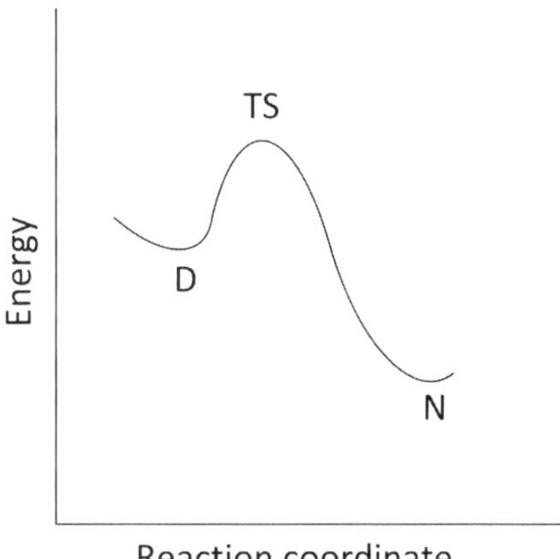

Fig. 2.7 Transition state picture of two-state protein folding. There are two stable states—the denatured state (D) and the native state (N). The reaction rate depends critically on the height of the transition state barrier relative to the denatured state well bottom. Under denaturing conditions the relative heights of the D and N wells are reversed and the denatured state lies at a lower energy than the native state

rate at which folding occurs. The first protein found that folds in this manner was the 64 amino acid residue chymotrypsin inhibitor 2. It was shown in 1991 to fold according to this simple prescription and since then a growing number of small proteins and protein modules has been found that fold rapidly in a manner consistent with this picture.

The two-state description of a chemical reaction in terms of an activation complex (transition state) is not a new one. It was introduced by the Swedish physicist and chemist Svante Arrhenius (1859–1927) in 1889. The reaction rate, k, for this process can be calculated in terms of the temperature by means of the Arrhenius equation:

$$k = k_0 \exp(-E / k_B T) \tag{2.3}$$

In this expression, k_0 is the rate constant, E is the activation energy, k_B is Boltzmann's constant, and T is the temperature. The key feature in this equation is the exponential factor in which $k_B T$ sets a scale for the energies. If the activation energy is much larger than the Boltzmann factor the reaction rate will become slow, and conversely, as the energy decreases below it the rate approaches its maximal value set by the rate constant. Similar arguments can be made as the temperature is varied.

Motivated by Einstein's theory of Brownian motion the Dutch physicist Hendrik Kramers (1894–1952) introduced a more detailed depiction of a chemical reaction in a famous paper published in 1940. In his article, he envisioned a diffusive process taking place along a one-dimensional reaction coordinate. In his mathematical formalism, he again utilized an exponential factor, but in this case it depends on the change in Gibb's free energy, and he introduced a diffusion-dependent preexponential factor:

$$k = k_D \exp(-\Delta G_{DTS} / k_B T) \tag{2.4}$$

with

$$k_D = \omega_D \omega_{TS} D / 2\pi k_B T \tag{2.5}$$

In Eqs. (2.4) and (2.5), ΔG_{DTS} is the difference in Gibb's free energies between the denatured and transition states, ω_D^2 and ω_{TS}^2 are the curvatures of the denatured state well and transition-state barrier, respectively, assumed for simplicity to have parabolic shapes, and D is the diffusion coefficient. Recall from Einstein's theory of Brownian motion that the diffusion coefficient is inversely proportional to the viscosity of the fluid through which the Brownian particles are moving. In protein folding, the underlying picture is that of the denatured/unfolded proteins moving through the solvent as they begin to fold. Another factor that enters later in the process is internal friction, which further slows the folding of the protein into its compact three-dimensional form. The effects of solvent viscosity on protein folding have been explored starting with myoglobin by the Frauenfelder and Eaton groups. Since then Kramers rate theory has been applied to the experimental analysis of proteins with growing attention to the role of internal friction.

2.8   Lessons from the Folding of Peptides and Small Globular Proteins

2.8.1   Folding of the Small Globular Proteins C2I and Barnase

Two small globular proteins that have been studied extensively experimentally and theoretically are chymotrypsin inhibitor 2 (CI2) mentioned in the previous section and barnase. These proteins fold from their denatured state to their native state in 10 ms and 50 ms, respectively. The sequence of steps through which CI2 folds is depicted in Fig. 2.8. This single-domain protein reversibly folds and unfolds, and serves as an archetypical example of two-state folding in which there is a single rate-limiting transition state. As can be seen in the figure, the denatured state is mostly but not completely unstructured. As it folds into a compact 3D form, the native-state secondary structure elements gradually coalesce. Many, if not most, of the secondary structure elements are already present in the transition-state structure.

The 110 residue bacterial enzyme barnase is a more complex protein than CI2. Its secondary structure is more extensive. It has several hydrophobic cores, and it possesses multiple modules that establish contact with one other. Its folding is depicted in Fig. 2.9. In contrast to CI2, a fairly long-lived intermediate state (I) is populated along the folding route prior to reaching the transition state. Folding occurs in several stages largely independent of one another. Lastly, the denatured state possesses several secondary structure elements and is fairly compact in comparison to CI2.

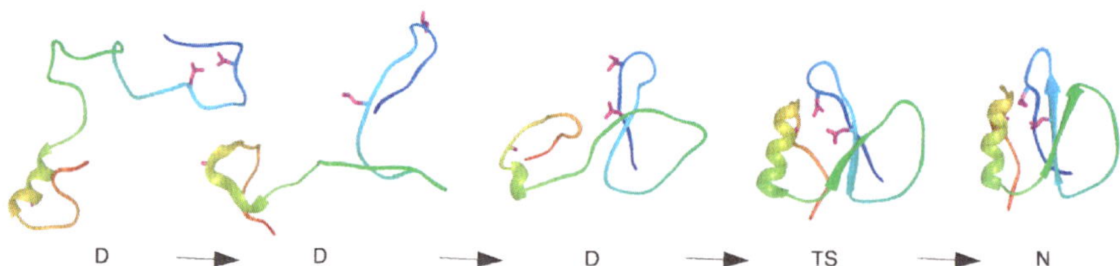

Fig. 2.8 Folding of CI2 determined through a combination of ϕ-value analysis, nuclear magnetic resonance, and molecular dynamics simulations (from Fersht *Cell* 108: 1 © 2002 Reprinted by permission from Elsevier)

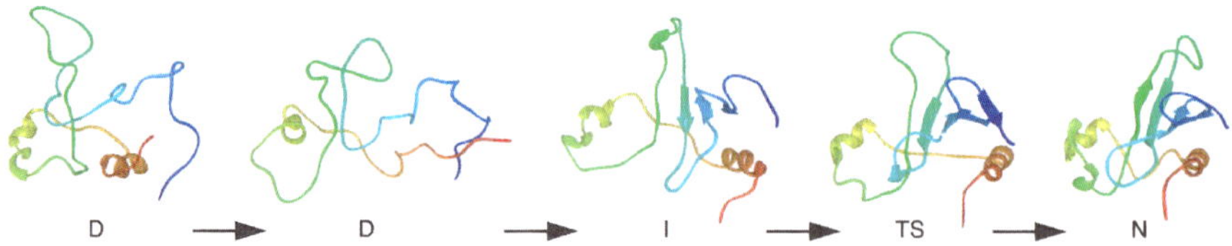

Fig. 2.9 Folding of barnase determined as in Fig. 2.8 (from Fersht *Cell* 108: 1 © 2002 Reprinted by permission from Elsevier)

2.8.2   *Cooperative Folding of Alpha Helices*

Considerable attention has been given to the folding of the secondary structural elements that come together to form the small globular proteins of the type just discussed as well as others, large and small. Prominent among the component systems belonging to all, or nearly all, of these proteins are (1) the alpha helix and its coalescence from a random coil, and (2) the formation and stabilization of small beta-hairpins that serve as basic units in beta sheet formation. The folding of an alpha helix from an open and unstructured conformation is known as the *helix-coil transition*, and its study produced two early models of protein folding, the *Zimm-Bragg model* and *Lifson-Roig model* introduced in 1959 and 1961, respectively.

In the Zimm-Bragg model, each residue in the chain is in either a c (coil) state or an h (helix) state, and the system evolves through sequences of local, nearest-neighbor interactions so that there are again two stable states—all c's or all h's. The states of the chain are then described by sequences of c's and h's such as ccccchhc-chhh…The folding of the chain of residues is then described by two parameters, σ and s. The first of these, σ, is a *nucleation* parameter. It describes the propensity of a chain (monomer) to occupy the helix state. The second parameter, s, is a propagation factor. It captures the nearest-neighbor interactions that bias a monomer towards the helix state if its neighbors are in that same state.

Results from the Zimm-Bragg mode are compared to experimental data in Fig. 2.10. As can be seen in the graphs, as the polymer length increases the data and

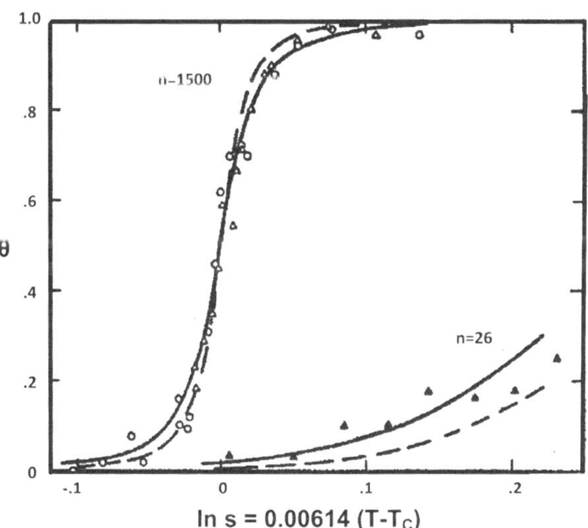

Fig. 2.10 Comparison of experimental and Zimm-Bragg model calculations for two different poly-γ-benzyl-L-glutamate polymer lengths (*n*). Experimental data are represented by *triangles*, *circles*, and *squares* while theoretical results are shown for two values of the nucleation parameter σ: *solid curves* correspond to $\sigma = 2 \times 10^{-4}$ and *dashed curves* to $\sigma = 1 \times 10^{-4}$. The critical temperature T_c is the value at which $\theta = 0.5$ (from Zimm *J. Chem. Phys.* 31: 526 © 1959 AIP Publishing LLC and reprinted with their permission)

curves become progressively more sigmoidal in shape. There is a lag phase, followed by a rapid rise that terminates in a flat plateau. Systems that behave in this manner are *cooperative*. *These are procedures that generate large-scale effects through sequences of small-scale, nearest-neighbor interactions*. This strategy is a common one and one of the ways that proteins bypass the Levinthal paradox.

2.8.3 Nucleation and Hydrophobic Cores in the Folding of Beta-Hairpins

The simplest example of how the several kinds of macromolecular forces work together to produce a stable fold is the *beta-hairpin*. An archetypical beta-hairpin is the 16-residue C-terminal fragment from protein G B1 depicted in Fig. 2.11. This structure, like all beta-hairpins, consists of *two antiparallel beta strands connected by a short turn*. This simple system has been studied extensively both experimentally and theoretically. From these studies it emerges that there are three stages in the folding.

1. Folding starts at the turn, which functions as a *nucleation* site; it brings the two strands into close contact.
2. A *hydrophobic core* develops built upon native-like contacts between nonpolar residues.
3. *Backbone hydrogen bonding* stabilizes the structure at the global free energy minimum.

Overall, folding is both rapid and consistent with the two-state model. Turn development is the rate-limiting step, and this depends on the amino acid composition in the turn and the relative position of the hydrophobic core.

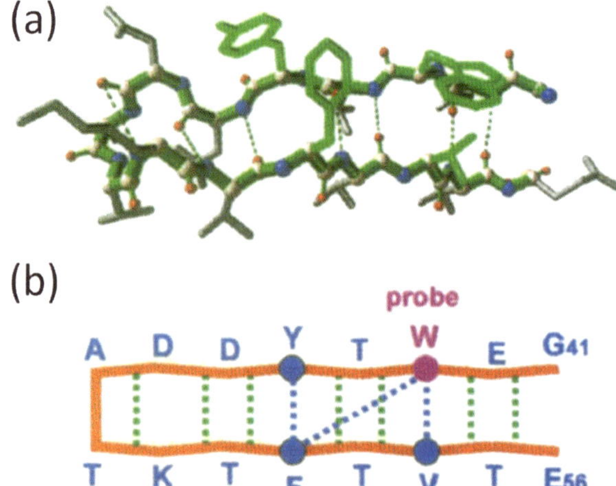

Fig. 2.11 Beta-hairpin structure of protein G B1, residues 41–56. (**a**) Structural representation and (**b**) schematic diagram of the hydrogen bonding and hydrophobic interactions involving residues Y, F, W, and V (from Muñoz PNAS 95: 5872 © 1998 National Academy of Sciences, U.S.A. and reprinted with their permission)

Folding times vary greatly with the size and topological complexity of the protein's native state. The helix-coil transition and beta-hairpin formation occur on a 0.1–$1.0\,\mu s$ timescale, beta-hairpin formation being about an order of magnitude slower than helix-coil transitions. Numerous ultrafast folders, that is, small protein motifs and supersecondary structure elements that fold on a similar timescale to these two, have been found. Fast folding proteins possess a considerable amount of local (e.g., helix and tight turn) structure while slower folders have a greater amount of nonlocal (e.g., beta sheet) structure. Studies of the folding times for structural elements such as these have led to the suggestion that there is a protein folding "speed limit". This limit can be represented by the simple expression $N/100\,\mu s$ for an N-residue single-domain protein lacking topological complexities such as disulfide bonds. Overall, protein folding times vary greatly from 10^{-7} s to 10^{2} s, with large multidomain proteins occupying the long folding times end of the scale. Oligomerization and amyloid fibril development take even longer, from 10^{3} s to more than 10^{6} s.

2.9   Several Highly Effective Folding Strategies Are Utilized

In examining the results of the folding studies of peptides, secondary structures, and globular proteins, a number of folding stratagems and mechanisms stand out. These act over a small scale (e.g., over helices and beta-hairpins), and in concert with one another, either simultaneously or sequentially, to fold entire proteins. These strategies are small in number and have led to the creation of a small set of mathematical/computer models, each encapsulating one or more specific core mechanisms and enabling their further study. These folding mechanisms and models are, as follows:

- Cooperativity and zippering
- Hydrophobic collapse
- Nucleation-growth
- Diffusion-collision
- Stabilization

2.9.1   Cooperativity and Zippering

Cooperative processes are widely encountered in physical and chemical systems and in biology at all levels or organization. Hemoglobin is a good example of the role of cooperativity in protein function. Recall that hemoglobin picks up, delivers, and releases oxygen. It contains four identical subunits each with its own oxygen binding site. Binding is cooperative—binding of oxygen at one binding site potentiates binding affinity of its neighboring subunits. Conversely, the release of oxygen at one binding site stimulates release at the other sites. The resulting oxygen dissociation curve (oxygen saturation versus oxygen partial pressure) has the sigmoidal shape that characterizes a cooperative process.

The classical example of a cooperative process in physics is that of ferromagnetism. In 1924, in his doctoral dissertation, Ernst Ising (1900–1998) introduced a simple one-dimensional nearest-neighbor model to describe how iron atoms placed in a rigid lattice align their spins and generate ferromagnetism. In his model, the interaction energy Φ_{ij} between adjacent, or nearest-neighbor, lattice sites, i and j, had the form

$$\Phi_{ij} = -Js_i \cdot s_{j=i+1} \tag{2.6}$$

where, the spin variables s can take on one of two values, either $+1$ or -1 depending on whether the spins are up or down. Like hemoglobin there are two lowest energy states corresponding to the cases where all spins are aligned and all point up or all point down (in hemoglobin—fully oxygenated or fully depleted).

This simple one-dimensional model lives on as a prototype for cooperative algorithms and phenomena throughout the engineering disciplines and in the sciences. In particular, it has guided the development of the helix-coil model just discussed. Proteins fold in a three-dimensional space and the notion of a protein nearest-neighbor is a far richer one than that of atoms fixed and immovable in a rigid lattice. Shown in Fig. 2.12 are the two types of nearest-neighbors—connected neighbors and topological ones—along with a depiction of local (short range) and nonlocal (long range) interactions appropriate for residues linked together in a flexible chain.

Zippering can be thought of as a limiting case where the transitions between stable states are rapid. There are several different kinds of zippers and these can be driven by either hydrogen bonding or hydrophobic interactions. One of these is the *tryptophan zipper*. This mechanism can drive the formation of beta-hairpins. In these situations, two pairs of hydrophobic tryptophan (W) residues form a hydrophobic core that stabilizes 11–16 residue long beta-hairpins. These beta-hairpins can fold and unfold reversibly in a highly cooperative manner, and are perhaps the smallest independently stable structures found to date. Still another kind of zipper is the *steric*

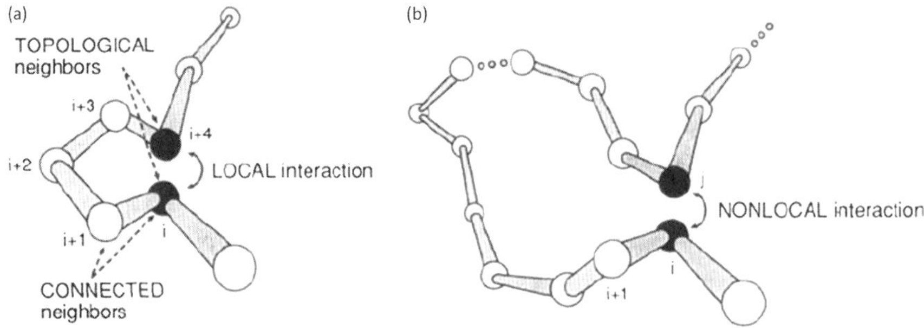

Fig. 2.12 Short- and long-range interactions. These terms refer to the distances separating residues of interest along the polypeptide chain. If they are near one another in sequence as in (**a**) the case of hydrogen bonding in an alpha helix, then they are short-range. If, on the other hand, as in (**b**) the two residues coming into contact are well separated in sequence, then they are long-range (Reprinted with permission from Dill *Biochem.* 29: 7133 © 1990 American Chemical Society)

zipper. It is generated through the inter-digitization of amino acid side chains, and promotes amyloid fibril formation. Lastly, *polar zippers* involving glutamate repeats were introduced by Max Perutz as a possible contributor to Huntington's disease.

2.9.2 Hydrophobic Collapse

Hydrophobic interactions that bring together and compact the polypeptide chain are a common theme in protein folding. They appear in the beta-hairpin descriptions, and more generally in a variety of folding scenarios. The term "molten globule" was coined by the Russian physicist Oleg Ptitsyn in 1973. In his depiction of folding, the protein would first fold along its backbone leading to formation of secondary structure elements. In the next stage, the protein would be driven by hydrophobic interactions and hydrogen bonding to form a more compact, *molten-globule state* that was situated along the folding pathway to the native state. In the *molten-globule model*, native-like secondary structure elements are present but the protein's tertiary structure is fluid and not yet fixed. Subsequent studies have provided experimental evidence for Ptitsyn's model, showing the emergence of the molten globule as a key intermediate. Other studies point to an initial hydrophobic collapse that can occur without the preceding or accompanying formation of any discernible secondary structure.

2.9.3 Nucleation

Nucleation is another pervasive mechanism in protein folding. It, like hydrophobic collapse, operates across organizational levels—from secondary structure to tertiary and quaternary. Hydrophobic collapse and nucleation along with remaining themes in the bullet list describe the overall series of mechanistic steps taken by all proteins as they fold as, for example, depicted in Figs. 2.8 and 2.9. These commonly occurring steps have been codified into several global protein folding models.

At the lowest level, nucleation was already encountered in the helix-coil model where one of the main parameters was a nucleation parameter, the other being a propagation factor. In larger-scale models the propagation step is referred to as a growth stage or, in the case of amyloid development, a polymerization phase. The *framework model* was introduced by Kim and Baldwin in 1982. In this hierarchical depiction of protein folding, a set of disconnected but stable secondary structure elements form first driven by local hydrogen bonding. These nucleating secondary structure elements then coalesce into the protein's tertiary structure.

In the two-state model of protein folding discussed earlier, there are no long-lived intermediate states. There is only the single-transition state. In examining the detailed evolution of the protein from initial to native state by means of ϕ-value analysis (to be discussed in Chap. 3) it emerged that secondary and tertiary structure elements can co-develop along with an overall collapse of the protein into a more compact form. This led to the introduction of the *nucleation-condensation model* of protein folding by Alan Fersht and his co-workers in 1995. In this picture, long-range tertiary and other hydrophobic interactions stabilize the nascent secondary structures present in the transition state. These diffuse secondary structure elements nucleate the folding process. This is an expansion of an earlier presentation of nucleation by Donald Wetlaufer in 1973 in which the notion of a domain was introduced. In his conceptual model, local interactions nucleate the independent formation of secondary structure elements in several regions of the polypeptide chain; these each seed the growth of additional secondary structure and the emergence of tertiary structures.

2.9.4　Diffusion-Collision

The *diffusion-collision model* of protein folding was introduced in 1976 by Martin Karplus and David Weaver. In their model, stable secondary structure elements (either alpha helices and beta sheets or hydrophobic clusters) form first. These structures, or microdomains, move about diffusively under the influence of internal and external random forces and come into contact (collide) with one another. When they do so the microdomains will sometimes coalesce, and through a series of these steps the protein will develop its tightly folded tertiary native structure.

2.9.5　Stabilization

Stability is crucial. The contributions to protein stability from the enthalpies associated with hydrogen bonding and hydrophobic interactions are almost completely matched by the loss configuration entropy occurring when a protein transitions from its denatured state to its native state. *The resulting $\Delta G_{D \rightarrow N}$ is quite small and ranges from -5 to -15 kcal/mol. Thus, a folded protein is only marginally stable and it takes only a few 1–2 kcal/mol mutations to destabilize it. The resulting loss of stability is a key trigger to the sequence of events leading from mutation-induced destabilization to partial unfolding to misfolding, aggregation and, neurodegeneration.*

One of the ways that peptides and small protein motifs are stabilized is through *capping*. Helices and other small structural elements are usually not stable without some attention being paid to the unpartnered N- and C-terminal sequences. In order to stabilize the ends, side chains are sometimes used to form terminal hydrogen bonds. In other situations, hydrophobic residues are present that cap these residues. In addition, and as will be discussed later, glycine and proline residues have special roles in stabilization. More generally, specific structural elements may not by themselves be stable but when these elements coalesce and assemble they are stabilized through contacts with the other structures. Lastly, and as will be discussed later in the text, in some cases, the final stabilizing steps in proteins folding may not occur until that protein comes into contact with its cognate binding partner.

2.10 Summary

1. Denaturation experiments in which proteins reversibly unfold and refold have provided essential insights into the thermodynamic properties of proteins. Starting with the experiments of Chick and Martin in the 1910–1912 timeframe and culminating in Anfinsen's experiments in the 1950s and 1960s these experiments led to the fundamental hypothesis of protein folding:

   - All the information necessary for folding a protein into its native state is contained in the primary sequence of amino acids, and
   - Protein folding obeys the laws of equilibrium thermodynamics; under normal conditions, the native state is a stable state of minimum Gibbs free energy.

2. Pioneering studies by Pauling, Corey, and Branson led to the discovery of the two main types of protein secondary structure—the alpha helix and the beta sheet. These structures are held together by hydrogen bonds between mainchain NH and CO groups. Combinations of these regular structures along with turns and loops comprise the higher layers of protein structure. There are six layers altogether (Table 2.2). These are:

   Alpha helices and beta strands and sheets are protein building blocks; by themselves they do not possess any functional specificity. The chief functional units are the protein domains. One or more of these comprise the protein's 3D native structure and determine its functional properties.

3. The application of X-ray crystallography to proteins by Perutz and others has enabled researchers to examine and study protein 3D native structure at the atomic level. Extension of X-ray crystallography to the study of DNA resulted in discovery of the double helix and to the central operating principle of the field, namely,

   - Genetic information is encoded in DNA, which is transcribed into RNA, which is then translated into the protein's primary sequence.
   - The primary sequence contains all the information needed to fold the protein into its native-state 3D structure.
   - The native-state 3D structure determines the cellular function.

4. Macromolecular forces—especially hydrogen bonding and hydrophobic interactions—drive the folding of a protein into its native state. These forces operate

Table 2.2 Hierarchical protein structure

Layer	Definition
Quaternary	Subunit associations of separate (folded) chains
Tertiary	3D native structure
Domain	Independent functional units
Supersecondary	Combinations of secondary structure elements
Secondary	Alpha helices, beta strands, loops, and turns
Primary	Covalent polypeptide chain structure

Table 2.3 Mechanistic models of protein folding

Folding model	Folding features emphasized by the model
Framework	Hierarchical assembly in which secondary structures develop first under the influence of local interactions; then followed by emergence of tertiary structure
Molten globule	Secondary structure elements form and then nonspecific hydrophobic interactions and hydrogen bonding drive the collapse of the protein into a compact, three-dimensional shape
Diffusion-collision	Secondary structure elements form and these elements subsequently collide with one another and compact the protein into its three-dimensional folded form
Nucleation condensation	Weak secondary structure elements form and these are assisted and stabilized by the emerging tertiary structure formation; both local and long-range interactions contribute
Zipping and assembly	Hydrophobic contacts initially form and are reinforced by other hydrophobic contacts and fragments folded in this manner then assemble to form the tertiary structure

together with evolutionary selection to produce primary sequences that fold into stable low energy states in physiological useful times. These times vary from micro- and milliseconds to minutes, hours, and longer depending on protein complexity. Peptides and small globular proteins can fold rapidly while large multidomain proteins and amyloids require far longer folding times.

Studies of models of secondary structure formation highlight the importance of cooperative processes in which local, nearest-neighbor interactions lead to formation of alpha helices. The beta-hairpin investigations extend these ideas and show how hydrogen bonding and development of a hydrophobic core lead to stability. These ideas are reinforced and extended through studies of two-state folding of small globular proteins. A common feature in protein folding is the emergence of a hydrophobic core consisting of nonpolar residues. Cooperative folding along with the other common folding strategies such as zippering, nucleation, hydrophobic collapse, diffusion, and stepwise assembly embody how the macromolecular forces enable proteins to fold as rapidly as they do. These themes are best summarized in terms of folding models that emphasize useful combinations of these strategies (Table 2.3).

5. The two-state folding of a protein into its 3D native conformation is best described by Kramers' formula, Eqs. (2.4) and (2.5). In these expressions, the rate-limiting step is passage from the denatured state over the transition-state barrier. It is conceptualized following Einstein's Brownian motion findings. It is pictured as a diffusive process in which the viscosity of the solvent plays a key role along with the path topology that is approximated by barriers and valley curvatures. In more recent studies, the importance of internal friction has emerged.

In the next chapter, energy landscapes and the various pathways through them from denatured state to native state will be examined. In doing so, more general situations will be explored in which larger and more complex proteins fold into their native state. A key property of the energy landscapes, roughness, will emerge; this property is closely related to internal friction. Another major consideration is the existence of one or more intermediate states that slow down folding. This too is related to roughness and leads to the kinetic control of folding.

Appendix 1. X-Ray Crystallography

In their work, the Bragg's noted that a three-dimensional crystal could be viewed as a set of equidistant parallel planes. In these situations, the conditions for maximal constructive interference are twofold. First the scattering from each plane must be a specular, or mirror, that is, reflection in which the angle of incidence equals the angle of reflection. This situation is depicted in Fig. 2.13. Second the X-ray wavelength λ, distance between parallel planes d, and angle of reflection θ obey the relationship known as Bragg's law:

$$2d \sin \theta = n\lambda \tag{2.7}$$

In Eq. (2.7), n is an integer that can take on the values 1, 2, 3, and so on. When $n=1$, the spots of light are known as first order reflections, and when $n=2$, they are called second order reflections. First order reflections are more intense than second order reflections, and similarly for third and higher order contributions.

In more detail, x-rays are produced whenever swiftly moving electrons strike a solid target. In an X-ray tube, a beam of electrons is generated that strikes a metallic anode (typically copper) to produce an X-ray beam. The X-rays in the beam are scattered by the electron clouds of atoms, particularly by tightly bound electrons near the center of the atoms. Light scattered from single atoms is too weak to observe, but the amount of light can be amplified using purified crystals. In a crystal, large numbers of identical molecules are arranged in a regular lattice. Light passing through a crystal will be scattered in a variety of directions. Spherical wavelets scattered by different atoms will interfere, some constructively and some

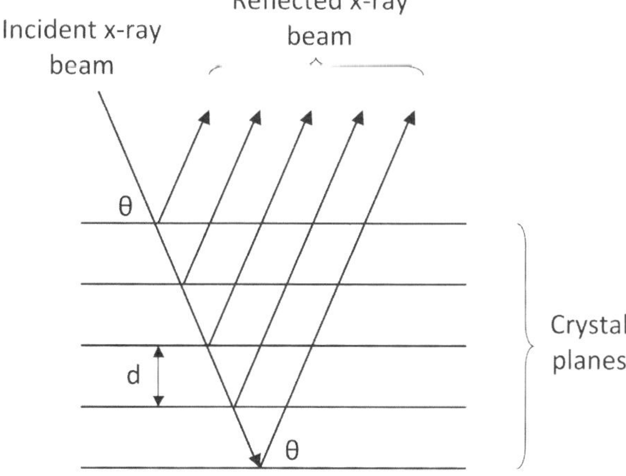

Fig. 2.13 The arrangement of crystal planes and the geometry in Bragg's law

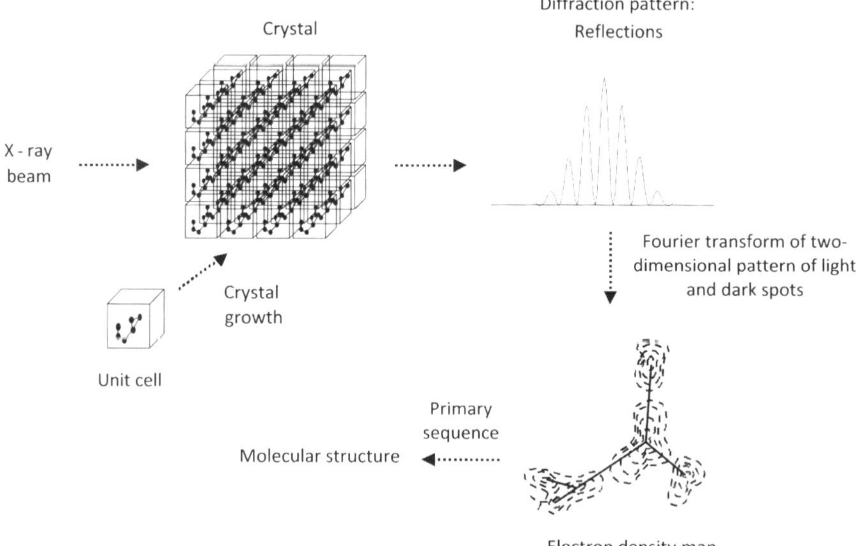

Fig. 2.14 Schematic depictions of the steps in X-ray crystallography in which an incident X-ray beam is directed at a crystal containing an array of unit cells, each cell containing the protein of interest. The two-dimensional diffraction pattern (here shown in 1D form with its peaks and valleys) is then Fourier-transformed (once the phase problem is solved). The resulting electron density map is further analyzed to produce the 3D protein structure

destructively. For certain wavelengths and scattering directions, the wavelets will be in phase to produce strong constructive interference. Constructive interference taking place between light waves scattered off of the atoms serves to amplify the light, producing a characteristic pattern of light spots and dark areas, a diffraction pattern that can be seen and analyzed to yield information on how the atoms are arranged.

The light spots are produced when the light waves arrive at the detector in phase with one another and thus constructively interfere. In an X-ray diffraction experiment, the intensities and positions of the light spots are recorded. The diffraction patterns are converted into electron density maps through application of a mathematical operation known as a Fourier transform. Several tens of thousands of reflections are collected in a typical X-ray diffraction experiment. Computer programs, taking as input the resulting electron density map and knowledge of the primary sequence, are used to deduce the three-dimensional arrangement of atoms in the protein (Fig. 2.14).

Today, once they are discovered, the three-dimensional coordinates (x, y, z) for each atom in a protein are routinely deposited in the Protein Data Bank (PDB) and made available to the research community. The PDB repository was established at Brookhaven National Laboratory in Long Island, and at several mirror sites throughout the world. Of the more than 84,000 structures for proteins and peptides that have been deposited in the PDB to-date, 75,000 were determined using X-ray crystallography and 9000 using nuclear magnetic resonance (NMR) spectroscopy (to be discussed in Chap. 3).

Bibliography

Denaturation and Heat Coagulation

Anson, M. L., & Mirsky, A. E. (1934). The equilibrium between active native trypsin and inactive denatured trypsin. *The Journal of General Physiology, 17*, 393–398.

Astbury, W. T., Dickinson, S., & Bailey, K. (1935). The X-ray interpretation of denaturation and the structure of the seed globulins. *The Biochemical Journal, 29*, 2351–2360.

Chick, H., & Martin, C. J. (1910). On the "heat coagulation" of proteins. *The Journal of Physiology, 40*, 404–430.

Mirsky, A. E., & Pauling, L. (1936). On the structure of native, denatured, and coagulated proteins. *Proceedings of the National Academy of Sciences of the United States of America, 22*, 439–447.

Wu, H. (1931). Studies on denaturation of proteins. XIII. A theory of denaturation. *The Chinese Journal of Physiology, 5*, 321–344.

The Alpha Helix and Beta-Pleated Sheet

Lovell, S. C. (2003). Structure validation by Cα geometry: φ, ψ and Cβ deviation. *Proteins: Structure, Function, and Genetics, 50*, 437–450.

Pauling, L., & Corey, R. B. (1951). The pleated sheet, a new layer configuration of polypeptide chains. *Proceedings of the National Academy of Sciences of the United States of America, 37*, 251–256.

Pauling, L., Corey, R. B., & Branson, H. R. (1951). The structure of proteins: Two hydrogen bonded helical configurations of the polypeptide chain. *Proceedings of the National Academy of Sciences of the United States of America, 37*, 205–210.

Ramachandran, G. N., Ramakrishnan, C., & Sasisekharan, V. (1963). Stereochemistry of polypeptide chain configurations. *Journal of Molecular Biology, 7*, 95–99.

The First Protein Structures

Bragg, W. L. (1912). The diffraction of short electromagnetic waves by a crystal. *Proceedings of the Cambridge Philosophical Society, XVII*(1), 43–57.

Kendrew, J. C., Bodo, G., Dintzis, H. M., Parrish, R. C., Wyckoff, H., & Phillips, D. C. (1958). A three-dimensional model of the myoglobin molecule obtained by X-ray analysis. *Nature, 181*, 662–666.

Perutz, M. F. (1956). Isomorphous replacement and phase determination in non-centrosymmetric space groups. *Acta Crystallographica, 9*, 867–873.

Perutz, M. F., Rossmann, M. G., Cullis, A. F., Muirhead, H., Will, G., & North, A. C. T. (1960). Structure of haemoglobin. A three-dimensional Fourier synthesis at 5.5-Å resolution, obtained by X-ray analysis. *Nature, 185*, 416–422.

Protein Folding and the Thermodynamic Hypothesis

Anfinsen, B. C. (1973). Principles that guide the folding of protein chains. *Science, 181*, 223–230.

Gutte, B., & Merrifield, R. B. (1969). The total synthesis of an enzyme with ribonuclease A activity. *Journal of the American Chemical Society, 91*, 501–502.

Hirschmann, R., et al. (1969). Studies on the total synthesis of an enzyme. V. The preparation of enzymatically active material. *Journal of the American Chemical Society, 91*, 507–508.

Levinthal, C. (1968). Are there pathways for protein folding? *Extrait du Journal de Chimie Physique, 65*, 44–45.

Protein Anatomy and Protein Folding Hierarchy

Lindorff-Larsen, K., Røgen, P., Paci, E., Vendruscolo, M., & Dobson, C. M. (2005). Protein folding and the organization of the protein topology universe. *Trends in Biochemical Sciences, 30*, 13–19.

Richardson, J. S. (1981). The anatomy and taxonomy of protein structure. *Advances in Protein Chemistry, 34*, 167–339.

Hydrophobic Forces

Chandler, D. (2005). Interfaces and the driving force of hydrophobic assembly. *Nature, 437*, 640–647.

Kauzmann, W. (1959). Some factors in the interpretation of protein denaturation. *Advances in Protein Chemistry, 14*, 1–63.

Pace, C. N., Shirley, B. A., McNutt, M., & Gajiwala, K. (1996). Forces contributing to the conformational stability of proteins. *The FASEB Journal, 10*, 75–83.

Tanford, C. (1962). Contribution of hydrophobic interactions to the stability of the globular conformation of proteins. *Journal of the American Chemical Society, 84*, 4240–4247.

Two-State Folding

Ansari, A., Jones, C. M., Henri, E. R., Hofrichter, J., & Eaton, W. A. (1992). The role of solvent viscosity in the dynamics of protein conformational change. *Science, 256*, 1796–1798.

Best, R. B., & Hummer, G. (2010). Coordinate-dependent diffusion in protein folding. *Proceedings of the National Academy of Sciences of the United States of America, 107*, 1088–1093.

Fersht, A. R., & Daggett, V. (2002). Protein folding and unfolding at atomic resolution. *Cell, 108*, 1–20.

Kramers, H. A. (1940). Brownian motion in a field of force and the diffusion model of chemical reactions. *Physica, 7*, 284–304.

Folding of Alpha Helices and Beta-Hairpins

Blanco, F. J., Rivas, G., & Serrano, L. (1994). A short linear peptide that folds in a native stable β-hairpin in aqueous solution. *Nature Structural Biology, 1*, 584–590.

Kubelka, J., Hofrichter, J., & Eaton, W. A. (2004). The protein folding 'speed limit'. *Current Opinion in Structural Biology, 14*, 76–88.

Lifson, S., & Roig, A. (1961). On the theory of helix-coil transition in polypeptides. *The Journal of Chemical Physics, 34*, 1963–1974.

Muñoz, V., Thompson, P. A., Hofrichter, J., & Eaton, W. A. (1997). Folding dynamics and mechanism of β-hairpin formation. *Nature, 390*, 196–199.

Plaxco, K. W., Simons, K. T., & Baker, D. (1998). Contact order, transition state placement and the refolding rates of single domain proteins. *Journal of Molecular Biology, 277*, 985–994.

Zimm, B. H., & Bragg, J. K. (1959). Theory of the phase transition between helix and random coil in polypeptide chains. *The Journal of Chemical Physics, 31*, 526–535.

Strategies and Models

Aurora, R., & Rose, G. D. (1998). Helix capping. *Protein Science, 7*, 21–38.

Cochran, A. G., Skelton, N. J., & Starovasnik, M. A. (2001). Tryptophan zippers: Stable, monomeric β-hairpins. *Proceedings of the National Academy of Sciences of the United States of America, 98*, 5578–5583.

Dill, K. A., Fiebig, K. M., & Chan, H. S. (1993). Cooperativity in protein folding kinetics. *Proceedings of the National Academy of Sciences of the United States of America, 90*, 1942–1946.

Fersht, A. R. (1997). Nucleation mechanisms in protein folding. *Current Opinion in Structural Biology, 7*, 3–9.

Itzhaki, L. S., Otzen, D. E., & Fersht, A. R. (1995). The structure of the transition state for folding of chymotrypsin inhibitor 2 analyzed by protein engineering methods: Evidence for a nucleation-condensation mechanism for protein folding. *Journal of Molecular Biology, 254*, 260–288.

Karplus, M., & Weaver, D. L. (1976). Protein folding dynamics. *Science, 260*, 404–406.

Karplus, M., & Weaver, D. L. (1994). Protein folding dynamics: The diffusion-collision model and experimental data. *Protein Science, 3*, 650–668.

Kim, P. S., & Baldwin, R. L. (1982). Specific intermediates in the folding reactions of small proteins and the mechanism of protein folding. *Annual Review of Biochemistry, 51*, 459–489.

Ozkan, S. B., Wu, G. A., Chodera, J. D., & Dill, K. A. (2007). Protein folding by zipping and assembly. *Proceedings of the National Academy of Sciences of the United States of America, 104*, 11987–11992.

Pace, C. N., et al. (2011). Contribution of hydrophobic interactions to protein stability. *Journal of Molecular Biology, 408*, 514–528.

Ptitsyn, O. B. (1987). Protein folding: Hypothesis and experiments. *Journal of Protein Chemistry, 6*, 273–293.

Wetlaufer, D. B. (1973). Nucleation, rapid folding, and globular intrachain regions in proteins. *Proceedings of the National Academy of Sciences of the United States of America, 70*, 697–701.

Chapter 3
Protein Folding: Part II—Energy Landscapes and Protein Dynamics

In thinking about how a protein might look in its three-dimensional fully folded form, Hsien Wu had envisioned it as forming a crystalline solid composed of repeated folded structural elements. That hope for simplicity by him and everyone else was effectively dashed by the pioneering studies of Kendrew and Perutz. Studies carried out later by Frederic Richards (1925–2009) [who solved the third ever protein structure in 1967, that of ribonuclease S] and others on packing densities confirmed part of Wu's depiction. There were few if any large voids in the protein interior, and the overall protein densities were indeed consistent with that of an organic compound in the crystalline state, but one without repeating regularities. However, when examined at a finer scale it turns out that the interiors are quite variable in their packing and do not resemble a tightly fit-together jigsaw puzzle so much as a randomly packed sets of nuts and bolts. The packing is; in fact, loose enough to permit a variety of movements.

This chapter will begin with protein motions and their importance for protein function. That discussion will set the stage for the modern landscape picture of protein folding which will follow. The landscape picture provides a conceptual framework and vocabulary for thinking about, and visualizing, protein folding. It encompasses not only the rapid two-state folding of small globular proteins but also the far slower folding of large multidomain ones and those that misfold and aggregate. Terms such as pathway diffusion and internal friction will find a simple interpretation within this picture as will kinetic control of folding and intermediate states.

Tools are important, and any field of science moves only as fast as its exploratory toolkit allows. In the case of protein folding, the field has advanced through creation of an ever-expanding body of experimental and theoretical/computational methods. This progression began with chemical denaturation techniques used to unfold and refold proteins. It then leapt forward through atomic level X-ray crystallography and NMR methods, and with chemical, spectroscopic, and computational methods that provide glimpses of the transition and intermediate states along the folding pathways. These exploratory tools will be introduced in the second part of this chapter.

© Springer International Publishing Switzerland 2015

M. Beckerman, *Fundamentals of Neurodegeneration and Protein Misfolding Disorders*,
Biological and Medical Physics, Biomedical Engineering,
DOI 10.1007/978-3-319-22117-5_3

3.1 Protein Motions Are Necessary for Protein Function

Proteins are dynamic, and not static, entities. This central fact has been long recognized and appreciated. According to Richard Feynman (1918–1988) proteins "jiggle and wiggle", while Gregorio Weber (1916–1997) characterizes them in an even more dramatic fashion as "kicking and screaming". The three-dimensional forms revealed by X-ray crystallography can be thought of as snapshots, frozen in time, of the proteins' average structure. Proteins undergo constant motions about these average conformations. Proteins motions range from small amplitude low-energy vibrational and rotational motions over femtosecond to nanosecond timescales to large-scale domain movements over a microsecond to second-plus timescales. The amplitudes of the motions vary from 0.01 Å to 100 Å, and their corresponding energies from 0.1 kcal to 100 kcal.

Motions are essential for protein function. In a series of landmark studies, Hans Frauenfelder, Robert Austin and their coworkers explored how motions regulate protein function. The protein examined by them was myoglobin (Mb), the 153 amino acid oxygen-storage protein whose structure was pioneered earlier by Kendrew and Perutz. The goal in the studies by Frauenfelder and Austin was to identify exactly how oxygen storage and release took place. What they found was that the penetration of dioxygen (O_2) and carbon monoxide (CO) into the heme binding sites was made possible by the ability of myoglobin to continually undergo small movements and equilibrium fluctuations (EFs). If myoglobin was completely fixed in its shape and unable to move the protein could not function. The picture that emerged was one in which the native and other long-lived states are not single unique conformations but instead consist of ensembles of substates arranged in a hierarchical manner to form an energy landscape. This central aspect of protein structure and function is depicted in Fig. 3.1.

3.1.1 Stability Against Perturbations

The most important property of a system in thermal equilibrium is its stability against random perturbations. The theory underlying this stability is summarized by a family of statistical physics findings known collectively as the *fluctuation-dissipation theorems*. These principles were first stated by Harry Nyquist (1889–1976) in 1928 and later proven by Herbert Callen (1919–1993) and Theodore Welton (1918–2010) in 1951. These theorems establish that the spontaneous fluctuations in a system at equilibrium with its surroundings are indistinguishable from deviations from equilibrium generated by small nonequilibrium perturbations. This happens because the regression (decay) of the spontaneous fluctuations in a system at equilibrium occurs through the same mechanisms that promote the relaxation towards equilibrium of (small) nonequilibrium perturbations in that system.

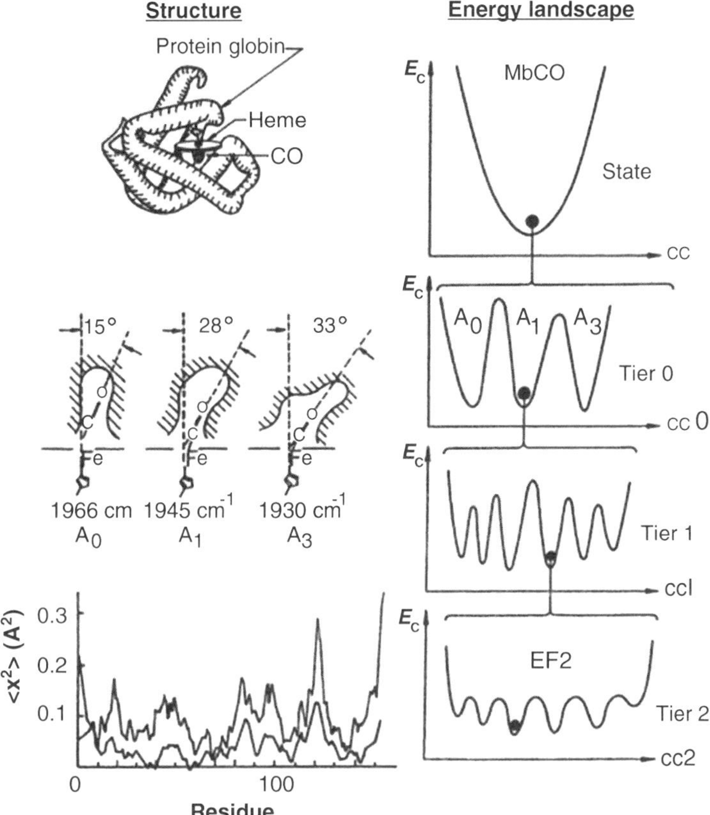

Fig. 3.1 Structure and free energy landscape of MbCO. *Left-hand panels*: Myoglobin structure. *Right-hand panels*: Hierarchy of conformational energies (E_c) as a function of a conformational coordinate (cc). Illustrated is the treelike arrangement of conformational states and substrates showing the progression from substrates separated by large barriers at the highest tiers to substates separated by small equilibrium fluctuation (EF) barriers at the lowest levels the hierarchy (from Frauenfelder *Science* 254: 1598 © 1991 Reprinted with permission from AAAS)

3.1.2   Motions Beget Function

The connection between motions and function are becoming clearer over time. For example, it has been found that motions enable an enzyme to carry out its function in a folded state in which the catalytic site is often buried, and underlie the ability of a protein to undergo allosteric regulation. In short, internal motions must be considered an intrinsic part of a protein's native-state 3D structure. They are essential for folding as they "lubricate" the process (to overcome internal friction) and enable escape from the myriad of small barriers that are encountered along the folding pathways no matter how smooth. Shown in Fig. 3.2 is a generalized depiction of a

Fig. 3.2 Hierarchical protein motion landscapes. Sates and substates have been arranged in a in three tiers according to the energies, barrier heights, and timescales involved in the states and transitions between them. Transition rates between states are determined by the barrier heights. The changes in color from *dark* to *light blue* illustrate how mutations and other actions can alter the relative positions of states A and B in the landscape (from Henzler-Wildman *Nature* 450: 964 © 2007 Reprinted by permission from Macmillan Publishers Ltd)

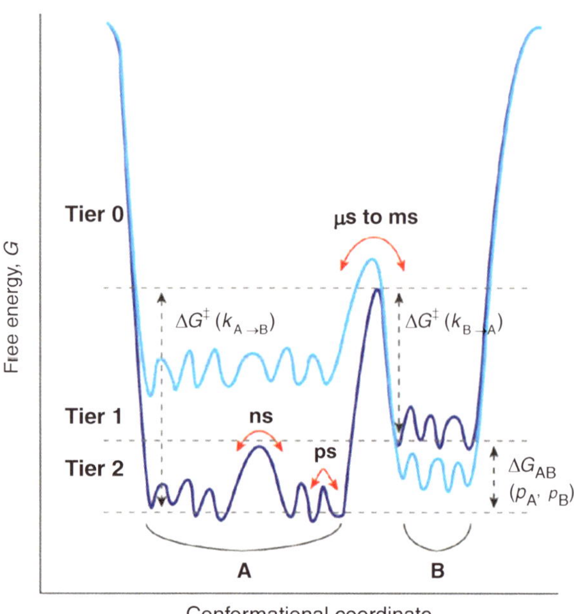

three-tiered, hierarchical arrangement of states and basins, and their substates. The lowest tiers encompass the fast picosecond and nanosecond timescales involved in small amplitude fluctuations—bond vibrations, side chain rotations, and loop motions—while the far slower microsecond, millisecond, and longer timescales are required for large-scale collective motions such as domain movements. This aspect will be discussed in greater detail in Sect. 3.6.

3.2 Role of Solvent Fluctuations in Protein Motions

The interactions between proteins and water have been a subject of study for over a hundred years. As early as 1913, Chick and Martin had looked at differences in density and volume between dry and hydrated caseinogen, egg- and serum albumin, and serum globulin. In examining their X-ray images of pepsin, Bernal and Crowfoot reported in 1934 that proteins were "relatively dense globular bodies…separated by relatively large spaces that contain water". That water filled and remained bound to proteins in solution was demonstrated by a series of measurements for hemoglobin in the 1930s by Gilbert Adair (1896–1979), the discoverer of protein quaternary structure and cooperative binding.

Given that water surrounds and permeates the native state of a protein it is perhaps not too surprising that water influences protein motions. Protein motions can be divided into two groups according to the influences of solvent fluctuations upon them:

- Nonslaved motions: These are independent of solvent fluctuations. These motions are determined by the protein conformation and vibrational dynamics.
- Slaved motions, in contrast, are tightly coupled to solvent fluctuations.

  - Primary, slow, α-fluctuations in the bulk solvent surrounding and permeating the protein drive and regulate the protein's first tier, consisting of large-scale motions and changes in conformation. These are controlled by the solvent viscosity.
  - Secondary, fast, β-fluctuations in the hydration layer drive and regulate the protein's smaller-scale internal motions taking place in the lower tiers in the hierarchy of protein substates.

3.3   The Energy Landscape Picture

Out of the universe of possible polypeptide chains, evolutionary pressure has sculpted a small number that fold into stable native shapes in physiologically useful times. The folding of these proteins is directed by the macromolecular forces in an energetically downhill fashion satisfying the thermodynamic requirements. These proteins utilize cooperativity (and all the other folding mechanisms discussed in the preceding chapter) in order to avoid becoming stuck in local minima surrounded by barriers too high to permit escape in reasonable time frames.

Proteins fold in ways that do not require exhaustive conformational searches. Instead, they fold along pathways that follow the contours of an energy surface in an overall downhill direction, taking them from an unfolded conformation to their native state. Each point in an energy landscape would represent a possible conformation of the protein. Similar conformations would be found near one another, and dissimilar ones further apart. Each state (point) in the energy landscape represents an ensemble of states and substates representing the intrinsic motions and influences of the solvent at the given temperature.

In this type of representation, the vertical axis denotes the sum of all contributions to the internal or potential energy of the protein while the entropic contribution to the Gibbs free energy is depicted by the width of the energy surface so generated. The horizontal axis itself gives the values of the various degrees of freedom, or reaction coordinates. Since there are too many coordinates to depict individually, one or two coordinates, or combinations thereof, are usually selected that capture the essential behavior of the protein as it folds.

The overall shape of the energy landscape is that of a funnel. Recall that the configuration entropy, $S(E)$, is a counting of the number of available states at a given energy, E. Mathematically, it is given by Ludwig Boltzmann's (1844–1906) famous ca. 1872 expression, cast into its present form by Max Planck (1858–1947) around 1900, namely,

$$S(E) = k_{\mathrm{B}} \ln \Omega(E) \tag{3.1}$$

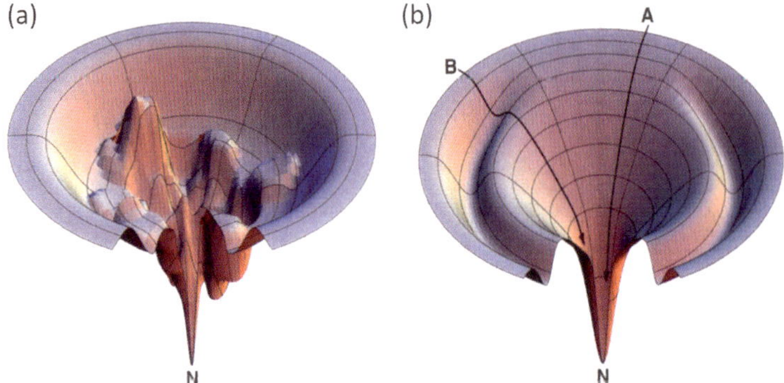

Fig. 3.3 Representative energy landscapes, or folding funnels. Depicted at the *right* (**b**) is a fairly smooth folding funnel. Pathway A is typical of that taken by a fast, two-state folding protein while a protein taking pathway B encounters a ridge that must be surmounted and consequently slows its folding. The folding funnel shown in (**a**) differs from the other (**b**): its energy landscape is quite rugged. A protein taking a pathway to the native state (N) in this landscape will take orders of magnitude longer to reach its destination (from Dill *Nat. Struct. Biol.* 4: 10 © 1997 Reprinted by permission from Macmillan Publishers Ltd)

In this formula, $\Omega(E)$ denotes the number of microstates (conformations) of the protein. In its denatured state, a protein may be in any of a large number of possible configurations. This freedom rapidly vanishes as the protein folds into a low-energy form that is far more compact. As a result the energy landscape is funnel shaped, broad at the top, and narrow at the bottom near the native state. A pair of stereotypic funnel-shaped energy landscapes is shown in Fig. 3.3.

The folding process can be depicted as a trajectory connecting many points on the landscape, denoting the sequence of small conformational changes that the protein undergoes as it folds. As shown in the figure, a folding trajectory starts out at a denatured state located at the top of the landscape at a high potential energy and ends at the native state located at the bottom of the landscape at a low potential energy. The ability of protein to undergo a variety of motions, large and small, enables the protein to sample ensembles of states and substates at each point along its folding pathways generated by the macromolecular forces.

The amount of time required for a protein to fold into its native state is an important aspect of the process. This is referred to as a kinetic requirement. Not only must a protein fold into its native state, but also it must do so in a physiologically reasonable time interval. The speed depends critically on the topography of the potential energy surface. If the surface is studded with deep minima separated by tall hills, and the folding trajectories pass close to them, the rate of folding will be slow. In these situations, the protein will fall into the minima and must escape before proceeding with its step-by-step movements towards its native state. The deep minima are called kinetic traps because of their slowing effects on the kinetics, or rates, of folding. Large and complex proteins, especially those involved in signaling pathways, tend to have rugged landscapes containing kinetic traps, or intermediate states, surrounded by high barriers. In contrast, small single-domain globular proteins

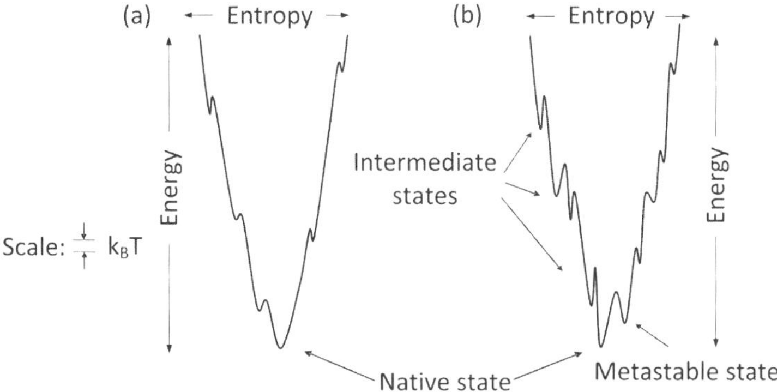

Fig. 3.4 One dimensional depiction of (**a**) smooth and (**b**) rugged folding funnels

often have landscapes that are fairly smooth. These proteins fold rapidly, lacking intermediate states and associated kinetic barriers that slow down the process. The difference between smooth and rough funnels is further highlighted in Fig. 3.4.

3.4   Metastable States

One of the features present in the rough funnel depicted in Fig. 3.4b is that of a low-lying intermediate states referred to as metastable states. Recall that one of the conditions for stability of the native state is that there be an appreciable energy gap between the native state and those lying above it. This condition is sometimes violated in a rough landscape. In these situations, a protein may dwell for a considerable amount of time in these low-energy states because of the high kinetic barriers. Although these are not states of minimal Gibbs free energy, proteins can still function in a useful fashion because of their long half-lives in those states.

In order to arrive at a state of minimal Gibbs free energy it must be kinetically accessible. In the case of many proteases, this condition has been exploited. These enzymes are synthesized as large proteins from which the active enzyme is subsequently cleaved. They typically contain a large amino-terminal proregion. This region is needed for kinetic accessibility and completion of the folding process to the native conformation. Once folding is complete and the protein is in its native state the proregion is removed. Removal of the proregion alters the energy landscape and the native state is no longer a state of minimal free energy. However, it is separated from the nearby lower-energy intermediate states by a high kinetic barrier and the enzyme is exceptionally stable. These kinetic situations are diagrammed in Fig. 3.5 for α-lytic protease.

Several examples of proteins that fold into functional metastable states rather than native states have been uncovered. Serpins (serine protease inhibitors) are a large family of protease inhibitors found in all kingdoms of life. These proteins regulate a variety of cellular processes; they help control complement cascades,

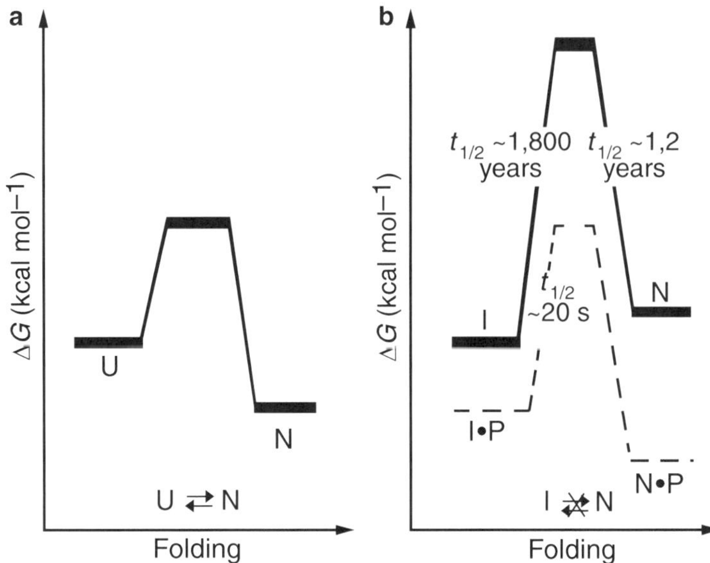

Fig. 3.5 Folding energetics (**a**) Thermodynamically controlled folding of a protein. U: unfolded state; N: native state. (**b**) Kinetic control of α-lytic protease folding and unfolding. The presence of the pro-region lowers the barrier for folding to the native state, and its absence raises the barrier thereby preventing its transition to one of several partially-unfolded intermediate states, I (from Jaswal *Nature* 415: 343 © 2002 Reprinted by permission from Macmillan Publishers Ltd)

angiogenesis, inflammation, tumor metastasis, apoptosis, axonal growth, and synaptic functions. Members of this large family include antithrombin which regulates blood coagulation, antitrypsin which mediates inflammation, PAI-1 which participates in tissue remodeling, and neuroserpin which regulates synaptic plasticity.

These proteins function as molecular mousetraps with an exposed reactive loop as the bait and the serine protease as the mouse. In its mousetrap-set form, the serpin resides in a metastable state. Binding of the serine protease to the reactive loop springs the trap and induces a major conformational enabling the serpin to transition to its native state.

3.5   Landscape Frustration and Spin Glasses

The energy landscape picture was introduced by Joseph Bryngelson and Peter Wolynes in a series of papers published in the late 1980s. The underlying inspiration for energy landscapes was provided in part by studies of spin glasses. Ordinary glasses are spatially disordered materials that lack the regularity of crystalline substances. These amorphous substances exhibit a "glass transition" from a solid and rather brittle form to a more liquid or molten state. Spin glasses are magnetic materials in which interactions between elements, rather than their positions, are disordered.

Mathematically, one can write an interaction term that closely resembles that of the Ising model, Eq. (2.6). To generate spin glass behavior the coupling factor J of the Ising model (discussed earlier in Chap. 2) is simply replaced by J_{ij} which then assumes a random set of values, some positive and some negative:

$$\Phi_{ij} = -J_{ij} s_i \cdot s_{j=i+1} \tag{3.2}$$

The result of this simple change can be dramatic. For many sets of coupling factors, spin orientations that produce good deep energy minima do not exist. Instead, the various constraints are in conflict with one another, and the energy landscape is fragmented into numerous shallow minima. Such systems are said to be "frustrated" because of the impossibility of simultaneously optimizing all the physical interactions and constraints with regard to the energy minimization.

Proteins are in several respects just like spin glasses. They too exhibit frustration, unable to fold without encountering situations where there are no good energy lowering steps available. Instead, there are too many conditions to be satisfied simultaneously and consequently the local energy surface possesses numerous suboptimal minima. Referring back to Fig. 3.3a, the energy landscape being depicted is not smooth, but instead corrugated with numerous hills and valleys. Landscapes convoluted with large numbers of hills and narrow valleys are said to be rugged. The fast folding proteins discussed in Chap. 2 possess energy landscapes that are smooth allowing for the rapid folding under complete thermodynamic control. In contrast, proteins that encounter rugged landscapes fold orders of magnitude more slowly under kinetic control and may or may not reach a state of global minimum in the Gibbs free energy. One of the key concepts emerging from the studies of folding pathways is the *principle of minimal frustration*. This principle states that protein primary sequences have been evolutionarily selected to fold via pathways in energy landscapes that are as smooth as possible thereby encountering a minimal number of kinetic traps.

Interestingly, hydrated protein motions largely cease as the temperature is lowered below a critical point, referred to in the literature as the "glass-transition" temperature. Below that temperature range, approximately 200 K, the only protein motions remaining are vibrations and the protein is said to be in a glassy state. As the temperature is raised above the transition value the protein becomes more liquid-like; it can now undergo large-scale motions and, most importantly, it can carry out its designed functions, which had ceased along with the loss of its proper range of motions, its jiggling and wiggling, at the lower temperatures.

3.6   Motions Enable Proteins to Carry out Their Cellular Tasks

To recap, proteins jiggle and wiggle; they are flexible and dynamic, populate an ensemble of states, and continually undergo transition from one conformation to another over multiple timescales. They undergo motions ranging from rigid body rotations of entire subunits, to side-chain and backbone movements, to local folding and unfolding. These biophysical properties are central to protein function and evolution. They enable proteins to recognize and bind their multiple partners, and may well provide the means whereby enzymes accelerate the rates of chemical reactions and allosteric effectors alter the functional properties of proteins. In the case of binding and recognition, increases in dwell-time in a sparsely-populated excited state may enable that state to act as a doorway to oligomerization and fibril formation. As a result, a protein's intrinsic flexibility can enable mutations and environmental factors to increase the likelihood that inappropriate aggregations occur and diseases emerge.

Protein recognition and binding: The classical model of ligand binding was introduced over a hundred years ago, in 1894, by the chemist Emil Fisher (1852–1919). Known as the *lock-and-key* mechanism of enzyme-substrate binding, the substrate (key) and enzyme (lock) are viewed as possessing complementary surfaces in terms of their shape and charge. These structures do not undergo major shape changes upon binding. Because their shapes fit into each other, the substrate is latched into the active site of the enzyme and this action initiates catalysis. To overcome shortcomings in Fisher's lock-and-key mechanism, Daniel Koshland (1920–2007) presented an alternative picture of enzyme catalysis in 1958. In his *induced-fit* model, the substrate causes (induces) a substantial change in the three-dimensional conformation of the amino acids at the enzyme's active site. These changes in shape, brought on by the substrate, properly align (uniquely fits) the catalytically active region of the enzyme with its substrate thereby enabling catalysis to take place.

The emergence of the protein folding landscape picture has led to an expanded view of how proteins bind one another. In the *conformational selection* model introduced in 1999, an ensemble of interconverting states exists prior to the interaction. When a protein comes into close proximity to its binding partner it forms an encounter complex that results in the selection (and stabilization) from the preexisting conformations those that best satisfy the geometric and electrostatic requirements for binding. These three mechanisms—static lock-and-key, semi-dynamic induced fit, and dynamic conformational selection—can operate either individually or in concert with one another to mediate protein recognition and binding.

Enzyme catalysis: Fast motions on a nanosecond timescale as well as slower motions spanning the microsecond to millisecond ranges underlie enzyme catalysis. That process has multiple steps. In addition to the chemical step, there are operations involving bringing together, aligning, opening and closing, and separating enzymes, substrates, cofactors, and products. Several steps in the reaction cycle are mediated by shifts in the equilibrium population of states in which a previously high-lying sparsely populated conformation becomes the new dominant low-lying state. If these shifts occur in a rate-limiting step they may explain the astonishing speedups observed in enzyme catalysis.

Allostery is dynamically driven: In an allosteric process, conformational perturbations at a particular site brought on by an effector generates functional changes at a distant, active site. Effectors are varied, ranging from ligands, to mutations and covalent modifications, to light and pH. The basic theory underlying "regulation (action) at a distance" was established 50 years ago in the mid-1960s by Monad, Wyman, and Changeux, formulated in terms of transitions between 'tensed' and 'relaxed' conformations, and by Koshland, Némethy, and Filmer based on their idea of an induced fit. The motivation for these studies was the desire to understand the sigmoidal (cooperative) binding of the multisubunit protein hemoglobin to molecular oxygen. This response property had been discovered in 1904 by Christian Bohr (1855–1911), father of Niels Bohr. Subsequent studies had provided details of the underlying hemoglobin physiology and Perutz and Kendrew had just produced a crystal structure upon which to anchor the theory. The resulting MWC and KNF models of allostery have been enormously influential and appear in all elementary textbooks on the subject.

In the intervening 50 years, the universe of proteins possessing allosteric properties has greatly expanded so that, today, allostery is regarded as a core biophysical property of most, if not all, dynamic (nonstructural), monomeric, as well as oligomeric, proteins. This expansion was made possible by (1) the development of solution NMR methods that enabled researchers to explore the dynamic processes underlying allosteric behavior and (2) by advances in theory, most notably the emergence of the energy landscape picture with protein motions and ensembles of interconverting conformational states serving as key unifying concepts.

Effector events such as mutations or covalent modifications or ligand binding alter the internal motions of the protein. They change the fast internal dynamics of the protein as well as its slow internal motions. In addition, they cause sifts in the conformational equilibrium, that is, they generate a reordering of the free energies within the ensemble of ground and excited states. As a consequence some of the higher lying and less stable conformations have their energies lowered and become the new stable (ground) states. By this means, the changes propagate through the protein to the active site, selectively turning them on and off to potential interaction partners, thereby altering their functional properties.

Thermodynamically, the effectors produce changes in the protein's configuration entropy and in its rotational and translational entropies. In more detail, the change in free energy of binding (ΔG_{bind}) is the sum of the enthalpy of binding (ΔH_{bind}) and the entropy of binding, which consists of contributions from the changes in protein, ligand, and solvent entropies:

$$\Delta G_{\text{bind}} = \Delta H_{\text{bind}} - T(\Delta S_{\text{prot}} + \Delta S_{\text{lig}} + \Delta S_{\text{solv}}) \tag{3.3}$$

In general, entropic penalties engendered by binding are closely matched by gains in enthalpy. As a result small changes in conformational entropy can have large effects in determining binding affinities. The influence of changes in solvent entropy with regard to folding was discussed earlier. It also plays an important role in binding processes, and is typically thought of in terms of the hydrophobic effect. The change in the protein's entropy consists of changes in its fast internal motions (its configuration entropy) and in its rotational and translational entropies. The nature and magnitudes of these quantities has become a subject of great interest because of their emerging roles in disease causation and their potential exploitation in drug intervention.

3.7 Experimental and Theoretical Methods of Exploring Protein Folding

Protein folding has been the subject of intensive studies for more than 60 years. During that time an ever-expanding suite of experimental and theoretical tools have been developed, starting with the previously discussed light microscopy, electron microscopy, and X-ray crystallography. These tools are often used synergistically with one another in order to explore the protein folding pathways and dynamic properties of the proteins as they fold and misfold. The most prominent of the theoretical/computational methods are

- Molecular dynamics (MD)
- Langevin/Brownian dynamics
- Simulated annealing (SA)

Molecular dynamics had its beginnings in the 1950s with the advent of modern electronic computers. The first molecular dynamics calculations were carried out by Alder and Wainwright in 1957. These calculations were aimed at exploring liquid–solid phase transitions of atoms that were treated as simple hard spheres. In 1964, Rahman carried out an MD simulation of a realistic system of argon atoms. That study was followed in 1971 with an exploration by Rahman and Stilinger of liquid water that brought out the importance of water's cooperative interactions and hydrogen-bonding network.

The first computer simulation of protein folding appeared in 1975 with the publication by Warshal and Levitt of their modeling and simulation study of the folding of bovine pancreatic trypsin inhibitor (BPTI), a small globular protein of known structure containing 58 amino acid residues. The first MD simulation study of BPTI dynamics was published shortly thereafter in 1977 by McCammon, Gelin, and Karplus. Their goal was to understand the dynamic fluctuations about the native conformation of the folded protein. Today, there are a number of widely used computer programs that enable users to carry out molecular dynamics simulations and protein structure prediction, dynamics, and design. These include CHARMM, AMBER, and GROMOS among others. Warshal, Levitt, and Karplus were awarded the 2013 Nobel Prize in Chemistry for their contributions to the field.

The folding of proteins is highly complex. As noted by Michael Levitt and Arieh Warshall in their 1975 paper, even a small protein of 50 residues has some 750 atoms and 200 degrees of freedom. When the solvent molecules are included the computational task becomes enormous. During the ensuing decades computer power has increased enormously and progressively more realistic and detailed simulations of protein folding have become possible, especially for the small globular proteins that fold rapidly.

Proteins fold through a sequence of states, most of which are transiently populated for a tiny fraction of a second. Partially folded intermediates are longer-lived for the reasons already discussed, and these can be studied to give valuable information on the routes taken by proteins as they fold and misfold. Three experimental methods—nuclear magnetic resonance (NMR), hydrogen-deuterium exchange, and

ϕ-value analysis are especially well suited for exploration of intermediate states and protein dynamics. These methods will be described following an examination of the theoretical, computational tools.

3.8 Molecular Mechanics (MM)

In molecular mechanics treatments of protein folding, the macromolecular forces between atoms are modeled in an empirical fashion. Two classes of forces are considered—covalent and noncovalent. Covalent bonds, in which electrons are shared, are the strongest of the bonds. They form the strong peptide linkages comprising the protein backbone. These bonds are thought of as operating in an elastic spring-like manner. In elastic springs, there is an equilibrium point and departures from that point due to stretching or compression build up potential energy. Once the perturbing force is removed the system returns to its equilibrium point.

$$U_{bonded}(r) = \sum_{bonds} \frac{1}{2} K_b \left(b - b_{eq} \right)^2 + \sum_{angles} \frac{1}{2} K_\theta \left(\theta - \theta_{eq} \right)^2$$
$$+ \sum_{torsions} \frac{1}{2} K_\varphi \left(1 + \cos\left(n\varphi - \delta \right) \right) \tag{3.4}$$

In molecular mechanics, three covalent force terms are defined. These describe the possible stretching, bending, and torsional (dihedral) motions about the bond axis.

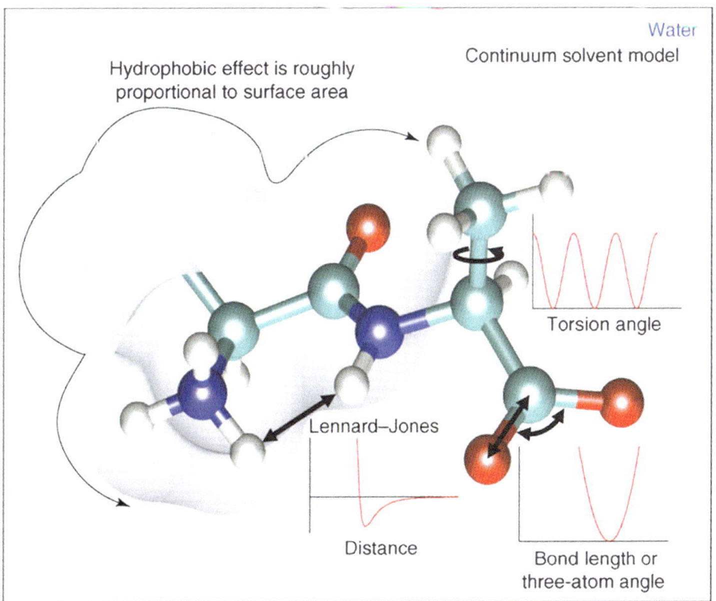

Fig. 3.6 Potential energy functions including the hydrophobic effect (from Boas *Curr. Opin. Struct. Biol.* 17: 199 © 2007 Reprinted by permission from Elsevier)

These three terms are given by Eq. (3.4), and are illustrated in Fig. 3.6. The length (bonds) and angles terms take the form of a harmonic potential that follows from Hooke's (in which the elastic spring force F is the product of the spring constant and the displacement from the equilibrium point). It has a minimum at the equilibrium position while the torsions term is a periodic function of the torsion angle.

The noncovalent forces include various combinations of point charge and dipole forces. These terms are jointly modeled as the sum of a Lennard-Jones, 6–12 potential, and a Coulomb-like point charge term as given by Eq. (3.5). These terms are summed over all atoms in the protein. The term in the Lennard-Jones (L-J) potential that varies as the sixth power of the radius is an attractive one while the other contribution, from the term varying as the twelfth power, is repulsive. The overall shape of the potential is depicted in Fig. 3.6. As can be seen in that figure the net effect being captured is strong repulsion at short distances corresponding to interpenetration of the electron clouds forbidden by the Pauli Exclusion Principle. There is a distance where the two terms just cancel one another. That distance is referred to as the van der Waals radius, and there is a minimum in the potential at the equilibrium radius.

$$U_{\text{non-bonded}}(r_{ij}) = \sum_{i<j}\{[\frac{a_{ij}}{r_{ij}^{12}} - \frac{b_{ij}}{r_{ij}^{6}}] + \frac{q_i q_j}{r_{ij}}\} \tag{3.5}$$

Because of their importance, hydrogen bonding and hydrophobic effects require additional attention. The hydrogen bonds in this model are handled in an approximate fashion by suitably adjusting the constants in the L-J and Coulomb potentials. In some implementations, an additional term similar to that presented in Eq. (3.5) is appended. This term is often taken to be of an angle-dependent 12-10 form rather than the standard, angle-independent 12-6 form. The angle in this term represents the donor-hydrogen-acceptor angle.

Hydrophobic interactions can be handled in one of two ways. In explicit solvation models, solvation energies are added that describe solvent-solvent and solvent-protein interactions. In the simpler continuum approaches, hydrophobic interactions are described in terms of the solvent accessible surface area (SASA) (Fig. 3.6). This quantity was first introduced by Lee and Richards in 1971 in their study of packing densities. Both approaches enable a more accurate treatment of the interactions with the water outside the protein and with water contained within the interior cavities being formed.

3.9 Molecular Dynamics (MD)

Molecular dynamics is the name given to a suite of computer simulation methods for studying how large systems of interacting atoms and molecules evolve over time. In this approach, Newton's equations of motion are numerically integrated using the potential functions from molecular mechanics to compute the forces. Introducing standard "dot notation" Newton's laws of motion are

$$\ddot{r}_i = \frac{\mathrm{d}}{\mathrm{d}t} v_i = a_i = \frac{F_i}{m_i} = -\frac{1}{m_i} \frac{\mathrm{d}}{\mathrm{d}r_i} U(r) \tag{3.6}$$

with

$$\dot{r}_i = \frac{\mathrm{d}}{\mathrm{d}t} r_i = v_i \tag{3.7}$$

In Eq. (3.6), the potential energy, $U(r)$, is the sum of the bonded and nonbonded potentials given by Eqs. (3.4) and (3.5), plus any additional terms that describe hydrogen-bonding contributions and hydrophobic effects. Once the potentials are specified the equations of motion are integrated. Numerical techniques known as *finite-difference methods* are used to convert the equations of motion into a form suitable for integration on a computer. The basic idea is to take the positions and momentum of each particle at a given time and compute how each quantity changes over a small time interval. One of the most widely used time stepping methods is the *Verlet algorithm*. The relevant expression is derived by first doing a Taylor's series expansion of the positions at time $t + \Delta t$ where Δt is the time step size, and keep only the first few terms:

$$r(t + \Delta t) = r(t) + \dot{r}(t)\Delta t + \frac{1}{2}\ddot{r}(t)(\Delta t)^2 \tag{3.8}$$

Upon carrying out some algebraic manipulations the Verlet algorithm can be produced:

$$r(t + \Delta t) = 2r(t) - r(t - \Delta t) + \ddot{r}(t)(\Delta t)^2 \tag{3.9}$$

In deriving this expression the velocities have been eliminated. The positions at time $t + \Delta t$ are computed from the positions at times t and the previous time $t - \Delta t$ and from the forces at time t through the acceleration term. In some cases, the velocities are important. In those situations another time stepping expression known as the *velocity Verlet algorithm* is used. It takes the form:

$$\dot{r}(t + \Delta t) = \dot{r}(t) + \frac{1}{2}[\ddot{r}(t) + \ddot{r}(t + \Delta t)]\Delta t \tag{3.10}$$

The step size, Δt, is a critical quantity. It is customary to use femtosecond (10^{-15} s) time steps in order to account for the fast motions—the atomic fluctuations, and side-chain and loop motions. However, even the most-rapidly folding motifs, super-secondary structures, and domains require microseconds (10^{-6}) to milliseconds (10^{-3} s) to fold. Thus, one has to integrate the equations of motion over 10^{12} time steps. Furthermore, the number of conformational states is enormous since there are perhaps tens of thousands of atoms present (especially when taking into account the attendant water molecules).

One of the ways to meet the computational challenge is to simplify the energy landscape by introducing a small number, one or two, of *order parameters* or effective coordinates. The most widely used of these is the order parameter usually designated by the symbol "Q" that represents the number of tertiary native contacts in a given conformation of the protein under study. If two residues are close to one another in space, that is, their α-carbons are within ~7 Å of one another, they are said to be in contact. If these contacts are found in the native state, then they are said to be a native contact and the order parameter Q counts the number of these, that is, it is a measure of nativeness. The unfolded state has few native contacts while in the native state Q is maximal. This type of order parameter is intended to capture the most salient features of the underlying energy landscape, that is, the prominent valleys, ridges, and basins encountered by the protein as its folds into its native state.

Shown in Fig. 3.7 is an energy landscape for hen lysozyme, a protein consisting of 129 amino acid residues organized into two domains, α and β. This protein can fold along two distinct pathways, one rapid and the other slow in which there is a prominent kinetically trapped folding intermediate. The figure combines theoretical simulations with experimental data acquired by means of nuclear magnetic resonance and hydrogen-deuterium exchange, and utilizes a pair of order parameters to simplify the landscape.

In the more recently developed approaches known as the *Markov state methods* (MSMs), the large computational problem is decomposed into an ensemble of smaller ones. In carrying out a molecular dynamics simulation, one has to sample the energy landscape. In an MSM, the landscape is pictured as containing a number of basins composed of states that rapidly interconvert into one another and separate from other basins by large free energy barriers. To determine the conformational dynamics (1) the long-lived metastable states visited by the protein as its folds along its pathways are found, and (2) one then uses MD or MC methods to calculate the transition rates between basins. This approach generates a far more detailed multipathway depiction of the folding process than obtained using one or two order parameters. As shown in Fig. 3.8, in place of a small number of long trajectories there are multiple shorter trajectories connecting the basins, and when combined these produce a multipathway description of the folding funnel.

3.10 Langevin Dynamics

In 1908, three years after Einstein's publication of his study of Brownian motion, Paul Langevin (1872–1946) published his own analysis of the phenomenon. Brownian motion was first described by the Scottish botanist Robert Brown (1773–1868) in 1826. When using a light microscope he observed that small specs of dust suspended in a fluid seem to be moving about in a random zigzag fashion. In his theoretical analysis, Einstein showed that the irregular motions of the dust specs arose from random collisions of the dust particles with the molecules of the fluid. In his treatment of the motion, Einstein considered the probability laws that must be obeyed by the dust particles, and then showed that these probabilities obeyed the

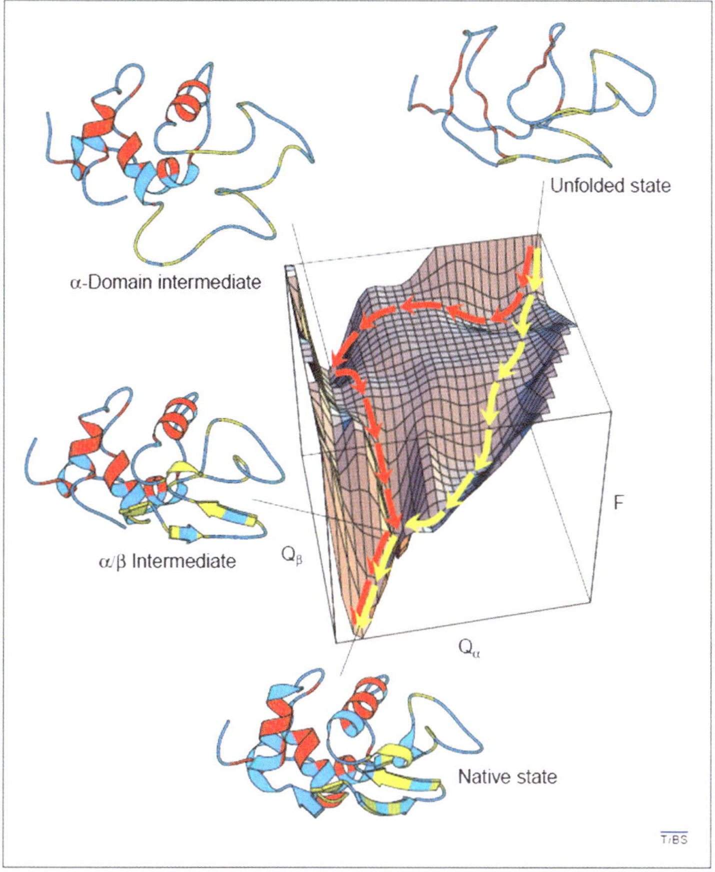

Fig. 3.7 Energy landscape describing the folding of hen lysozyme. Plotted along the vertical axis is the free energy while the horizontal axes represent the number of native contacts. The *yellow trajectory* represents the fast pathway in which the α and β domains form concurrently and only transiently populate the α/β intermediate state. The red trajectory passes through a long-lived intermediate state is which only the α domain has formed its secondary structure. The system must then pass over a high energy barrier and may partially unfold in order to complete its passage to the native state (from Dinner et al *Trends Biochem. Sci.* 25: 331 © 2000 Reprinted by permission from Elsevier)

diffusion equation. Langevin took a different approach and produced a description in terms of a first order (stochastic) differential equation now called the Langevin equation. This equation is

$$m\frac{dv}{dt} = -\alpha v + F(t) \tag{3.11}$$

In this expression, there are two force terms. One of these is a friction term with friction coefficient α, and the second term, $F(t)$, is a random fluctuating force.

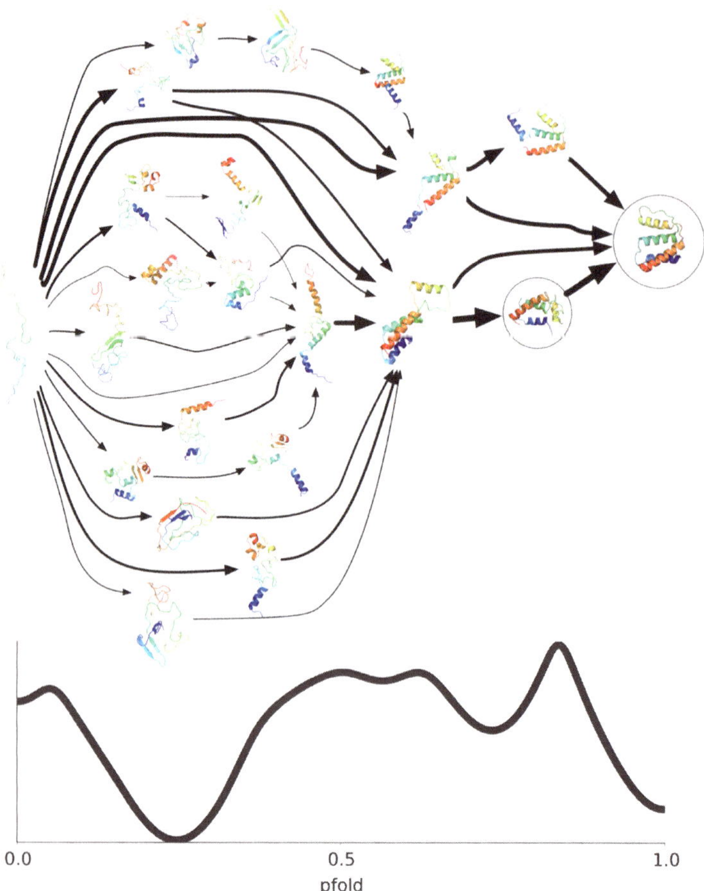

Fig. 3.8 Markov state method generated energy landscape and folding funnel for Acyl-CoA binding protein (ACBP). The *upper frame* depicts the flow through the various basins of the landscape as the protein folds into its native state while the lower plot shows a one-dimensional projection (from Lane *Curr. Opin. Struct. Biol.* 23: 58 © 2013 Reprinted by permission from Elsevier)

A key result of solving this equation is a pair of limiting expressions for the mean square displacements, $\langle x^2 \rangle$, of the Brownian particles. His results, identical to those found by Einstein, are

$$x^2 = \begin{cases} (\dfrac{k_B T}{m})t^2, & t \ll \dfrac{m}{\alpha} \\[3mm] (\dfrac{2k_B T}{\alpha})t, & t \gg \dfrac{m}{\alpha} \end{cases} \tag{3.12}$$

The first limiting expression pertains to small time intervals with m/α serving as the relevant scale factor. In this case, the motion is inertial with particles moving at a constant thermal velocity $(k_\mathrm{B}T/m)^{1/2}$. The second limiting case is applicable to long time intervals. In this limit, the random character of the movements comes into play; the particles are moving diffusively, and the net displacements are far smaller than that would be expected from movement at a constant thermal velocity. These theoretical predictions were verified experimentally by Jean Perrin (1870–1942) in the 1908–1909 time period. Taken together, these findings had an immediate important consequence—they dispelled still lingering doubts in the science community as to the existence of atoms first postulated a century earlier by John Dalton (1766–1840).

The Langevin equation can be used as an alternative to molecular dynamics by incorporating the MM force term, $F(x)$, into the Langevin equation:

$$m\frac{\mathrm{d}v}{\mathrm{d}t} = F(x) - \alpha v + F(t) \tag{3.13}$$

In this simulation approach, the Langevin equation becomes an alternative to Newton's second law of motion. In applying this formalism to protein folding, the friction term takes into account interactions between protein and solvent and the second term models random movements resulting from the various collisions taking place.

The friction coefficient, α, controls how fast equilibrium is approached; the greater the friction the faster the approach to equilibrium. The fluctuations are generated by the coupling interactions between Brownian particles and the molecules of the heat bath, and the greater the coupling the faster the relaxation to equilibrium. As a consequence of the fluctuation-dissipation theorem there is a direct, inverse relationship between the diffusion coefficient, D, and the friction coefficient, namely:

$$D = \frac{k_\mathrm{B}T}{\alpha} \tag{3.14}$$

This expression relates the mobility or inverse of the friction coefficient to the diffusion constant and thus to the fluctuations in velocity of the Brownian motion. Lastly, the friction (drag) coefficient and viscosity, η, of the solvent are related through Stokes' law first presented in 1851 by the mathematician/physicist George Gabriel Stokes (1819–1903). This law states that:

$$\alpha = 6\pi\eta r_0 \tag{3.15}$$

where r_0 is the radius of the diffusing particle.

3.11   Monte Carlo Methods

3.11.1   The Metropolis Algorithm

The Monte Carlo method is the name given to a random sampling technique introduced by John von Neumann and Stanislaw Ulam in 1947. They had noted that in many instances it is far easier to obtain a numerical solution to a set of equations by finding a stochastic process whose probability distributions or parameters satisfied the equations of interest, and then studying its statistical output, than it was to directly solve those equations. It was quickly recognized that this method provides a means of obtaining precise results for complex systems of multiple interacting elements. It rapidly became the method of choice during the 1950s for simulating nuclear collision processes and a multitude of other physical and chemical processes involving large numbers of atoms and molecules.

A landmark paper during this period was the study in 1953 by Nicholas Metropolis, Arianna and Marshall Rosenbluth, and Augusta and Edward Teller. They wanted to understand certain properties of liquids and dense gases, and introduced a Monte Carlo random sampling technique known to this day as the Metropolis algorithm. Their intent in introducing the sampling procedure was to avoid having to directly evaluate every possible configuration of the large system of N atoms. Instead they chose to carry out importance sampling—that is, avoid sampling unlikely configurations of the system and instead concentrate on those states that were likely to occur. (Note the similarity of their reasoning to that of protein folding along pathways in an energy landscape!)

Their sampling procedure, or algorithm, is, as follows. Select a pair of states, x and y, to consider in an unbiased way. Then calculate the energy difference between states y and x, $\Delta E_{xy} = \Delta E_y - \Delta E_x$. If state "$y$" has lower energy than state "x" the energy difference is negative and the transition from the current state "x" to the new state "y" is always allowed. If, on the other hand, the energy difference is positive, then select a random number ξ between 0 and 1. If ξ is less than $\exp(-\Delta E_{xy}/kT)$ then allow the move. Otherwise reject the move and choose another candidate transition to consider. To summarize:

$$t_{xy} = \begin{cases} 1, & \text{if } \Delta E_{xy} \leq 0 \\ \exp(-\dfrac{\Delta E_{xy}}{k_B T}), & \text{if } \Delta E_{xy} > 0 \end{cases} \tag{3.16}$$

3.11.2   Simulated Annealing

The ratio of ΔE_{xy} to the $k_B T$ factor has an important effect on the transition probabilities. If the ratio of the two factors is large and positive then the transition probability will be small and conversely, if, for example, the temperature is elevated, then the same ΔE will not induce such a large penalty on the potential move. This behavior is exploited in the simulated annealing extension to the Metropolis algorithm. The term "simulated annealing" is borrowed from metallurgy. It refers to a process of heating and cooling a metal in order for the metal's domains to optimally orient and configure themselves. If, in contrast, the metal is cooled rapidly, or quenched, the domains will be frozen in nonoptimal configurations (local minima) resulting in a far more brittle metal.

In simulated annealing, the temperature is gradually reduced from a large starting value to its final, lower one. Thus, at the outset, the penalties for increasing the energy are small and the algorithm can sample large parts of the space of possible configurations. Later on in the simulation, the penalties will grow and the tendency will increase to only accept energy decreasing (downhill) moves. This approach was introduced in 1984 in a famous paper by Kirkpatrick, Gelatt, and Vecchi, and independently by Cerny in another well-known paper. Beginning in the 1980s and 1990s this generalization of the Metropolis algorithm was applied to finding optimal, low-energy stable or metastable conformations of proteins.

3.12 Nuclear Magnetic Resonance

3.12.1 Protein Structure and Dynamics

Nuclear magnetic resonance (NMR) is the second major experimental method (after X-ray crystallography) for determining three-dimensional, atomic level protein structures. It is also a key method for exploring protein dynamics. In NMR, one utilizes radiation in the radiofrequency portion of the electromagnetic spectrum. More specifically, frequencies are typically in the range 300–600 MHz corresponding to wavelengths from 100 to 50 cm. Photons of these energies can induce transitions between nuclear (proton and neutron) spin states. The subsequent relaxation of the nuclear spins back to their equilibrium distributions is sensitive both to the electron clouds surrounding the nuclei (shielding effect) and to neighboring nuclear spins (spin-spin, or J, couplings). As a result this technique can provide detailed information on molecular structure and connectivity.

Recall that electrons, protons, and neutrons have an intrinsic angular momentum, or spin. Because they have a (nonzero) spin, they have a magnetic dipole moment and can interact with an external magnetic field. Electrons, protons, and neutron have a spin value of $\frac{1}{2}$. Nuclei with an odd number of neutrons and an even number of protons, or alternatively, an odd number of protons and an even number of neutrons will have a net spin of $\frac{1}{2}$. Spin $\frac{1}{2}$ nuclei encountered in biomolecules are ^{1}H, ^{13}C, ^{15}N, ^{19}F, and ^{31}P. This is not the only nonzero spin value possible. Nuclei with an odd number of proteins and also an odd number of neutrons, like ^{2}H, have a spin of 1.

Spin is a fundamental property of particles. The existence of spin was first postulated in the 1920s by Wolfgang Pauli (1900–1958), George Uhlenbeck (1900–1988), Samuel Goudsmit (1902–1978), and others; it was demonstrated experimentally in a famous 1922 experiment by Otto Stern (1888–1969) and Walther Gerlach (1889–1979), and occupies a central place in the formulation of quantum mechanics. The phenomenon of nuclear magnetic resonance was first observed in 1937 by Isidor I. Rabi (1898–1988) in his measurements of magnetic moments using molecular beams.

The utilization of NMR as a diagnostic tool in protein science and medicine took place in several stages spanning four decades. The transition from physics lab to practical application began with the demonstration of NMR in fluids by Felix Bloch (1905–1983) and in solids by Edward Purcell (1912–1997). This occurred in 1946 at the end of World War II when radar and associated radiofrequency technology had become available.

That NMR could be used to characterize atomic species and molecules developed through a series of discoveries by several groups in the 1950–1951 time periods. The first of these events was the discovery of chemical shifts by Proctor and Yu. The negatively charged electron clouds surrounding nuclei reduce the magnitude of the magnetic field experienced by the spin ½ protons and neutrons residing in the atomic nucleus. Each atomic species has a uniquely different electron cloud, and thus different atoms will undergo different chemical shifts in the presence of the same external magnetic field. Nearby atoms will influence the chemical shifts through the shielding effects, as well. The subject of Proctor and Yu's experiment was ammonium nitrate, NH_4NO_3. In contrast to their expectation of a single nitrogen peak, they observed two peaks, one from NH_4 and the other from NO_3 clearly showing the influence of the local chemical environment on the chemical shifts. Similar results were reported from several other groups confirming the effect, and illustrating that nonequivalent chemical environments within the same molecule will induce different shifts, as well.

At around the same time a second key discovery was made. Herbert Gutowsky, David McCall, and Charles Slichter observed the splitting of a single chemical shift peak into multiple peaks, or fine structure, arising from spin-spin (J-J) coupling between different atoms in the same molecule. By exploiting this additional effect NMR spectroscopy is further able to give valuable information on the molecular structure and connectivity. An example of the chemical shifts and spin-spin line splittings produced by CH_3CH_2 is presented in Fig. 3.9.

Finally, the further extension of NMR to uncovering three-dimensional protein structure and dynamics became possible with pioneering efforts by Richard Ernst and Kurt Wüthrich in the 1970s and 1980s. These researchers developed (1) the 2D NMR method known as COSY (two-dimensional correlated spectroscopy) to detect line splitting arising from spin-spin couplings among bonded atoms, and (2) NOE (nuclear Overhauser effect) that can be used to measure line splitting arising from spins of nonbonded atoms located in close spatial proximity to one another.

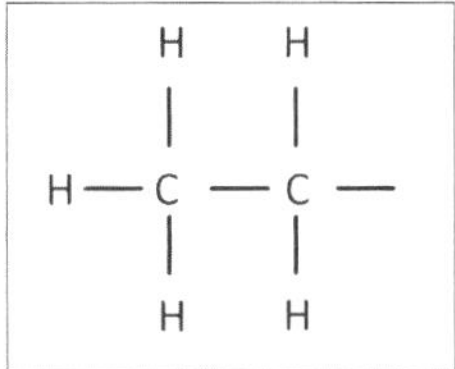

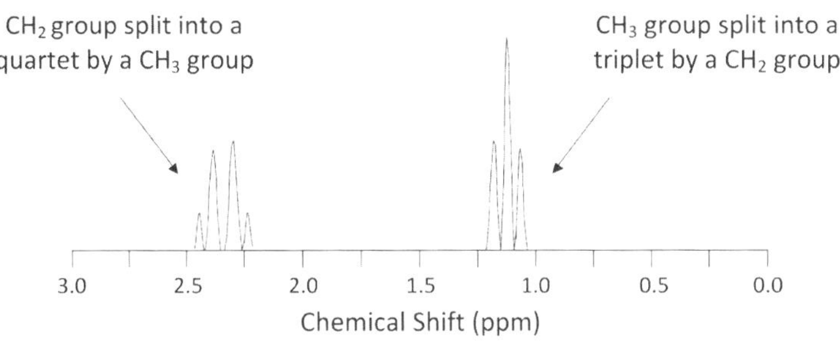

Fig. 3.9 Chemical shifts and spin-spin line splittings for CH_3CH_2

3.12.2   Medical Diagnostics

Magnetic resonance imaging (MRI) was invented by Paul Lauterbur (1929–2007). He had become interested in using NMR to find tumors in tissue and came up with the idea that he could spatially localize an NMR signal emanating from complex tissue by using a second magnetic field. The presence of the second field restricts the interaction of the object with the first magnetic field to a small spatial region. Thus, its use bypasses the resolution requirement, that is, that the wavelength be shorter than the dimensions of the object being imaged. That this could be done was initially greeted with disbelief by many in the field. When he submitted his paper to *Nature* it was initially rejected, and when it was finally published the editors deleted the sentence at the end noting the potential applications to medical diagnostics!

Today MRI is a major diagnostic tool in medicine. Its use in diagnosing neurodegenerative conditions was greatly advanced with the invention of functional MRI (fMRI) by the Japanese researcher Seiji Ogawa in 1990. In fMRI, differences in the magnetic properties of hemoglobin in its deoxygenated and oxygenated states, first noted by Linus Pauling and Charles Coryell (1912–1971) in 1936, are exploited. This difference plus the spatially localized vascular blood oxygenation levels, which reflects neuronal metabolic activity (and thus brain activity), are central to the imaging method. The two phenomena give rise to the use of blood-oxygenation-level-dependent (BOLD) contrast as a measure of brain activity.

3.13 Mass Spectrometry and Ion-Mobility Spectrometry Are Used to Probe Amyloid States

In *mass spectrometry*, electric and/or magnetic fields are used to filter, disperse, and separate charged particles according to their charge (z) to mass (M) ratios. All mass spectrometers have three main components, an ion source, a mass analyzer, and an ion detector. The source is responsible for producing a beam of ionized particles of the material to be mass analyzed. The analyzer separates the beam ions according to their M/z values, and the detector is responsible for their detection. The main breakthrough leading to the use of mass spectrometry in studying proteins was the development of techniques for producing beams of charged gaseous proteins. Two techniques are widely used. The first is electrospray ionization (ESI) and the second is matrix-assisted laser desorption ionization (MALDI). These methods can be employed for biomolecules with masses up to 50 kDa (ESI) and more than 300 kDa (MALDI).

A second method, *ion-mobility spectrometry*, is often used in conjunction with mass spectrometry. In ion-mobility spectrometry, gas-phase ions are separated according to their mobility (drift time) through a drift gas. The mobility of the ions depends on the collision cross-section of the ions, which, in turn, depends on physical properties of the ions—size and shape. The ion species of interest are accelerated through the gas-filled chamber or tube by an applied electric field. Large elongated ion will take longer than small compact ions of the same mass to traverse a given distance. That happens because the larger more extended ones collide more frequently with the buffer gas molecules than the small compact ions. The joint application of the two methods has enabled researchers to study oligomer and amyloid formation, composition, and assembly.

3.14 Hydrogen–Deuterium (H/D) Exchange

If hydrogen atoms in protein backbones and side chains are exposed to the solvent they will continuously exchange places with those of nearby water molecules. These exchanges will take place for amide hydrogen atoms in the backbone and for hydrogen atoms bonded to nitrogen, oxygen, and sulfur atoms in side chain polar groups. As the name implies, in hydrogen/deuterium exchange, protein hydrogen atoms are replaced by solvent deuterium (^{2}H, or D) atoms. This is usually accomplished by diluting the H_2O concentration with D_2O. The exchange rates are dependent on a number of factors. For example, the rate will be reduced by orders of magnitude if hydrogen bonding is present. Thus, the occurrence of hydrogen bonding, which is highly indicative of secondary structure elements being formed, is reflected by fewer deuterium atoms being present. In addition, portions of the protein that are buried within the core and thereby shielded from the solvent will be unlikely to exchange and thus will possess fewer deuterium atoms that solvent-exposed elements.

This method of exploring protein structure and dynamics was invented by Kai Linderstrøm-Lang in Copenhagen in the 1950s shortly after Pauling's discovery of the α-helix and β-sheet. The basic model of the exchange kinetics introduced by

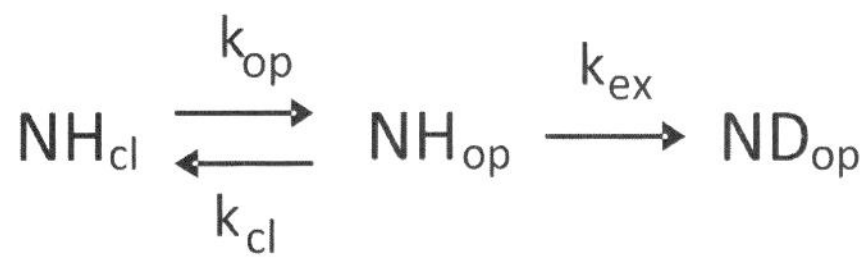

$$NH_{cl} \underset{k_{cl}}{\overset{k_{op}}{\rightleftarrows}} NH_{op} \xrightarrow{k_{ex}} ND_{op}$$

Fig. 3.10 H-D exchange kinetics. There are two useful limiting situations. In the more common EX2 limit, the folded protein is highly stable, $k_{cl} \gg k_{op}$ and $k_{cl} \gg k_{ex}$. In the other limiting case, termed the EX1 limit, $k_{ex} \gg k_{cl}$

Linderstrøm-Lang and formalized by Hvist and Nielsen in their 1966 paper is that of a two-step process illustrated in Fig. 3.10 in terms of the backbone amide hydrogens.

In the closed state, the amide hydrogen atoms are protected against deuterium exchange as a result of their helix and sheet-forming hydrogen bonding, but can exchange in the nonhydrogen-bonded open state. There are two ways that the open state can be populated. In the first of these, the protein is in its native state, and undergoes thermal equilibrium motions spanning multiple timescales. While undergoing these motions the amide hydrogens transiently break and reform their hydrogen bonds. These transient breakages may occur in a noncooperative manner that allows the solvent to penetrate to amides buried in the protein interior. Alternatively, local unfolding may occur in which portions of helices unfold in a cooperative manner thereby exposing multiple hydrogen bonds to the solvent. Finally, the exchanges can occur from first excited states that have slightly higher free energies but greatly altered conformations through either a penetration mechanism or a locally unfolded mechanism.

These native-state mechanisms are referred to in the literature as the penetration model (due to Woodward 1982), the local unfolding model (due to Englander 1984), and the Miller-Dill model (due to Miller 1995) [see Bibliography]. In addition to native-state mechanisms, hydrogen exchange can take place from proteins that have become unfolded. By exposing the proteins to denaturing conditions, various intermediate states along the protein folding landscape can be explored. The decomposition of cytochrome c into its foldons, and their stepwise assembly discussed in the last chapter (Sect. 2.4.2 and Fig. 2.6) was established by means of hydrogen exchange measurements that utilized urea [$CO(NH_2)_2$] as a denaturant.

3.15 Phi (ϕ)-Value Analysis

Phi (ϕ)-value analysis is a protein engineering technique that enables researchers to explore the 3D structure of a protein's transition state and folding intermediates. It was developed by Alan Fersht and his coworkers in the late 1980s and since then has been a major method for determining the structures of these elusive but critical states along the folding pathways. In this approach, mutations are introduced in a site-specific manner throughout the protein, and the changes in energetics between wild-type and mutated proteins are measured. In more detail, mutations are introduced that alter the interaction of a side chain with a ligand or other residues in the protein. These alterations change the energetics of that particular reaction. Changes in rates of the reaction and equilibrium constants brought on by the substitution are then measured. From these results one infers whether or not that reaction participates in the transition or intermediate state of the protein as it folds. This is accomplished by computing the quantity ϕ, as follows:

$$\phi = -\frac{\Delta\Delta G_{\ddagger-D}}{\Delta\Delta G_{N-D}} \tag{3.17a}$$

where

$$\Delta\Delta G_{\ddagger-D} = \Delta G_{\ddagger-D} - \Delta G_{\ddagger'-D'} \tag{3.17b}$$

and

$$\Delta\Delta G_{N-D} = \Delta G_{N-D} - \Delta G_{N'-D'} \tag{3.17c}$$

In these expressions, $\Delta\Delta G_{\ddagger-D}$ is change brought on by the substitution in the free energy of activation, and similarly $\Delta\Delta G_{N-D}$ is the corresponding change in the free energy of folding. The definitions of the various energy differences appearing in these equations are depicted in Fig. 3.11 for two illustrative cases. In the situation

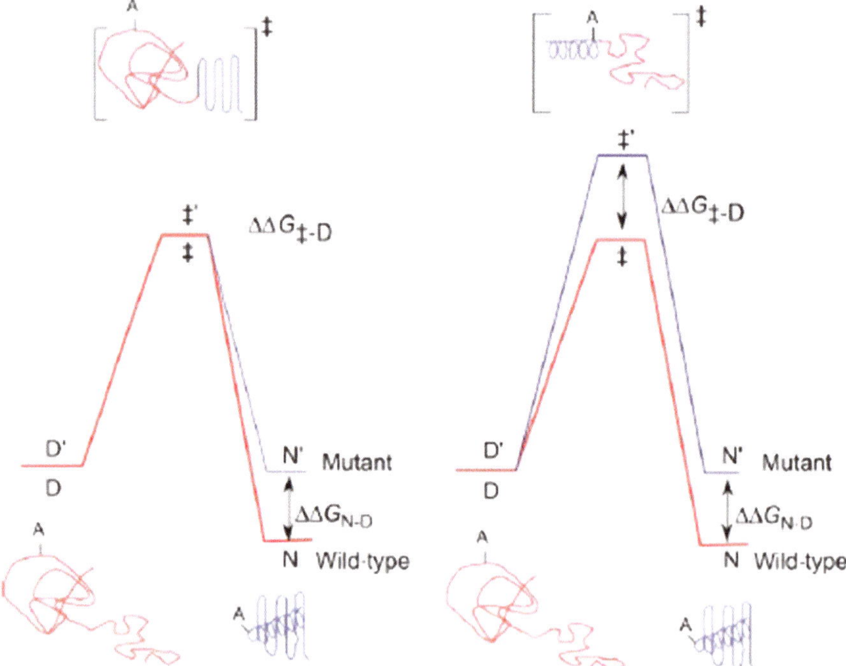

Fig. 3.11 Free energy profiles and ø-value analysis. *Left-hand panel*: The site of the mutation at the transition state lies in a region that has the same structure as the denatured state. This is reflected in the energetics with $\Delta\Delta G_{\ddagger-D}=0$ and therefore $\phi=0$. *Right-hand panel*: Here, the site of the mutation lies in a fully-developed helical region identical to that of the native state. As a result $\Delta\Delta G_{\ddagger-D}=\Delta\Delta G_{N-D}$ and $\phi=1$ (from Fersht *Cell* 108: 1 © 2002 Reprinted by permission from Elsevier)

shown on the left, ϕ is 0 and thus the transition state's structure at the site of the substitution resembles the denatured state, whereas in the situation on the right, ϕ is 1 and therefore the transition state's structure at that site is native-like.

This technique has been applied to characterizing transitions state of small globular proteins especially those that fold rapidly through two-state kinetics and have fairly smooth energy landscapes. ϕ-value analysis used together with NMR and molecular dynamics simulations made possible the detailed results for barnase and C2I presented in the previous chapter, in Figs. 2.8 and 2.9. In this combined experimental and theoretical approach, simulations were carried out at elevated temperatures for extended, nanosecond timescales in order to follow the entire reaction process. Experiments provided data on the denatured, transition, and intermediate states, and for checking and validating the molecular dynamics simulations. These experiments, like those utilizing hydrogen exchange, typically employ denaturants such as urea and guanidinium chloride.

3.16 Summary

1. Proteins are dynamic entities and their motions are essential for proper function.

   - Proteins continually undergo several different kinds of motions among which are:

     - Vibrational motions on the femtosecond timescale
     - Side chain rotations on a picosecond to nanosecond timescale
     - Large-scale domain movements on a microsecond to second timescale
     - Misfolding and aggregation on a second plus timescale

   - Proteins do not populate single unique conformations, but instead populate ensembles of states and substates.

     - The proteins continually transition from one to another as they undergo equilibrium fluctuations and functionally important motions
     - The ensembles of states and substates are arranged in a hierarchical manner to form an energy landscape

   - Proteins and the solvent in which they are embedded interact with one another and many of their motions are coupled together.

     - Solvent α-fluctuations in the bulk solvent regulate protein large-scale motions, while
     - β-fluctuations in the hydration layer drive protein small-scale internal motions.

2. Protein folding is directed by the macromolecular forces in an energetically downhill fashion satisfying the thermodynamic requirements. By utilizing cooperativity and other folding strategies proteins avoid becoming stuck in deep local minima. Proteins fold in ways that do not require exhaustive conformational searches. Instead, they fold along pathways that follow the contours of an energy surface in an overall downhill direction, taking them from an unfolded conformation to their native state.

Table 3.1 Terms and concepts used to describe protein folding

Terms and concepts	Meaning and significance
Denatured state	Name given to a large number of high-energy configurations of a newly synthesized or an unfolded protein
Energy landscape	A graphical representation of the number of states available to a protein at each value of the potential energy as a function of a few significant degrees of freedom
Fast folding	Submillisecond folding of simple proteins, whose energy landscapes have few barriers and traps
Folding funnel	The overall shape of the potential energy landscape. With many high-energy states and few low-energy ones, the surface narrows as the potential energy (or enthalpy) is reduced
Frustration	The inability of simultaneously optimizing all physical interactions and constraints at each site
Intermediate state	Long-lived, kinetically trapped, partially folded state
Internal friction	Generic term for intrachain collisions / landscape ruggedness that slows down the rate of protein folding
Kinetic trap	A local minimum in the energy landscape enclosed by energy barriers large compared to the thermal energy
Metastable state	Low-energy (native-like) intermediate state
Native state	Name given to the small number of low-energy configurations of the biologically active protein
Principle of minimal frustration	The primary structure of a protein has been evolutionarily selected to enable folding into a stable native state along pathways with few kinetic traps

Table 3.2 Experimental methods for exploring protein folding and structure

Experimental method	Brief description
X-ray Crystallography	High resolution, atomic level detail, of protein 3D structure using X-ray diffraction
Nuclear Magnetic Resonance (NMR)	High resolution, atomic level detail, of protein 3D structure using nuclear spin flips
Mass Spectroscopy	Precise determination of protein masses using mass and charge separation
Hydrogen-Deuterium (H/D) Exchange	Provides information on secondary and tertiary structure, and folding intermediates using hydrogen-deuterium exchange
Phi (ϕ)-Value Analysis	Provides information on the 3D structure of transition states and folding intermediates using protein engineering

In an energy landscape representation, the vertical axis denotes the sum of all contributions to the internal or potential energy of the protein while the entropic contribution to the Gibbs free energy is depicted by the width of the funnel-shaped energy surface so generated. The horizontal axis itself gives the values of one or two reaction coordinates that capture the essential behavior of the protein as it folds. Table 3.1 summarizes some of the key terms and concepts associated with protein folding and energy landscapes.

3. Listed in Table 3.2 are some of the most prominent methods being used to probe how proteins fold into their native state and misfold into forms that

produce neurodegenerative diseases. The first two methods in the table, X-ray crystallography and nuclear magnetic resonance, provide atomic level details of protein structures.

Proteins fold through a sequence of states, most of which are transiently populated for a tiny fraction of a second. Partially folded intermediates are longer-lived for the reasons already discussed, and these can be studied to give valuable information on the routes taken by proteins as they fold and misfold. The next two methods, hydrogen-deuterium exchange and ϕ-value analysis, along with nuclear magnetic resonance (NMR), are especially useful for exploration of intermediate states and protein dynamics, and mass spectroscopy is increasingly used in conjunction with H/D exchange.

These are not the only experimental methods used to explore protein structure and dynamics. For example, several methods are particularly useful for exploring secondary structure. Circular dichroism provides information on a protein's secondary structure using left and right circularly polarized UV light. Infrared (IR) spectroscopy provides information on a protein's secondary structure using IR absorption by vibrational chemical bonds, and Raman spectroscopy provides information on a protein's secondary structure using inelastically scattered IR light to populate vibrational bands. During the last few years single-molecule biophysics has moved to the fore to supply new means of exploring protein structure and dynamics. These methods provides structural dynamics and folding information at the level of single biomolecules using (a) fluorescence methods, most notably, Förster resonance energy transfer (FRET); (b) atomic force microscopy (AFM), and (c) optical tweezers. These methods will be examined in Chap. 5.

4. Molecular mechanics and the three entries that follow in Table 3.3 are key computational and theoretical tools. They are not only crucial for exploring theoretical issues but also are essential component of the acquisition and analysis of the experimental data. Molecular dynamics, Langevin/Brownian dynamics, and simulated annealing have their beginnings in the earliest days of computers and have been used to explore protein folding for several decades. As computers have become progressively more powerful, and dedicated protein folding machines have been developed, the frontiers of what can be explored computationally have been pushed back considerably.

Table 3.3 Theoretical methods for exploring protein folding and structure

Theoretical method	Brief description
Molecular Mechanics	Empirical modeling of the macromolecular forces that hold proteins together and drive protein folding
Molecular Dynamics	Theoretical, deterministic method for exploring protein structure and dynamics of folding carried out by numerically solving Newton's equations of motion
Langevin/Brownian Dynamics	Theoretical method based on solving the Langevin equation in which diffusion, friction, and fluctuating forces are emphasized
Simulated Annealing	Theoretical, stochastic optimization method for exploring transitions between unfolded and folded states in proteins

Bibliography

Liang, J., & Dill, K. A. (2001). Are proteins well-packed? *Biophysical Journal, 81*, 751–766.
Richards, F. M. (1974). The interpretation of protein structures: Total volume, group volume distributions and packing density. *Journal of Molecular Biology, 82*, 1–14.

Motions of Proteins

Austin, R. H., Beeson, K. W., Eisenstein, L., Frauenfelder, H., & Gunsalus, I. C. (1975). Dynamics of ligand binding to myoglobin. *Biochemistry, 14*, 5355–5373.
Frauenfelder, H., Sligar, S. G., & Wolynes, P. G. (1991). The energy landscapes and motions of proteins. *Science, 254*, 1598–1603.
Henzler-Wildman, K., & Kern, D. (2007). Dynamic personalities of proteins. *Nature, 450*, 964–972.
Weber, G. (1972). Ligand binding and internal equilibria in proteins. *Biochemistry, 11*, 864–878.

Protein Dynamics, Solvent Slaving, and the Glass-Transition

Adair, G. S., & Adair, M. E. (1936). The densities of protein crystals and the hydration of proteins. *Proceedings of the Royal Society B, 120*, 422–446.
Bernal, J. D., & Crowfoot, D. (1934). X-ray photographs of crystalline pepsin. *Nature, 133*, 794–795.
Chick, H., & Martin, C. J. (1913). The density and solution volume of some proteins. *The Biochemical Journal, 7*, 92–96.
Fenimore, P. W., Frauenfelder, H., McMahon, B. H., & Parak, F. G. (2002). Slaving: Solvent fluctuations dominate protein dynamics and function. *Proceedings of the National Academy of Sciences of the United States of America, 99*, 16047–16051.
Frauenfelder, H., Chen, G., Berendzen, J., Fenimore, P. W., Jansson, H., McMahon, B. H., et al. (2009). A unified model of protein dynamics. *Proceedings of the National Academy of Sciences of the United States of America, 106*, 5129–5134.
Rasmussen, B. F., Stock, A. M., Ringe, D., & Petsko, G. A. (1992). Crystalline ribonucleaese A loses function below the dynamical transition at 220 K. *Nature, 357*, 423–424.
Ringe, D., & Petsko, G. A. (2003). The 'glass transition' in protein dynamics: What it is, why it occurs, and how to exploit it. *Biophysical Chemistry, 105*, 667–680.

Energy Landscape Picture

Bryngelson, J. D., Onuchic, J. N., Socci, N. D., & Wolynes, P. G. (1995). Funnels, pathways and the energy landscape of protein folding: A synthesis. *Proteins, 21*, 167–195.
Bryngelson, J. D., & Wolynes, P. G. (1987). Spin glasses and the statistical mechanics of protein folding. *Proceedings of the National Academy of Sciences of the United States of America, 84*, 7524–7528.
Dill, K. A., & Chan, H. S. (1997). From Levinthal to pathways to funnels. *Nature Structural Biology, 4*, 10–19.
Leopold, P. E., Montal, M., & Onuchic, J. N. (1992). Protein folding funnels: A kinetic approach to the sequence-structure relationship. *Proceedings of the National Academy of Sciences of the United States of America, 89*, 8721–8725.

Metastability

Jaswal, S. S., Sohl, J. L., Davis, J. H., & Agard, D. A. (2002). Energetic landscape of α-lytic protease optimizes longevity through kinetic stability. *Nature, 415*, 343–346.

Jaswal, S. S., Truhlar, S. M. E., Dill, K. A., & Agard, D. A. (2005). Comprehensive analysis of protein folding activation thermodynamics reveals a universal behavior violated by kinetically stable proteases. *Journal of Molecular Biology, 347*, 355–366.

Lomas, D. A., & Mahadeva, R. (2002). α_1-Antitrypsin polymerization and the serpinopathies: Pathobiology and prospects for therapy. *The Journal of Clinical Investigation, 110*, 1585–1590.

Whisstock, J. C., & Bottomley, S. P. (2006). Molecular gymnastics: Serpin structure, folding and misfolding. *Current Opinion in Structural Biology, 16*, 761–768.

Spin Glasses

Binder, K., & Young, A. P. (1986). Spin glasses: Experimental facts, theoretical concepts, and open questions. *Reviews of Modern Physics, 58*, 801–976.

Molecular Mechanics

Boas, F. E., & Harbury, P. B. (2007). Potential energy functions for protein design. *Current Opinion in Structural Biology, 17*, 199–204.

Brooks, B. R., Bruccoleri, R. E., Olafson, B. D., States, D. J., Swaninathan, S., & Karplus, M. (1983). CHARMM: A program for macromolecular energy, minimization, and dynamic calculations. *Journal of Computational Chemistry, 4*, 187–217.

Cornell, W. D., Cieplak, P., Bayly, C. I., Gould, I. R., Merz, K. M., Jr., Ferguson, D. M., et al. (1995). A second generation force field for the simulation of proteins, nucleic acids, and organic molecules. *Journal of the American Chemical Society, 117*, 5179–5197.

Schueler-Forman, O., Wang, C., Bradley, P., Misura, K., & Baker, D. (2005). Progress in modeling of protein structures and interactions. *Science, 310*, 638–642.

Recognition and Binding

Boehr, D. D., Nussinov, R., & Wright, P. E. (2009). The role of dynamic conformational ensembles in biomolecular recognition. *Nature Chemical Biology, 5*, 789–796.

Csermely, P., Palotai, R., & Nussinov, R. (2010). Induced fit, conformational selection and independent dynamic segments: An extended view of binding events. *Trends in Biochemical Sciences, 35*, 539–546.

Fischer, E. (1894). Einfluss der Configuration auf die Wirkung der Enzyme. *Berichte der Deutschen Chemischen Gesellschaft, 27*, 2984–2993.

Koshland, D. E., Jr. (1958). Application of a theory of enzyme specificity to protein synthesis. *Proceedings of the National Academy of Sciences of the United States of America, 44*, 98–104.

Ma, B., Kumar, S., Tsai, C. J., & Nussinov, R. (1999). Folding funnels and binding mechanisms. *Protein Engineering, 12*, 713–720.

Tokuriki, N., & Tawfik, D. S. (2009). Protein dynamism and evolvability. *Science, 324*, 203–207.

Tsai, C. J., Kumar, S., Ma, B., & Nussinov, R. (1999). Folding funnels, binding funnels, and protein function. *Protein Science, 8*, 1181–1190.

Enzyme Catalysis

Boehr, D. D., McElheny, D., Dyson, H. J., & Wright, P. E. (2006). The dynamic energy landscape of dihydrofolate reductase catalysis. *Science, 313*, 1638–1642.

Eisenmesser, E. Z., Millet, O., Labeikovsky, W., Korzhnev, D. M., Wolf-Watz, M., Bosco, D. A., et al. (2005). Intrinsic dynamics of an enzyme underlies catalysis. *Nature, 438*, 117–121.

Henzler-Wildman, K. A., Thai, V., Lei, M., Ott, M., Wolf-Watz, M., Fenn, T., et al. (2007). Intrinsic motions along an enzymatic reaction trajectory. *Nature, 450*, 838–844.

Allostery and Conformational Entropy

Ferreon, A. C. M., Ferreon, J. C., Wright, P. E., & Deniz, A. A. (2013). Modulation of allostery by protein intrinsic disorder. *Nature, 498*, 390–394.

Frederick, K. K., Marlow, M. S., Valentine, K. G., & Wand, A. J. (2007). Conformational entropy in molecular recognition by proteins. *Nature, 448*, 325–329.

Hilser, V. J., & Thompson, E. B. (2007). Intrinsic disorder as a mechanism to optimize allosteric coupling in proteins. *Proceedings of the National Academy of Sciences of the United States of America, 104*, 8311–8315.

Koshland, D. E., Jr., Némethy, G., & Filmer, D. (1966). Comparison of experimental binding data and theoretical models in proteins containing subunits. *Biochemistry, 5*, 365–385.

Monod, J., Wyman, J., & Changeux, J. P. (1965). On the nature of allosteric transitions: A plausible model. *Journal of Molecular Biology, 12*, 88–118.

Motlagh, H. N., Wrabi, J. O., Li, J., & Hilser, V. J. (2014). The ensemble nature of allostery. *Nature, 508*, 331–339.

Nussinov, R., & Tsai, C. J. (2013). Allostery in disease and in drug discovery. *Cell, 153*, 293–305.

Tzeng, S. R., & Kalodimos, C. G. (2012). Protein activity regulated by conformational entropy. *Nature, 488*, 236–240.

Molecular Dynamics

Adcock, S. A., & McCammon, J. A. (2006). Molecular dynamics: Survey of methods for simulating the activity of proteins. *Chemical Reviews, 106*, 1589–1615.

Alder, B. J., & Wainwright, T. E. (1959). Studies in molecular dynamics: I. General methods. *The Journal of Chemical Physics, 31*, 459–466.

Chodera, J. D., Singhal, N., Pande, V. S., Dill, K. A., & Swope, W. C. (2007). Automatic discovery of metastable states for the construction of Markov models of macromolecular conformational dynamics. *The Journal of Chemical Physics 126*: art. 155101.

Dinner, A. R., Šali, A., Smith, L. J., Dobson, C. M., & Karplus, M. (2000). Understanding protein folding via free-energy surfaces from theory and experiment. *Trends in Biochemical Sciences, 25*, 331–339.

Freddolino, P. L., Harrison, C. B., Liu, Y., & Schulten, K. (2010). Challenges in protein folding timescale, representation, and analysis. *Nature Physics, 6*, 751–758.

Gō, N. (1983). Theoretical studies of protein folding. *Annual Review of Biophysics and Bioengineering, 12*, 183–210.

Kohlhoff, K. J., Shukla, D., Lawrenz, M., Bowman, G. R., Konerding, D. E., Belov, D., et al. (2014). Cloud-based simulations on Google Exacycle reveal ligand modulation of GPCR activation pathways. *Nature Chemistry, 6*, 15–21.

Kubelka, J., Hofrichter, J., & Eaton, W. A. (2004). The protein folding 'speed limit'. *Current Opinion in Structural Biology, 14*, 76–88.

Levitt, M., & Warshal, A. (1975). Computer simulation of protein folding. *Nature, 253,* 694–698.

Lindorff-Larsen, K., Piana, S., Dror, R. O., & Shaw, D. E. (2011). How fast-folding proteins fold. *Science, 334,* 517–520.

McCammon, J. A., Gelin, B. R., & Karplus, M. (1977). Dynamics of folded proteins. *Nature, 267,* 585–590.

Noé, F., Schütte, C., Vanden-Eijnden, E., Reich, L., & Weikl, T. R. (2009). Constructing the equilibrium ensemble of folding pathways from short off-equilibrium simulations. *Proceedings of the National Academy of Sciences of the United States of America, 106,* 19011–19016.

Piana, S., Lindorff-Larsen, K., & Shaw, D. E. (2013). Atomic level description of ubiquitin folding. *Proceedings of the National Academy of Sciences of the United States of America, 110,* 5915–5920.

Rahman, A. (1964). Correlations in the motions of atoms in liquid argon. *Physical Review A, 136,* 405–411.

Rahman, A., & Stillinger, F. H. (1971). Molecular dynamics study of liquid water. *The Journal of Chemical Physics, 55,* 3336–3359.

Shaw, D. E., Maragakis, P., Lindorff-Larsen, K., Piana, S., Dror, R. O., Eastwood, M. P., et al. (2010). Atomic-level characterization of the structural dynamics of proteins. *Science, 330,* 341–346.

Voelz, V. A., Bowman, G. R., Beauchamp, K., & Pande, V. S. (2010). Molecular simulation of *ab initio* protein folding for a millisecond folder NTL9(1–39). *Journal of the American Chemical Society, 132,* 1526–1528.

Langevin Dynamics

Einstein, A. (1905). Über die von der molekular-kinetischen Theorie der Wärmer geforderte Bewegung von der ruhenden Flüssigkeiten suspenierten Teilchen. *Annals of Physics (Leipzig), Series 4, 17,* 549–560.

Langevin, P. (1908). Sur la théorie du mouvement brownien. *Comptes Rendus de l'Académie des Sciences (Paris), 146,* 530–533.

Perrin, J. (1909). Mouvement brownien et réalité moléculaire. *Anneles de Chemie et de Physique, Series 8, 18,* 1–114.

Monte Carlo Methods

Cerny, V. (1985). Thermodynamic approach to the traveling salesman problem: An efficient simulation algorithm. *Journal of Optimization Theory and Applications, 45,* 41–51.

Kirkpatrick, S., Gelatt, C. D., & Vecchi, M. P. (1983). Optimization by simulated annealing. *Science, 220,* 671–680.

Metropolis, N., Rosenbluth, A. W., Rosenbluth, M. N., Teller, A. H., & Teller, E. (1953). Equation of state calculations by fast computing machines. *The Journal of Chemical Physics, 21,* 1087–1092.

Ulan, S. M., & von Neumann, J. (1947). On combination of stochastic and deterministic processes. *Bulletin of the American Mathematical Society, 53,* 1120.

Nuclear Magnetic Resonance

Arnold, J. T., Dharmatti, S. S., & Packard, M. E. (1951). Chemical effects on nuclear induction signals from organic compounds. *The Journal of Chemical Physics, 19,* 507.

Aue, W. P., Bertholdi, E., & Ernst, R. R. (1976). Two dimensional spectroscopy: Application to nuclear magnetic resonance. *The Journal of Chemical Physics, 64*, 2229–2246.

Block, F., Hansen, W. W., & Packard, M. (1946). Nuclear induction. *Physics Review, 69*, 127.

Gutowsky, H. S., McCall, D. W., & Slichter, C. P. (1953). Nuclear magnetic resonance multiplets in liquids. *The Journal of Chemical Physics, 21*, 279–292.

Lauterbur, P. C. (1973). Image formation by induced local interactions: Examples employing nuclear magnetic resonance. *Nature, 242*, 190–191.

Ogawa, S., Lee, T. M., Kay, A. R., & Tank, D. W. (1990). Brain magnetic resonance imaging with contrast dependent on blood oxygenation. *Proceedings of the National Academy of Sciences of the United States of America, 87*, 9868–9872.

Ogawa, S., Tank, D. W., Menon, R., Ellermann, J. M., Kim, S. G., Merkle, H., et al. (1992). Intrinsic signal changes accompanying sensory stimulation: Functional brain mapping with magnetic resonance imaging. *Proceedings of the National Academy of Sciences of the United States of America, 89*, 5951–5955.

Proctor, W. G., & Yu, F. C. (1950). The dependence of a nuclear magnetic resonance frequency upon chemical compound. *Physics Review, 77*, 717.

Purcell, E. M., Torrey, H. C., & Pound, R. V. (1946). Resonance absorption by nuclear magnetic moments in a solid. *Physics Review, 69*, 37–38.

Williamson, M. P., Havel, T. F., & Wüthrich, K. (1985). Solution conformation of proteinase inhibitor IIA from bull seminal plasma by 1H nuclear magnetic resonance and distance geometry. *Journal of Molecular Biology, 182*, 295–315.

Mass Spectrometry

Benesch, J. L. P., & Ruotolo, B. T. (2011). Mass spectrometry: Come of age for structural and dynamic biology. *Current Opinion in Structural Biology, 21*, 641–649.

Bernstein, S. L., Dupuis, N. F., Lazo, N. D., Wyttenbach, T., Condron, M. M., Bitan, G., et al. (2009). Amyloid-β protein oligomerization and the importance of tetramers and dodecamers in the aetiology of Alzheimer's disease. *Nature Chemistry, 1*, 326–331.

Hydrogen-Deuterium Exchange

Englander, S. W., & Kallenbach, N. R. (1984). Hydrogen exchange and structural dynamics of proteins and nucleic acids. *Quarterly Reviews of Biophysics, 16*, 521–655.

Hvidt, A., & Nielsen, S. O. (1966). Hydrogen exchange in proteins. *Advances in Protein Chemistry, 21*, 288–386.

Miller, D. W., & Dill, K. A. (1995). A statistical mechanical model for hydrogen exchange in globular proteins. *Protein Science, 4*, 1860–1873.

Woodward, C., Simon, I., & Tüchsen, E. (1982). Hydrogen exchange and the dynamic structure of proteins. *Molecular and Cellular Biochemistry, 48*, 135–160.

Phi-Value Analysis

Fersht, A. R., Matouschek, A., & Serrano, L. (1992). The folding of an enzyme: I. Theory of protein engineering analysis of stability and pathway of protein folding. *Journal of Molecular Biology, 224*, 771–782.

Matouschek, A., Kellis, J. T., Jr., Serrano, L., & Fersht, A. R. (1989). Mapping the transition stte and pathway of protein folding by protein engineering. *Nature, 340*, 122–126.

Chapter 4
Protein Misfolding and Aggregation

The reversible folding and unfolding of proteins uncovered by Anfinsen and Pauling was a landmark event. It established the primacy of the amino acid sequence in determining the functional, native state of a protein. It further set the stage for the equally remarkable discovery that proteins can possess more than one stable conformation; and that these alternative conformations, brought on by mutations or abnormal environmental conditions, can promote aggregation and a host of neurodegenerative diseases.

This second fundamental idea took over 30 years to emerge. Several key discoveries led the way. Chief among these were the 1959 electron microscope experiments of Cohen and Calkins followed by the 1968 X-ray diffraction studies by Glenner and Wong. These experiments established that amyloids were composed of sets of β-sheets oriented parallel to the amyloid fibril axis with each strand in each β-sheet oriented perpendicular to the fibril axis thereby producing a characteristic cross-β pattern.

Additional studies repeated over several years with different proteins established that many if not most proteins had the capability to adopt this alternative form if their environment was suitably manipulated. Remarkably, the proteins that underwent this change in conformation had nothing in common. These changes occurred whether the proteins were large or small, globular or extended, predominately beta-sheet or alpha-helical. The conclusion drawn from these studies was that the ability to adopt the cross-β-generating conformation was a basic property of the protein backbone. Under normal conditions the adoption of an alternative, potential-disease-causing conformation does not occur. In stark contrast, in the neurodegenerative disorders, one or more proteins fail to properly fold and remain in their native state and instead adopt the alternative cross-β conformation and collect in insoluble extracellular and intracellular deposits.

A major challenge in the study of neurodegenerative disorders has been, and remains, to firmly establish the causal chain linking the observed amyloids and other kinds of deposits of misfolded proteins to the disease severity and progression.

© Springer International Publishing Switzerland 2015 95
M. Beckerman, *Fundamentals of Neurodegeneration and Protein Misfolding Disorders*,
Biological and Medical Physics, Biomedical Engineering,
DOI 10.1007/978-3-319-22117-5_4

Table 4.1 Prominent disease-causing proteins and the visible deposits they form

Protein	Deposit	Site of action	Disease
Aβ	Amyloid plaques	Temporoparietal	AD
α-Synuclein	Lewy bodies	Midbrain	PD
	Lewy bodies	Frontotemporal	LBD
tau	Tangles	Temporoparietal	AD
	Pick bodies	Frontotemporal	FTLD
Huntingtin	NIIs	Basal ganglia	Huntington's
SOD1	Bunina bodies	Motor cortex	ALS
TDP-43, FUS	Inclusion bodies	Motor cortex	ALS
	Inclusion bodies	Frontotemporal	FTLD

In many of these disorders, several morphologically distinct types of deposits are formed, not just the ones presented in this short table. As will be seen in the later chapters this list of diseases is a highly abbreviated one containing only the most prominent of the neurological disorders uncovered to-date. *AD* Alzheimer's disease, *PD* Parkinson's disease, *LBD* Lewy body disease, *FTLD* frontotemporal lobar degeneration, a general heading encompassing a number of disorders, *ALS* amyotrophic lateral sclerosis, *tangles* neurofibrillary tangles, *NIIs* neuronal intranuclear inclusions

In the case of the systemic amyloidoses discussed in Chap. 1, the broad features if not the details are becoming clear. The misfolded proteins form intrusive deposits in cells, tissues, and organs leading to their malfunction, failure, and death. Not so in the case of Alzheimer's disease or Parkinson's disease or any of the other neurodegenerative disorders. The correlations between the visible deposits (Table 4.1) and disease are far more complex. They are made so by the presence of multiple types of aggregates that can and do interconvert into one another. A minimal list of the types of structures formed by misfolded proteins is, as follows:

- Monomers
- Soluble oligomers
- Amorphous aggregates
- Amyloid fibrils

Monomers, single molecules that alter their conformation from that of the native state, are the natural starting point in development of a protein misfolding disorder. These may append to a growing fibril or alternatively aggregate into small, soluble oligomeric complexes containing two, three, four, or more misfolded proteins that may or may not lie on the pathway leading to amyloid fibrils. There is emerging evidence that in many instances these small assemblies may be more toxic than the large insoluble deposits found in inclusion bodies and extracellular spaces. In addition, not all misfolded proteins generate amyloids. Misfolded huntingtin and TDP-43 do not do so. Instead, they accumulate in inclusion bodies as amorphous aggregates with varying degrees of amyloid-like properties.

A major difficulty in looking deeper into the structure of the amyloids was their large insoluble character. A way around this difficulty presented itself through the realization that the core structure of the amyloids might be generated from short fibril-generating sequences of just six or seven or eight amino acid residues. These findings led in 2005 to the first atomic level three-dimensional amyloid fibril crystal structures built from these short sequences. This chapter will begin with an examination of amyloid microcrystals and with the two leading models of how fibrils they might naturally develop.

There are numerous as yet unanswered questions on the road to effective therapeutics. One paramount set of questions has as its focus the identity of the toxic species responsible for each of the neurodegenerative diseases, and how these species might be produced. A key concept is of their development through loss of stability and partial unfolding; another is the contributing role of protein fluctuations and perturbations. An examination of oligomer formation by these means will follow the exploration of fibril structure and formation.

Many proteins especially those involved in cell regulation and signaling, the *sine qua non* of neurons, contain large unstructured regions and remain unfolded in their native state. These polypeptides are referred to as intrinsically unstructured proteins (IUPs) or even more commonly as intrinsically disordered proteins (IDPs). Numbered among this group are many of the proteins with prominent roles in neurodegeneration, including most, if not all, of the entries in Table 4.1. These proteins contain large regions lacking any discernable secondary structure. This chapter will conclude with a first look at this class of proteins.

4.1   Structure of the Amyloid Fibril

The first microcrystal structure solved in 2005 was that of the yeast protein Sup35. This structure is presented in Fig. 4.1. In this figure, one observes the cross-β spine of the amyloid fibrils formed by the seven-residue sequence GNNQQNY that is rich in glutamine (Q/Gln) and asparagine (N/Asn) amino acids. Repeating segments of this sequence assembled within several hours in solution into the pair of parallel beta-sheets shown in Fig. 4.1a. Each sheet was erected from identical seven-residue strands exactly in register, stacked one on top of another.

An extensive network of hydrogen bonds is formed within each beta-sheet. Some of these hydrogen bonds are between backbone residues, while the others are amide-amide hydrogen bonds formed between identical pairs of polar Asn and Gln residues in adjacent molecules. Recall that glutamine and asparagine side chains end with a hydrogen-acceptor oxygen atom and a hydrogen-donor amide group. These side chains enable the formation of "amide stacks" within each sheet. This arrangement of parallel beta-sheet-forming residues was given the name "polar zipper" by Max Perutz in a 1993 study of how long polyglutamine repeats prevalent in Huntington's disease and other polyQ disorders might form parallel beta-sheet-like structures.

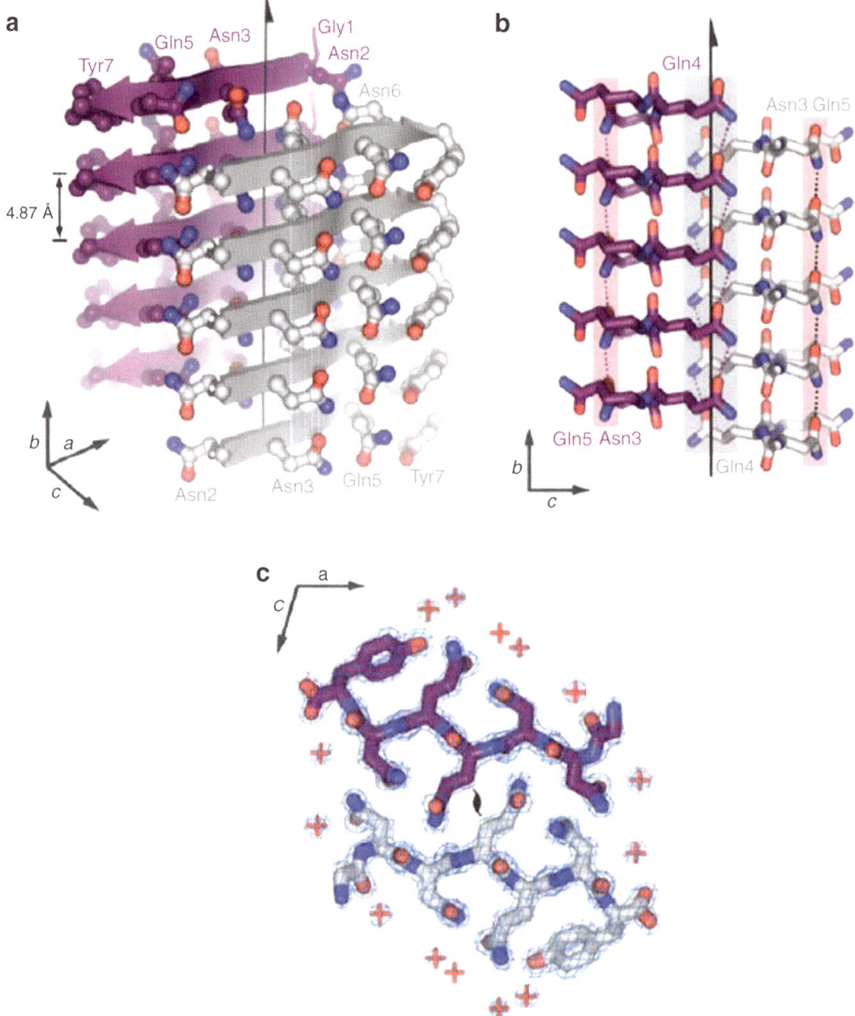

Fig. 4.1 Amyloid fibrils formed by the seven-residue sequence GNNQQNY. (**a**) Stacking of the monomers one on top of the other along the fibril axis (indicated by the *vertical arrow*). (**b**) The two parallel sheets are shifted relative to one another to permit the interdigitization of the side chains. (**c**) *Top-down view* showing the locations of the water molecule indicated by *red plus signs* showing that water molecules are absent from the interface (from Nelson *Nature* 435: 773 © 2005 Reprinted by permission from Macmillan Publishers Ltd)

The second important feature of this structure is that the two beta-sheets are held together not by hydrogen bonds but instead by van der Waals forces between polar side chains on beta-strands belonging to adjacent β-sheets. These side chains are interdigitized with one another (Fig. 4.1b). This arrangement was given the name "steric zipper" in analogy to Perutz's polar zipper. The third key feature of the structure is that water molecules although readily present elsewhere are expelled from the interface, which is tight and dry (Fig. 4.1c).

One of the most significant set of findings preceding the atomic level structures were the apparent universality of the amyloid fibril structure. A large number of polypeptides, irrespective of their primary sequences, could be made to refold into amyloid-forming conformations. Some of the proteins were small; others were large. Some folded normally into tight globular shapes while others assumed more extended forms. One conclusion drawn from these observations is that the ability of these proteins, or portions thereof, to fold into amyloid-templating structures is a general property of the polypeptide backbone. The sequence dependence, which under normal circumstances would play a key role in determining the fold, still plays a role albeit a different one. Its role in this alternative folding universe manifests itself at the atomic level through the existence of more than one kind of microcrystalline structure.

This point was brought out in studies that examined the spines of amyloid fibrils generated by different short peptide sequences. It was found, for example, that eight different kinds of steric zippers can be generated by short peptide amyloidogenic sequences. Examples studied included sequences from Aβ, tau, Sup35, and insulin. The eight ($2 \times 2 \times 2$) different classes are defined by (1) orientation of their faces (face-to-face or face-to-back); (2) orientation of their strands (up-up or up-down); and (3) whether the strands within each sheet were oriented in parallel or antiparallel.

Further variability in the fibril structures arises from their dependence upon their surroundings. The same sequences can assemble into several different microcrystalline forms as the growth and environmental conditions are altered. This dependence may include influences of the full chains. There is evidence that the microcrystalline forms represent reasonably well the actual in vivo fibrillar structures, but this remains to be established firmly. What can be said is that without question the forms generated by amyloid-competent sequences are highly polymorphic and responsive to their surroundings.

A second 2005 amyloid fibril atomic structure was that of the amyloid-β peptide generated from H/D-exchange NMR. It is presented in Fig. 4.2. Whereas the sequence shown in Fig. 4.1 was rich in polar residues, the amyloid-β fibril is dominated by hydrophobic residues. The Aβ fibrils have as their core two dissimilar sequences connected by a flexible loop resulting in U-turn heterozipper. The fibril-generating core consists of the hydrophobic segment containing residues 18–26 that comprise the β1 strand and residues 31–42 that form the β2 strand.

Yet another atomic level three-dimensional fibril structure is shown in Fig. 4.3. This structure, determined by means of solid state NMR, was of the prion protein HET-s (218–289). In this case, the β-sheets form a solenoid-like structure in which each monomer contributes two windings. The structure possesses a compact triangular core of hydrophobic residues along with two asparagine ladders, several salt bridges, and a large number of hydrogen bonds. The core contains three strands per winding resulting in parallel in-register β-sheets aligned along the fibril axis. The overall structure is a highly stable one and far less subject to polymorphic variability than many of the other structures.

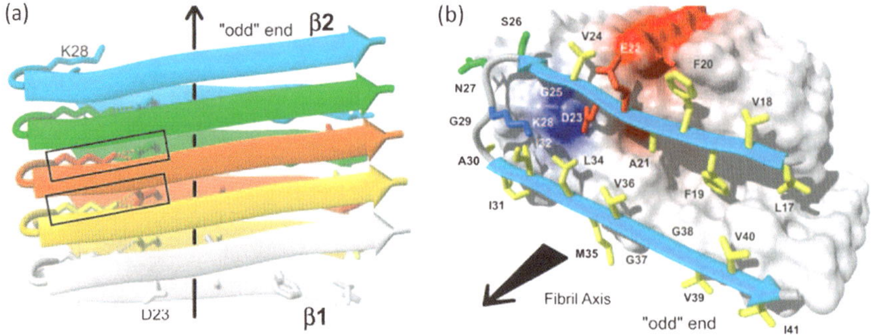

Fig. 4.2 Amyloid fibrils formed by residues 18–26 and 31–42 from Aβ (1–42) as determined by means of hydrogen-deuterium exchange and NMR. *Arrows* denote the fibril axis. (**a**) Parallel, in-register stacked strands form two β-sheets along the fibril axis indicated by the *arrow*. (**b**) Hydrophobic, polar, negatively charged and positively charged amino acid side chains are colored *yellow*, *green*, *red*, and *blue*, respectively; all others are depicted as *white* (from Lührs *PNAS* 102: 17342 © 2005 National Academy of Sciences, U.S.A. and reprinted with their permission)

Fig. 4.3 Amyloid fibrils from the fungal prion protein HET-s (218–289), which forms a triangular-shaped β-solenoid. The *grey arrow* denotes the fibril axis. The five winding pairs are color coded. The figure was prepared using Jmol with atomic coordinates deposited in the PDB under accession code 2NMR

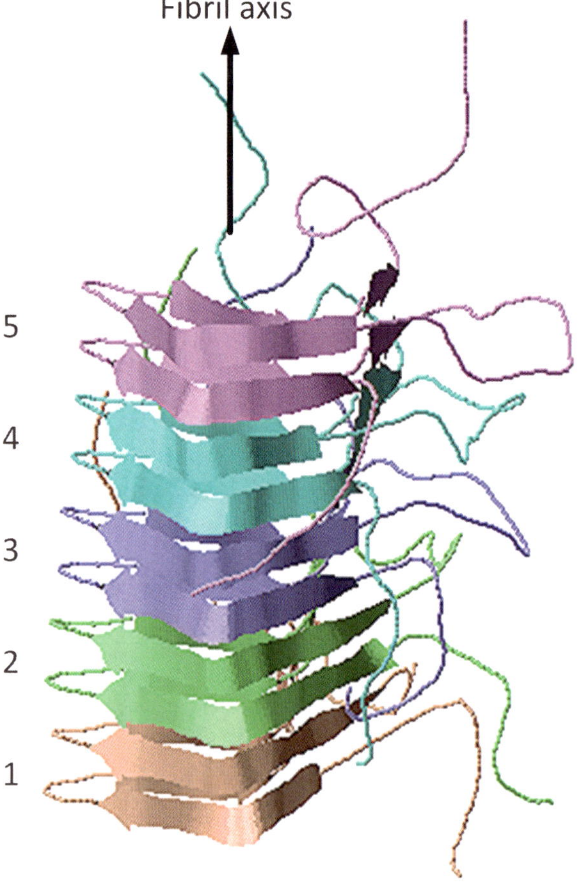

4.2 Amyloids as Biomaterials

William Astbury coined the term "molecular biology" in 1950, in his *Harvey Lectures*, to describe the new emerging field concerned with the forms of biological materials, that is, with their three-dimensional structure and their genesis and function. This field had as its cornerstone the X-ray crystallography studies of the Bragg's, of Bernal and Pauling and Kendrew and Perutz, along with Astbury, and all their essential collaborators and colleagues too numerous to mention. Now, some 80 years after Astbury's examinations of keratin, amyloids in their various aggregated forms are seen as the causes of neurodegenerative diseases, and another new field that of engineered biomaterials for tissue and organ repair is in its early stages of development.

Spider silk fibrils are remarkable structures. They, like amyloids, are assembled from beta-strands oriented perpendicular to the fibril axis and held together by hydrogen bonds. Although technically not amyloids they are similar, both belonging to a large class of structures erected from simple polypeptide building blocks. Like amyloids the spider silks are held together by cooperative networks of hydrogen bonds. Although these bonds are individually weak far more so than covalent bonds the overall effect of hydrogen bonding is to generate materials that are remarkably strong. In addition, whereas covalent bonds once broken stay broken, hydrogen bonds that break can reestablish themselves and are thus capable of self-repair. However, there are geometric/size limits to the stiffness and stability of hydrogen bonded amyloids and spider silk. Once individual strands and fibrils exceed a critical length they become weaker, more brittle, and less resilient to mechanical stresses. Fragmentation can then occur; this secondary process has important consequences not just for nucleating additional fibril growth but also for prion-phenomena as will be discussed later.

Amyloids, dragline silk, and other hydrogen bonded materials have high bending rigidity. That point is illustrated in Fig. 4.4 in which bending rigidity is plotted against the moment of inertia for a wide variety of materials. Recall that Young's modulus, Y, also known as the elastic modulus, E, is a material-specific measure of stiffness. It is related to bending rigidity, B, and the (area-specific, cross-sectional) moment of inertia, I, through the expression $B = Y \cdot I$. The data points shown in the figure represent values for the Young's modulus. These fall in several bands, according to the nature of the intermolecular forces holding the fibers together. Data points for a variety of amyloidogenic proteins are found in the orange band; their values are displayed in Fig. 4.4 as blue dots with error bars.

4.3 Amyloids in Normal Physiology

Given their remarkable physical properties it is perhaps not surprising that cross-β amyloids fibrils have a role in normal physiology. They are utilized across multiple kingdoms—in bacteria, in yeasts, and in humans. In bacteria, they are secreted and

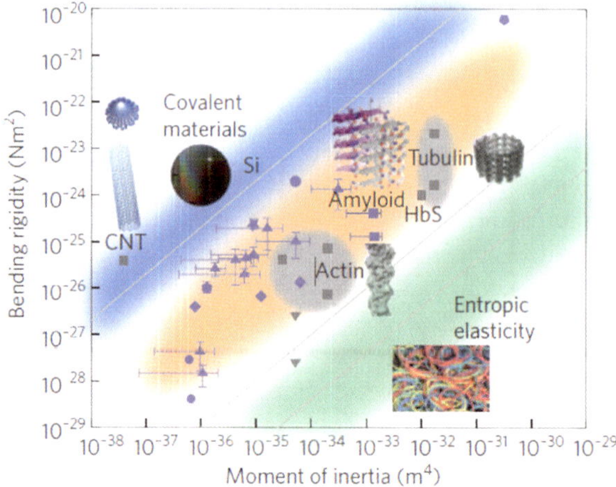

Fig. 4.4 Plot of bending rigidity versus the moment of inertia for different classes of materials. The *blue band* shows the range of values for metallic and covalently bonded substances; the *orange band* encompasses representative values for strong non-covalently bonded (e.g., hydrogen-bonded) structures. The *green band* contains values of weak non-covalently bonded materials. *Blue symbols* represent amyloidogenic substances; other types of materials are depicted in *grey* (from Knowles *Nat. Nanotechnol.* 6: 469 © 2011 Reprinted by permission from Macmillan Publishers Ltd)

function as an essential component of biofilms. Bacteria such as *Escherichia coli* (*E. coli*) are highly adaptable and can reversibly switch between a free-swimming planktonic lifestyle and a surface-attached, communal existence. Under stressful conditions they switch their gene expression programs to support construction and participation in communal single or multispecies biofilms. The biofilms contain not only the bacteria but also a protective, extracellular matrix of secreted molecules that facilitate attachment to surfaces and provides protection against agents in the environment potentially harmful to the bacteria. In many instances, amyloidogenic fibers are a major component of the biofilm extracellular matrix. Like the amyloids in neurodegenerative diseases, these fibers bind Congo red, exhibit birefringence, and possess an underlying cross-beta-sheet organization.

Bacterial amyloids are not a result of protein misfolding but instead are produced through a highly regulated assembly process, and, when desired, can be systematically disassembled to enable the bacteria to resume a planktonic lifestyle. The *E. coli* amyloidogenic materials are referred to as Curli fibers. Two Curli subunits—CsgA and CsgB—are centrally involved. The fibers themselves are composed of repeating CsgA units while the CsgB subunit directs their surface attachment, nucleation, and polymerization. The CsgA subunits contain an amyloid core domain consisting of a series of repeats enriched in glutamine and asparagine residues that generate the cross-beta-structure. These residues create a hydrogen bonded network that provides stability to the assemblage as they do for Sup35 discussed earlier in this chapter.

Functional amyloids are also encountered in yeasts where they facilitate adhesion to solid surfaces, cells, and the extracellular matrix, and promote formation of biofilms. One particularly well-studied example is that of the human opportunistic fungal pathogen *Candida albicans*. In these studies, atomic force microscopy was used to explore how the cell surface Als adhesins mediate fungal attachment. A key finding is the presence of force-activated partial unfolding that exposes seven-residue amyloidogenic sequence IVIVATT. This sequence is highly enriched in hydrophobic (I, V) and C_β-branched amino acids (I, V, T) that because of their particular side chains are ill-disposed to form alpha-helices and instead strongly favor formation of beta-strands. The beta-strands are stacked in amyloid fashion on top of one another to supply strong adhesive forces along with a stretching ability.

A question that naturally arises is how are the toxic effects of the functional amyloids being avoided? One answer is that in humans the amyloids are sequestered in membrane-bound compartments thereby limiting their potential toxic effects. One example of this is the storage of peptide hormones in secretory granules, another is the melanosome matrix protein (Pmel17) fibers critical for melanosome maturation. In the former instance, endocrine hormones form stable amyloids while they are stored in secretory granules. Once they are released the large structures rapidly dissociate into monomers and are secreted from the cell. Thus, the potentially toxic amyloids are not only being encapsulated and sequestered, but they also readily dissociate unlike, for example, the Aβ peptides. In the second example, the Pmel17 amyloid fibers are also packaged in membrane-bound organelles, and in this case the melanosomes are subject to fast kinetics, both aspects limiting any toxic effects from oligomers and other potentially harmful intermediate sized aggregates.

4.4 Amyloid Growth Through Nucleated Polymerization

How repeating structures such as three-dimensional metallic crystals and one-dimensional amyloid fibrils self-assemble has been the subject of study for a century or more. Two models, or pictures, of how these structures develop have been put forth over the years. The older of the two is the *nucleated polymerization* picture inspired by how crystal growth occurs and how actin or tubulin fibers develop. This model was proposed as the basis for Aβ and prion amyloid growth by Jarrett and Lansbury in 1992. The other, *nucleated conformational conversion*, was put forth by Serio in 2000; it was suggested by the presence of significant numbers of structurally fluid prion oligomers.

In crystal growth, atoms undergo a phase change from a three-dimensional disordered ensemble to a highly ordered spatially regular arrangement. This process takes place in several stages. In the, initial, slow stage a seed had to form that serves as the nucleus for the subsequent and far more rapid crystal growth stage. In a series of papers published in the 1939–1941 time period, Melvin Avrami derived a simple expression, the Avrami equation, for the crystallization kinetics:

$$X(t) = 1 - \exp(-Kt^n) \tag{4.1}$$

In this expression, $X(t)$ is the fraction of atoms that have transitioned to the crystalline phase at the given time t, K is a geometry-dependent growth rate, and n is an integer whose value varies with the type of nucleation and structure of the crystal. Similar expressions to this one have been applied to the growth of amyloid fibrils.

Sickle cell anemia is a disorder brought on by a single point mutation in the gene coding for the beta-chain of hemoglobin. This mutation results in the substitution of valine for glutamic acid at position 6; this alteration has no effect on the fully oxygenated state of hemoglobin, but results in a reduced solubility of deoxygenated hemoglobin, HbS. As a result HbS forms aggregates referred to as a gel that deforms the normal red blood cells into a crescent or sickle shape and results in vascular occlusion and decreased blood flow through the microcirculation. Under an electron microscope the fibrils comprising the gel are found to be organized into two helical layers—an inner layer of four polymeric strands surrounded by a second helical layer consisting of ten strands. The result is a 21 nm diameter structure consisting of 14 strands twisted together to form long fibrils.

Sickle cell disease was first described in 1910 by the physician James Bryan Herrick (1861–1954) who had noted the odd-shaped cells. Further studies by Herrick and others over the next 15 years or so further established the connection between the sickle-shaped red blood cells and the anemia. That set the stage in 1949 for the landmark paper by Linus Pauling that identified an aberrant form of hemoglobin as the causal agent of the disease and introduced the concept of "molecular medicine", and the study by James Neel (1915–2000) that established its genetic origins.

Another set of early pioneering studies focused on normal growth in the cell of tubulin and actin filaments. Those considerations led to the adaptation of crystal growth ideas into the realm of one-dimensional protein polymerization by the Japanese physicists Fumio Oozawa and Sho Asakura. Their investigation was followed a short time later by a series of studies by William A. Eaton and coworkers in the 1970s and 1980s on HbS polymerization and the further expansion of Oozawa and Asakura's model of nucleated polymerization to include secondary polymerization. The overall joint result is a central model of how both normal and abnormal filaments and amyloids develop. The key mechanisms underlying protein nucleated polymerization are illustrated in Fig. 4.5.

This model describes several features of significance. First, the initial stage, referred to as a lag phase, is a slow one since a number of nuclei or seeds have to form first, and this is kinetically unfavorable. Once nuclei have formed the process speeds up with the addition of monomers to either end of the growing one-dimensional fiber. As noted in the section on amyloids as a biomaterial, once the fiber length exceeds a critical value it becomes increasingly brittle and prone to shed pieces. Several studies have been carried out that establish that fragmentation and secondary nucleation are important contributors to amyloid pathology. Lastly, in the nucleated polymerization model, the concentration of amyloid-competent monomers must exceed a threshold; there are few oligomers smaller than a nucleus, and

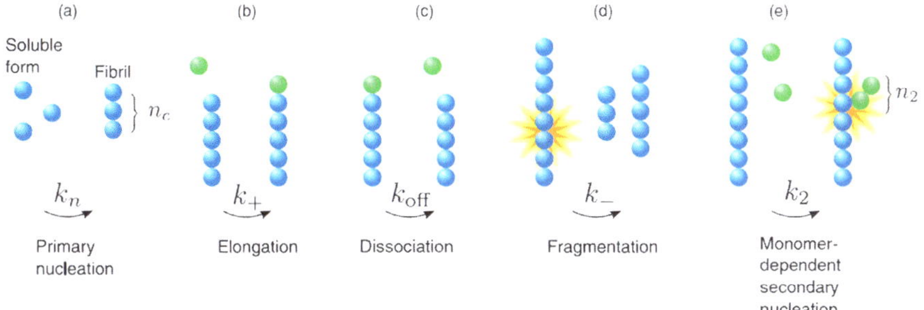

Fig. 4.5 Nucleation-polymerization mechanisms. (**a**) Primary polymerization leading to formation of a filament of length n_c; (**b**) linear growth from either end of the elongating filament; (**c**) loss of monomers from either end of the filament; (**d**) fragmentation, and (**e**) secondary nucleation leading to formation of branch (secondary) filament of length n_2 (from Cohen 2011 *J. Chem. Phys.* 135: 065106 © 2011 AIP Publishing, LLC and reprinted with their permission)

the lag phase can be shortened and perhaps eliminated if seeds are externally supplied at the outset.

4.5 Amyloid Growth through Nucleated Conformational Conversion

Oligomers are non-covalently-attached assembles of several identical monomers. These small aggregates are quire varied in their numbers and in their geometric arrangements (morphology). Some form pore-like water-bearing rings, others are cylindrical in shape but have a dry interior. Pathway destinations are of considerable importance. Some oligomers are amyloidogenic and serve as aggregation intermediates on the pathway towards amyloid assembly. Others are, in themselves, aggregation endpoints and may be toxic or not.

Oligomers are the main components in an alternative model of how amyloid fibrils may develop. In this second model oligomers serve as the locus of the conversion from non-amyloid-competent to amyloid-competent units. This process is one in which monomers initially and rapidly associate into amorphous oligomers. The units within these aggregates then undergo a slow nucleated or templated conformational conversion through their interactions with one another into ordered beta-strands and sheets that self-assemble into the fibrils as depicted in Fig. 4.6.

This model has a different kinetics from that of the nucleated polymerization model and a different dependence on concentration and associated thresholds. For example, unlike the nucleated polymerization model a significant number of soluble oligomers are present early in the fibril-forming process. Several recent studies of amyloid formation have been carried out that support this more recent oligomer-centric model emphasizing as they do the appearance of significant numbers of oligomers before the first fibrils appear.

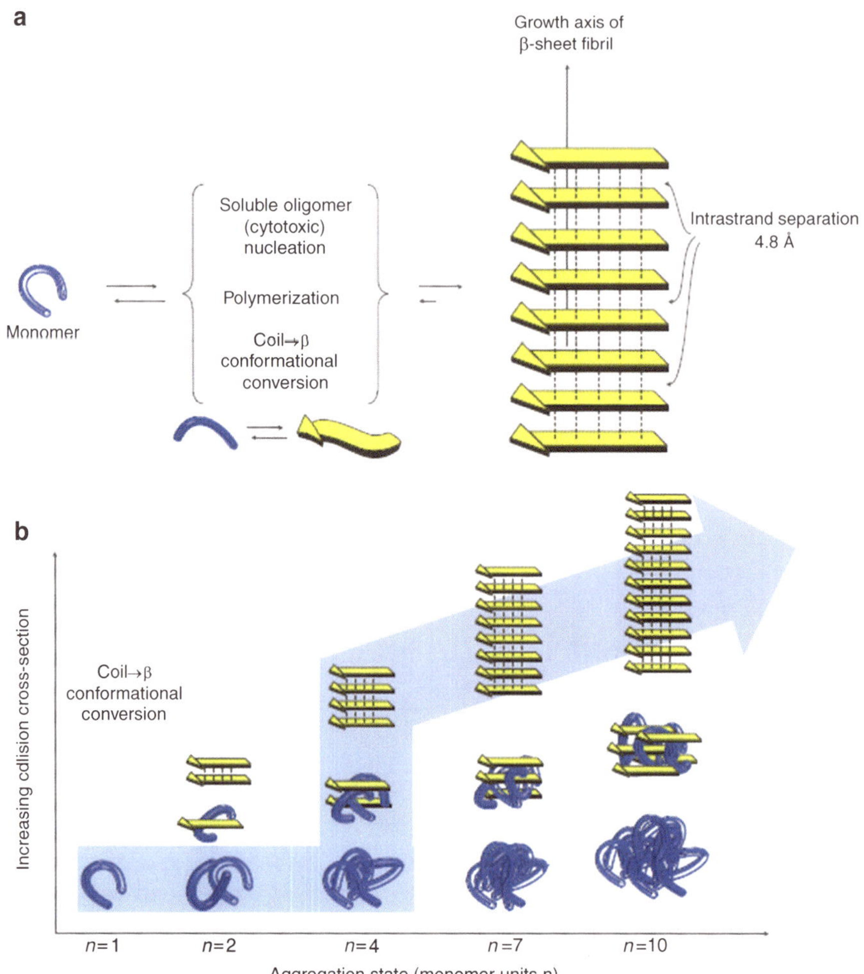

Fig. 4.6 Nucleated conformational conversion. *Blue bent* monomers aggregate into oligomers which then serve as a "reactor" for the conversion of the oligomers into beta strands depicted as *yellow arrows* (from Blieholder *Nat. Chem.* 3: 172 © 2011 Reprinted by permission from Macmillan Publishers Ltd)

4.6 The Toxic Oligomer Hypothesis

The "amyloid cascade hypothesis" appeared in a mature form in a 1992 paper by Hardy and Higgins. This hypothesis posits that deposits of Aβ peptide in the brain are the crucial events leading to Alzheimer's disease. This has been the basic operating principle behind much of the activity in the field during that past 20-plus years. However, identification of the specific form or forms of Aβ responsible for causing dementia is not clear. The primary object of interest has been the extracellular amyloid plaques, but clinical correlations between the extracellular deposits and the disease progression are not convincing and, in many cases, are absent. The presence of tau and the tangles added another level of complication as does the natural conversion of one form to the other through growth and fragmentation.

In an update and revision to the amyloid cascade hypothesis, toxic oligomers have been hypothesized to be the primary causative agents in Alzheimer's disease and other forms of neurodegeneration. These smaller soluble species are far more mobile and expose a greater fraction of hydrophobic surfaces than do the large insoluble amyloids and amorphous aggregates. Support for this shift in attention is provided by investigation of several different diseases. One of these is type 2 diabetes. In this disorder, islet amyloid polypeptide (IAPP) accumulates in the pancreas. These deposits were initially believed to be the main cause of β-cell failure paralleling the finding for most if not all the systemic amyloidosis. But difficulties arose with ascribing the cause of the disease to the extracellular deposits. The main event in the disease progression is death of β-cells through apoptosis. However, that this was due to the extracellular deposits of IAPP seemed to be implausible. Evidence that oligomers might be capable of permeating the cell membrane thereby triggering a series of events leading to apoptosis soon followed. The idea that oligomers might be the toxic agent in type 2 diabetes was further strengthened by the lack of a similar effect on membranes by components of the extracellular IAPP deposits.

A series of reports that appeared in the decade from 2002 to 2011 provided further evidence for a major role of toxic oligomers in Alzheimer's disease and other neurological disorders. Two biophysical features of these oligomers—their expose of hydrophobic surfaces and the presence of regions of intrinsic disorder in the candidate toxic proteins and their interaction partners—were centrally implicated in their ability to cause harm to and kill cells. Complicating the issue of the identity of specific toxic oligomers is their tremendous variability in oligomer size, shape, density, and hydrophobicity. Heterogeneous mixtures of monomers, dimers, trimers, tetramers, and so on up to and including large amorphous aggregates and fibrillary deposits can occur. The fractional contribution of each type of small oligomer and large aggregate to a given mixture at any time depends critically on the particular species and its biophysical properties.

The identification of certain features of the oligomers as being important was made possible by the creation of agents that recognize specific-binding epitopes on those proteins. Among these are antibodies called OC and A-11, and the fluorescence probe 8-anilino-1-naphthalenesulfonic acid (ANS). As noted above, a variety of oligomeric assembles are encountered; these small soluble oligomers vary in size and morphology. Some of them, referred to as *prefibrillar oligomers*, bind A-11 while at the same time are negative to binding the fibril-specific antibody OC. Others, termed *fibrillar oligomers*, bind OC but not A-11. Interestingly, the A-11 antibody can select out oligomers of diverse sequence structure apparently recognizing a common epitope. Similarly, ANS, which is used to detect conformational changes, recognizes a common hydrophobic epitope in proteins possessing unrelated primary sequences. Furthermore, the strength of the binding of ANS to these proteins correlates well with their increasing toxicity.

There is yet another layer of complexity beyond that of the differing sizes and morphologies of the aggregates. That layer is associated with variations in the primary sequence from which oligomers and fibrils are generated. These differences arise in several ways: First, there are variations in primary sequence arising from alternative splicing. Secondly, cleavage of the full-length protein often occurs resulting in the generation of protein fragments. Thirdly, and obviously, mutations alter the primary sequence. Leaving aside for the moment the mutations, the following are examples of alterations in primary sequence associated with the aggregation-prone proteins listed in Table 4.1:

- **Aβ**: Peptides vary in length from 39 to 43 amino acids; in addition, shortened peptides are produced in which a number of the N-terminal-most residues have been removed.
- **tau**: Six isoforms exist resulting in proteins ranging in length from 352 to 441 amino acids; shortened proteins are formed through truncation of N- and C-terminal residues.
- **huntingtin**: PolyQ tract lengths vary; full-length and N-terminal fragments are produced.
- **TDP-43**: C-terminal fragments are generated.

These changes in primary sequence can profoundly alter the tendency to misfold, aggregate, and cause disease. Even small differences can have large effects. Shown in Fig. 4.7 are the results of a study of Aβ1–40 and Aβ1–42 oligomer formation. Of the two forms, Aβ1–42 is regarded as being far more toxic. As the figure shows there are differences in the sizes and morphologies of the corresponding oligomeric species formed, and these distinctions are thought to be linked to their varying disease-causing propensities.

In many situations, the conformational variability of the putative toxic proteins creates challenges in identifying the exact toxic species responsible for a particular disorder. Those protein containing extensive disordered regions are particularly sus-

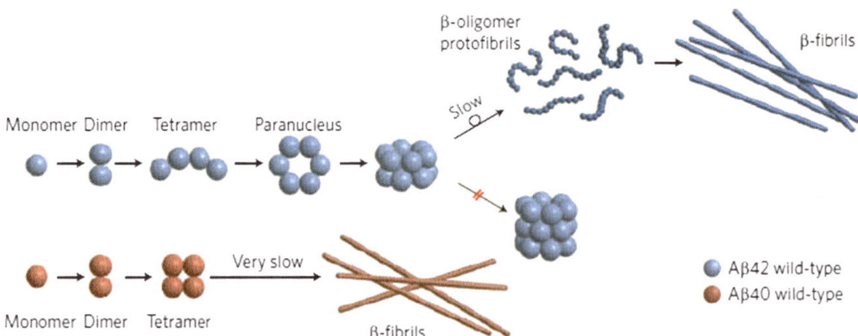

Fig. 4.7 Aβ1–40 (Aβ40) versus Aβ1–42 (Aβ42) oligomer and fibril formation (from Clemmer *Nat. Chem.* 1: 257/Bernstein *Nat. Chem.* 1: 326 © 2009 Reprinted by permission from Macmillan Publishers Ltd)

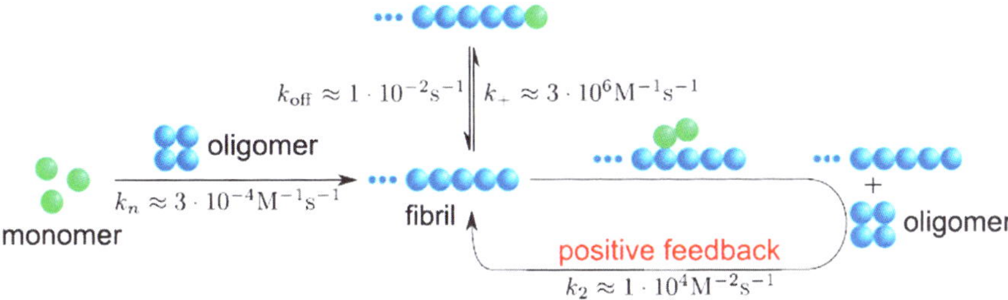

Fig. 4.8 Role of secondary nucleation in Aβ aggregation (from Cohen *PNAS* 110: 9758 © 2013 Reprinted with permission from Tuomas Knowles)

ceptible to these complications. This class of proteins, referred to as intrinsically disordered proteins, is heavily implicated as causative agents in neurodegeneration. Prominent examples include alpha-synuclein, tau, and huntingtin. Portions of SOD1, TDP-43, and FUS implicated in the frontotemporal dementias and ALS are unstructured, as well. As notable example of this flexibility is α-synuclein, which justifies adding another entry to the above-mentioned bullet-list:

- **α-Synuclein**: Called a chameleon because of its conformational diversity, this protein exhibits varying amounts of α-helical and β-sheet secondary structure.

Lastly, the relationship between oligomers and fibrils is a complex one that depends on the biophysical and biochemical details of the involved species. As seen in studies of Aβ aggregation, amyloid fibrils are variable in structure and may be generated from monomers and other small oligomers in more than one way. An added dimension to their complex interrelations is the existence of fragmentation and secondary nucleation pathways, processes, illustrated schematically in Figs. 4.5 and 4.8 that link together oligomers and fibrils. An important recent observation in this regard is that new oligomers may be generated, when both monomers and fibrils are present in sufficient numbers, by means of fibril surface-catalyzed (secondary) nucleation. For all of these reasons fibrils may yet play an important role in the pathogenesis of Alzheimer's disease and the other amyloidogenic disorders.

4.7   Coding Strategies that Prevent Misfolding and Aggregation

Physics and genetics work together to ensure that proteins that are expressed can fold into their native conformations in physiologically useful timeframes. As discussed in Chap. 2, they do not carry out exhaustive and time-consuming conformational searches; instead, they utilize a number of highly efficient macromolecular-force-driven folding stratagems. The interplay between physics and genetics goes much further. Sequences that are especially aggregation-prone are normally prevented from doing so by a variety of coding strategies. First and foremost, aggregation-prone portions of the protein tend to be buried in the core of the natively folded protein and only upon destabilization become exposed.

Secondly, certain patterns of residues—large blocks of hydrophobic residues, or patterns of alternating polar and no-polar residues—are encoded with low frequency, far lower than would be expected on purely statistical considerations. The highest frequency amino acid residues in sequences associated with Alzheimer's disease peptide Aβ42, tau, and α-synuclein are tryptophan, phenylalanine, cysteine, tyrosine, isoleucine, and valine while proline residues appear least often. These findings are similar to those obtained in examinations of other amyloid-forming sequences. In those situations where aggregation-prone sequences are present, additional coding strategies are utilized to prevent aggregation. Prominent among these are the presence of negative design elements, gatekeepers, and tight control of expression levels.

Negative design: Peripheral strands on beta-sheets can easily attach to similar peripheral strands on adjacent proteins, and initiate edge-to-edge aggregation. Beta-strands on protein fragments are similarly positioned to form aggregates, and cellular stresses causing unfolding greatly increases such dangers. However, proteins possess not only helices and beta-sheets but also various loops, turns, bulges, shielding strands, and capping helices. These function as "negative design elements", a term introduced by the Jane and David Richardson in 2002. These elements flank potentially dangerous peripheral, hydrophobic-residue-rich beta-strands and impeded their ability to form aggregates. Especially prominent among these design features are the use of proline and glycine in protective turns and loops.

Gatekeepers: Another aggregation-limiting strategy is to place residues lying near or at the bottom of the hydrophobicity scale within or flanking long stretches of hydrophobic residues. These residues—proline, lysine, arginine, glutamate, and aspartate—have been given the name "gatekeepers" by Otzen and coworkers in 1999–2000 and more recently by Rousseau and coworkers in 2006. The gatekeeper residues utilize charge-charge repulsion, entropic penalty, and low hydrophobicity to prevent the clumping together of beta-strands. Mutations that replace these critical residues will increase the aggregation propensity of the proteins.

Expression levels: The list of misfolded proteins associated with neurodegenerative diseases contains a small number of proteins. One question that naturally arises is what do they have in common? One possible answer to this question was obtained from the studies of HbS aggregation by Hofrichter in the 1970s. It turns out that the propensity of HbS to aggregate is concentration dependent. This idea has been extended to other proteins. There is evidence that one common feature shared by many neurodegenerative disease-causing proteins is that they are expressed at high levels when taking into account their intrinsic capability to form aggregates in the presence of cellular stresses. Rigid control of expression levels of these proteins is therefore another way that proteins try to limit misfolding and aggregation. In addition to these (largely passive) strategies that maintain solubility, cells invest considerable resources in active maintenance of proper protein folding. Those processes will be the subject of the next two chapters.

4.8   Loss of Stability Occurs through Changes in Environmental Conditions and Mutations

4.8.1   Changes in Environmental Conditions

The key event leading to formation of potentially damaging oligomers and amyloids is loss of native state stability. A method used by Anfinsen, Tanford, and many others since then in protein folding studies is to lower the pH of the aqueous environment of the protein. In some experiments, the goal was to reproduce the acidic conditions present in lysosomes; in others and more generally, it was aimed at inducing the denaturation of the protein in order to study alternative conformations leading to aggregation and fibril formation. The chief underlying process is the protonation of the proteins leading to positive charge-charge repulsion between amino acid residues. The consequences of doing this depend on hydrophobicity and other specifics of the amino acid composition. Those proteins that do successfully undergo pH-induced denaturation lose the bulk of their tertiary structure and little, some, or most of their secondary structure, while the backbone remains intact.

Another environmental factor of interest is ionic strength (salt concentration). This factor influences protein solubility and stability, and can either accelerate or impede fibril growth. The influences are diverse and complex because proteins bear many charges, both positive and negative. The effects of salts on proteins were famously reported by Franz Hofmeister (1850–1922) in 1888, and the ordered effects of different salts on protein solubility are known to this day as the Hofmeister series. In brief, some salts stabilize the native conformation and maintain solubility while others stimulate unfolding and intermediates on the pathway to forming aggregates. The salts can modulate electrostatic and hydrophobic interactions in several ways, for example, through (electrostatic) Coulomb screening and direct protein binding, or alternatively by their influences on the hydrogen bonding network of the water molecules.

4.8.2   Effects of Mutations

Missense mutations, that is, point mutations resulting in the substitution of one amino acid for another affect protein solubility and aggregation propensity. Most notably, solubility is reduced by mutations that disrupt the hydrophobic core of a protein, replacing, for instance, a nonpolar residue by a polar one, or by introducing residues with bulky side chains. Other, lesser reductions in stability can be brought on by mutations that disrupt hydrogen bonding and electrostatic interactions. The most important way aggregation is prevented is by maintenance of native state stability.

There are important interplays between mutations and biophysical stability in a protein. These interplays arise because of the limited stability of the native state(s) - they may be stable but they are not so by large amounts. As a consequence, many missense mutations destabilize proteins so much that they unfold at least partially.

This is a far more common consequence of a mutation than those leading to a gain or loss of function. It has been estimated that more than 80 % of disease-causing mutations destabilize the native state. These mutations affect protein stability by 1–3 kcal/mol. This may be compared to the thermodynamic stability measure of a protein, which is usually taken as ΔG between unfolded and native (folded) states of a protein and falls in the range −3 to −10 kcal/mol. For example, mutations to the hydrophobic core can destabilize a protein by more than 5 kcal/mol resulting in complete loss of the folded structure. Interestingly, not all mutations decrease protein stability. Instead, there are some mutations (about 5 %) that increase protein stability. These mutations can serve in a compensatory role during protein evolution. The basic idea is that a mutation that confers a new and beneficial function to a protein, but is destabilizing, can be preserved by a second compensatory mutation that restores stability to the native fold.

Mutations can alter a protein's energy landscape. An example of this type of change is presented in Fig. 4.9. The protein under consideration is lysozyme, an enzyme that functions within the innate immune system and attacks bacterial cell walls. That enzyme was given the name "lysozyme" by Alexander Fleming (1881–1955) in a paper published in 1922 entitled "On a remarkable bacteriolytic element found in tissues and secretions". In 1965, lysozyme became the first enzyme to have

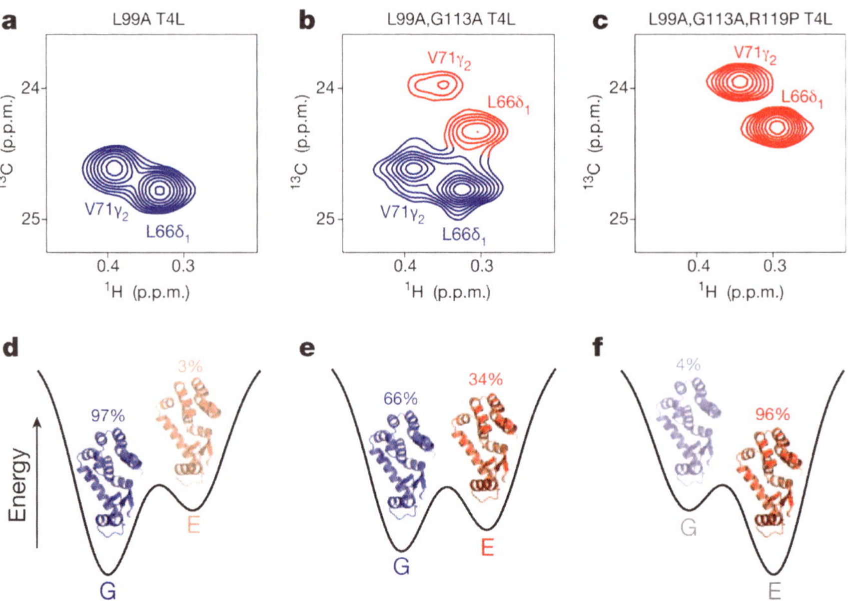

Fig. 4.9 Effect of mutations on the energy landscape of lysozyme proteins from phage T4 (T4L). The mutations under investigation are shown at the *top* in panels **a**, **b**, and **c**. The numbers above the corresponding ground (G) and excited (E) state structures in energy landscape panels **d**, **e**, and **f** denote the fractional populations (from Bouvignies *Nature* 477: 111 © 2011 Reprinted by permission from Macmillan Publishers Ltd)

its 3D structure determined by means of X-ray crystallography. Its structure was determined by David Chilton Phillips (1924–1999), and enabled him to propose a first mechanism for how enzymes worked. As shown in the figure the specific point mutations cause a population inversion. Normally, the protein rarely and transiently populates an excited state (E). However, the presence of the mutations causes a shift in equilibrium whereby the state E becomes the new stable ground state and the original ground state (G) is shifted upward and is only rarely populated.

4.9 Formation of Soluble Oligomers and Larger Aggregates

A number of oligomeric forms have been discovered in the last few years. These differ from one another in terms of their size and shape and destiny. A variety of questions that need answering accompany these findings. Among these are: what events trigger their formation? Which ones are toxic, under what conditions, and in which way? And, once these questions are addressed, at least initially, how do the various and well-established risk factors listed in Chap. 1 fit into the picture, if the oligomers or the fibrils, or both, are indeed the culprits?

The energy landscape picture presented in the last chapter can be expanded to include the presence of pathways leading to formation of different types of oligomers and large aggregates. The expanded representation has as its main feature the presence of additional kinetic barriers and intermediate states and an entire new set of folding funnels ending in deeper minima than those of the native state. While the native state is stable with respect to folding it is not so with respect to aggregation and amyloid formation. Instead, it is metastable. Under normal cellular conditions the folding funnel is sequestered from the oligomerization and aggregation pathways by high kinetic barriers that are not surmounted. That is, folding and aggregation pathways are kinetically isolated from one another. A start in addressing the series of questions posed in the preceding paragraph is to understand how this kinetic partitioning might break down.

SH3 domains: A first example of a breakdown in kinetic partitioning resulting in increased aggregation is presented in Fig. 4.10. Src homology 3 (SH3) domains are small globular protein modules, 60–85 residues in length, that function in cellular signaling pathways where they specifically recognize proline-rich peptide sequences (generally) of the form PxxP. These domains fold into a compact 3D structure composed of a pair of antiparallel beta-sheets plus a hydrophobic pocket that functions as the ligand-binding site. Under normal conditions these modules do not form amyloids and are not associated with any known neurodegenerative disease, but when subjected to low pH conditions the modules aggregate, forming a gel consisting of amyloid fibrils. In lowering the pH, the protein partially unfolds. The subsequent aggregation is mediated by a short amino acid stretch that becomes exposed.

Figure 4.10 illustrates another important feature of the SH3 domain and, by implication, other small globular proteins. The SH3 domain subjected to A39V, N53P, V55L mutations possesses a native-like intermediate state. This figure shown

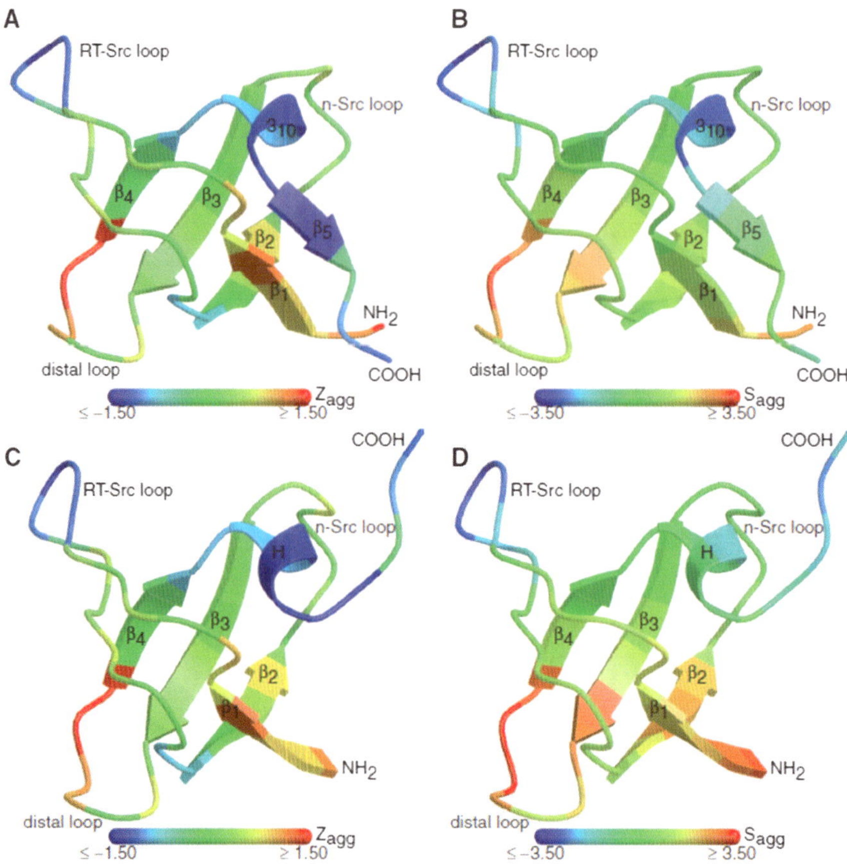

Fig. 4.10 Aggregation-prone regions of the SH3 domain. Z_{agg} is the color-coded, sequence-specific aggregation-propensity score for the native state (**a**) and folding intermediate (**c**). S_{agg} is the color-coded, surface aggregation propensity score for the native state (**b**) and folding intermediate (**d**) (from Neudecker *Science* 336: 362 © 2012 Reprinted with permission from AAAS)

that this low-lying intermediate state is far more aggregation-prone than the ground state. It is also accessible by means of thermal fluctuations. In this case under study, the mutations produce a partial (local) unfolding of the β_5 strand. Ordinarily the β_5 strand shields the aggregation-prone β1 strand. (This is one of the protective coding strategies discussed in Sect. 4.7.) However, this protective measure is deactivated by the mutations. As a result the aggregation-prone β1 strand is exposed resulting in formation of SH3 fibrils.

FF Four-helix bundle: The partial unfolding, removal and protective native interactions, and aggregation are not the sole provenance of polypeptides rich in beta-strands and sheets. As Kendrew showed myoglobin folds into a compact globular shape rich in alpha-helical secondary structure. Later studies by Dobson and colleagues further established that all partially folded structures along myoglobin's folding pathway are alpha-helical, as well. It is therefore doubly significant that under partial denaturing conditions this protein refolded in beta-sheet-rich conformations and formed amyloid fibrils.

The FF domain is a four-helix bundle protein. It contains three α-helices arranged in an orthogonal bundle, plus a 3–10 helix situated in a loop between the second and third helices. This domain, like myoglobin, is quite different structurally from the SH3 domain (alpha-helical rather than beta-sheet dominated). However, as shown in Fig. 4.11, mutations can increase the dwell time in a transiently populated excited state functioning as a doorway to aggregation and amyloid formation. Specifically, there is an increase in the dwell time in the partially folded intermediate from 1 % to 4 %.

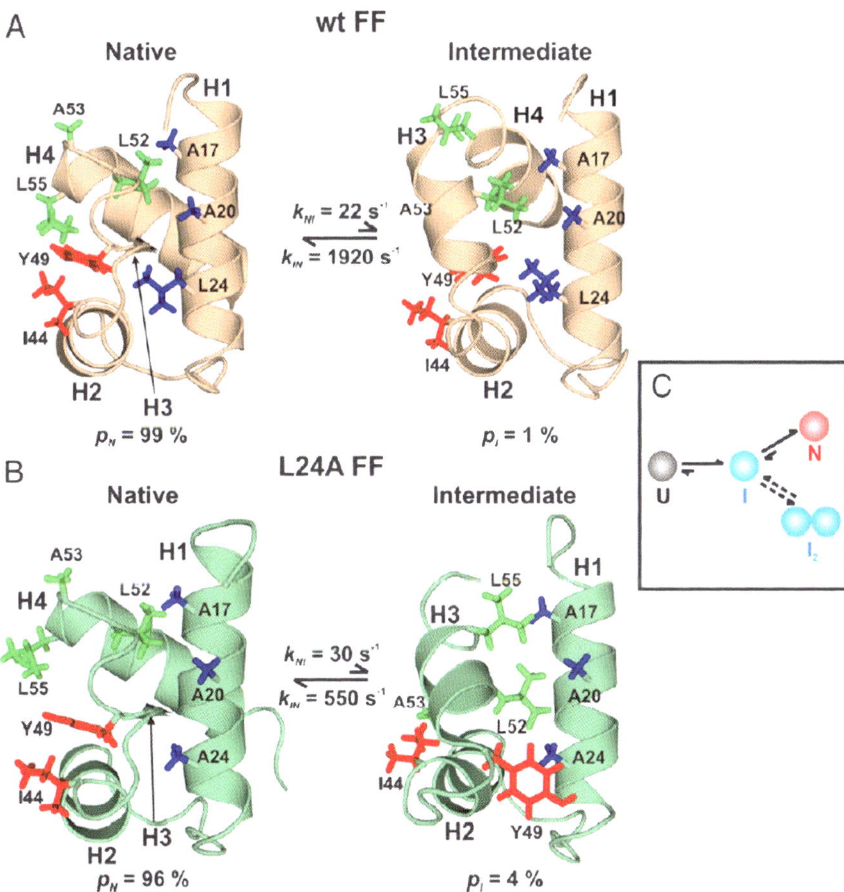

Fig. 4.11 Atomic structures for wild-type (wt) and L24A mutated states of the four-helix bundle FF domain. (**a**) wt native and transient intermediate structures; (**b**) L24A native and transient intermediate structures; (**c**) depiction of the pathways for folding and dimerization (from Sekhar *PNAS* 110: 12867 © 2013 Reprinted with permission from Lewis Kay)

The net effect of the mutation in the presence of the intrinsic protein motions is to promote dimerization and, by implication, aggregation. That possibility is illustrated in Fig. 4.11c, which shows two pathways branching out from the intermediate state. One leads to the native state while the other promotes dimerization. The appearance of this alternative route is highly suggestive of how an aggregation pathway might develop. Three distinct timescales are involved. The $U \rightarrow I$ transition occurs rapidly on a microsecond timescale; the $I \rightarrow N$ transition is far slower, taking place over a millisecond time frame, and the $I \rightarrow I_2$ transition is still slower, requiring seconds for its completion.

β2-Microglobulin: Dialysis-related amyloidosis (DRA) is brought on by partial unfolding and aggregation of β2 microglobulin. This protein normally functions as a 12 kDa subunit that together with the heavy-chain alpha-subunits forms the class-I major histocompatibility complex (MHC-I). The β2 microglobulin ($β_2$m) subunit maintains stability of the membrane-spanning alpha-chain, but these subunits are released extracellularly during MHC-I turnover and migrate to the kidneys. In diabetic individuals, they are no longer catabolized efficiently by the kidneys and as a result the concentration of $β_2$m in the circulation increases by as much as a factor of 60. The resulting buildup of $β_2$m leads to DRA, characterized by the buildup of amyloid fibrils in the joints of affected individuals. This informative example is discussed further in an Appendix to this chapter.

4.10 Intrinsically Disordered Proteins (IDPs)

Proteins responsible for signaling and regulation tend to be far larger than these carrying out, for example, metabolic tasks. They typically possess multiple domains, one or more of which will have little secondary structure and remain largely unfolded under physiological conditions. The lack of compact, unique tertiary structure is the defining property of these proteins. These proteins differ in their amino acid composition from proteins that fold into compact three-dimensional shapes. They tend to have a smaller complement of bulky hydrophobic residues and a greater fraction of polar and charged residues. This complement of amino acids enables the IDPs to remain in an open conformation.

These large, unstructured proteins have a number of interrelated properties important for their normal cellular functions. The most important of these are:

- Conformational flexibility
- Binding versatility
- Environmental responsiveness

First, these proteins are highly exposed to the solvent and can sample a broad spectrum of conformational states and substates. In some instances, these multidomain proteins possess unstructured linkers; in other cases, one or more of their domains remains largely disordered. Motions both small and large, involving domain movements, can occur. These proteins have energy landscapes differing from those of the small globular proteins. As depicted in Fig. 4.11 they do not utilize folding funnels but instead their landscapes are fractured into numerous shallow minima as is the case of the frustrated systems discussed in Chap. 3.

Secondly, their conformational flexibility enables many of them to function as signaling platforms, or scaffolds, or as signaling hubs, or as transcription regulators. In doing so, their open conformations greatly facilitate their interactions with multiple signaling partners and coregulators. These versatile proteins organize and control multiple cellular activities, and as a result their misfolding and malfunctions have severe cellular consequences. Thirdly, conformation and binding selectivity is not arbitrary but instead is sensitive to environmental conditions. These properties are important for the cellular functions of these proteins, but can go awry under stressful conditions and cause disease.

As noted above, these proteins tend to be composed of amino acids possessing low hydrophobicity along with nonzero net charge. For example, α-synuclein contains an N-terminal region of net charge -13 and a C-terminal region of total charge $+3$ resulting in an overall charge for the protein of -10. This protein can assume a variety of conformations depending on environmental conditions. It can be natively unfolded, or it can assume an aggregation-prone partially folded conformation. It may exhibit one or more regions of alpha-helical or beta-sheet secondary structure, organize into small oligomeric structures, or generate large fibrillar aggregates.

The tauopathies are so-named for the common presence of misfolded and aggregated tau protein. Like α-synuclein this is an intrinsically disordered protein. It is highly enriched in hydrophilic residues and although nominally a microtubule-binding protein it may interact with many structurally diverse binding partners. Like other IDPs, the open conformation provides a far greater surface area than the equivalent globularly folded protein. In the case of tau, it is estimated that its open conformation enables it to sweep out 27 times the surface area of its globular equivalents.

IDPs are widely viewed as populating an ensemble of low-energy conformational substates and have multiple-binding partners. As depicted in Fig. 4.12 the low-energy states are separated by barriers that are comparable to or smaller than the Boltzmann energy $k_{\mathrm{B}}T$. In these situations, the substates are weakly populated and the proteins continually transition from one state to another in a reversible manner. When a potential binding partner comes into close proximity to the target protein there is a population shift towards those states that recognize and bind the partner protein, and these states are then stabilized through the interaction. This flexible-binding situation is depicted in terms of one-dimensional energy landscapes in this figure.

Evidence for this type of conformational selection mechanism either by itself or in concert with the induced-fit model has been uncovered. Situations have been observed where first contacts between potential binding partners occur when they are far apart and are weakly coupled, but greatly strengthen as they come into closer contact and settle into a more structured and tightly bound conformation. This depiction has been given the name *fly-casting* because of its resemblance to a reeling-in process. Another important observation is that disorder-to-order transition may occur in which nominally unstructured regions develop helical or beta-sheet structures resulting in deeper landscape minima.

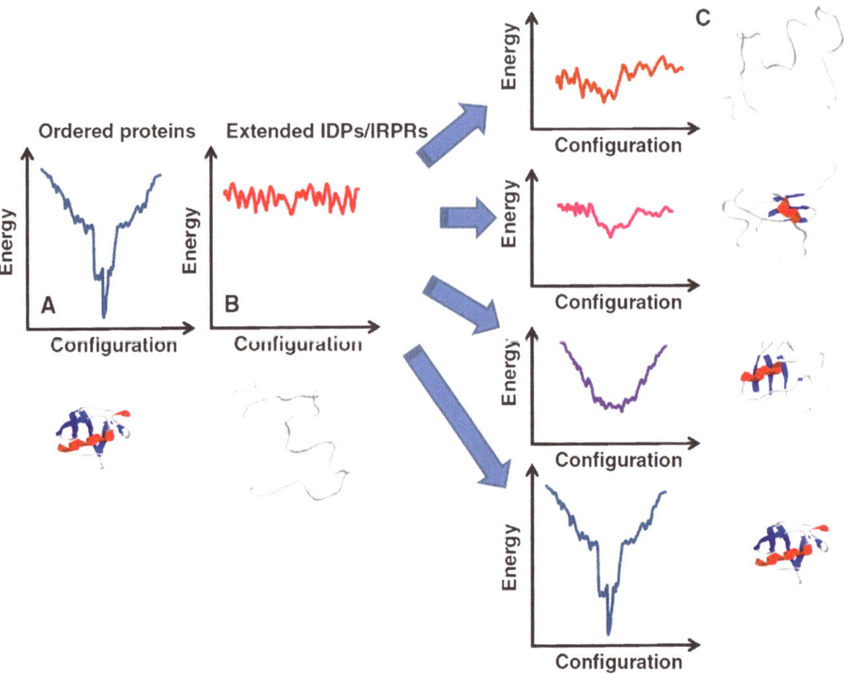

Fig. 4.12 Contrasting energy landscapes of globular proteins and IDPs. (**a**) Typical deep-valley of a small globular protein. (**b**) Multiple shallow minima characteristic of IDPs. (**c1–c3**) Landscapes accompanying different binding partners (from Uversky *Biochim. Biophys. Acta* 1834: 932 © 2013 Reprinted by permission from Elsevier)

Lastly, turning to thermodynamics, the role of entropic contributions in the binding of IDPs is particularly informative. At first thought, their entropic penalties should be exceptionally large and difficult to overcome, yet these proteins are promiscuous binders. At least part of the answer to this conundrum is that considerable entropy remains after binding; that is, the protein is not as rigidified by the binding as might be expected. This residual configuration entropy, referred to by some as the protein's "*dark energy*", is fine-tuned by allosteric effectors thereby increasing or decreasing the various binding affinities.

The appellation "protein misfolding disorder" when applied to these proteins takes on a different meaning from the standard one. Rather than capturing the notion that a protein has largely or partially unfolded from its initial well-folded native state and then refolded into a disease-causing state, IDPs strongly populate one or more aberrant, disease-associated conformational states as a result of mutations or abnormal cellular conditions or interactions. In some well-studied instances, essential cellular proteins that possess intrinsic disordered regions but are not mutated or altered end up sequestered in large amyloidogenic aggregates because of their binding promiscuity.

4.11 Summary

1. Amyloid fibrils are elongated structures built from multiple beta-strands oriented perpendicular to the long axis of the assemblage. These fibrils, and their characteristic cross-beta form, can be generated from many different polypeptide sequences implying that this capability may be a general property of the protein backbone.

   For many years the association of amyloid proteins in large insoluble aggregates served as a bottleneck to atomic level investigations. That impediment was overcome by the discovery that amyloid fibril formation was driven primarily by short stretches of aggregation-prone amino acids. Two kinds of amyloidogenic stretches—Q/N-rich sequences and hydrophobic stretches—were particularly prominent in amyloid-generating proteins. By focusing on these stretches amyloid crystal structures could be built and analyzed by X-ray diffraction and other advanced physical techniques.

   In brief, these structures are held together (1) through networks of hydrogen bonds within each beta-sheet and (2) by means of van der Waals interactions between polar side chains belonging to strands on adjacent beta-sheets. To facilitate the van der Waals interactions the side chains are interdigitized with one another.

   The resulting structures are known as steric zippers. The amyloid fibrils seen in steric zippers exhibit considerable conformational diversity at the atomic level; different kinds of side chains generate different classes of steric zippers.
2. Amyloids are of considerable interest as a nanomaterial. It is counterintuitive but significant that these structures held together as they are by multiple weak interactions possess the strength that they do. Another structure with a similar physics and a remarkable strength is spider silk. However, there is a length threshold beyond which individual amyloid fibers become increasingly brittle and prone to fragmentation.

   About 20 amyloid species are associated with human diseases, but others are not. Instead, they can be found carrying out normal physiological functions across multiple kingdoms. They are utilized in bacteria and yeasts where they contribute to surface attachment and biofilm formation. In humans, they assist in storage. To prevent toxic side effects they are sequestered in cellular compartments and controlled through fast kinetics.
3. There are two leading models of how these structures assemble. The nucleated polymerization model is monomer centric. Monomers slowly assemble into a nucleus that serves as a seed for the subsequent rapid growth phase of fibril formation. The concentration of monomers must exceed a threshold value, and the initial lag phase can be shortened if seeding nuclei are supplied. Fragmentation can occur that promotes secondary fibril formation and generates oligomers.

   In the second, conformational conversion model, oligomers are the key elements. In contrast to nucleated polymerization, there is an early buildup of oligomers. Monomers once they are incorporated into oligomers convert into amyloid-competent conformations and shed internal water molecules.

4. Physics and genetics work together through a variety of coding strategies to promote efficient folding of proteins into their native state ensembles in physiologically useful timeframes. The utilization of folding mechanisms such as cooperativity, zippering, and hydrophobic collapse were discussed in Chap. 2. That these physics-driven mechanisms work is a direct consequence of the evolutionary sculpting of the primary sequences.

   A number of other coding strategies were discussed earlier in the present chapter. These can be discerned from computer analyses of aggregation-prone sequences by taking into account properties such as hydrophobicity, charge, pH, expression levels, and responses to stresses. Chief among the aggregation-avoiding coding strategies that emerge from these studies are:

   - Burial of hydrophobic residues in the protein core
   - Avoidance of aggregation-promoting stretches of amino acids
   - Negative design
   - Gatekeepers
   - Tight control over expression levels

5. Not all proteins fold into compact shapes. Some proteins remain in a largely unfolded conformation while others contain large unstructured regions. This property enables them to carry out key signaling and regulatory functions. These proteins are referred to as intrinsically disordered proteins, or IDPs. That grouping contains a number of proteins centrally involved in neurodegenerative disorders.

6. Amyloid fibrils are not the only kind of beta-sheet-rich assemblages present in neurodegenerative diseases. In some disorders, misfolded proteins aggregate into insoluble amorphous aggregates rather than ordered fibrils. Both the ordered fibrils and the amorphous aggregates are frequently accompanied by a variety of smaller soluble oligomers that vary in size and morphology. These assemblages are more mobile and potentially expose a greater fraction of hydrophobic surface than the large insoluble aggregates. Recent studies point towards these small assemblages as the primary toxic agents. The following set of questions naturally arises:

   - Which species of aggregates are toxic?
   - Under what cellular conditions do these proteins partially unfold and become toxic?
   - What is (are) the cellular mechanism(s) of their toxicity?
   - How do the various and well establish risk factors listed in Chap. 1 exert their effects?
   - Why does the likelihood of neurodegeneration increase so dramatically with age?

   The last two questions was not posed earlier in this chapter but will be examined beginning in the next chapter. In addition to possessing multiple coding strategies, cells invest a considerable fraction of their total resources in order to carry out protein quality control, or *proteostasis*. The multiple systems that cells utilize ordinarily ensure that proteins fold properly, and do not at any time misfold and aggregate. Evidence of their involvement in neurodegeneration comes directly from examining the composition of the inclusion bodies invariably present in the disorders. The success and possible failure of these systems (and their connections to the risk factors and to aging) will be the subject of the next chapter.

Appendix 1. β₂-Microglobulin in Dialysis-Related Amyloidosis

β_2 microglobulin is a small 99 amino acid proteins that folds into a compact three-dimensional beta-sandwich in which two beta-sheets are linked by a central disulfide bond. The critical rate-limiting step in conversion from a well-folded soluble protein into an aggregation-prone one is a Pro32 cis-trans isomerization resulting in formation of an amyloidogenic intermediate state, I_T. As shown Fig. 4.13 the conformational change results in the repackaging of the two beta-sheets leading to an increased exposure of amyloid-promoting residues. The exact cause of this switch is still an area of active research. In studying this process, several striking features have been uncovered. That protein concentration is important is one such feature. It was discussed earlier in this chapter in the discussion of intrinsically disordered proteins and will be encountered several times more in later chapters.

Another factor that can helps drive the aggregation process is high concentrations of metal cations, specifically, Cu^{2+}, which are elevated in dialysis patients.

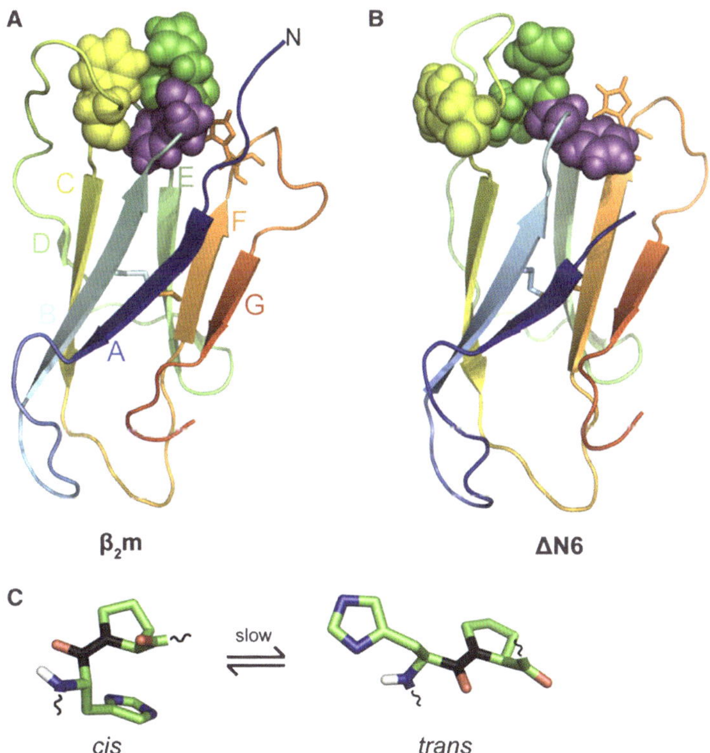

Fig. 4.13 Structural conversion of β2-microglobulin. (**a**) Normal and (**b**) ΔN6 form showing the effect of the cis-trans isomerization (**c**) (from Gierasch *Mol. Cell* 41: 129 © 2011 Reprinted by permission from Elsevier)

These ions create transient bridges between pairs of partially unfolded β_2m molecules that stabilize the formation of a β_2m dimer. This assemblage then serves as a nucleus for growth and formation of β_2m amyloid. Similarly, elevated amounts of metal divalent cations such as Zn^{2+} and Ni^{2+} are suspected of having a facilitating role in promoting aggregation and amyloid formation in neurodegenerative diseases such as Parkinson's disease.

Perhaps the most striking of the findings deals with the formation and aggregation of a truncated form of β_2m in which the first six residues have been removed. This form, called ΔN6, is seen in DRA patients. Once the protein has shed its first six residues it undergoes Pro32 cis-trans isomerization characteristic of the intermediate state (Fig. 4.13). Interactions between truncated ΔN6 and native β_2m stimulate the additional conversion from native form to the I_T intermediate. Not only concentration but also pH has an important effect on the conversion process with mild acidic conditions enhancing ΔN6's ability to form aggregates. As a next step, these findings point towards searches for a proteolytic enzyme responsible in DRA individuals.

In summary, many themes that run through protein misfolding science are recapitulated in dialysis-related amyloidosis. In DRA, there is a buildup of amyloid deposits in the joints of people undergoing kidney dialysis. These deposits are composed of misfolded and aggregated β_2m (light chain) subunits of the MCH-I complex. These subunits are normally shed into the circulation and removed by the kidneys but this removal does not occur in diabetic patients. Instead, there is a buildup in β_2m concentration; increased amounts of divalent copper ions are present, acidification occurs in certain areas associated with amyloid deposits, and (although not discussed) certain mutations promote misfolding and the disease state. In addition, some the β_2m subunits are truncated having lost their N-terminal residues. As a consequence of these conditions the subunits undergo a critical cis-trans isomerization at Pro32 thereby forming a long-lived amyloid-prone intermediate. Although the details may vary from protein to protein, a general series of steps is expected: partial unfolding followed by misfolding into an intermediate state that nucleates the formation of insoluble aggregates. This is similar to what is observed with transthyretin, lysozyme, and other small globular protein amyloid diseases. Lastly, in the interactions between ΔN6 and β_2m, prion-like behavior takes place, an emerging new theme that will be explored in Chap. 7.

Bibliography

Structure of the Amyloid Fibril

Lührs, T., Ritter, C., Adrian, M., Riek-Loher, D., Bohrmann, B., Döbeli, H., et al. (2005). 3D structure of Alzheimer's amyloid-β (1–42) fibrils. *Proceedings of the National Academy of Sciences of the United States of America, 102*, 17342–17347.

Nelson, R., Sawaya, M. R., Birbinie, M., Madsen, A. Ø., Riekel, C., Grothe, R., et al. (2005). Structure of the cross-β spine of amyloidlike fibrils. *Nature, 435*, 773–778.

Perutz, M. (1993). Glutamine repeats as polar zippers: Their possible role in neurodegenerative diseases. *Proceedings of the National Academy of Sciences of the United States of America, 91*, 5355–5358.

Petkova, A. T., Leapman, R. D., Guo, Z. H., Yau, W. M., Mattson, M. P., & Tycko, R. (2005). Self-propagating, molecular-level, polymorphism in Alzheimer's β-amyloid fibrils. *Science, 307,* 262–265.

Petkova, A. T., Ishii, Y., Balbach, J. J., Antzutkin, O. N., Leapman, R. D., Delaglio, F., et al. (2002). A structural model for Alzheimer's β - amyloid fibrils based on experimental constraints from solid state NMR. *Proceedings of the National Academy of Sciences of the United States of America, 99,* 16742–16747.

Sawaya, M. R., Sambashivan, S., Nelson, R., Ivanova, M. I., Sievers, S. A., Apostol, M. I., et al. (2007). Atomic structures of amyloid cross-β spines reveal varied steric zippers. *Nature, 447,* 453–457.

Sunde, M., Serpell, L. C., Bartlam, M., Fraser, P. E., Pepys, M. B., & Blake, C. C. (1997). Common core structure of amyloid fibrils by synchrotron X-ray diffraction. *Journal of Molecular Biology, 273,* 729–739.

Wasmer, C., Lange, A., Van Melckebeke, H., Siemer, A. B., Riek, R., & Meier, B. H. (2008). Amyloid fibrils of the HET-s(218–289) prion form a β-solenoid with a triangular hydrophobic core. *Science, 319,* 1523–1526.

Amyloids as Nanomaterials

Keten, S., Xu, Z. P., Ihle, B., & Buehler, M. J. (2010). Nanoconfinement controls stiffness, strength and mechanical toughness of β-sheet crystals in silk. *Nature Materials, 9,* 359–367.

Knowles, T. P. J., & Buehler, M. J. (2011). Nanomechanics of functional and pathological amyloid materials. *Nature Nanotechnology, 6,* 469–479.

Knowles, T. P., Fitzpatrick, A. W., Meehan, S., Mott, H. R., Vendruscolo, M., Dobson, C. M., et al. (2007). Role of intermolecular forces in defining material properties of protein nanofibrils. *Science, 318,* 1900–1903.

Amyloids in Normal Physiology

Blanco, L. P., Evans, M. L., Smith, D. R., Badtke, M. P., & Chapman, M. R. (2012). Diversity, biogenesis and function of microbial amyloids. *Trends in Microbiology, 20,* 66–73.

Fowler, D. M., Koulov, A. V., Alory-Jost, C., Marks, R. S., Balch, W. E., & Kelly, J. W. (2006). Functional amyloid formation within mammalian tissue. *PLoS Biology, 4,* e6.

Lipke, P. N., Garcia, M. C., Alsteens, D., Ramsook, C. B., Klotz, S. A., & Dufrène, Y. F. (2012). Strengthening relationships: Amyloids create adhesion nanodomains in yeasts. *Trends in Microbiology, 20,* 59–65.

Maji, S. K., Perrin, M. H., Sawaya, M. R., Jessberger, S., Vadodaria, K., Rissman, R. A., et al. (2009). Functional amyloids as natural storage of peptide hormones in pituitary secretory granules. *Science, 325,* 328–332.

Nucleated Polymerization

Avrami, M. (1939). Kinetics of phase change: I. General theory. *The Journal of Chemical Physics, 7,* 1103–1112.

Cohen, S. I. A., Vendruscolo, M., Dobson, C. M., & Knowles, P. J. (2012). From macroscopic measurements to microscopic mechanisms of protein aggregation. *Journal of Molecular Biology, 421,* 160–171.

Cohen, S. I. A., Linse, S., Luheshi, L. M., Hellstrand, E., White, D. A., Rajah, L., et al. (2013). Proliferation of amyloid-β 42 aggregates occurs through a secondary nucleation mechanism. *Proceedings of the National Academy of Sciences of the United States of America, 110*, 9758–9763.

Dykes, G., Crepeau, R. H., & Edelstein, S. J. (1978). Three-dimensional reconstruction of the fibres of sickle cell hemoglobin. *Nature, 272*, 506–510.

Ferrone, F. A., Hofrichter, J., Sunshine, H. R., & Eaton, W. A. (1980). Kinetic studies on photolysis-induced gelation of sickle cell hemoglobin suggest a new mechanism. *Biophysical Journal, 32*, 361–380.

Hofrichter, J., Ross, P. D., & Eaton, W. D. (1974). Kinetics and mechanism of deoxyhemoglobin S gelation: A new approach to understanding sickle cell disease. *Proceedings of the National Academy of Sciences of the United States of America, 71*, 4864–4868.

Neel, J. W. (1949). The inheritance of sickle cell anemia. *Science, 100*, 64–66.

Oosawa, F., & Asakura, S. (1975). *Thermodynamics of the polymerization of protein*. London: Academic.

Pauling, L., Itano, H. A., Singer, S. J., & Wells, I. C. (1949). Sickle cell anemia, a molecular disease. *Science, 110*, 543–548.

Powers, E. T., & Powers, D. L. (2008). Mechanisms of protein fibril formation: Nucleated polymerization with competing off-pathway aggregation. *Biophysical Journal, 94*, 379–391.

Nucleated Conformational Conversion

Bleiholder, C., Dupuis, N. F., Wyttenbach, T., & Bowers, M. T. (2011). Ion-mobility mass spectrometry reveals a conformational conversion from random assembly to β-sheet in amyloid fibril formation. *Nature Chemistry, 3*, 172–177.

Chimon, S., Shaibat, M. A., Jones, C. R., Calero, D. C., Aizezi, B., & Ishii, Y. (2007). Evidence of fibril-like β-sheet structures in a neurotoxic amyloid intermediate of Alzheimer's β-amyloid. *Nature Structural & Molecular Biology, 14*, 1157–1164.

Lee, J., Culyba, E. K., Powers, E. T., & Kelly, J. W. (2011). Amyloid-β forms fibrils by nucleated conformational conversion of oligomers. *Nature Chemical Biology, 7*, 602–609.

Morris, A. M., Watsky, M. A., & Finke, R. G. (2009). Protein aggregation kinetics, mechanism, and curve-fitting: A review of the literature. *Biochimica et Biophysica Acta, 1794*, 375–397.

Reddy, G., Straub, J. E., & Thirumalai, D. (2010). Dry amyloid fibril assembly in a yeast prion peptide is mediated by long-lived structures containing water wires. *Proceedings of the National Academy of Sciences of the United States of America, 107*, 21459–21464.

Serio, T. R., Cashikar, A. G., Kowal, A. S., Sawicki, G. J., Moslehi, J. J., Serpell, L., et al. (2000). Nucleated conformational conversion and the replication of conformational information by a prion determinant. *Science, 289*, 1317–1321.

Loss of Stability

Chiti, F., Stefani, M., Taddei, N., Ramponi, G., & Dobson, C. M. (2003). Rationalization of the effects of mutations on peptide and protein aggregation rates. *Nature, 424*, 805–808.

DePristo, M. A., Weinreich, D. M., & Hartl, D. L. (2005). Missense meanderings in sequence space: A biophysical view of protein evolution. *Nature Reviews. Genetics, 6*, 678–687.

Tokuriki, N., & Tawfik, D. S. (2009). Protein dynamism and evolvability. *Science, 324*, 203–207.
Wang, X., Minasov, G., & Shoichet, B. K. (2002). Evolution of an antibiotic resistance enzyme constrained by stability and activity trade-offs. *Journal of Molecular Biology, 320*, 85–95.
Yue, P., Li, Z. L., & Moult, J. (2005). Loss of protein structure stability as a major causative factor in monogenic disease. *Journal of Molecular Biology, 353*, 459–473.

Coding Strategies

Ciryam, P., Tartaglia, G. G., Morimoto, R., Dobson, C. M., & Vendrusculo, M. (2013). Widespread aggregation and neurodegenerative diseases are associated with supersaturated proteins. *Cell Reports, 5*, 1–10.
Minor, D. L., Jr., & Kim, P. S. (1996). Context-dependent secondary structure formation of a designed protein sequence. *Nature, 380*, 730–734.
Monsellier, E., & Chiti, F. (2007). Prevention of amyloid-like aggregation as a driving force of protein evolution. *EMBO Reports, 8*, 737–742.
Otzen, D. E., Kristensen, O., & Oliveberg, M. (2000). Designed protein tetramer zipped together with a hydrophobic Alzheimer homology: A structural clue for amyloid formation. *Proceedings of the National Academy of Sciences of the United States of America, 97*, 9907–9912.
Pawar, A. P., DuBay, K. F., Zurdo, J., Chiti, F., Vendruscolo, M., & Dobson, C. M. (2005). Prediction of "aggregation-prone" and "aggregation-susceptible" regions in proteins associated with neurodegenerative disease. *Journal of Molecular Biology, 350*, 379–392.
Richardson, J. S., & Richardson, D. C. (2002). Natural β-sheet proteins use negative design to avoid edge-to-edge aggregation. *Proceedings of the National Academy of Sciences of the United States of America, 99*, 2754–2759.
Rousseau, F., Serrano, L., & Schymkowitz, J. W. (2006). How evolutionary pressure against protein aggregation shaped chaperone specificity. *Journal of Molecular Biology, 355*, 1037–1047.
Tartaglia, G. G., Pawar, A. P., Campioni, S., Dobson, C. M., Chiti, F., & Vendruscolo, M. (2008). Predication of aggregation-prone regions in structured proteins. *Journal of Molecular Biology, 380*, 425–436.

Toxic Oligomer Hypothesis

Bolognesi, B., Kumita, J. R., Barros, T. P., Esbjorner, E. K., Luheshi, L. M., Crowther, D. C., et al. (2013a). ANS binding reveals common features of cytotoxic amyloid species. *ACS Chemical Biology, 5*, 735–740.
Bucciantini, M., Giannoni, E., Chiti, F., Baroni, F., Formigli, L., Zurdo, J., et al. (2002). Inherent toxicity of aggregates implies a common mechanism for protein misfolding diseases. *Nature, 416*, 507–511.
Cleary, J. P., Walsh, D. M., Hofmeister, J. J., Shankar, G. M., Kuskowski, M. A., Selkoe, D. J., et al. (2005). Natural oligomers of the amyloid-β protein specifically disrupt cognitive function. *Nature Neuroscience, 8*, 79–84.
Glabe, C. G. (2008). Structural classification of toxic amyloid oligomers. *The Journal of Biological Chemistry, 283*, 29639–29643.
Haataja, L., Gurlo, T., Huang, C. J., & Butler, P. C. (2008). Islet amyloid in Type 2 diabetes, and the toxic oligomer hypothesis. *Endocrine Reviews, 29*, 303–316.
Kayed, R., Head, E., Thompson, J. L., McIntire, T. M., Milton, S. C., Cotman, C. W., et al. (2003). Common structure of soluble amyloid oligomers implies common mechanism of pathogenesis. *Science, 300*, 486–489.

Lambert, M. P., Barlow, A. K., Chromy, B. A., Edwards, C., Freed, R., Liosatos, M., et al. (1998). Diffusible, nonfibrillar ligands derived from Aβ are potent central nervous system toxins. *Proceedings of the National Academy of Sciences of the United States of America, 95*, 6448–6453.

Lashuel, H. A., & Lansbury, P. T., Jr. (2006). Are amyloid diseases caused by protein aggregates that mimic bacterial pore-forming toxins? *Quarterly Reviews of Biophysics, 39*, 167–201.

Lesné, S., Koh, M. T., Kotilinek, L., Kayed, R., Glabe, C. G., Yang, A., et al. (2006). A specific amyloid-β protein assembly in the brain impairs memory. *Nature, 440*, 352–357.

Shankar, G. M., Li, S., Mehta, T. H., Garcia-Munoz, A., Shepardson, N. E., Smith, I., et al. (2008a). Amyloid-β protein dimers isolated directly from Alzheimer brains impair synaptic plasticity and memory. *Nature Medicine, 14*, 837–842.

Walsh, D. M., Klyubin, I., Fadeeva, J. V., Cullen, W. K., Anwyl, R., Wolfe, M. S., et al. (2002). Naturally secreted oligomers of amyloid β protein potently inhibit hippocampal long-term potentiation in vivo. *Nature, 416*, 535–539.

Formation of Soluble Oligomers

Baldwin, A. J., Knowles, T. P. J., Tartaglia, G. G., Fitzpatrick, A. W., Devlin, G. L., Shammas, S. L., et al. (2011). Metastability of native proteins and the phenomenon of amyloid formation. *Journal of the American Chemical Society, 133*, 14160–14163.

Chiti, F., & Dobson, C. M. (2009). Amyloid formation by globular proteins under native conditions. *Nature Chemical Biology, 5*, 15–22.

Chiti, F., Taddei, N., Baroni, F., Capanni, C., Stefani, M., Ramponi, G., et al. (2002). Kinetic partitioning of protein folding and aggregation. *Nature Structural Biology, 9*, 137–143.

Korzhnev, D. M., Vernon, R. M., Religa, T. L., Hansen, A. L., Baker, D., Fersht, A. R., et al. (2011). Nonnative interactions in the FF domain folding pathway from an atomic resolution structure of a sparsely populated intermediate: An NMR relaxation dispersion study. *Journal of the American Chemical Society, 133*, 10974–10982.

Lim, K. H., Dyson, H. J., Kelly, J. W., & Wright, P. E. (2013). Localized structural fluctuations promote amyloidogenic conformations in transthyretin. *Journal of Molecular Biology, 425*, 977–988.

Neudecker, P., Robustelli, P., Cavalli, A., Walsh, P., Lundström, P., Zarrine-Afsar, A., et al. (2012). Structure of an intermediate state in protein folding and aggregation. *Science, 336*, 362–366.

Intrinsically Disordered Proteins

Dunker, A. K., Lawson, J. D., Brown, C. J., Williams, R. M., Romero, P., Oh, J. S., et al. (2001). Intrinsically disordered proteins. *Journal of Molecular Graphics & Modelling, 19*, 26–59.

Jensen, M. R., Markwick, P. R. L., Meier, S., Griesinger, C., Zweckstetter, M., Grzesiek, S., et al. (2009). Quantitative determination of the conformational properties of partially folded and intrinsically disordered proteins using NMR dipolar couplings. *Structure, 17*, 1169–1185.

Mukrasch, M. D., Bibow, S., Korukottu, J., Jeganathan, S., Biernat, J., Griesinger, C., et al. (2009a). Structured polymorphism of 441-residue Tau at single residue resolution. *PLoS Biology, 7*, e1000034.

Olzscha, H., Schermann, S. M., Woerner, A. C., Pinkert, S., Hecht, M. H., Tartaglia, G. G., et al. (2011). Amyloid-like aggregates sequester numerous metastable proteins with essential cellular functions. *Cell, 144*, 67–78.

Shoemaker, B. A., Portman, J. J., & Wolynes, P. G. (2000). Speeding molecular recognition by using the folding funnel: The fly-casting mechanism. *Proceedings of the National Academy of Sciences of the United States of America, 97*, 8868–8873.

Sugase, K., Dyson, H. J., & Wright, P. E. (2007). Mechanism of coupled folding and binding of an intrinsically disordered protein. *Nature, 447*, 1021–1025.

Tompa, P. (2002). Intrinsically unstructured proteins. *Trends in Biochemical Sciences, 27*, 527–533.

Uversky, V. N., Gillespie, J. R., & Fink, A. L. (2000). Why are "natively unfolded" protein unstructured under physiological conditions? *Proteins: Structure, Function, and Genetics, 41*, 415–427.

Wright, P. E., & Dyson, H. J. (1999). Intrinsically unstructured proteins: Re-assessing the structure-function paradigm. *Journal of Molecular Biology, 293*, 321–331.

β_2-*Microglobulin*

Eakin, C. M., Berman, A. J., & Miranker, A. D. (2006). A native to amyloidogenic transition regulated by a backbone trigger. *Nature Structural & Molecular Biology, 13*, 202–208.

Eichner, T., Kalverda, A. P., Thompson, G. S., Homans, S. W., & Radford, S. E. (2011). Conformational conversion during amyloid formation at atomic resolution. *Molecular Cell, 41*, 161–172.

Jahn, T. R., Parker, M. J., Homans, S. W., & Radford, S. E. (2006). Amyloid formation under physiological conditions proceeds via a native-like folding intermediate. *Nature Structural & Molecular Biology, 13*, 195–201.

Kameda, A., Hoshino, M., Higurashi, T., Takehashi, S., Naiki, H., & Goto, Y. (2005). Nuclear magnetic resonance characterization of the refolding intermediate of β_2-microglobulin trapped by non-native prolyl peptide bond. *Journal of Molecular Biology, 348*, 383–397.

Chapter 5
Protein Quality Control: Part I—Molecular Chaperones and the Ubiquitin-Proteasome System

The 1930s, 1940s, and 1950s were a time of change for how proteins were viewed. Results from X-ray crystallography and denaturation experiments guided the way from the idea that proteins were the repository of genetic information, and DNA provided structural support, to one in which DNA was the data repository. In the revised modern picture, proteins provided structure support and were the workhorses of the cell, performing most of the necessary cellular tasks.

Notions on the permanence of proteins also underwent major changes during this period. At the outset, it was believed that there were two kinds of proteins—those that were derived from foodstuffs and were degraded to provide fuel for the body and proteins that were used for structural and functional roles in the body. The latter were essentially permanent and required replacement only when damaged. The view of the permanence of proteins was widely accepted awaiting the advent of tools that could be used to follow them in space and time. The needed tool was provided by the discovery of heavy isotopes of stable elements by Harold C. Urey (1893–1981). As a result Rudolf Schoenheimer (1898–1941) and David Rittenberg (1906–1970) were able to use ^{15}N radiolabeled amino acids to demonstrate that structural and functional proteins underwent cycles of synthesis and destruction.

The striking notion of dynamic protein turnover gained traction with the discovery of a compartment where cellular components could be safely degraded. The lysosome was uncovered by Christian de Nuve in the 1953–1955 time-period. Further studies over the next several decades established the full extent of the lysosomal/vacuolar system for transporting and degrading exogenous and endogenous proteins and organelles. However, it was also found that not all proteins were degraded in lysosomes. Instead, many if not most cellular proteins were degraded somewhere else. Furthermore, it was found that this other, nonlysosomal degradation system required metabolic energy (that is, it was ATP-dependent).

The long-sought-for system was the ubiquitin-proteasome system. It took another 20 years or so for that to be established. The road to those discoveries led through studies of how abnormal hemoglobin was degraded in reticulocytes, newly

© Springer International Publishing Switzerland 2015

M. Beckerman, *Fundamentals of Neurodegeneration and Protein Misfolding Disorders*, Biological and Medical Physics, Biomedical Engineering, DOI 10.1007/978-3-319-22117-5_5

produced red blood cells that, importantly, do not contain lysosomes. These findings were reinforced by the discoveries on how unwanted proteins were degraded in bacteria. It was found that the nonlysosomal degradation system required ATP and involved conjugation of the small protein ubiquitin to the target substrate leading to degradation by the as yet unknown protease. Those discoveries were followed by the identification of the unknown protease as a large multisubunit molecular machine—the 20S proteasome in 1980 followed by the discovery of the 26S proteasome shortly thereafter. Those studies culminated with the award of the 2004 Nobel Prize in chemistry to the main discoverers of the ubiquitin-proteasome system (UPS), Israeli biochemists Avram Hershko and Aaron Ciechanover, and American biophysicist Irwin Rose.

The existence of protein turnover and the lysosomal and proteasomal degradation systems was just one or several major developments following the discovery of the cellular roles of proteins, RNA, and DNA. A second major development was the emergence of a molecular chaperone system that assisted in protein folding, unfolding, disaggregation, and oligomeric assembly. These nanomolecular machines are active in bacteria, yeasts, plants, and animals where they play essential roles in maintaining protein homeostasis, or *proteostasis*.

The unveiling of the molecular chaperones began in the early 1960s in Pavia where Ferruccio Ritossa was exploring the effects of a brief exposure to elevated temperature on *Drosophila* gene expression. What he found was that the elevated temperatures triggered a change in the overall gene expression pattern—some genes were expressed more strongly while others had their expression tuned down. That finding was followed in 1974 by the study by Tissières, Mitchell, and Tracy in which *Drosophila* subjected to heat shock triggered a pronounced increase in RNA transcription and protein synthesis of a particular set of polypeptides logically termed heat-shock proteins (hsps). The most pronounced of these increases was of a protein of molecular weight of approximately 70,000 Da and thereafter called Hsp70.

The uncovering of Hsp70 was followed a few years later, in 1978, with a report by Ron Laskey that hsps were required in the assembly of histones into nucleosomes. This finding went against the prevailing belief that subunit assembly was an autonomous process. In addition, Laskey had discovered that while the hsp nucleoplasmin assisted in assembling the nucleosomes it was not part of the final structure. For that reason he coined the term "chaperone" to describe this unexpected role.

Those findings were followed a few years later by a convergence of two independent lines of investigation on the existence of a second major family of heat-shock proteins—the chaperonins. One line of research centered on the assembly of the chloroplast protein Rubisco; the other involved the replication of bacteriophages within *Escherichia coli*. In brief, John Ellis had discovered the existence of a Rubisco large subunit binding protein that assisted in assembly of newly synthesized Rubisco subunits. Another protein, named GroEL was identified and studied by several groups. This protein, needed for phage-λ morphogenesis in *E. coli*, consisted of 14 subunits organized into two stacked heptameric rings forming a cage that encapsulated its substrate. It was accompanied by a co-chaperone, named

Table 5.1 Heat-shock protein inducers (after Schlesinger, *J. Biol. Chem.* 265: 12111, 1990)

Environmental stress conditions	Pathological stress conditions	Normal conditions
Heat	Microbial infections	Cell cycle
Heavy metals	Tissue trauma	Embryonic development
Organics	Genetic lesions	Cell differentiation
Oxidants		Hormonal stimulation
		Microbial growth

GroES in bacteria (Hsp10 in eukaryotes). Once GroEL was sequenced and compared to the chloroplast protein the similarity between them led to establishment of the groEL/Hsp60 chaperonin family essential for the folding of newly synthesized proteins and their assembly into oligomeric complexes.

By the early 1980s the Hsp60 and Hsp70 families of heat-shock proteins had been uncovered in yeasts, plants, and man, the bacterial counterpart to Hsp70, called dnaK, having been found by the late 1970s. Their investigation had established that heat-shock proteins/molecular chaperones were (1) highly conserved across multiple kingdoms; (2) were essential, and (3) were induced under both abnormal and normal conditions. A short table from a 1990 review by Milton Schlesinger makes the latter point (Table 5.1).

Those discoveries led John Ellis to reintroduce the concept of "molecular chaperones" in 1987 following a suggestion to that effect by Hugh Pelham in a 1986 paper. In that paper, Pelham summarized the possible role of these proteins in addition to a role in assembly. Specifically they (1) could assist in handling denatured or abnormal proteins' (2) disrupt hydrophobic aggregates through their use of ATP, and (3) become upregulated in response to overloading of the protein degradation system.

5.1 Macromolecular Crowding Is Present in the Cell

The cellular environment is a crowded place; the fraction of the cellular volume occupied by dry matter is roughly 30 %, most of which is composed of proteins. That point is illustrated in Fig. 5.1 in which portions of typical eukaryotic and prokaryotic cells are shown using realistic representations of the resident macromolecule shapes and sizes. Crowding results in the steric inhibition of one macromolecule by another. One obvious consequence of the excluded volume is that it reduces diffusion rates. Beyond that, the effects of crowding are both less obvious and more interesting. Several studies have been carried out of how crowding affects protein folding. One conclusion is that crowding promotes the folding of globular proteins and enhances their refolding capabilities. It does so by reducing the conformation freedom of the unfolded chain far more than that of the compact native state, thus increasing the stability of the latter.

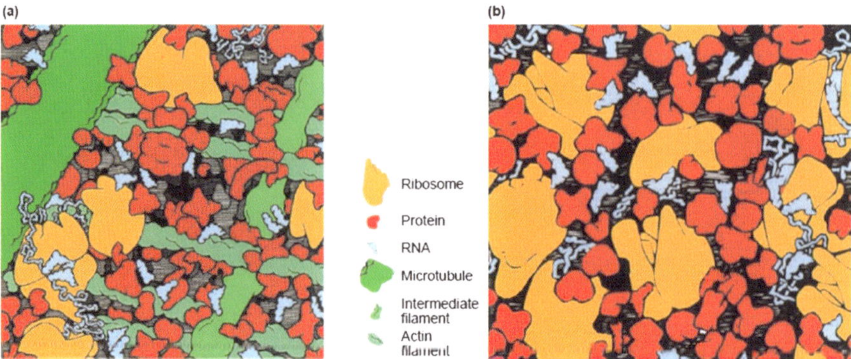

Fig. 5.1 Macromolecular crowding. Typical cytoplasm (**a**) in a eukaryotic cell and (**b**) in a bacterial (*E. coli*) cell. *Squares* depict the face of a cube of cytoplasm 100 nm in length on a side (from Ellis *Curr. Opin. Struct. Biol.* 11: 114 © 2001 Reprinted by permission from Elsevier)

The situation for intrinsically disordered proteins discussed in the last chapter is more complex and depends on the structural details and the ruggedness of the energy landscape. Significantly, it does not appear to impede the ability of these proteins to carry out their cellular tasks. Lastly, while crowding can enhance the speed at which a polypeptide chain can collapse, it can also increase the rate of aggregate formation especially of larger or more complex proteins that fold slowly, possess kinetic traps, and/or do not rapidly bury hydrophobic patches in their interior.

A broad spectrum of tasks, all related to maintenance of protein health, that is, to *proteostasis*, is performed by molecular chaperones. As noted by Pelham and Ellis some time ago, molecular chaperones assist in the folding of nascent polypeptide chains in the crowded cellular environments. These chains are susceptible to forming inappropriate contacts with other proteins and with other domains of the same chain. Acting further downstream other molecular chaperones help to refold misfolded proteins, disaggregate proteins that have clumped together, and direct proteins that cannot be recovered to either the UPS or, if unavailable, to alternate disposal pathways. They also unfold proteins that have to cross intracellular membranes and refold them once they have transited through the membranes.

Not all proteins require the assistance of molecular chaperones. The fast folders, those proteins that fold rapidly through two-state kinetics in microseconds to milliseconds, do not, but others cannot fold without their help. Kinetic traps are a key impediment to proper folding and for these proteins the assistance of chaperones is essential. As illustrated in Fig. 5.2 one way the chaperones promote folding is by destabilizing the kinetic traps.

The ubiquitin-proteasome system and molecular chaperones occupy the upper layer of a multilayered defense system that protects against potentially toxic proteins. Molecular chaperones assist in folding of nascent polypeptide chains, the refolding of misfolded proteins, and the unclumping of aggregated ones. Any remaining malformed protein is directed either to the ubiquitin-proteosome system

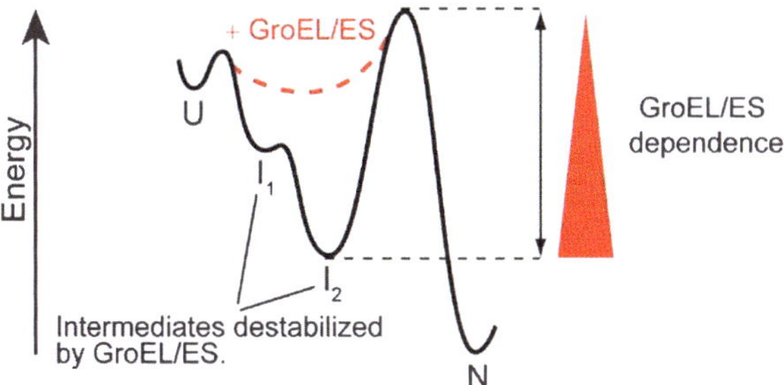

Fig. 5.2 Chaperone modification of the energy landscape for folding. GroEL/ES function destabilizes kinetic traps, thereby enabling a protein to exit folding intermediates and reach its native state (from Kerner *Cell* 122: 209 © 2005 Reprinted by permission from Elsevier)

or undergo autophagic transport to lysosomes. Importantly, if these systems fail cells sequester the misfolded/aggregated/amyloidogenic proteins in inclusion bodies, the amyloidogenic deposits discussed in earlier chapter, and may even ship them out of the cell as will be discussed later in the text.

Inclusion bodies such as Lewy bodies, Pick bodies, and Bunina bodies contain ubiquitinated deposits of the misfolded proteins along with molecular chaperones and other components of the protein quality control system. In the remainder of this chapter and in the following one, the protein quality control system will be examined. One of the features that may characterize the proteins that aggregate and collect in IBs is that they are found at concentrations that are too high. Recall from the discussion of β2-microglobulin at the end of the last chapter that expression levels are important, and there is a need to maintain aggregation-prone proteins at low levels. The machinery responsible for the maintenance of proper concentration levels will be examined in the later, disease-specific chapters.

5.2   Ubiquitination Tags Proteins for Degradation by Proteosomes

Ubiquitin is a small protein, consisting of 76 amino acid residues with a mass of 8.5 kDa. It gets its name because of its ubiquitous presence in the cytosol and nucleus of bacteria, yeasts, higher plants, and animal cells. It was discovered in 1975 by Goldstein and coworkers and given the provisional name UBIP (ubiquitous immunopoietic polypeptide). The reversible attachment of this molecule to proteins has emerged as a major theme in cellular signaling since its role in degradation of short-lived species was first explored in the early 1980s by Ciechanover, Hershko, and Rose. The importance of this covalent modification was lent further support by the discoveries in just the last few years of the deubiquitinating enzymes.

Ubiquitin-dependent proteolysis is carried out in two stages. In the first step, the protein to be degraded is tagged by covalent attachment of a string of four or more

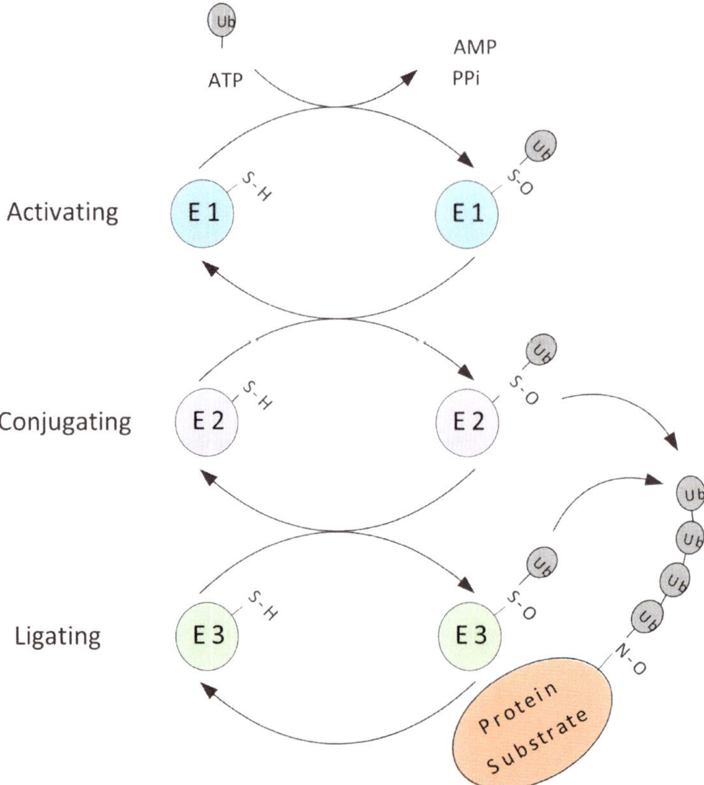

Fig. 5.3 Protein ubiquitination illustrating the three-stage process of ubiquitin attachment and chain growth. Ubiquitin possesses seven lysine residues, and can be conjugated on any of a number of these among which are K48 and K63. These two alternative sites for ubiquitination lead to different outcomes. In the case of K48 ubiquitination, the chain so produced leads to degradation of the tagged proteins by the 26S proteasome. In situations where K63 is the site for ubiquitination, the chain serves not as a tag for degradation but rather as a platform for assembly of complexes that activates downstream signaling

ubiquitin molecules. In the second stage, the protein so-tagged is degraded within a large chambered multisubunit assemblage called the 26S proteasome. Three types of enzymes operating in a sequential fashion are responsible for attachment of the ubiquitin molecules to the target protein (Fig. 5.3). The first enzyme to act is an ubiquitin-activating, or E1 enzyme. It is followed by an ubiquitin-conjugating, or E2 enzyme, and finally by an ubiquitin-ligating, or E3 enzyme.

To begin the process the E1 catalyzes the attachment of an ubiquitin molecule to the sulfhydryl group on a catalytic cysteine residue within the E1. A high-energy thioester bond is formed between the sulfhydryl group and the carboxyl group of an ubiquitin C-terminus glycine residue. This is done in an ATP-dependent way as depicted in Fig. 5.3. Next, the E2 catalyzes the transfer of the ubiquitin protein from the E1 to an E2 catalytic cysteine residue to which it is again linked by means of a thioester bond. Lastly, the E3 catalyzes the transfer of the ubiquitin to an amino group of a side chain substrate lysine residue and to the growing multiubiquitin chain. Transfer to the substrate can occur either using the E3 as a transfer intermediate or directly from the E2. The ubiquitin-handling enzymes, the E1s, E2s, and E3s, form a hierarchy. A small number of E1s interact with a larger number of E2s and these interact with a yet larger number of E3s. The E3s confer substrate specificity to the degradation process acting either by themselves or together with the E2s.

5.3   Proteasomes Are Proteolytic Machines that Cleave the Tagged Proteins into Small Pieces

The 20S and 26S proteasomes are cylindrical-shaped proteolytic machines that degrade intracellular proteins. The 20S core unit (Fig. 5.4) is composed of four heptameric rings stacked one on top of the other with a total molecular mass of 750 kDa. The two inner rings consist of seven different β-subunits and the two outer rings contain seven different α-subunits. The β-subunits comprise the catalytic core of the proteasome while the α-subunits have a structural role and assist in controlling the entrance and exit of the substrate protein from the inner chamber. Three catalytic β-subunits positioned on the inner surface of each of the catalytic rings are responsible for cleaving peptide bonds. These units chop the client proteins into short peptide fragments ranging in size from 3 to 15 amino acid residues. These fragments are further degraded into amino acids by peptidases outside the proteasome.

The 26S proteasome consists of a 20S catalytic core plus one or two 19S regulatory particles (RPs) that cap one or both ends of the core unit. The 19S RP contains two types of subunits—regulatory particles of AAA-ATPases (Rpts) and regulatory parti-

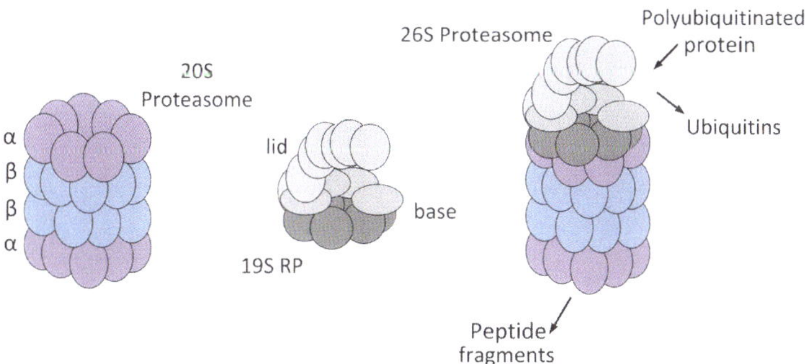

Fig. 5.4 Schematic diagrams of the 20S and 26S proteasomes based on electron microscopy and X-ray crystallography data. 20S proteasome: The outer α-subunits are shown in *purple* and the inner β subunits are depicted in *blue*. 19S regulatory particle (RP): The hexameric AAA ATPase ring that forms the base is shown in *dark grey*; the four Rnp subunits that belong to the base as well and the nine or so Rnps that comprise the lid are depicted in *light grey*. 26S proteasome: The structure shown is of a 20S proteasome capped by a single 19S RP

cles of non-ATPases (Rpns). The base subcomplex contains the six Rpts, which form a ring, plus four ancillary Rnps. The lid contains nine or more Rnps for a total of at least 19 subunits resulting in a 26S proteasome with a molecular mass of approximately 2 MDa. The proteolytic core chops substrate proteins in a nonspecific manner, but to accomplish this task the proteins must be inserted in an unfolded form through the narrow opening of the barrel-shaped 20S core. The lid of the 19S RP removes the ubiquitins from the protein. The base recognizes polyubiquitinated proteins, and prepares the protein for degradation by trapping and unfolding it. It then gates open the pore to allow entry of the unfolded protein into the 20S degradation chamber.

As just noted, in order for the 26S proteasome to degrade client proteins it must first unfold them and then feed the chain into the proteolytic chamber. In the 26S proteasome, the base of the 19S RP contains a hexameric AAA-ATPase, an unfoldase that handles the first task. The 20S proteasome, the peptidase, is then able to carry out the second task, degradation of the polypeptide chain. Shown in Fig. 5.5 is an example of the sequential action by ClpX(P), a heavily studied molecular machine that helps maintain protein quality control by degrading unwanted and misfolded proteins. In this example, the hexameric AAA-ATPase, ClpX, unfolds proteins, which are then fed into a double-ringed proteolytic chamber, ClpP. This molecular machine is present in bacteria and in mitochondria where proteins bearing a degradation tag, or degradon, are targeted for destruction. In these machines, like the proteasome, energy supplied by ATP hydrolysis is converted into mechanical work.

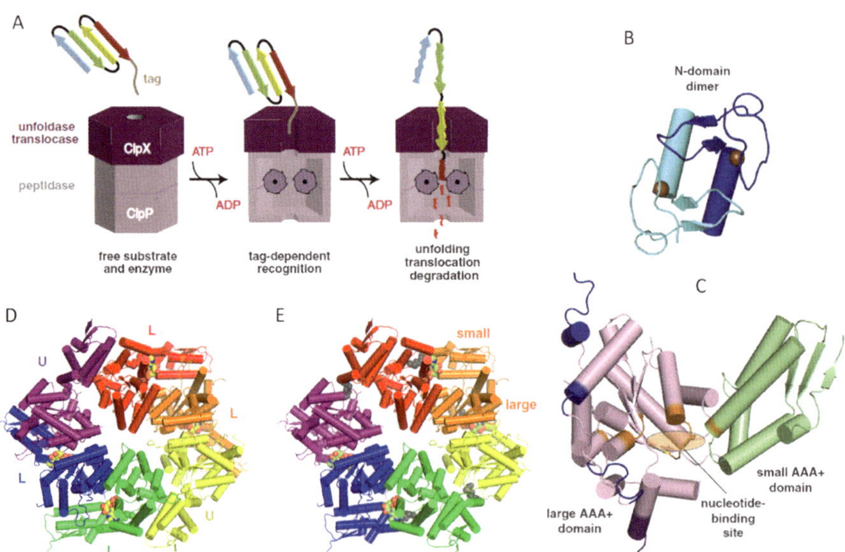

Fig. 5.5 The ClpXP molecular machine. (**a**) Schematic representation of the sequential unfolding and degradation operations. (**b**) Domain structure of ClpX N-domain dimer. (**c**) Domain structure of the AAA+ large and small domains within a subunit with L and U denoting the nucleotide loadable and unloadable subunits. (**d**) ClpX hexamer colored by subunit. (**e**) ClpX hexamer colored by rigid-body units (from Baker *Biochim. Biophys. Acta* 1823: 15 © 2012 Reprinted by permission from Elsevier)

5.4   Hsp60 Family Chaperonins Encapsulate Proteins in Folding Cages

Molecular chaperones are among the most highly expressed proteins in human cells. Recent analyses of proteomic data have revealed that some 10 % of the total protein mass is composed of molecular chaperones. These proteins can be grouped into a number of families each designated by their molecular weights—Hsp60, Hsp70, Hsp90, and Hsp100 plus sHSP, a family of small heat-shock proteins. Each family specializes in a particular set of proteostasis tasks. The chaperones form proteostasis networks in which a member of one family will hand off a problematic client protein to a member of a second family for further triage or to a protease for destruction if unrepairable.

Chaperonins directly participate in the folding process, supplying energy through ATP hydrolysis to the substrate protein. The additional energy supplied to the protein enables that protein to surmount energy barriers and escape from kinetic traps. The client protein is able to fold far more rapidly into its native state since it can avoid getting stuck in nonoptimal and nonfunctional states for long periods of time. The subunits that form the folding cage repeatedly bind and release the substrate protein. They disrupt the misfolded structures that are kinetically trapped, and release the protein in a less folded configuration through the use of mechanical stretching forces. At the end of each round of binding and release, the protein is free to continue to fold and search for its low-energy native state.

There are two kinds of chaperonins. Group I chaperonins of which Hsp60 and GroEl are typical members are found in bacteria, mitochondria, and chloroplasts. These chaperonins require the presence of a co-chaperone, Hsp10 (GroES in bacteria), which functions as a lid over the Hsp60 (GroEL). Group I folding chambers are built from 16 identical subunits arranged in two rings, one stacked on top of the other, and are widely referred to as "Anfinsen cages". The operation of the GroEL/GRoEs folding chamber is shown in Fig. 5.6. Group II chaperonins, of which TRiC [t complex polypeptide 1 (TCP1) ring complex] is perhaps the best known member, are found in archaea and the eukaryotic cytosol. As was the case for Group I chaperonins, the Group II chaperonins are composed of two rings stacked on top of one another. However, in this class, the subunits possess an additional segment that serves as a lid and they do not require the assistance of a co-chaperone. Also, the type-II chaperonin rings are built from eight different subunits, each with its own unique role.

The encapsulation of a substrate protein within dedicated folding chambers serves several purposes. First, it sequesters the misfolded proteins from potential harmful interactions with other macromolecules in the crowded cellular environment. The folding chamber is, in fact, rather small, and this close confinement has a positive effect on the folding. As discussed in the preceding section on crowding, limiting the physical space that can be sampled by the polypeptide chain can by itself help collapse the protein and drive it towards the native state. Both of these effects are passive ones. In addition, the inner surface of the cage is lined with hydrophobic residues, and the subunits undergo large-scale movements driven by

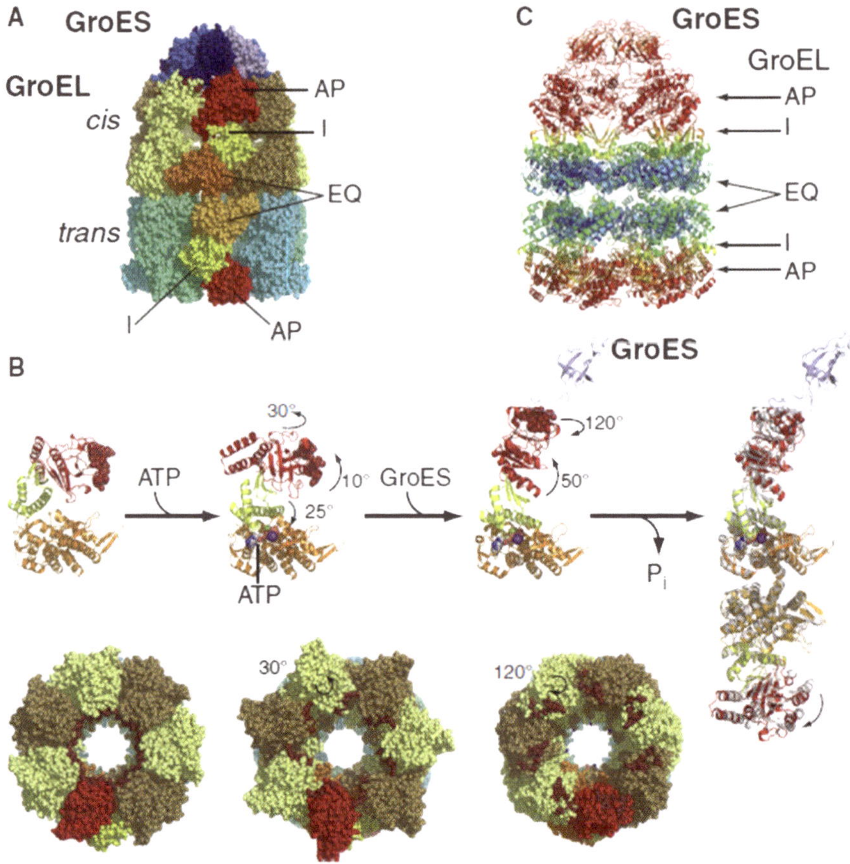

Fig. 5.6 Structure and changes in conformation that accompany ATP binding and hydrolysis. (**a**) Structure of GroEL/GroES complex. GroEL subunits are colored *brown/yellow* (cis-ring) and *cyan/green-cyan* (trans-ring); Equatorial (EQ), intermediate (I), and (AP) apical domains are colored *orange*, *yellow*, and *red*, respectively. GroES subunits are shaded in *blue*. (**b**) Conformational changes in a single GroEL subunit accompanying ATP binding and hydrolysis. (**c**) Debye-Waller factor (also called the thermal or b factor) for the GroEL/GroES complex colored *blue-to-red* for *low-to-high*. This factor provides a measure of the thermally driven conformational activity taking place within each structural element in the complex (from Mayer *Mol. Cell* 39: 321 © 2010 Reprinted by permission from Elsevier)

ATP hydrolysis. The chaperonin cage is much more than a passive chamber. Instead, it functions as a molecular machine that applies forces to the misfolded protein and does work on the proteins through repeated cycles of binding and release. Mechanistically, if these cycles are short, on the order of 1 s, the process can be regarded as occurring through *iterative annealing*. If, on the other hand, the cycles take longer, ~10–15 s, then the process may be thought of as one of *forced unfolding*.

5.5 Hsp70 Chaperones Maintain Cellular Proteostasis

Hsp70 family members assist in the folding of nascent proteins and proteins that have been exposed to cellular stresses resulting in their partial unfolding. They bind hydrophobic patches on newly synthesized chains, and bind and help transport proteins across organelle membranes and into the endoplasmic reticulum and mitochondria. They not only assist newly synthesized proteins in their folding to the native stage, but also prevent their aggregation and, if aggregation occurs, they help solubilize and refold the clumped proteins. For those proteins presenting difficulties, the Hsp70s will hand off proteins either to (1) Hps60s that form Anfinsen cages to further refold kinetically trapped and difficult to fold proteins, or to (2) Hsp90s to maintain their stability and solubility and enable then to carry out their cellular functions. Alternatively, if the damage to the proteins is unrecoverable the Hsp70s will convey/send them to the UPS or via autophagy to lysosomes for degradation.

The aforementioned families of molecular chaperones operate as a highly integrated Hsp70/Hsp60/Hsp90 network. The Hsp70/HSPAs chaperones reside at the network center with Hsp60/chaperonins and Hsp90/HSPCs interconnected to them along with their co-chaperones and small heat-shock proteins (HSPBs). A schematic representation of this network is presented in Fig. 5.7. Several groups of co-chaperones assist in the above-mentioned activities. The Hsp70/Hsp90 organizing protein (HOP) facilitates the transfer from Hsp70s to Hsp90s. Chaperones belonging to the J-domain protein family such as Hsp40 and DnaJ (bacteria) provide substrate specificity to the Hsp70s; some J-proteins bind client proteins while others target the Hsp70s to the clients.

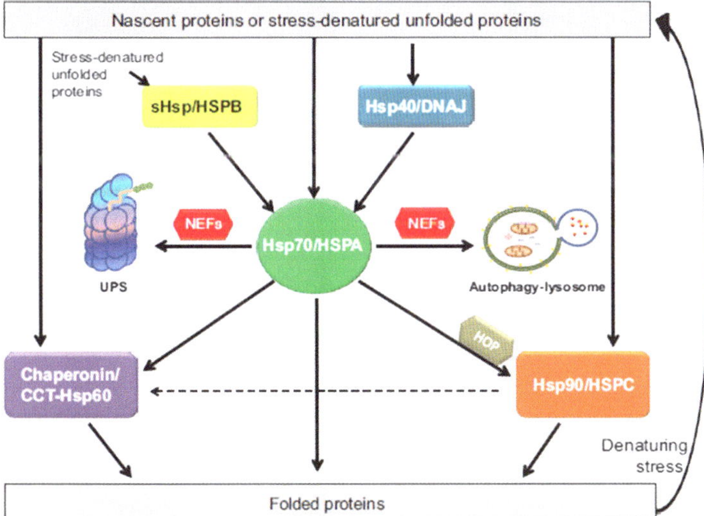

Fig. 5.7 The molecular chaperone network operating under acute stress conditions within the cell requiring their assistance in folding nascent proteins and in refolding stress-denatured unfolded proteins (from Kakkar *Dis. Models Mech.* 7: 421 © 2014 Reprinted by permission of the authors)

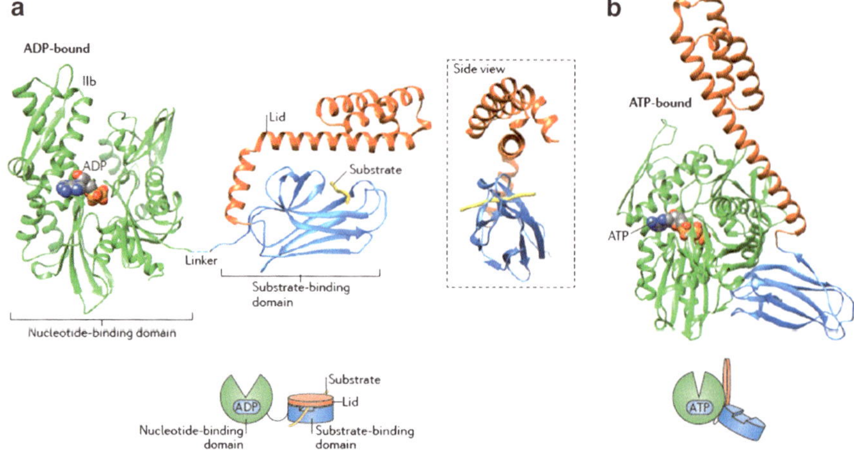

Fig. 5.8 Hsp70 domain structure and cycle of binding and release. (**a**) ADP-bound nucleotide-binding domain (NBD) is connected to the substrate-binding domain (SBD) plus lid by a flexible linker. (**b**) In the ATP-bound state the cleft in the NBD has narrowed and the lid and SBD undergo large conformational changes enabling release and exchange of portions of the substrate. Once hydrolysis occurs the lid closes back down onto the substrate and SBD (from Saibil *Nat. Rev. Mol. Cell Biol.* 14: 630 © 2013 Reprinted by permission from Macmillan Publishers Ltd)

Hsp70 chaperones undergo multiple ADP/ATP cycles of binding and release. In this process, the J-domain proteins assist Hsp70 by working together with the substrate to lower the activation energy for hydrolysis. As depicted in Fig. 5.8, once bound, the substrate and co-chaperone stimulate hydrolysis of ATP leading to the closing of the substrate binding cavity about the substrate. In the next step, the ADP molecule is released to be rapidly replaced by another ATP molecule leading to opening of the substrate binding cavity, and the cycle is ready to be repeated. Another set of cofactors, the nucleotide exchange factors (NEFs), control the duration of the substrate binding by facilitating the exchange of ADP for ATP and triggering the release of the client protein.

5.6 Hsp90 Chaperones Stabilize Proteins

Each family of molecular chaperones has a distinct set of roles in preserving protein homeostasis. Members of the Hsp90 family play a key role in stabilizing proteins in nonnative intermediate states. To do so they form stable complexes with Hsp70, its co-chaperone Hsp40, and HOP. Once the Hsp90 binds and stabilizes its client, another cofactor, AHA1, triggers ATP hydrolysis resulting in client folding and release. Hsp90 chaperones play special roles in cellular signaling networks where they stabilize proteins that tend to be, at best, only weakly stable (or metastable) until they contact their signaling partners. Once these proteins establish contact they dissociate from the Hsp90s, and undergo stability-promoting conformational changes. The Hsp90 cycle of binding, stabilization, and activation is depicted in Fig. 5.9.

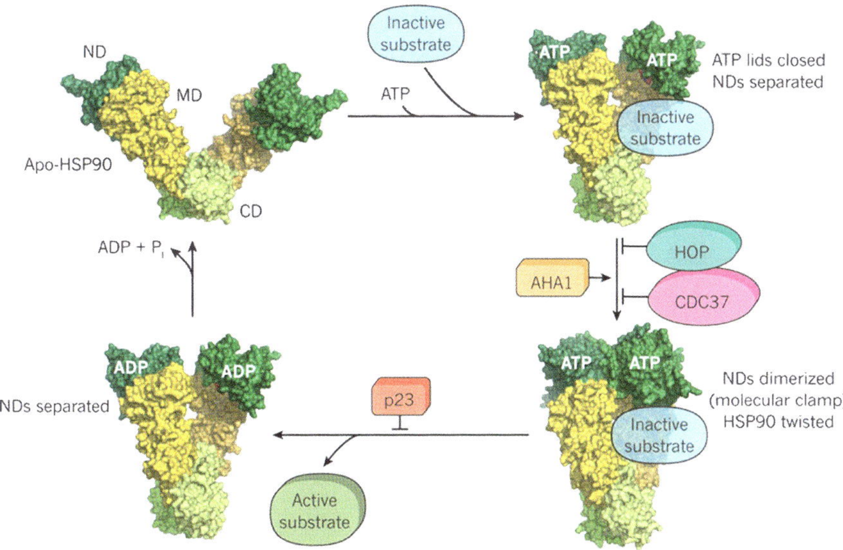

Fig. 5.9 The Hsp90 cycle. *Upper-left panel*: Hsp90 contains an N-terminal domain (ND), middle domain (MD), and C-terminal domain (CD), and functions as a dimer. *Upper-right panel*: ATP binding induces conformational changes and closure of the ATP lid. *Lower-right panel*: An ND dimer forms and the assemblage can function as a molecular clamp with its twisted subunits. *Lower-left panel*: ATP hydrolysis occurs resulting in activation and release of the substrate (from Hartl *Nature* 475: 324 © 2011 Reprinted by permission from Macmillan Publishers Ltd)

One of Hsp90s client proteins is the heat-shock stimulating factor Hsf1. This protein is the most conserved stress responder from yeasts to humans. It serves as a master transcription factor that binds one or multiple short sequences termed heat-shock elements (HSEs) located in promoters of a large number of target genes. Hsf1 not only regulates the transcription of heat-shock proteins but also regulates the transcription of proteins involved in related activities such as energy generation and protein trafficking. Under low-stress conditions Hsf1 is maintained in an inactive state through binding to Hsp90 and its cofactors. It detaches from Hsp90 in response to the buildup of ubiquitinated proteins indicative of protein folding-related stresses. Among the proteins upregulated by Hsf1 are Hsp70, Hsp40, the small heat-shock protein Hsp27, and Hsp90.

Interestingly and significantly, Hsp90 has many clients and it preserves not only partially-folded beneficial proteins but also misfolded, disease-causing ones. Among the Hsp90 clients are mutated proteins contributing to all of the hallmarks of cancer. In the case of neurodegeneration, something similar occurs with regard to preservation of aggregates of the primary contributors in each of the disease classes. This has given rise to the hope that by developing Hsp90 inhibitors, Hsf1 would be released, as it would in the case of cellular stresses, and upregulate transcription of the beneficial Hsp70s and Hsp40s.

5.7 Molecular Chaperones Handle Difficult-to-Treat Misfolded, Intrinsically Disordered, and Amyloidogenic Proteins

The handling of misfolded/intrinsically disordered, amyloidogenic proteins such as Aβ, α-synuclein, and Htt is especially challenging to the chaperone network. These proteins do not respond easily to refolding endeavors. Instead, they need to be sent either to the UPS or to the autophagic-lysosomal system for disposal. In these situations, the small heat-shock proteins send badly folded proteins directly to the degradation systems and not to the Hsp70s. This alternative routing by the HSPBs is depicted in Fig. 5.10. In addition, it is surmised that the Hsp40s/DNAJ co-chaperones may do the same. Proteostasis is maintained in these instances by rapidly disposing difficult-to-handle misfolders before they form toxic oligomers and large potentially dangerous aggregates. A comparison of Fig. 5.10 to 5.7 shows that the situation is far more complex in the case of amyloidogenic and other difficult-to-handle proteins.

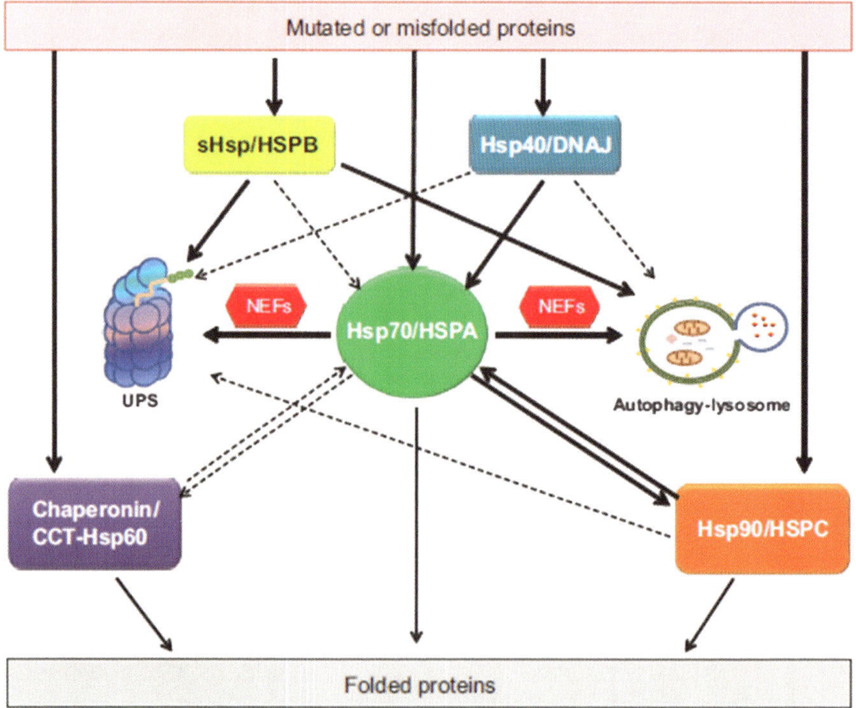

Fig. 5.10 The molecular chaperone network operating under chronic stress conditions within the cell focuses on directing the mutated or misfolded proteins to the UPS and to the autophagic-lysosomal pathway for degradation (from Kakkar *Dis. Models Mech.* 7: 421 © 2014 Reprinted by permission of the authors)

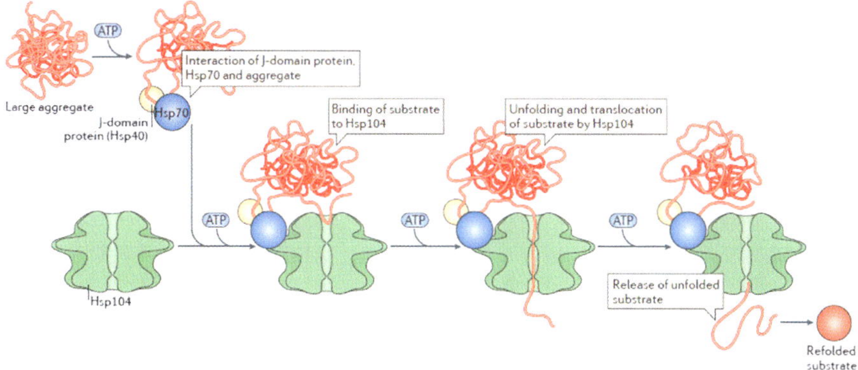

Fig. 5.11   Action of the Hsp104/Hsp70/Hsp40 disaggregation network in yeast (from Doyle *Nat. Rev. Mol. Cell Biol.* 14: 617 © 2013 Reprinted by permission from Macmillan Publishers Ltd)

Two families of proteins—the small heat-shock proteins and the Hsp104s—have been a focus of attention because of their ability to disaggregate proteins that have clumped together and/or formed amyloid fibrils. Heat-shock protein 104 is a yeast chaperone that is able to resolubilize a large variety of protein aggregates ranging from amorphous assemblages to fibrillary deposits. These molecular chaperones as well as their bacterial homolog ClpB are AAA-ATPases, hexameric molecular machines that help unfold proteins, disassemble/disaggregate protein complexes, and resolubilize their components (AAA: ATPase associated with various cellular activities). Members of this family of chaperones include ClpX and the 26S proteasomal unfoldase discussed earlier and p97/VCP to be discussed later. Hsp104 does not operate alone; instead, it functions in close association with Hsp70 and Hsp40, forming an Hsp104/Hsp70/Hsp40 disaggregation network (Fig. 5.11).

Interestingly, there does not seem to be a metazoan counterpart to Hsp104. One possible counterpart is Hsp100. This protein is variously regarded as an Hsp70 NEF or as forming a subfamily of Hsp70 chaperones. In support of its having an analogous role in metazoans, Hsp110/Hsp70/Hsp40 systems are present in metazoans, and these networks are capable of disassembling some amorphous aggregates. However they cannot disassemble and resolubilize either Aβ or α-synuclein fibrillar deposits, while Hsp104-based networks, when presented with these aggregates, can handle them. It seems that when encountering difficult-to-treat misfolded, intrinsically disordered, and amyloidogenic proteins metazoans primarily attempt to dispose of them through the UPS and autosomal-lysosomal pathway.

The small heat-shock proteins (sHsps/HSPBs) are a diverse family of proteostasis triage specialists ranging in size from 15 to 43 kDa. First discovered in 1974, these chaperones become activated rapidly in response to elevated cellular stresses. Once activated, they bind exposed hydrophobic patches on proteins populating non-native conformations and through those actions prevent misfolded proteins from clumping together and forming aggregates. There are ten members of this family (HSPB1–HSPB10), each characterized by the presence of a conserved domain of approximately 100 amino acid residues called as *α-crystallin* domain. These proteins carry out their tasks independent of ATP hydrolysis, forming large oligomeric complexes containing varying numbers of subunits. These complexes exist as ensembles of conformers that continually interconvert into one another over time thereby varying their quaternary structure. These proteins act in concert with other chaperones to potentiate their actions and greatly improve their disaggregation and amyloid-handling capabilities.

5.8 Protein Folding Can Be Studied through Single-Molecule Biophysics

The multisubunit ATP-dependent proteases and molecular chaperones function as molecular machines. These machines exert pulling forces on the client proteins to unfold them, to thread them into degradation chambers, and to facilitate their folding into their native conformational ensembles. They anneal the proteins, supply energy, and use force to wiggle them, modify their energy landscapes, promote their escape from kinetic traps, stabilize them, and disaggregate them. Single-molecule biophysics provides structural dynamics and folding information at the level of single biomolecules. As noted in the brief comments in Chap. 3, this term encompasses three distinct methods: Förster resonance energy transfer, atomic force microscopy, and optical tweezers.

5.8.1 Atomic Force Microscopy

The atomic force microscope (AFM) was invented in 1986 by Binnig, Quate, and Gerber. In a typical AFM experiment, the protein of interest (or assembly of proteins) is attached to a sharp tip mounted on a flexible cantilever. The other end of the protein assembly is attached to a surface that can be withdrawn away from the tip thereby applying a pulling force. The protein extension that results will generate a restoring force (much like in the stretching of a spring) that will bend the cantilever and this deflection can be measured precisely using laser light and a photodetector as shown in Fig. 5.12.

The restoring force is an entropic one. The stretched protein resides in conformations that are less probable than the unstressed ones and has undergone a decrease in entropy. The restoring force resets the protein into a more probable state and consequently the springing back of stretched materials is referred to as entropic elasticity. One model invented to describe entropic restoring force is called the worm-like chain model. This model was previously applied to DNA, which behaves in a similar manner. The equation for that restoring force is presented in Fig. 5.12 (C). Rubber is the canonical example of a material that derives its elastic properties from entropic forces.

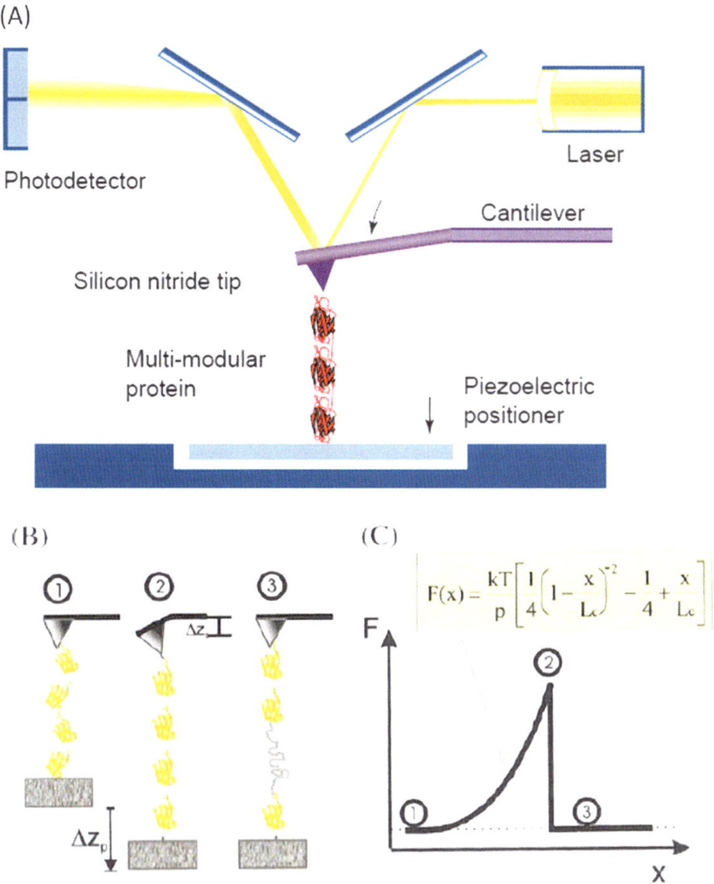

$$F(x) = \frac{kT}{p}\left[\frac{1}{4}\left(1 - \frac{x}{L_c}\right)^{-2} - \frac{1}{4} + \frac{x}{L_c}\right]$$

Fig. 5.12 Measurements of protein unfolding using an atomic force microscope. (**A**) Schematic diagram of the experimental setup used in the measurements. Force F is measured by deflection of the cantilever. Extension Z_p is determined from the position of the piezoelectric actuator (from Fisher *Trends Biochem. Sci.* 24: 279 © 1999 Reprinted by permission from Elsevier). (**B**) Unfolding of a protein through application of a pulling force. As the protein is stretched from states 1 to 2, the reduction in entropy generates a restoring force that bends the cantilever. When the domain unfolds in state 3, the contour length of the protein increases and the force on the cantilever is reduced to near zero. The cycle then repeats. (**C**) Calculation of the entropic restoring force $F(x)$ using a worm-like chain model (WLC). L_c denotes the protein's contour length, p its persistence length, and x its extension. ((**B, C**) from Carrion-Vazquez *Prog. Biophys. Mol. Biol.* 74: 63 © 2000 Reprinted by permission from Elsevier)

5.8.2   *Optical Tweezers*

The basic idea underlying optical tweezers is that light carries momentum and because of this can push and move particles. In optical tweezers experiments, the goal is not to push particles with light but rather to trap them with it. This is

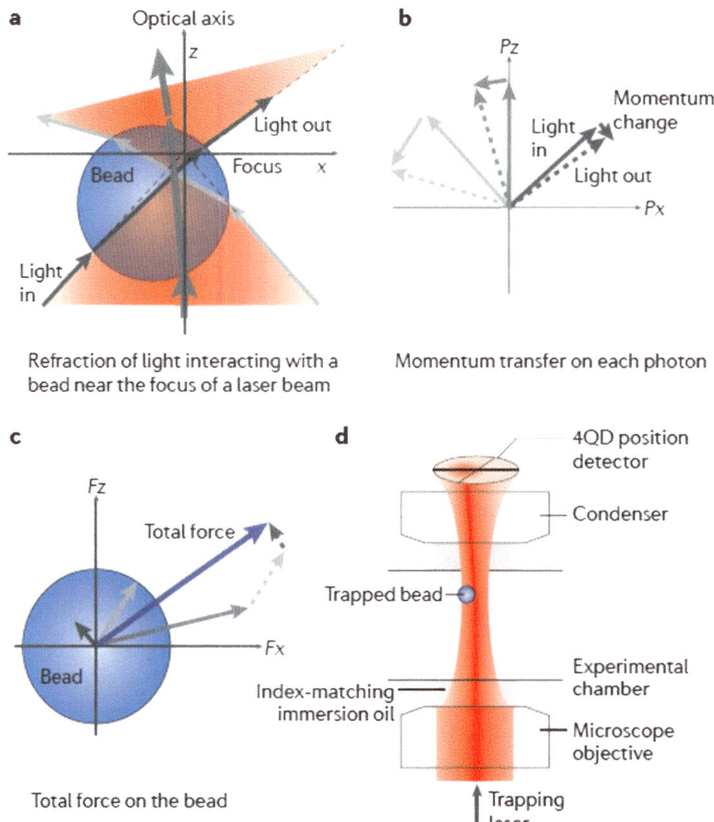

Fig. 5.13 Physics underlying the use of optical tweezers. (**a**) Refraction of light showing the passage of light rays through the bead positioned slightly off-center from the optic axis. (**b**) Resulting momentum changes and transfer onto the bead for the three (*color-coded*) rays depicted in (**a**). (**c**) Total force on the bead produced by the sum of all incident rays. (**d**) Experimental setup (from Veigel *Nat. Rev. Mol. Cell Biol.* 12: 163 © 2011 Reprinted by permission from Macmillan Publishers Ltd)

accomplished by (a) passing laser light through a dielectric material usually in the form of a spherical bead suspended in a fluid medium of lower refractive index, and (b) designing the laser beam so it has steep Gaussian lateral and axial (using a focusing lens) intensity gradients. When these conditions are met the bead will be optically trapped laterally and axially in three dimensions. This occurs because the intensity gradients push displaced particles back towards the center where the peaks in the intensities occur (Fig. 5.13).

In a typical optical tweezers experiment, the laser beam is focused by a microscope objective lens. The biological object of interest is attached at one end to a small dielectric bead, which is manipulated inside a sample chamber. The other end of the biological object is attached to the chamber or to another bead in a way that

Fig. 5.14 Photograph of Steven Chu and Art Ashkin taken in 1986 in the laboratory shortly after the first optical trap experiment was performed (from Chu *Rev. Mod. Phys.* 70: 685 © 1998 Reprinted by permission from American Physical Society)

allows for its manipulation. By using near infrared and visible light biological objects can be studied noninvasively. The forces at work during protein folding and misfolding can be probed along with the underlying energy landscapes.

Lasers play key roles in the single-molecule methods. Shown in Fig. 5.14 is a photo of the laser apparatus used by Art Ashkin and Steven Chu and their coworkers in the first optical trap experiments. These efforts were preceded by studies by Ashkin in which small dielectric particles were trapped, and were followed by the expanded utilization of this setup in the form of optical tweezers to study biological systems. As important aspect of laser trapping is that atoms can be cooled through interaction with radiation. This aspect was exploited by Chu, Claude Cohen-Tannoudji, and Phillips, who combined optical with magnetic confinement to achieve near absolute zero temperatures. Those experiments led to the development of atomic clocks and attainment of Bose-Einstein condensation.

5.8.3   *Pulling Forces and Protein Unfolding*

One notable finding that emerged from early studies on pulling forces was the observation that bonds strengths rather than being fixed quantities are, in fact, variable and could be weakened through the application of external forces. A particularly dramatic example of this property was in the transient adhesive bonds formed and broken by leukocytes as they roll along the inner surface of blood vessels. These and similar studies led to the development of a simple formula by GI Bell in 1978 that modeled the phenomena in terms of the changes in the energy landscape brought on by the applied forces.

In Bell's approach, bond lifetimes are exponentially reduced when subjected to pulling forces. This dependence can be described by expressions of the form:

$$k_{\mathrm{off}}(F) = k_{\mathrm{off}}^{0} \exp(-E(F)/k_B T) \tag{5.1}$$

In this expression, the usual dependence of the rate on the barrier height E has been replaced by a force-dependent energy $E(F)$. The reduction in the barrier height by the applied force-dependent barrier height can be described by an expression of the form $E(F) = E - F \cdot x$, where x is the appropriate reaction coordinate. Inserting this quantity back in Eq. (5.1) yields an expression for the exponential reduction in bond lifetime, the inverse of k_{off}:

$$k_{\mathrm{off}}(F) = k_{\mathrm{off}}(0) \exp(F \cdot x / k_B T) \tag{5.2}$$

where

$$k_{\mathrm{off}}(0) = k_{\mathrm{off}}^{0} \exp\left(-\frac{E}{k_B T}\right) \tag{5.3}$$

is the off-rate in the absence of the applied force. This simple picture is perhaps too simple; recent experiments demonstrate that the effects of applied forces on bonds are more complex, but this expression does highlight the important idea that bond strengths depend on how their observation/measurement is conducted.

5.9 Molecular Machines Convert Chemical Energy into Mechanical Work

The emergence of the ATP-dependent proteases and chaperones as key players in protein folding and neurodegenerative disease prevention has led to a major thrust to understand how these systems physically operate using single-molecule biophysics. Shown in Fig. 5.15 are the results of pulling on molecules of titin, a prominent component of the cytoskeleton. As seen in part (a) of this figure, a plot of extensions as a function of the applied force has a sawtooth pattern. This type of pattern is a characteristic feature of pulling measurements. The zigzags correspond to the successive unfolding of portions of the molecule. The extension of the protein generates an elastic, spring-like restoring force that is entropic in character as just discussed.

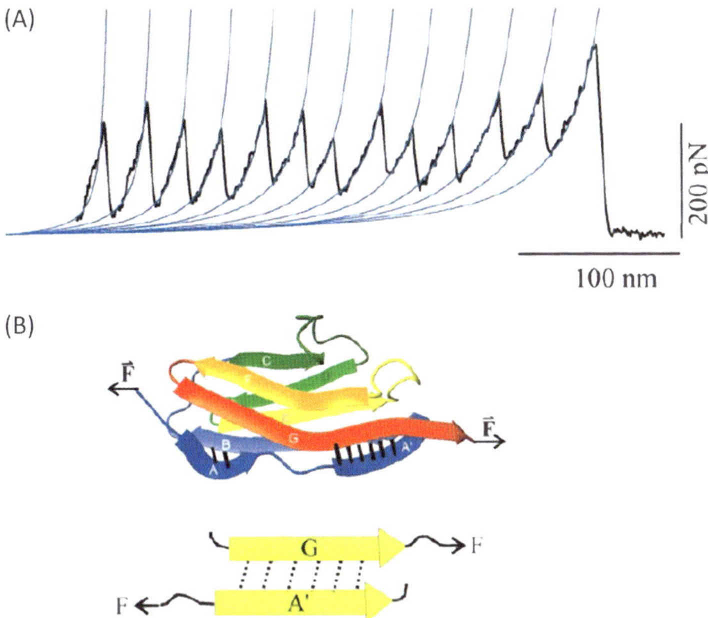

Fig. 5.15 Unfolding of single titin immunoglobulin domains using an atomic force microscope. (**a**) Characteristic sawtooth force-extension pattern. (**b**) Titin beta sandwich formed by seven beta strands organized into two beta-sheets, and including some of the critical hydrogen bonds as indicated by the *dashed lines*. *Black arrows* denote direction and sense of the applied pulling force *F* (from Carrion-Vazquez *Prog. Biophys. Mol. Biol.* 74: 63 © 2000 Reprinted by permission from Elsevier)

Ring ATPases and chaperone chambers are molecular machines that utilize energy derived from ATP hydrolysis to perform useful work. Other machines and mechanical processes also require energy to do work but derive that energy from random thermal fluctuations. These large protein complexes utilize the enormous reservoir of energy in a cell that manifests itself as thermal fluctuations. At first glance such a process would appear to be impossible. After all, the random thermal fluctuations are just that—random, so how could they be exploited to perform directional, useful tasks? That would be like getting something for nothing—a perpetual motion machine. That subject, in the form of the impossibility of violating the second law of thermodynamics, was discussed by the Scottish physicist James Clerk Maxwell (1831–1879) in his famous 1871 tome on heat. In his discussion, he introduced a little demon who attempted to violate the second law. That discussion has persisted to this day with a most interesting recent example introduced by American physicist Richard P. Feynman (1918–1988) is his 1963 *Lectures*. Maxwell's demon and Feynman's ratchet are described further in the Appendix to this chapter.

In short, the second law is safe, there are no perpetual motion machines, and biological system utilizes remarkably diverse set of ways of biasing the physical processes so that thermally driven Brownian motion can be utilized to do work. For example, in the *Brownian ratchetmodel* of how Hsp70 might facilitate transit of a protein through a membrane pore, chaperones allow motion in the desired direction while preventing motion in the opposite direction. On one side of the pore a chaperone unfolds the protein while on the other side another chaperone refolds any chains that have diffused through the pore. Because the protein undergoes a refolding upon exiting the pore it cannot pass back to the other side and this step rectifies the Brownian motion to give left-right or up/down directionality to an otherwise completely random diffusion.

This mechanism is not the only one capable of explaining this capability of Hsp70. In the alternative, *power-stroke model*, the chemical energy released through ATP hydrolysis is directly converted to mechanical work—the chaperones undergo conformational changes that variously push and pull the protein through then pore. These two models are illustrated schematically in Fig. 5.16. In many instances, a combination of these two limiting cases may best explain the operation of the molecular motor under consideration.

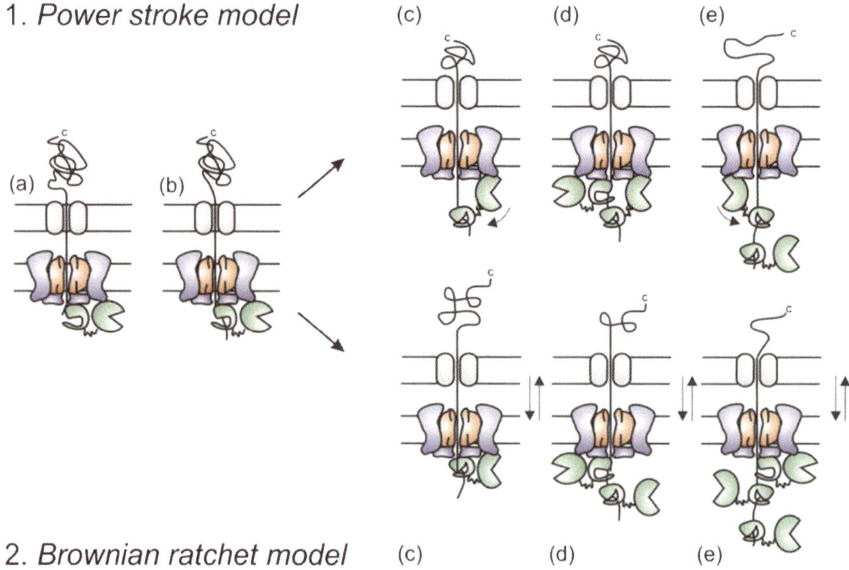

Fig. 5.16 Two models of Hsp70-dependent protein import into mitochondria. Translocation channels (*red*), motor proteins (*green*), and accessory proteins (*blue*) are depicted. In the power stroke model, the motor protein ATP binding and hydrolysis induce conformational changes that produce a pushing force that unfolds the client protein on the *cis*-side and pulls it to the *trans*-side. In the Brownian ratchet model, the protein undergoes random oscillations back and forth in the channel. ATP hydrolysis enables the Hsp70 binding pocket to close about the client protein thereby preventing its movement back to the *cis*-side and refolding, and trapping it on the *trans*-side (from Tomkiewicz *FEBS Lett* 581: 2820 © 2007 Reprinted by permission from Elsevier)

5.10 Summary

1. Step by step, piece by piece, an understanding of neurodegeneration is emerging. The central element in this theory is that of protein misfolding. This primary event leads to several sets of consequences. First, large deposits of misfolded proteins accumulate in the affected cells. These deposits not only contain the misfolded and aggregated proteins but also ubiquitin chains, ubiquitinating enzymes, and molecular chaperones involved in protein repair and disposal. Secondly, a variety of smaller oligomers are present that vary in size and morphology, and in the composition of the primary chain from which they are derived.
2. During the past 20 years or so the existence of a multilayered protein quality control system has been uncovered. The highest level in this system consists of multiple cooperating families of molecular chaperones, ATP-consuming molecular machines that protect cells against misfolded proteins. These versatile proteins network with one another. They assist newly synthesized proteins in their initial folding, their refolding and stabilization, their disaggregation, and in their disposal when triage fails. The following describes the main actions taken by each of these families of molecular chaperones:

   - Hsp60/GroEL: Assists in completing the folding of kinetically trapped and difficult to fold proteins.
   - Hsp70/DnaK: Assists in the folding of newly synthesized proteins, in the refolding of misfolded proteins, and in preventing aggregation.
   - Hsp90: Stabilizes complex proteins especially those involved in signaling and regulation, maintains their solubility.
   - Hsp104/Hsp110: Prevents protein aggregation, disaggregates protein assemblages.
   - sHsps: Prevents protein aggregation and like the Hsp110 proteins forms protein networks with other families of molecular chaperones.

3. The maintenance of protein quality control is a major endeavor in the cell. About 10 % of the protein mass is taken up with chaperones. These molecular machines consume a considerable amount of ATP as do the machines that chop up and dispose of unwanted and damaged proteins. In order to degrade the misfolded proteins, they must first be unfolded and then threaded into the proteolytic chamber. This is accomplished by placement of an unfoldase, typically a hexameric ring-shaped AAA-ATPase, on top of a multiringed, multisubunit peptidase (degradase). Both operations require ATP. The physical properties of proteins that make these operations possible, and underlie their ability to function as molecular machines, are being explored using single-molecular biophysics techniques such as atomic force microscopy and optical tweezers.
4. In examining the molecular machines, two sources of energy are utilized—energy released through ATP hydrolysis and thermal energy locked up in Brownian motion. The latter can be utilized if the random motion is biased,

ratcheted, or rectified in some manner. The way in which this is accomplished varies from machine to the next. Two limiting cases, in which one energy source or the other is used exclusively, are referred to as the Brownian ratchet and power stroke. Whether energy is derived from the hydrolysis of ATP or harnessed from Brownian motion, chemical energy is converted to mechanical work by the molecular machines.

5. The second layer of protein quality control consists of the ubiquitin-protease system and the autophagic-lysosomal pathway. If for any reason the ubiquitin-proteasome system is overwhelmed, misfolded, and damaged, proteins are disposed by the autophagic-lysosomal system. Two compartments receive special attention—the protein-synthesizing endoplasmic reticulum (ER) and the energy-generating mitochondria. The examination of protein quality control will continue in the next chapter with an exploration of autophagy and mitophagy (mitochondrial autophagy) and quality control in the ER. These discussions will bring into play a number of risk factors for developing a neurodegenerative disorder. By far the predominant risk factor is age with its accompanying decline in the effectiveness of protein quality control. A second emerging major factor is overnutrition. These threads, explored in the laboratory through caloric restriction (CR), will be examined in the next chapter.

Appendix. Maxwell's Demon and Feynman's Ratchet

The notion that random fluctuations can be used to perform useful work is not an obvious one. After all Maxwell famously showed in his 1871 tome on heat that such a process by itself would violate the second law of thermodynamics. The guilty party in his thought experiment was a little fellow known to this day as Maxwell's demon. His job was to sit by a trapdoor between two chambers. The chambers are filled with particles undergoing thermal motions at equilibrium. Whenever a particle moving faster than the average approaches the favored chamber the demon opens the door to permit passage. And, whenever a particle moving slower than the average approaches the door from the other chamber he again opens the door. Otherwise, the door remains closed. Such a process would decrease the entropy of the two-chamber system and thus violates the second law.

However, that bit of reasoning is flawed—it does not take into account an essential part of the overall system—the demon. If the entropic cost to the demon and his information-requiring decision making is taken into account the apparent violation does not occur. That this is so has been discussed in different ways for the past 140 plus years. The subject was revisited by the Polish physicist M. von Smoluchowski (1872–1917) in 1912 and by many others since then. In 1963, the demon reemerged in a lecture by Feynman in which he reformulated the second law violation question as one involving a motor containing a ratchet and pawl as depicted in Fig. 5.17. The resulting analysis presented in his *Lectures* illustrated in a novel way the impossibility of using thermal motions to build a perpetual motion machine.

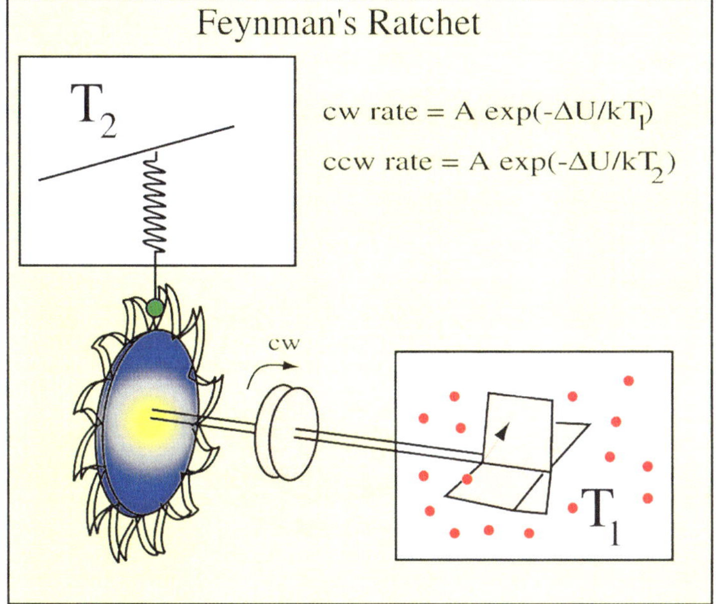

Fig. 5.17 Feynman's ratchet. The ratchet arises from the asymmetry in the teeth of the device giving rise to the (incorrect) notion that by this means thermal energy could be harnessed to do work. However, the spring is subject to thermal vibrations. In the down position of the pawl (spring plus cog), thermal collisions impinging on the paddlewheel will turn the device in the desired (cw) direction. However, in the up and disengaged pawl position, movement in either direction is possible, and only a small movement is needed to send the device back one tooth. In contrast, a far greater movement is needed to drive the device forward another tooth. When $T_1 = T_2$ there is no net motion and no perpetual motion either (from Astumian *Eur. J. Biophys.* 27: 474 © 1998 Reprinted by permission from Springer)

Bibliography

Discovery of the Ubiquitin-Proteasome System

Arrigo, A. P., Tanaka, K., Goldberg, A. L., & Welch, W. J. (1988). Identity of the 19S 'prosome' particle with the large muntifunctional protease complex of mammalian cells (the proteasome). *Nature, 331*, 192–194.

Etlinger, J. D., & Goldberg, A. L. (1977). A soluble ATP-dependent proteolytic system responsible for the degradation of abnormal proteins in reticulocytes. *Proceedings of the National Academy of Sciences of the United States of America, 74*, 54–58.

Goldstein, G., Scheid, M., Hammerling, U., Boyse, E. A., Schlesinger, D. H., & Niall, H. D. (1975). Isolation of a polypeptide that has lymphocyte-differentiating properties and is probably represented universally in living cells. *Proceedings of the National Academy of Sciences of the United States of America, 72*, 11–15.

Hershko, A., Ciechanover, A., & Rose, I. A. (1979). Resolution of the ATP-dependent proteolytic system from reticulocytes: A component that interacts with ATP. *Proceedings of the National Academy of Sciences of the United States of America, 76*, 3107–3110.

Hershko, A., Heller, H., Elias, S., & Ciechanover, A. (1983). Components of ubiquitin-protein ligase system. *The Journal of Biological Chemistry, 258*, 8206–8214.

Hough, R., Pratt, G., & Rechsteiner, M. (1987). Purification of two high molecule weight proteases from rabbit reticulocyte lysate. *The Journal of Biological Chemistry, 262*, 8303–8313.

Discovery of the Molecular Chaperones

Ang, D., Liberek, K., Skowyra, D., Zylicz, M., & Georgopoulos, C. (1991). Biological role and regulation of the universally conserved heat shock proteins. *The Journal of Biological Chemistry, 266*, 24233–24236.

Ellis, J. (1987). Proteins as molecular chaperones. *Nature, 328*, 378–379.

Goloubinoff, P., Christeller, J. T., Gatenby, A. A., & Lorimer, G. H. (1989). Reconstitution of active dimeric ribulose biphosphate carboxylase from an unfolded state depends on two chaperonin proteins and Mg-ATP. *Nature, 342*, 884–889.

Hemmingsen, S. M., Woolford, C., van der Vies, S. M., Tilly, K., Dennis, D. T., Georgopoulos, C. P., et al. (1988). Homologous plant and bacterial proteins chaperone oligomeric protein assembly. *Nature, 333*, 330–334.

Laskey, R. A., Honda, B. M., Mills, A. D., & Finch, J. T. (1978). Nucleosomes are assembled by an acidic protein which binds histones and transfers them to DNA. *Nature, 275*, 416–420.

Neupert, W., Hartl, F. U., Craig, E. A., & Pfanner, N. (1990). How do polypeptides cross the mitochondrial membranes? *Cell, 63*, 447–450.

Pelham, H. R. B. (1986). Speculation on the functions of the major heat shock and glucose-regulated proteins. *Cell, 46*, 959–961.

Macromolecular Crowding

Cheung, M. S., Klimov, D., & Thirumalai, D. (2005). Molecular crowding enhances native state stability and refolding rates of globular proteins. *Proceedings of the National Academy of Sciences of the United States of America, 102*, 4753–4758.

Cino, E. A., Kartunnen, M., & Choy, W. Y. (2012). Effects of molecular crowding on the dynamics of intrinsically disordered proteins. *PLoS ONE, 7*, e49876.

Ciryam, P., Tartaglia, G. G., Morimoto, R. I., Dobson, C. M., & Vendruscolo, M. (2013). Widespread aggregation and neurodegenerative diseases are associated with supersaturated proteins. *Cell Reports, 5*, 781–790.

Ellis, R. J. (2001). Macromolecular crowding: An important but neglected aspect of the intracellular environment. *Current Opinion in Structural Biology, 11*, 114–119.

Minton, A. P. (2005). Models for excluded volume interaction between an unfolded protein and rigid macromolecular cosolutes: Macromolecular crowding and protein stability revisited. *Biophysical Journal, 88*, 971–985.

Olzscha, H., Schermann, S. M., Woerner, A. C., Pinkert, S., Hecht, M. H., Tartaglia, G. G., et al. (2010a). Amyloid-like aggregation sequester numerous metastable proteins with essential cellular functions. *Cell, 144*, 67–78.

Szasz, C., Alexa, A., Toth, K., Rakacs, M., Langowski, J., & Tompa, P. (2011). Protein disorder prevails under crowded conditions. *Biochemistry, 50*, 5834–5844.

Proteasomes

Groll, M., Ditzel, L., Löwe, J., Stock, D., Bochtler, M., Bartunik, H. D., et al. (1997). Structure of 20S proteasome from yeast at 2.4 Å resolution. *Nature, 386*(6624), 463–71.

Lander, G. C., Estrin, E., Matyskiela, M. E., Bashore, C., Nogales, E., & Martin, A. (2012). Complete subunit architecture of the proteasome regulatory particle. *Nature, 482,* 186–191.

Lasker, K., Förster, F., Bohn, S., Walzthoeni, T., Villa, E., Unverdorben, P., et al. (2012). Molecular architecture of the 26S proteasome holocomplex determined by an integrative approach. *Proceedings of the National Academy of Sciences of the United States of America, 109,* 1380–1387.

Unno, M., Mizushima, T., Morimoto, Y., Tomisugi, Y., Tanaka, K., Yasuoka, N., et al. (2002). The structure of the mammalian 20S proteasome at 2.75 Å resolution. *Structure, 10,* 609–618.

AAA-ATPases

Aubin-Tam, M. E., Olivares, A. O., Sauer, R. T., Baker, T. A., & Lang, M. J. (2011). Single-molecule protein unfolding and translocation by an ATP-fueled proteolytic machine. *Cell, 145,* 257–267.

Baker, T. A., & Sauer, R. T. (2012). ClpXP, an ATP-powered unfolding and protein-degradation machine. *Biochimica et Biophysica Acta, 1823,* 15–28.

Maillard, R. A. (2011). ClpX(P) generates mechanical force to unfold and translocate its protein substrates. *Cell, 145,* 459–469.

Meyer, H., Bug, M., & Bremer, S. (2012). Emerging functions of the VCP/p97 AAA-ATPase in the ubiquitin system. *Nature Cell Biology, 14,* 117–123.

Hsp60 Chaperonins

Clare, D. K., Vasishtan, D., Stagg, S., Quispe, J., Farr, G. W., Topf, M., et al. (2012). ATP-triggered conformational changes delineate substratebinding and -folding mechanics of the GroEL chaperonin. *Cell, 149,* 113–123.

Lin, Z., Madan, D., & Rye, H. S. (2008). GroEL stimulates protein folding through forced unfolding. *Nature Structural & Molecular Biology, 15,* 303–311.

Thirumalai, D., & Lorimer, G. H. (2001). Chaperonin-mediated protein folding. *Annual Review of Biophysics and Biomolecular Structure, 30,* 245–269.

Todd, M. J., Lorimer, G. H., & Thirumalai, D. (1996). Chaperonin-facilitated protein folding: Optimization of the rate and yield by an iterative annealing mechanism. *Proceedings of the National Academy of Sciences of the United States of America, 93,* 4030–4035.

Zhang, J., Baker, M. L., Schröder, G. F., Douglas, N. R., Reismann, S., Jakana, J., et al. (2010). Mechanism of folding chamber closure in a Group II chaperonin. *Nature, 463,* 379–383.

Hsp70 Chaperones and Network Operation

Kakkar, V., Meister-Broekema, M., Minoia, M., Carra, S., & Kampinga, H. H. (2014). Barcoding heat shock proteins to human diseases: Looking beyond the heat shock response. *Disease Models and Mechanisms, 7,* 421–434.

Kampinga, H. H., & Craig, E. A. (2010). The HSP70 chaperone machinery: J proteins as drivers of functional specificity. *Nature Reviews. Molecular Cell Biology, 11*, 579–592.

Mayer, M. P., & Bukau, B. (2005). Hsp70 chaperones: Cellular functions and molecular mechanism. *Cellular and Molecular Life Sciences, 62*, 670–684.

Shorter, J. (2011). The mammalian disaggregase machinery: Hsp110 synergizes with Hsp70 and Hsp40 to catalyze protein disaggregation and reactivation in a cell-free system. *PLoS ONE, 6*, e26319.

Hsp90 Chaperones

Luo, W., Sun, W., Taldone, T., Rodina, A., & Chiosis, G. (2010). Heat shock protein 90 in neurodegenerative diseases. *Molecular Neurodegeneration, 5*, 24.

Rutherford, S. L., & Lindquist, S. (1998). Hsp90 as a capacitor for morphological evolution. *Nature, 396*, 336–342.

Whitesell, L., & Lindquist, S. L. (2005). Hsp90 and the chaperoning of cancer. *Nature Reviews. Cancer, 5*, 761–772.

Hsp104/Hsp110 Chaperones and Disaggregation

Doyle, S. M., Genest, O., & Wickner, S. (2013). Protein rescue from aggregates by powerful chaperone machines. *Nature Reviews. Molecular Cell Biology, 14*, 617–629.

Goloubinoff, P., Mogk, A., Ben Zvi, A. P., Tomoyasu, T., & Bukau, B. (1999). Sequential mechanism of solubilization and refolding of stable protein aggregates by a bichaperone network. *Proceedings of the National Academy of Sciences of the United States of America, 96*, 13732–13737.

Mattoo, R. U. H., Sharma, S. K., Priya, S., Finka, A., & Goloubinoff, P. (2013). Hsp110 is a bona fide chaperone using ATP to unfold stable misfolded polypeptides and reciprocally collaborate with Hsp70 to solubilize protein aggregates. *The Journal of Biological Chemistry, 288*, 21399–21411.

Parsell, D. A., Kowal, A. S., Singer, M. A., & Lindquist, S. (1994). Protein disaggregation mediated by heat-shock protein Hsp104. *Nature, 372*, 475–478.

Saibil, H. (2013). Chaperone machines for folding, unfolding and disaggregation. *Nature Reviews. Molecular Cell Biology, 14*, 630–642.

Small Heat Shock Proteins

Basha, E., O'Neill, H., & Vierling, E. (2012). Small heat shock proteins and α-crystallins: Dynamic proteins with flexible functions. *Trends in Biochemical Sciences, 37*, 106–117.

Duennwald, M. L., Echeverria, A. L., & Shorter, J. (2012). Small heat shock proteins potentiate amyloid dissolution by protein disaggregases from yeast to humans. *PLoS Biology, 10*, e1001346.

Hilton, G. R., Lioe, H., Stengel, F., Baldwin, A. J., & Benesch, J. L. P. (2013). Small heat shock proteins: Paramedics of the cell. *Topics in Current Chemistry, 328*, 69–98.

Mchaourab, H. S., Godar, J. A., & Stewart, P. L. (2009). Structure and mechanism of protein stability sensors: The chaperone activity of small heat shock proteins. *Biochemistry, 48*, 3828–3837.

Atomic Force Microscope

Binnig, G., Quate, C. F., & Gerber, C. (1986). Atomic force microscopy. *Physical Review Letters, 56*, 930–933.

Bustamante, C., Marko, J. F., Siggia, E. D., & Smith, S. (1994). Entropic elasticity of λ-phage DNA. *Science, 265*, 1599–1600.

Rief, M., Gautel, M., Oesterhelt, F., Fernandez, J. M., & Gaub, H. E. (1997). Reversible unfolding of individual titin immunoglobulin domains by AFM. *Science, 276*, 1109–1112.

Optical Tweezers

Ashkin, A., Dziedzic, J. M., Bjorkholm, J. E., & Chu, S. (1986). Observation of a single-beam gradient force optical trap for dielectric particles. *Optics Letters, 121*, 288–290.

Finer, J. T., Simmons, R. M., & Spudich, J. A. (1994). Single myosin molecule mechanics: Piconewton forces and nanometer steps. *Nature, 368*, 113–119.

Pulling Forces and Molecular Motors

Bell, G. I. (1978). Models for the specific adhesion of cells to cells. *Science, 200*, 618–627.

Evans, E., & Ritchie, K. (1997). Dynamic strength of molecular adhesion bonds. *Biophysical Journal, 72*, 1841–1855.

The Brownian Ratchet and Power Stroke

Astumian, R. D. (1997). Thermodynamics and kinetics of a Brownian motor. *Science, 276*, 917–922.

Dimroth, P., von Ballmoos, C., & Meier, T. (2006). Catalytic and mechanical cycles in F-ATP synthases. *EMBO Reports, 7*, 276–282.

Okuno, D., Iino, R., & Noji, H. (2011). Rotation and structure of F_0F_1-ATP synthase. *Journal of Biochemistry, 149*, 655–664.

Oster, G., & Wang, H. (2003). Rotary protein motors. *Trends in Cell Biology, 13*, 114–121.

Peskin, C. S., Odell, G. M., & Oster, G. F. (1993). Cellular motions and thermal fluctuations: The Brownian ratchet. *Biophysical Journal, 65*, 316–324.

Simon, S. M., Peskin, C. S., & Oster, G. F. (1993). What drives the translocation of proteins? *Proceedings of the National Academy of Sciences of the United States of America, 89*, 3770–3774.

Freyman's Ratchet

Astumian, A., & Derényi, I. (1998). Fluctuation driven transport and models of molecular motors and pumps. *European Biophysics Journal, 24*, 474–489.

Bennett, C. H. (1987). Demons, engines and the second law. *Scientific American, 257*, 108–116.

Feynman, R. P., Leighton, R. B., & Sands, M. (1963). *The Feynman lectures on physics*. Reading, MA: Addison-Wesley.

Chapter 6
Protein Quality Control: Part II—Autophagy and Aging

Protein quality control is carried out continuously throughout the cell—in the cytosol, in the endoplasmic reticulum, in the mitochondria and everywhere else. It is an intensive endeavor requiring the participation of multiple sensing, signaling, and degradation pathways. The focus in the last chapter was on the extensive network of molecular chaperones and on the ubiquitin-proteasome system that forms a first line of defense against misfolded and aggregated proteins. In that chapter, several different kinds of protein machines were introduced. Among these were ATP-dependent molecular chaperones and degradases that carry out folding and unfolding and separation and degradation operations on proteins. In this chapter, the autophagic-lysosomal pathway will be described along with another class of protein machines, the molecular motors that transport vacuoles and organelles and their contents to distant locales where they are needed. The autophagic-lysosomal pathway has been the subject of great interest in the field due to its ability to remove damaged organelles and proteins that have become misfolded and aggregated because of mutations or other causes and cannot be handled by the UPS.

Cells belonging to the immune, endocrine, and nervous systems signal extensively. They secrete large quantities of signaling molecules and position numerous receptors on their cell surface. Proteins destined for secretion and membrane localization are translocated into the endoplasmic reticulum where they are modified and subjected to protein quality control. Proteins found to be misfolded by resident chaperones are retranslocated from the ER lumen to the cytosol and degraded by the 26S proteasome by a process termed *endoplasmic reticulum-associated degradation (ERAD)*. One of the widespread characteristics of neurodegeneration is the accumulation of misfolded and aggregated proteins in the ER. That buildup impairs the normal function of the ER, a phenomenon termed *ER stress*. In response, cells initiate a set of programmatic changes referred to as the *unfolded protein response (UPR)*. Protein quality control in the ER along with the maintenance of mitochondrial health critical for maintaining an adequate energy supply will accompany the exploration of autophagy. The chapter will then conclude with a look at current

© Springer International Publishing Switzerland 2015

M. Beckerman, *Fundamentals of Neurodegeneration and Protein Misfolding Disorders*, Biological and Medical Physics, Biomedical Engineering, DOI 10.1007/978-3-319-22117-5_6

theories of aging, which have as their main thesis the progressive decline in protein quality control. In that discussion, special attention will be paid to caloric restriction and its ability to promote longevity through a variety of ways.

6.1   Autophagy and the Active Transport of Cargo Are Discovered

By 1963 electron micrographs had revealed the presence in cells of vesicles encapsulated by single or double membranes containing portions of the cytoplasm and organelles in the process of being digested. Earlier, in 1955, Christian de Duve (1917-2013) had discovered the lysosome using a high-speed centrifuge. At a conference on lysosomes, de Duve coined the term "autophagy", or self-eating, to denote the notion of self-directed degradation of unwanted and damaged cellular components. Today, this disposal route is recognized as playing a major role whenever the ubiquitin-proteasome system is overloaded or cannot handle the materials due to size or composition limitations. Cytosolic materials, misfolded, aggregated, and excess proteins, parts of the ER, the Golgi, and the nucleus, and entire organelles, especially mitochondria, are disposed through autophagic transport and digestion by hydrolytic enzymes in lysosomes.

Autophagy takes one of three forms. In *macroautophagy*, phagophores are formed that encapsulate the cytoplasmic materials to be degraded. As illustrated in Fig. 6.1 these become autophagosomes when encirclement is completed. These compartments subsequently fuse with lysosomes to form autolysosomes. Autophagic and endocytic pathways converge prior to reaching the lysosomal compartments,

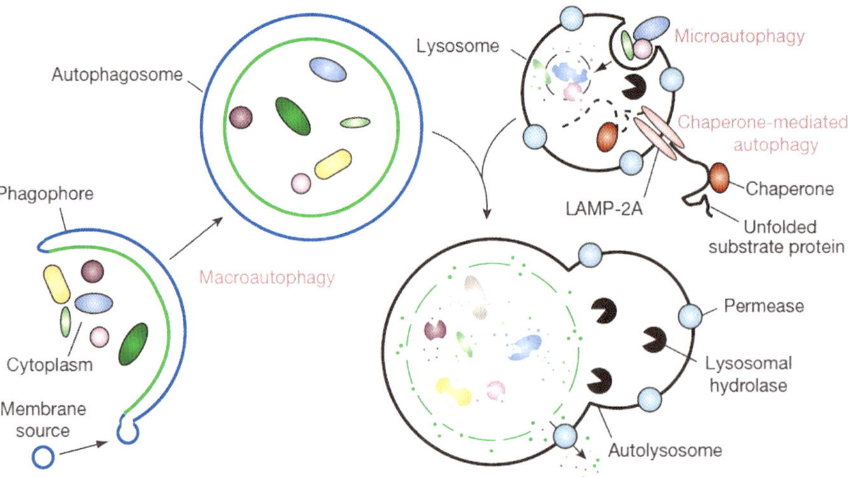

Fig. 6.1   The three different kinds of autophagy. Shown are macroautophagy, microautophagy and chaperone-mediated autophagy (from Mizushima *Nature* 451: 1069 © 2008 Reprinted by permission from Macmillan Publishers Ltd)

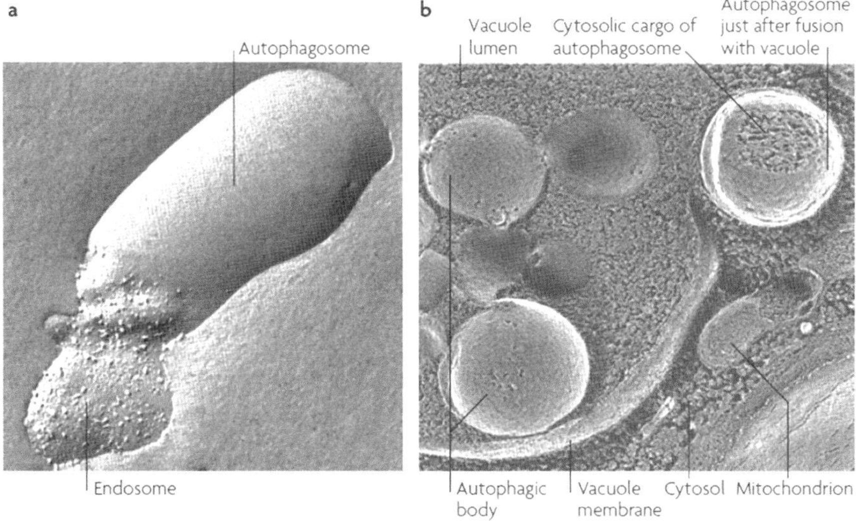

Fig. 6.2 Autophagic events visualized by means of freeze-fractured electron microscopy. (**a**) Fusion of an autophagosome with an endosome, and (**b**) fusion of an autophagosome with a vacuole (from Klionsky *Nat. Rev. Mol. Cell Biol.* 8: 931 © 2007 Reprinted by permission from Macmillan Publishers Ltd)

and several examples of fusion events of this type captured using electron microscopy are presented in Fig. 6.2. In the second form of autophagy, *microautophagy*, lysosomes directly engulf small components of the cytoplasm. Lastly, in *chaperone-mediated autophagy*, substrate proteins are chaperoned one-by-one across the lysosomal membrane.

Macroautophagy is the predominant mechanism, and will hereafter be referred to as autophagy. It can be further broken down into two subtypes. *Nonselective autophagy* supplies materials for macromolecular synthesis and energy production by recycling the amino acid, free fatty acid, and glucose building blocks. This process is especially valuable during starvation conditions and even takes place under normal conditions, as well. It is estimated that 1 to 1.5 % of cellular proteins are catabolized every hour through nonselective (basal) autophagy. In *selective autophagy*, specific tagged proteins and organelles are degraded.

Neurons present special challenges due to their morphology. Materials need to be transported over large distances out to axonal and dendritic terminals and back again to the soma, the primary sites of protein synthesis and lysosomal degradation. This requirement necessitates the use of an active transport system. That such a system might exist was first posited by Ramon y Cajal as early as 1928. Evidence in favor of that hypothesis was initially obtained in studies of peripheral nerve regeneration by Weiss and Hiscoe in the years just after the end of World War II, motivated by limb injuries to war veterans. Some years later, with the advent of radiolabelling techniques, the flow of materials out to the periphery from the cell body was directly observed.

By 1980 it was established that two-way flows regularly occurred and with varying rates depending on the the cargoes being shuttled. The cargoes undergoing active transport included organelles, vesicles, cytoskeleton components, receptors, and other signaling proteins. Rates for anterograde transport of neurotransmitters, membrane proteins, and lipids range from 200 to 400 mm/day, while rates for retrograde transport of lysosomal vesicles vary from 200 to 300 mm/day. Mitochondria are transported at reduced rates, 50–100 mm/day, punctuated by stops and reverses designed to uniformly distribute mitochondria throughout the axon. The aforementioned movement processes are referred to as *fast axonal transport*. That is in contrast to the transport of cytoskeleton components such as microfilaments and microtubules that move at rates up to 100 times slower. The shipping of those cargoes is called *slow axonal transport*.

6.2 Cargo Is Transported by Motor Proteins along Actin and Microtubule Rails

Active transport is made possible by molecular motors that travel along tracks constructed out of actin filaments and microtubules. Recall that microtubules are polarized structures composed of polymers of α- and β-tubulin. Their fast-growing plus end is directed outward, away from the microtubule organizing center (MTOC), towards the axonal terminal and its slower-growing minus end is directed inward towards the soma and MTOC. Three kinds of motors are utilized in active transport (Table 6.1). Myosins transport cargoes along actin filaments while kinesins and dyneins shuttle cargoes along microtubules. Kinesins mediate the anterograde transport from the cell body out along axons and dyneins move cargo in the opposite, or retrograde, direction back to the cell body. The motor proteins appearing in Table 6.1 belong to large families that carry out ATP-dependent, force-producing mechanical tasks ranging from muscle contraction to ciliary and flagellar movement to walking along rails while conveying cargo.

Myosins, the first of the motor proteins to be studied, have been known for some time because of their role in muscle contraction. Early studies on muscle contraction culminated with the discovery in 1942 that contraction was brought about through the interaction of ATP with two proteins—actin and myosin. A key development in this arena was the development over more than a decade of the *sliding filament model* and an accompanying picture of force generation by Andrew Fielding Huxley (1917–2012) and others. In their model, actin-myosin linkages form, move, and

Table 6.1 Axonal transport and its motor proteins

Motor protein	Type of transport	Rail system
Kinesins	Long-distance anterograde	Microtubules
Dyneins	Long-distance retrograde	Microtubules
Myosins	Short-distance	Actin filaments

generate a sliding force when ATP is hydrolyzed by the enzymatic actions of myosin. The discovery of other myosin motors was advanced by the exploration by Pollard and Korn of how amoeboid movement comes about. Those investigations led to the discovery in 1986 of myosin I, initially referred to as an "unconventional" myosin, and in 1993 Rayment solved the first three-dimensional structure of a myosin motor protein.

The discovery of dynein, the first microtubule-associated protein to be found, is intimately connected to explorations of how cilia and flagella move. Those efforts began in the 1950s with the use of electron microscopy to examine the structural core and culminating in 1965 with the isolation of the dynein ATPase by Gibbons and Rowe. The name, dynein, given to the protein by them was taken from "dyne" the Greek word for force. Dyneins are structurally far more complex than either myosins or kinesins, and it was not until 2011 that the three-dimensional crystal structure of its motor domain was solved. Kinesin, the second microtubule-based motor protein, was discovered by several groups in the period from 1982 to 1985. Those efforts were inspired by the desire to find the elusive motor protein responsible for the fast axonal transport of organelles and other cargo. The name given to the new protein, kinesin, was taken from the Greek "kinein", meaning to move. It is the smallest of the three cargo-transporting motor proteins, and its three-dimensional crystal structure was solved in 1996.

6.3   Structure and Function of the Motor Proteins

Members of the three families of motor proteins are found in multiple kingdoms and carry out a broad spectrum of functions. Of these, five members can be singled out for their prominent roles as cytoplasmic cargo transporters. As illustrated in Fig. 6.3 they each have a pair of heads that make contact with the microtubule and actin rails and contain the ATPase activity. Kinesins and myosins have one ATP binding site per head while dynein possesses a hexameric AAA-ATPase ring with four potential ATP binding sites and two putative regulatory sites. Cargo is carried by the feet of the kinesins and dyneins and in the equivalent location on the dyneins through the dynactin partner complex.

Single molecule biophysics methods have been used extensively to probe the structure and dynamics of the walking motors. That includes atomic force microscopy and optical tweezers. It also includes FRET and the application of quantum dots, fluorescent semiconductor nanoparticles that overcome many of the limitations of conventional fluorophores. Using these single molecule biophysical methods the manner in which these three classes of motor protein move their cargo over actin and microtubule rail systems has been explored.

Kinesins and myosins are stepping motors. As can be seen in Fig. 6.3, kinesin is much smaller than myosin and has been called the world's smallest biped. In spite of their lack of similarities in physical size, amino acid sequences, and enzymatic activities, the walking motors utilize similar power-generation and walking strate-

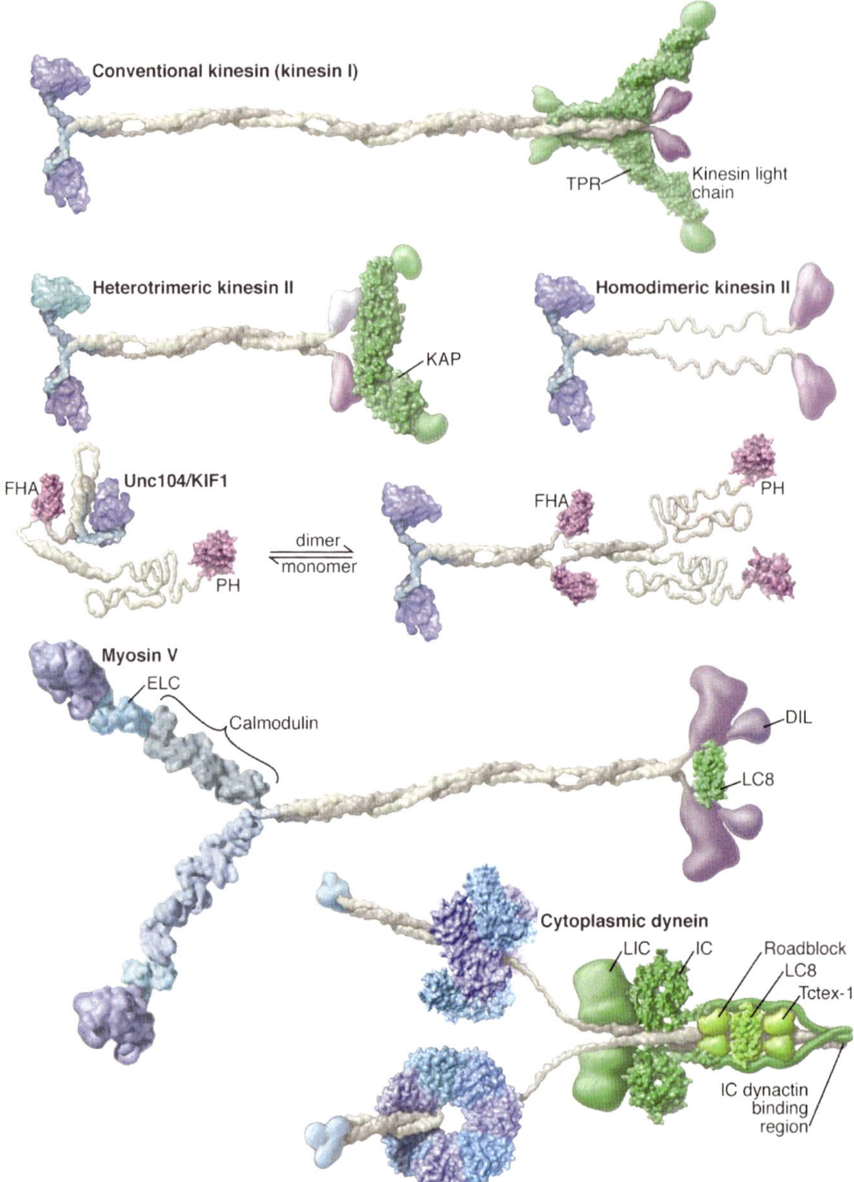

Fig. 6.3 Motor proteins with prominent roles in cargo transport. The kinesins and myosin each contain a head region shown in *blue* on the *left*, a long stalk colored *grey*, and a tail region highlighted in *purple* on the *right*. The long structures connecting the myosin heads to the stalk are referred to as the lever arms. The structure of dynein differs in two main ways—it contains a hexameric ring structure connected to a microtubule-binding stalk and in place of the feet it possesses a dynactin-binding region that serves as the interface to the cargo (from Vale *Cell* 112: 467 © 2003 Reprinted by permission from Elsevier)

gies. Using single molecule biophysics and associated techniques the motor proteins have been found to function in the following manner:

- They hydrolyse one ATP molecule for each 8 nm step in the case of kinesins and myosins. Dynein steps are more variable and range up to 20 nm or so. In some instances, a pair of motor proteins may attach to a single cargo, and move that cargo either forward or backward as the need dictates.
- The term *processive* denotes the ability of some motor proteins to take many steps before releasing from the actin filaments and microtubules. These motor proteins may take more than 100 steps before detaching. Homodimeric kinesin 1, myosin V, and cytoplasmic dynein undergo processive movement. However, others such as muscle myosin II move nonprocessively. That is, they undergo isolated movements in which forces are exerted transiently followed by detachment. The movement of both kinds of motor proteins along microtubule and actin tracks is depicted in Fig. 6.4.
- The movements of the two heads are coordinated, and enable them to move in a *hand-over-hand* manner. There are three models for the gait—symmetric hand-over-hand, inchworm, and asymmetric hand-over-hand. The asymmetric gait is supported by the majority of the evidence at present. The three ways of walking are illustrated in Fig. 6.5.

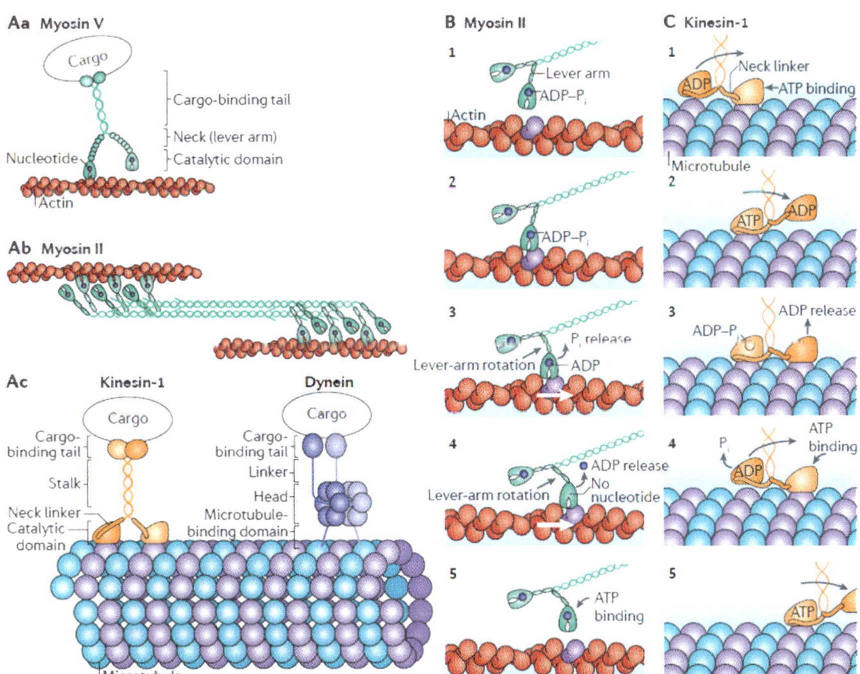

Fig. 6.4 Cytoskeletal motor protein form and function. (**Aa**) Dimeric myosin V processive movement along actin filaments. (**Ab**) Dimeric myosin II undergoing non-processive movement along actin filaments. (**Ac**) Kinesin-1 and Dynein movement along microtubules. (**B, C**) Illustration of the Myosin II and Kinesin-1 power strokes (from Veigel *Nat. Rev. Mol. Cell Biol.* 12: 163 © 2011 Reprinted by permission from Macmillan Publishers Ltd)

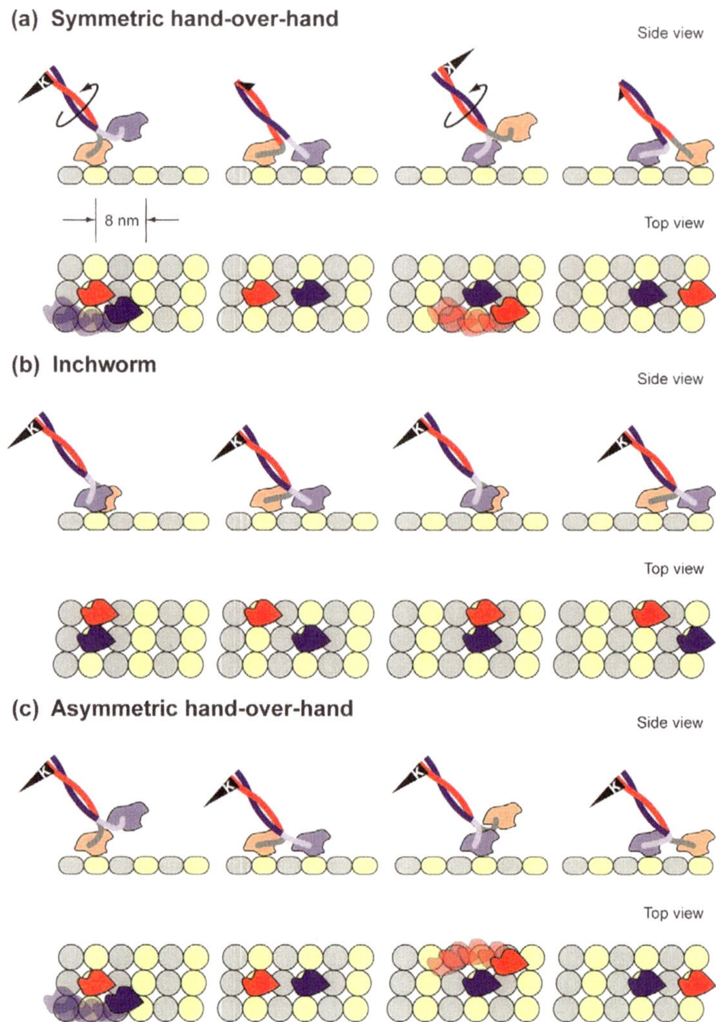

Fig. 6.5 Gaits that can be used by walking motor proteins. (**a**) Symmetric hand-in-hand in which the two heads are completely synchronized. (**b**) Inchworm in which one head is always in the lead. (**c**) Asymmetric hand-over-hand, or limping, in which the proteins alternate between two different conformations as they step (from Asbury *Curr. Opin. Cell Biol.* 17: 89 © 2005 Reprinted by permission from Elsevier)

These three families of motor proteins have important roles in maintaining cellular homeostasis and protein quality control. The short-range, actin-based myosins (e.g., Myosins Va and Vb) move glutamergic and GABAergic receptors as well as recycling endosomes into dendritic spines, and haul transport vesicles into presynaptic terminals. At the same time, the long-range microtubule-based haulers move organelles and granules, and receptors and neurotrophic factors, into and out of axons and dendrites. The kinesins and cytoplasmic dyneins transport mitochondria, RNA granules, endosomes, synaptic vesicle precursors, and NMDA and AMPA receptors. Most importantly, from the viewpoint of protein quality control cytoplasmic dynein hauls misfolded, damaged, and aggregation-prone proteins in autophagic vesicles to lysosomes for fusion and degradation.

The key element in making walking possible is the generation of force, of power strokes in this case, by the motor proteins. The mechanism underlying this process is remarkably similar to the one used in muscle contraction as identified by Huxley in 1969. Known as the *swinging cross-bridge mechanism* it has as its main tenet the transduction of chemical energy from ATP hydrolysis into mechanical power strokes. When applied to motor proteins it enables them to move along actin filaments or microtubules. A key observation underlying the mechanism is that small movements generated in the vicinity of the catalytic site are amplified by the motor protein's light-chain binding region, which acts as a lever arm.

The *swinging lever arm mechanism* is a more modern name for Huxley's swinging cross-bridge. The various motor proteins are built on the same broad principles, but differ in a number of ways from one another in their details. These differences are reflected by variations in, for example, precise angles through which the lever arms swing and the step size generated as a result. As shown in Fig. 6.6, the myosin II lever arm swings through 70° thereby generating a 10 nm stroke. Myosin V also swings through 70°, but has a longer lever arm that produces a 20 nm stroke while Myosin VI's lever arm swings through a full 180° and generates a 30 nm stroke (the latter two not shown). Illustrations of how "walking" is generated through cycles of ATP/ADP binding and release by the catalytic heads of myosins, kinesins, and dyneins are presented in Fig. 6.4.

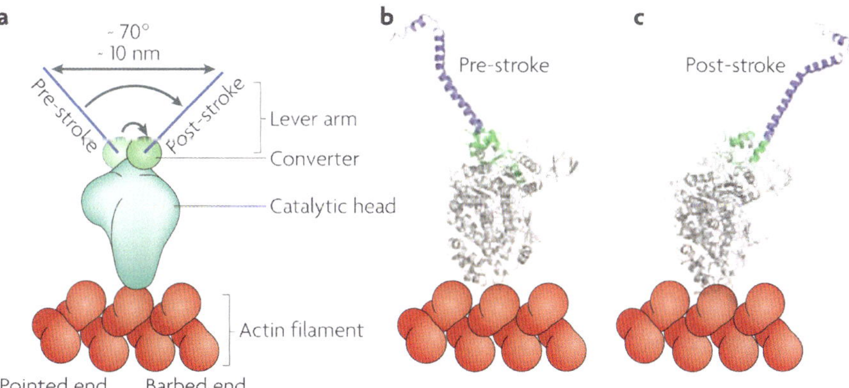

Fig. 6.6 The swinging lever arm mechanism. (**a**) Schematic representation of the myosin II head and its pre- and post-stroke orientations. (**b, c**) Corresponding crystal structures illustrating the 70° rotation of the converter and its amplification by the lever arm (from Spudich *Nat. Rev. Mol. Cell Biol.* 11: 128 © 2010 Reprinted by permission from Macmillan Publishers Ltd)

There are many as yet unanswered questions and unresolved mysteries. One mystery is: where does the energy to drive the dynein, kinesin, and myosin motors come from? Some of the axons that have to be traversed are as long as a meter. Typical step lengths are about 8 nm, and each step requires the hydrolysis of an ATP molecule. The canonical answer is that the needed energy is supplied by mitochondria that are distributed uniformly throughout the axons and dendrites. However, mitochondria are not distributed uniformly but rather are concentrated in synaptic terminals and other high use areas and less so in other regions. So, in those regions devoid of mitochondria, where does the energy come from? A possible answer, and one that is certainly satisfying, is that the motors carry their own energy generators. Specifically, they carry the glycolytic enzyme GAPDH along with them and continuously generate ATP by breaking down sugars. Evidence that this is, in fact, what they may do was uncovered recently.

6.4   Complexes and Aggregates of Misfolded Proteins, and Damaged Organelles, Are Removed through Selective Autophagy

6.4.1   Formation of the Phagophore

The formation of an autophagosome that encapsulates misfolded and aggregated proteins, and malfunctioning organelles, in a double membrane is a remarkable process that begins with initiation and nucleation. That step is followed by an elongation phase and then by capture of the appropriate selective substrates. The phagophore is then closed up to form an autophagosome ready for transport to the lysosome to which it fuses to generate the final structure, the autolysosome (Fig. 6.7). A number of complexes have been identified that orchestrate it by bringing together and helping assemble the lipids needed for the membranes, the protein machinery required to identify and capture the correct substrates, and the proteins that interface to the transport motors.

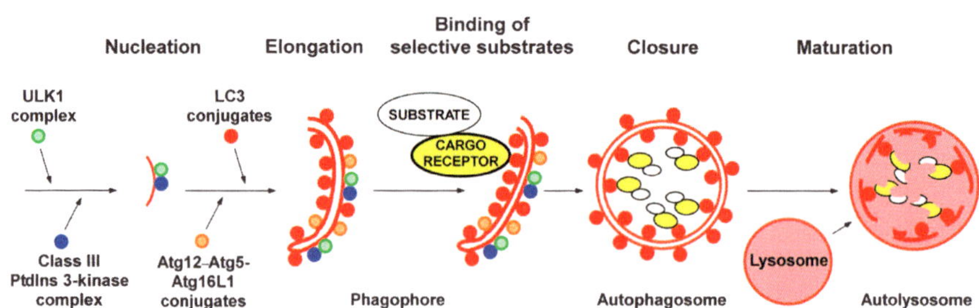

Fig. 6.7 Selective autophagy—steps in the assembly of autophagosomes and autolysosomes (from Johansen *Autophagy* 7: 279 © 2011 Landes Biosciences reprinted with permission from Taylor & Francis)

Initiation and nucleation: The first complex to act is the Ulk1 complex. This resident complex awaits activating nutrient, energy, and other signals and, once received, begins the recruitment process. It contains in addition to Ulk1 (a serine/threonine kinase) three other proteins—two members of the Atg (autophagy-related) protein family, Atg13 and Atg101, and FIP200 (focal adhesion kinase family interacting protein of 200 kDa). Ulk1 is the mammalian homolog of the yeast Atg1 protein and this complex is often referred to as the Atg1 complex.

Mutual interactions between the Ulk1 and Class III PtdIns 3-kinase complex (Fig. 6.7) occur next leading to nucleation of the phagophore. The latter has two principle members—Beclin 1, a Bcl2 family member, and Vsp34, a Class III phosphoinositide (PtdIns) 3-kinase. Bcl2 proteins are best known as stress sensors and apoptosis signal transducers. The Bcl2 proteins can be grouped into three subfamilies; some, like Bcl-X_L and Bcl-2, inhibit apoptosis, while others belonging to the so-called BH3-only subfamily act as stress sensors and when activated promote apoptosis. Beclin 1 is a member of the BH3-only subfamily, but promotes autophagy rather than apoptosis. Under normal growth and nutrient conditions this protein is bound by Bcl-2 or Bcl-X_L and remains inactive. Starvation and other stress conditions lead to its dissociation from Bcl-2 thereby enabling it to associate with autophagic proteins and function as a signal integrator. The other member of the complex is Vps34 along with its regulatory subunits Atg14 and Vps15. Vps34 phosphorylates phosphatidylinositol to generate phosphatidylinositol 3-phosphate [PI(3)P] essential for membrane trafficking.

Elongation: The next step, elongation, makes use of a set of operations remarkably similar to the ones used to tag protein for degradation by the proteasome. Like the ubiquitin-protease, a set of E1, E2, and E3 enzymes facilitate the attachment of ubiquitin-like proteins (Ubls) Atg8 and Atg12 to selected substrates. Two sets of conjugates are established. Atg12 is conjugated to its substrate Atg5 to form the Atg12-Atg5-Atg16L1 (E3) complex by the successive actions of Atg7 (E1) and Atg10 (E2) and, similarly, Atg8/LC3 (ubiquitin-like light-chain 3) is conjugated sequentially by Atg7 and Atg3 (E2) to phosphatidylethanolamine (PE) to create lipidated LC3-II complexes.

6.4.2 Substrate Binding and Autophagosome Formation

Upon completion of these steps the LC3-conjugates occupy the inner surface of the nascent phagophore where they provide binding sites for cargo receptors. For many years, disposal via the ubiquitin-proteasome system and disposal through the autophagic-lysosomal pathway were thought to be independent, nonoverlapping processes. That idea was abandoned when selective autophagy emerged as its own process distinct from basal autophagy and in which cargo receptors recognize ubiquitinated substrates and interface them to the LC3s.

One of the key elements (and potential points of failure) in the autophagic-lysosomal pathway is the p62 protein. One of its first properties uncovered in the 1990s was its ability to aggregate proteins for temporary storage. For that reason, its gene was named

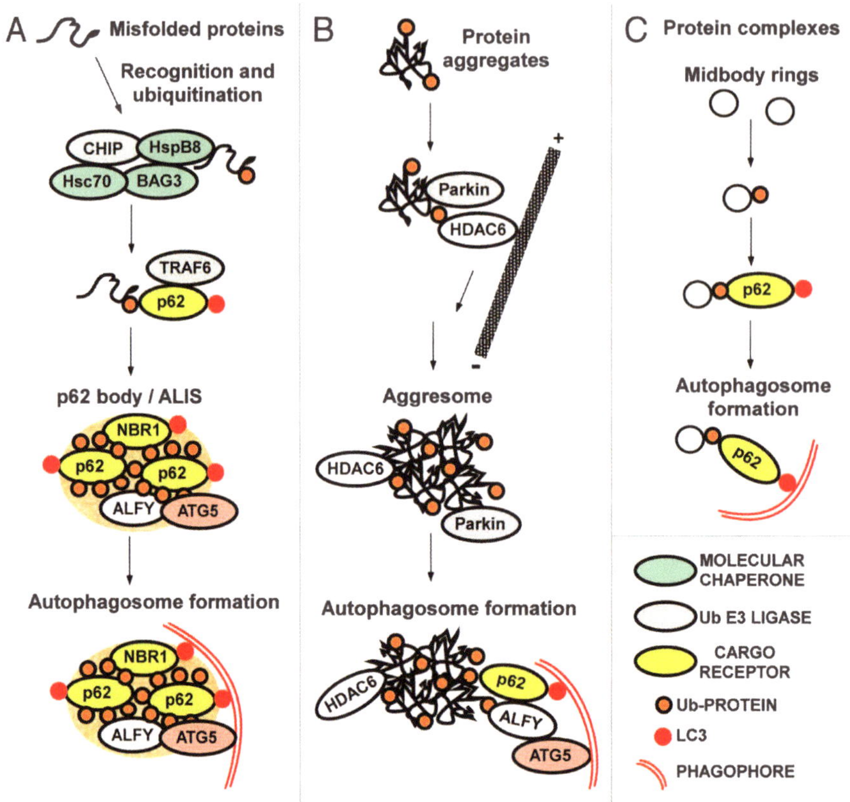

Fig. 6.8 Receptors and adapters involved in selective autophagy. (**a**) Formation of p62 bodies containing misfolded ubiquitinated proteins. *ALIS:* aggresome-like induced structure. (**b**) Formation of aggresomes containing protein aggregates. (**c**) Putting the pieces together (from Johansen *Autophagy* 7: 279 © 2011 Landes Biosciences reprinted with permission from Taylor & Francis)

sequestosome 1 (SQSTM1). Its collection activity is quite selective; this protein binds polyubiquitinated (misfolded) proteins and stores them for later disposal via the autophagic-lysosomal pathway. This protein is a cargo receptor as is NBR1; they bind to ubiquitinated substrates and to the LC3s. The p62-enriched intracellular inclusions depicted in Fig. 6.8 are referred to as *p62 bodies*. They contain besides p62 and the misfolded protein substrates, molecular chaperones and ubiquitin E3 ligases. Alfy is a large scaffolding protein found in aggregates of misfolded proteins such as (polyQ) mutant huntingtin and α-synuclein. It functions as an adapter mediating interaction between p62-bound cargo and the Atg12-Atg5-Atg16 complex.

Lastly, autophagosomes are closed, undergo maturation, and are transported by motor proteins along the rail system to where the lysosomes reside in the soma and finally the two compartments fuse. These too are nontrivial operations, and present an impressive number of potential points of failure. However, there are size limitations to entry into the proteasome, and for many large proteins and aggregates, the only available disposal route is by means of autophagy. The disposal of misfolded, aggregated, and ubiquitinated proteins through autophagy is further illustrated in Fig. 6.8.

Aggresomes are inclusion bodies that form in the vicinity of MTOCs/centrosomes. Aggregates are transported to these central locales along microtubule rails by minus-end directed dynein motor proteins. These IBs are characterized by a cage of intermediate filament (IF) proteins, most notably, vimentin. Aggresomes are dynamic entities; they are situated near and recruit components of the ubiquitin-proteasome system and molecular chaperones to deal with the misfolded proteins. The histone deacetylase-6 (HDAC6) protein has a key role in the pathway leading from aggregates to aggresomes to autophagic clearance in lysosomes. This protein is a microtubule-associated protein that regulates microtubule acetylation and the recruitment of dynein motor complexes in a p62-dependent manner. It has an ubiquitin-binding motif that enables it to bind mono-and polyubiquitinated proteins. By possessing this set of properties HDAC6 is able to function as an adapter between ubiquitinated proteins and microtubule-associated dynein motor complexes. Parkin, another component of the aggresome, is an E3 ubiquitin ligase.

Lipids and an alternative pathway: One question that naturally arises is: where do the lipids used in autophagosome biogenesis come from? Several answers have been put forth, each with supporting evidence. The vast majority of the answers postulate that the requisite lipids are derived from a pre-existing membrane rather than being synthesized de novo. Evidence in support of several candidate sources has been found. These include: (1) the endoplasmic reticulum; (2) the Golgi apparatus; (3) mitochondria, and (4) the plasma membrane. Most of these pathways utilize Atg5-Atg7 and LC3 but interestingly, an Atg5-Atg7-LC3-independent-pathway has been uncovered, as well. These observations are not mutually exclusive and it seems that several sources of membrane lipids that can be drawn upon when needed to support autophagy.

6.5 Mitochondria Are Dynamical Entities

Mitochondria are dynamic entities. They readily change their morphology, and continually undergo fission and fusion with each other. In fusion, there is an exchange of materials between the fusing organelles. In this diffusive mixing between healthy and damaged mitochondria, the healthy ones can rescue damaged ones thereby reducing mitochondrial stress. Fusion can increase their size while fission can increase their number and reduce their size. Mitochondrial dynamics and transport have emerged is the last few years as a major subject of research into aging and neurodegeneration. To see why this might be so consider the following:

• Neurons have large energy demands. Not only do the molecular machines that maintain protein heath require continual ATP input, but synapses are huge energy consumers, as well. That aspect was pointed out over 50 years ago by Sanford Palay (1918–2002) who discovered huge concentrations of mitochondria in axon terminals. There, they supply the energy needed for endocytosis, exocytosis, vesicle release, and the mobilization of vesicle reserves. The postsynaptic density is highly enriched in signaling molecules, and neurons maintain a high concentration of mitochondria in dendritic spines and postsynaptic terminals.

- Neurons have a unique morphology that puts special demands on the transport system. An adequate energy supply must be maintained over large distances, especially in axons, to the points where consumption occurs. In order to satisfy this need, there is a continual trafficking and recycling of mitochondria from soma to synaptic terminals and back again to the soma for disposal.
- Mitochondria are sensitive to buildups in intracellular stress. In response, these organelles may decrease ATP synthesis, increase reactive oxygen species (ROS) production, fail to properly handle Ca^{2+}, and release apoptosis- and necrosis-promoting factors. Declining ATP generation by the electron transport chain and increased oxidative damage correlate with chronological aging.
- Mitochondria, like the cell at large, contain a complete multilayer quality control system consisting of an ensemble of mitochondrial chaperones and a damaged-mitochondria transport and macroautophagic disposal network, termed *mitophagy*. The latter prevents untimely release of death-promoting factors from damaged mitochondria and maintains a proper balance between metabolic adjustments, clearance, recovery, and cell death.
- At some point, repair of damaged mitochondria through fusion with healthy ones may fail. Those damaged mitochondria are then removed by the abovementioned mitophagy. Fission may trigger mitophagy by testing daughter membrane potential for functionality. If they fail this stress test they are isolated from the fusion machinery to prevent their contaminating the healthy mitochondrial network, and are shipped via the mitophagy for lysosomal disposal.

6.6 Endoplasmic Reticulum-Associated Degradation and the Unfolded Protein Response

6.6.1 ERAD

Endoplasmic reticulum-associated degradation requires the participation of not only ER-resident chaperones but also E1, E2, and E3 ubiquitin ligases and a variety of recognition, solubility, and bridging factors. In brief, there are four main steps in ERAD. These are:

- **Recognition** in which the misfolded protein is recognized and tagged.
- **Ubiquitination** in which the tagged protein acquires a chain of four or more ubiquitin molecules.
- **Retranslocation** in which the ubiquitinated protein is unfolded and extracted from the ER.
- **Degradation** in which the protein is conveyed to the 26S proteasome for degradation.

One of the most prominent participants in ERAD is BiP, an ER-resident member of the Hsp70 family of molecular chaperones. This protein, like other family members, recognizes exposed hydrophobic patches on the misfolded proteins. Another important participant is p97. The p97 protein is also known as valosin-containing protein (VCP) while in yeast it is referred to as Cdc48. This abundant and versatile protein belongs to the AAA-ATPase family of heptameric ring-shaped proteins discussed in the last chapter. This arrangement enables the enzyme to use the energy of ATP hydrolysis to unfold and extract proteins from membranes, unwind DNA, and disassemble proteins complexes. In addition, p97 protein complexes escort extracted proteins to 26S proteasome for degradation.

6.6.2   *The Unfolded Protein Response (UPR)*

As noted at the beginning of this chapter, some cell types—neurons, endocrine, and immune cells, in particular—secrete large numbers of proteins. A prominent example is the pancreatic beta cell that secretes insulin. Such proteins utilize disulfide bond formation and glycosylation, and these and other finishing or maturation operations are carried out in the ER's membranous network. Failure the carry out these modifications adequately because of a buildup in unfolded, misfolded, and/or unfinished proteins is termed ER stress. In order to avoid this condition, cells sense buildups in malformed proteins and make programmatic adjustments intended to return the ER to normal operating conditions. If these protective responses, including autophagy, all fail, that is, if the stresses become chronic, the cells undergo apoptosis.

In order to restore the ER to normal operations signals are sent to both the transcription and translation machinery. These actions not only stimulate increases in the production of chaperones to deal with an increased load, but also slow protein synthesis thereby reducing the growing accumulation of unfolded proteins in the ER. In more detail: (1) Signals are sent that increase expression of genes that remove un- and misfolded proteins either to the UPS or by means of autophagy. (2) Other signals increase the expression of genes that alleviate the stresses (e.g., those encoding chaperones and antioxidants). (3) Still other signals tune down the translational machinery thereby reducing the folding burden placed on the organelle, and lastly, (4) if the stresses are severe and persistent so that protein homeostasis cannot be restored, signals trigger apoptosis leading to destruction of the overburdened cell.

The abovementioned recovery measures are orchestrated beautifully by the three sensors—inositol requiring enzyme 1 (IRE1), PKR-like ER kinase (PERK), and activating transcription factor 6 (ATF6). In the absence of ER stresses, these proteins are inactive residents of the ER membrane. Once they sense excessive stresses, either directly or indirectly, they coordinately mediate the unfolded protein response. The signaling pathways activated by these sensors ER are illustrated in Fig. 6.9. A brief summary of what each one does is, as follows:

- IRE1: This protein kinase and endoribonuclease is activated either directly or indirectly by the stressors, in the former case through direct binding and in the latter case through release and relocation of a blocking chaperone such as BiP. In response to these activating events, IRE1 undergoes dimerization and autophosphorylation that creates docking sites. These sites then recruit signaling proteins and initiate multiple downstream actions. These include reduction of the protein folding burden through mRNA degradation and splicing/activation of XBP-1s (X-box protein 1 splice factor), a transcription factor that upregulates chaperones, lipid synthesis, ERAD proteins, and factors needed for autophagy.

- PERK: Upon activation by stressors this protein kinase, like IRE1, dimerizes and autophosphorylates. It subsequently phosphorylates eukaryotic translation initiation factor eIF2α thus reducing the protein burden by tuning down most protein synthesis while activating ATF4, a transcription factor that upregulates expression of chaperones, redox enzymes, apoptosis-promoting transcription factors such as CHOP (C/EBP homology protein), and factors required for autophagy.
- ATF6: This sensor contains a transcription factor cytosolic domain. Under normal, unstressed conditions it resides in the ER. Stressors induce its translocation to the Golgi apparatus where it is processed by S1P and S2P (site 1 and site 2 proteases) resulting in release of a transcription-competent fragment ATF6f. This transcription factor upregulates ER chaperones that assist in folding such as BiP and Grp94, a Hsp90 family member, along with ERAD components and XBP1s.

Cells take multiple actions to alleviate ER stress. In response, they reroute nascent secretory and membrane proteins from the ER translocon to the cytosol where they are degraded. (The ER translocon is a multiprotein complex built upon the Sec61 channel protein that transports polypeptides into the ER lumen and integrates them into membranes.) This protective process, referred to as *co-translocational degradation*, is a pre-emptive form of protein quality control that

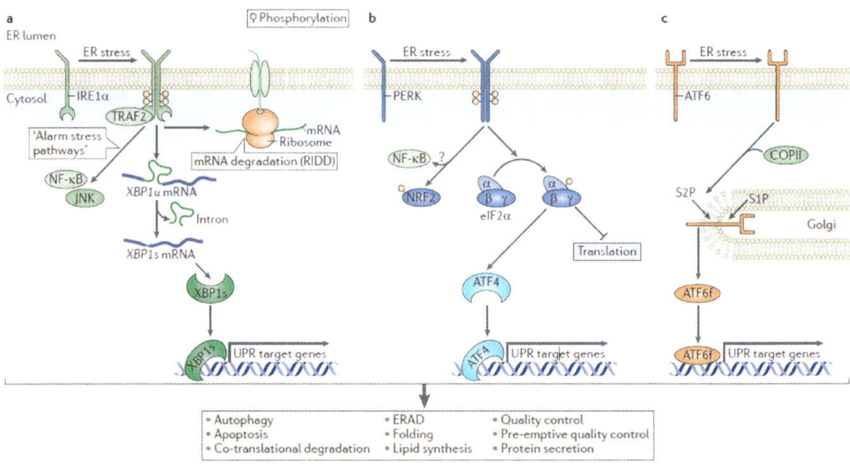

Fig. 6.9 The unfolded protein response. (**a**) The IRE1 pathway. (**b**) The PERK pathway. (**c**) The ATF6 pathway. See text for details (from Hetz *Nat. Rev. Mol. Cell Biol.* 13: 89 © 2012 Reprinted by permission from Macmillan Publishers Ltd)

reduces the ER burden. Proteins that require finishing in the ER contain hydrophobic localization signals recognized by the translocon. *Mislocalization* of these polypeptides to the cytosol on the order of 10–20 % can occur as a result of mutations in their targeting sequences. Aggregation-prone mislocalized proteins are handled by a pre-insertional protein quality control pathway that recognizes and directs them to the proteasome for degradation. Lastly, protein quality control is carried out independently at the inner nuclear membrane (INM). [The outer nuclear membrane is contiguous with the ER.] An ubiquitin ligase complex called Asi (aminoacid signaling independent) has been found in yeast to mediates degradation of INM mislocalized membrane proteins.

Protein quality control is carried out in mitochondria. PQC in mitochondria is a challenging endeavor. Mitochondria not only have a complex architecture, but also continually generate protein- and DNA-damaging reactive oxygen species, and must assemble an electron transport chain whose components are encoded by both nuclear and mitochondrial genes. In these organelles, stresses may be generated within the matrix and within the intermembrane spaces. Like the ER, mitochondria possess a dedicated ensemble of chaperones, proteases, and accessory factors that maintain protein homeostasis. Many of these are upregulated by the translocation of signaling agents from the mitochondria to the nucleus resulting in, for example, upregulation of CHOP, which in associated with another factor, C/EBPβ, that stimulates the transcription of the chaperonin Hsp60 and the protease ClpP.

6.7 Aging and Caloric Restriction

The likelihood of acquiring a cancer or type 2 diabetes or a neurodegenerative disorder rapidly increases late in life. Aging is, in fact, the number one risk factor for coming down with a neurodegenerative disorder. Aging itself is a mystery. As George C. Williams (1926–2010) noted in 1957:

> It is remarkable that after a seemingly miraculous feat of morphogenesis a complex metazoan should be unable to perform the much simpler task of merely maintaining what is already formed.

That is not to say that no one has thought about the subject. On the contrary a great amount of thought has been given to aging and a number of insightful observations have been made. One of the most notable of these was made by Peter Medawar (1915–1987). In his 1951 lecture published in 1952, he introduced the notion of *mutation accumulation*. According to Medawar late acting germline gene mutations may accumulate during aging. These genes are not negatively selected for since they only become active after decline in reproduction has set in. A few years later, Williams introduced the concept of *antagonistic pleiotropy* in which pleiotropic genes exert their positive effects early in life and exhibit deleterious effects later in life. Timing of these opposing effects is crucial as the positive ones, but not the negative ones, occur during reproductive years where natural selection is operative.

At a more mechanistic level, Thomas Kirkwood proposed the *disposable soma theory* in 1977. In his viewpoint, the body partitions the amount of available energy (in the form of ATP) between support of reproduction (preservation of its germline) and repair and maintenance of its soma (components not involved in reproduction). According to Kirkwood aging is a consequence of the random accumulation of unrepaired damage brought on by built-in limitations in repair and maintenance. Earlier, in 1961, Hayflick and Moorehead posited that there exists a cellular factor whose loss through successive cellular divisions limits the replicative capacity of those cells. The result is *cellular senescence*, an irreversible growth arrest. Cellular senescence has possible contributions not only from telomere shortening but also from the accumulation of DNA damage, and activation of senescence-promoting pathways.

Today, *caloric restriction* is arguably the most firmly established theme in the field, giving rise to a useful experimental protocol, and pointing to possible contributing factors and underlying mechanisms. It dates back to a landmark paper by McCay, Crowell and Maynard that appeared in 1935. In it, they reported that rats fed a diet restricted in its caloric content lived longer than rats fed a standard diet. In the intervening years, evidence mostly positive for this phenomenon has been acquired in organisms ranging from yeast to worms to flies to primates. The mechanism underlying this effect is not, as the name might suggest, a consequence of dieting, *per se*. Instead, caloric restriction activates a set of evolutionarily ancient (starvation) stress responses and it is the tuning up of these responses that are responsible for the observed increase in longevity in the laboratory test subjects. Central to these responses are the upregulation of genes encoding molecules chaperones and autophagy.

6.8 Signaling Pathways that Regulate CR and Induce Autophagy

One of the most significant set of findings into the mechanisms underlying CR was the discovery that reduced insulin signaling together with a reduction in plasma glucose levels recapitulates the essential features of the CR phenotype. This finding was quite general, and both reducing insulin growth factor signaling and CR extended lifespan in test animals ranging from fruit flies to rodents. As expected from the very nature of caloric restriction, mechanisms that sense nutrient status would be expected to become active, as well. One of these is cytosolic AMPK, another is the sirtuins. These three nutrient-sensing routes—insulin/insulin-like, AMPK, and sirtuins—execute the CR program.

In more detail, insulin signaling is initiated when insulin produced and released by pancreatic β-cells in response to elevated glucose levels binds an insulin receptor embedded in a target cell membrane. This action, and similar ones in which growth factors bind insulin-like receptors, triggers a cascade of signaling events through a cellular network containing a number of critical nodes (Fig. 6.10). The insulin/insulin receptor substrate (IRS) node is in itself quite important. For example, feedback signals onto the IRS triggered as a result of inflammation and

Fig. 6.10 Schematic depiction of the insulin signaling "backbone." Node 1 (*blue*): Insulin receptor (IR) and insulin receptor substrate (IRS); Node 2 (*green*): Phosphoinositide-3-OH kinase (PI3K)-akt/protein kinase B; Node 3 (*yellow*): Tuberous sclerosis complex (TSC); (*orange*): Ras homolog enriched in brain (Rheb); Node 4 (*purple*): mammalian target of rapamycin (mTOR)/Raptor (TORC1) complex. *Blue arrows*: stimulatory actions; *red flattops*: inhibitory

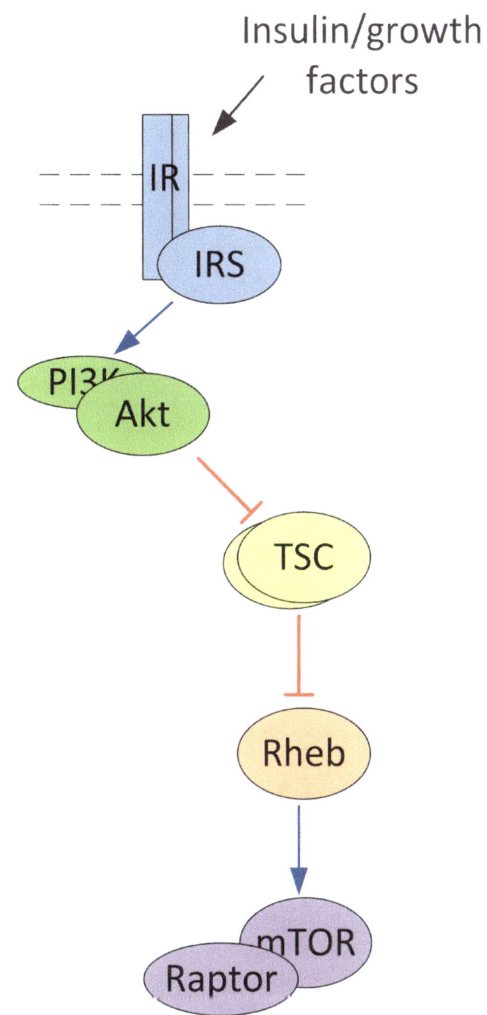

excessive metabolites result in insulin resistance, the hallmark of type 2 diabetes. The serine/threonine kinase Akt phosphorylates a number of substrates. By that means it regulates fatty acid synthesis, glycolysis, glycogen synthesis, and translocation of the glucose transporter, GLUT4. Shown in Fig. 6.10 is its inhibition of output signals from the tuberous sclerosis complex (TSC). The TSC, in turn, inhibits Rheb, an activator of the mTOR cassette. The mTOR cassette coordinates and regulates cellular metabolism and growth in cells receiving insulin and other growth signals. As long as these signals are received and transduced, Akt inhibits TSC and mTOR remains active, promoting growth-related activities such as protein synthesis.

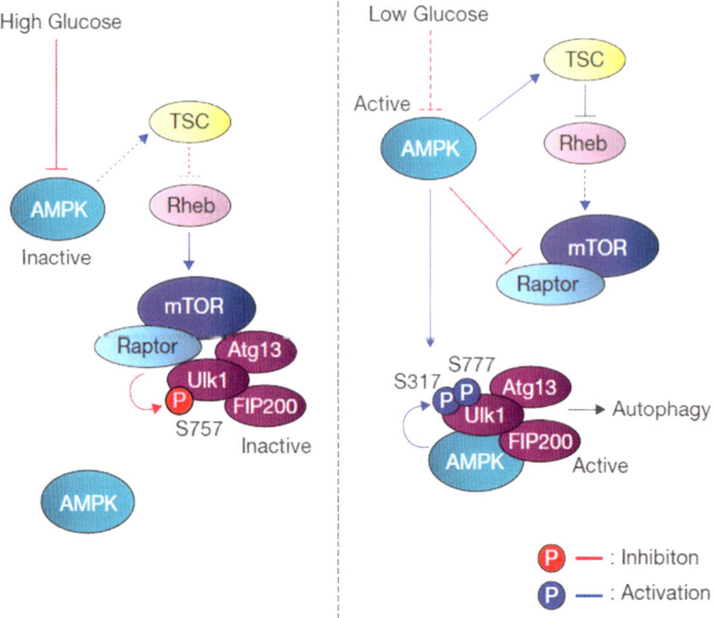

Fig. 6.11 AMPK—mTOR signaling at high and low glucose levels. *Left-hand side* (high glucose): AMPK is inactive, and activation of mTOR keeps the Ulk1 complex bound and inactivated. *Right-hand side* (low glucose): AMPK is activated and phosphorylates the TSC, Raptor, and Ulk1 resulting in dissociation of the Ulk1 complex from the mTOR cassette (from Kim *Nat. Cell Biol.* 13: 132 © 2011 Reprinted by permission from Macmillan Publishers Ltd)

Nodes are central loci where signals from multiple pathways are integrated. Both the TSC and mTOR are nodes that integrate multiple signals. One of the pathways working together with the insulin signaling pathway to provide this information to mTOR is AMP activated protein kinase (AMPK), the master regulator of energy balance in the cell. This signaling molecule is activated when energy supplies are depleted; it restores energy balance by stimulating energy-producing catabolic activities, tuning back energy consuming anabolic processes, and halting growth. This pathway conveys signals to both the TSC and mTOR. As shown in Fig. 6.11 phosphorylation of the TSC by AMPK activates that complex, thereby opposing Akt's inhibitory effects. Signals conveyed to mTOR are also crucial. There, AMPK not only inhibits Raptor but also phosphorylates Ulk1, and the combination of these signaling actions leads to the release of several factors belonging to the Ulk1-complex needed to begin the building of a phagophore (as discussed in Sect. 6.4).

The third sensing route stimulated by CR has as its central element the sirtuins. There are seven mammalian members of this family (Table 6.2), all requiring the metabolite nicotinamide adenine dinucleotide (NAD^+) as a cofactor. Because of their dependence on NAD^+ the sirtuins are sensitive to metabolite status and are activated whenever energy supplies become depleted. Interest in the sirtuins was kindled by

Table 6.2 Mammalian sirtuins, their subcellular locations, and principal enzymatic activities

Sirtuin	Site(s) of action	Catalytic function
SIRT1	Nucleus	Deacetylase
SIRT2	Nucleus, cytoplasm	Deacetylase
SIRT3	Mitochondria	Deacetylase
SIRT4	Mitochondria	ADP-ribotransferase[a]
SIRT5	Mitochondria	Deacetylase
SIRT6	Nucleus	ADP-ribotransferase[a]
SIRT7	Nucleolus	Deacetylase

[a]These enzymes catalyze the transfer of adenosine diphosphate ribose (ADP-ribose) derived from NAD^+ to acceptor proteins while the deacetylases hydrolyze one NAD^+ molecule for each lysine side chain undergoing deacetylation

studies carried out in the late 1990s in the budding yeast *Saccharomyces cerevisiae* that pointed to a possible connection between histone deacetylation by the yeast sirtuin, Sir2p (silent information regulator 2), and lifespan extension. In those studies, Sir2p acting on certain lysine in the NH_2 termini of histones H3 and H4 silenced rDNA chromatin. The resulting extensions in lifespan mimicked the effects of caloric restriction and were abolished by deactivating mutations in the *SIR2* gene. In more detail, Sir2p silencing of rDNA results in reduction in recombination and production of extrachromosomal rDNA circles (self-replicating circles of ribosomal DNA) a known cause of replicative senescence in aging*Saccharomyces cerevisiae* cells.

AMP-activated protein kinase (AMPK) is a key sensor and master regulator of metabolic status in the cell. In response to diminishing energy supplies, AMPK tunes up catabolic ATP-generating processes (e.g., oxidative phosphorylation and mitochondrial biogenesis) and tunes down anabolic ATP-depleting ones. To do so it senses changes in the AMP/ATP ration that functions as the key measure of metabolic status. Similarly, many metabolic enzymes are regulated by the ratio of oxidized to reduced forms of NAD, that is, by the $NAD^+/NADH$ ratio and by that means NAD^+ serves as a second metabolic indicator. Significantly, the two sensors—the sirtuins and AMPK—operate in a single integrated pathway to jointly modify transcription factors (e.g., PGC-1α by deacetylation [SIRT1] and phosphorylation [AMPK]) and proteins in several cellular compartments thereby adjusting energy metabolism.

The possible use of sirtuins in ameliorating protein aggregation and neurodegeneration has and is being explored. Studies carried out in cell culture and in laboratory test subjects are encouraging. For example, there is evidence for the reduction in aggregation brought on through sirtuin-catalyzed deacetylation (i) of transcription factors and coactivators that influence aggregation, (ii) of toxic protein species leading to their increased clearance, and (iii) of Atg5, Atg7, and Atg8 required for autophagosome formation leading to an improved clearance of damaged proteins and organelles, and especially, mitochondria. One of the ways of activating the sirtuins in these studies was through the addition of resveratrol. This natural phenol is produced by a number of plants in response to injury and attack by pathogens, and has been brought to public attention through its presence in red wine. Resveratrol is an example of a *sirtuin-activating compound, or STAC*. A number of these agents have been discovered, many of which are far more potent activators of sirtuins than resveratrol. The mechanism of activation by these compounds is a complex one and requires mutually assisted binding between the STAC and an acetylated substrate containing hydrophobic residues in two critical locations.

6.9 Aging, Breakdown in Protein Quality Control, and Neurodegeneration

Studies in *C. elegans* point to *proteostasisfailure* as a major event accompanying aging. As this organism reached adulthood increasing levels of misfolded proteins are observed along with a loss of function of a broad group of proteins containing missense mutations. These negative changes are accompanied by a steep reduction in the heat shock response and the unfolded protein responses.

These and other studies point to a buildup of protein aggregates with increasing age as a general phenomenon. They may produce a cascading effect in which one kind of misfolded protein ties up other proteins, especially ones that are metastable, in the same aggregate. The maintenance of protein solubility is essential, and for that reason expression levels of aggregation-prone proteins are tightly regulated. Proteins whose cellular concentrations exceed a safe level with regard to solubility are referred to as being *supersaturated*.

The decline in protein quality control with age greatly increases the likelihood of developing a neurodegenerative disorder. Support for this assertion is provided by numerous examples where point mutations in components of the proteostasis network generate a neurodegenerative disorder. Here are some of the best-established examples where a mutated component of the proteostasis network results in a disease situation.

HSPB8 and Bag3: HSPB8 codes for the small heat shock protein Hsp27. Mutations in this gene as well as in HSPB1 which codes for Hsp22 are associated with familial forms of ALS, hereditary distal motor neuropathy (DMN), and Charcot-Marie-Tooth neuropathy type 2L (CMT2L). Recall from the last chapter that sHSPs do not refold proteins; instead they chaperone and maintain them in soluble, folding-competent states. They also work together with other molecular chaperones and co-chaperones to promote the autophagic removal of misfolded proteins. One of its key partners is the co-chaperone Bag3 (Bcl-2-associated athanogene 3). This protein together with Bag1 constitutes a switch in aged individual from proteasomal degradation to autophagy. This switch is thrown when there is a buildup of misfolded/aggregated proteins that the proteasomal system cannot efficiently handle. In these situations, Bag3 working together with Hsp27 and Hsp70 mediates the targeting and transport of the misfolded and aggregated proteins into the aggresomes.

Dynein and dynactin: Dyneins require the presence of a binding partner, the large multisubunit transporter protein dynactin, in order to transport cargo. The dynactin protein binds to the dynein motor protein, to the microtubules, and to the cargo requiring transport. Mutations in the DCTN1 gene encoding the p150 subunit of dynactin lead to loss of motor neurons and the onset of motor neuron disease. Specifically the G59S point mutation has been found to give rise to ubiquitin-positive cytoplasmic inclusions, defects in vesicular trafficking in the cell bodies, and loss of motor neurons. Mutations in the dynein motor protein also generate a buildup of inclusion bodies and indications are that the autophagosome-lysosome fusion stage is impaired. In situations where mutant huntingtin is present, its toxicity is enhanced by the mutations in dynein that prevent its autophagic clearance.

VCP/p97/Cdc48: This is a highly expressed, versatile protein. Perhaps its most prominent role is in extracting targeted proteins from the ER membrane and chaperoning them to the 26S proteasome for disposal. More generally, it provides the force to extract polyubiquitinated proteins from complexes to which they are bound and from membranes within which they are embedded, and in removing damaged mitochondria. This protein also has an important role in autophagy where it helps seal ubiquitin-containing autophagosome membranes. In carrying out its diverse activities, it forms partnerships with different substrate-recruiting cofactors. Mutations in the VCP gene are associated with inclusion body myopathy associated with Paget's disease of bone and frontotemporal dementia (IBMPFD) and ALS.

SQSTM1/p62: Recall that p62 plays a critical role in autophagy by functioning as a cargo receptor that interfaces with ubiquitin-bearing misfolded proteins and with LC3. Mutations in the SQSTR1 gene that encodes p62 are associated with Paget's disease of bone, and with familial and sporadic instances of ALS. P62-positive inclusions (p62-bodies) are observed in tauopathies and in α-synucleinopathies. The protein ordinarily undergoes rapid autophagic removal and buildups of p62 is often used in neurodegenerative disease diagnoses and taken as a sign of autophagic failure.

Parkin and PINK1: Mutations in PINK1 and parkin are responsible for autosomal recessive forms of Parkinson's disease as will be discussed in Chap. 9. These two proteins have prominent roles in mitophagy. The participation of parkin, an E3 ubiquitin ligase involved in autophagic removal of protein aggregates, was noted in Fig. 6.8. A landmark development in understanding PD was the finding that parkin and PINK1 cooperate in a single to mitophagic pathway with PINK1-sensing mitochondrial damage and recruiting parkin to the depolarized and dysfunctional mitochondria.

There are several links between the failure to maintain mitochondrial quality control and neurodegenerative diseases. Mitochondrial fission is mediated by a cytosolic dynamin family member Drp1; fusion is by three GTPases—OPA1, and Mfn1, 2. Mutations in the latter three proteins result in Charcot-Marie-Tooth disease type 2A (CMT2A) and dominant optic atrophy (DOA). Mutations in the former produce deficits in ATP production and impaired synaptic transmission.

6.10 Summary

1. Protein quality control is a major cellular endeavor. It is carried out continuously by an integrated network of molecular chaperones, proteolytic enzymes and chambers, ubiquitin-handling enzymes, dynein/dynactin motor complexes, autophagy proteins, and signal transducers. These systems handle the normal turnover and recycling of proteins and organelles, and prevent excessive build-ups of misfolded proteins and inappropriate assemblages. The two main systems are the ubiquitin-proteasome system and the autophagic-lysosomal system. Molecular chaperones play a major role in orchestrating and managing the UPS. These proteins constitute about 10 % of the protein mass of a cell. They monitor and assist in the folding of nascent polypeptide chains and refold mis-folded proteins (Hsp70), assist in completing the folding of kinetically trapped proteins (Hsp60), prevent their aggregation (sHsp and Hsp100), and maintain marginally stable proteins in soluble "ready-states" (Hsp90). They also direct irrevocably misfolded cytosolic proteins to the proteasome for degradation, and assist in ERAD.

2. The autophagic-lysosomal system is the partner system to the UPS. It handles basal turnover and recycling of cellular components, the degradation of mis-folded proteins and protein aggregates that the UPS is unable to handle because of size and composition limitations, and the breakdown of unwanted and dam-aged organelles. Parts of the ER, the Golgi, and the nucleus, damaged mitochon-dria, and other organelles are degraded by this system.

   This second protein quality control system is far more complex than the UPS. Macroautophagy, the main form, requires formation of autophagosomes that encapsulate misfolded proteins, aggregates, and organelles. A series of proteins complexes acting one after the other orchestrate the various stages in their assembly. These temporary storage compartments are then transported by dynein molecular motors along rail system from the axons and dendrites to the soma, and upon arrival they undergo fusion with lysosomes containing the hydrolytic enzymes. There are two other, simpler, forms of autophagy—microautophagy in which the lysosomes form about the material to be digested and chaperone-mediated autophagy in which chaperones convey substrates one-by-one across lysosomal membranes.

3. The ER and mitochondria require special attention. The former, involved as it is in finishing proteins destined for the plasma membrane and secretion is suscep-tible to buildup in unfolded, misfolded, and unfinished proteins (ER stress). It possesses its own set of chaperones and disposal system, ERAD. If and when that fails, an unfolded protein response is launched by three sensors of ER stress—ATF6, PERK, and IRE1. These stimulate a greater production of molec-ular chaperones, reduce the cellular protein burden, and increase production of proteins needed for autophagy. Mitochondria receive special attention due to their primary role in energy generation, their susceptibility to environmental stresses, and their complex nuclear plus mitochondrial genome regulation.

Mitochondrial autophagy gets is own name—mitophagy, and its own set of autophagic agents.

4. The effectiveness of the protein quality control system declines with age. This decline is accompanied by an increase in the number of intracellular inclusion bodies (IBs). These structures, of which there are many different kinds, are composed of various combinations of misfolded proteins, amyloids and other types of protein aggregates, and damaged organelles. These are more often than not polyubiquitinated. Also present in the IBs are molecular chaperones, E3 ubiquitin ligases, and components of the autophagic network.

   Aging has been and is a subject of great interest. Early ideas on the subject emphasized that genetic mutations may well accumulate with age. There might also be a decline in the electron transport chain's efficiency resulting in an increased production of ROS. That increase along with the presence in the environment of other molecules that can damage DNA and other cellular components are thought to contribute as well. A particularly important observation is that caloric restriction increases longevity in a broad spectrum of test subjects. In examining why this is so, attention is drawn to a set of signaling pathways activated by starvation and CR. These include the insulin/growth factor pathway leading from membrane receptors to the TSC and mTOR cassete, major signaling nodes where signals are integrated. Two other pathways, containing AMPK and sirtuins as the central nutrient and energy sensors are integrated with one another and with the insulin pathway. These pathways coordinately upregulate molecular chaperones and proteins needed for autophagy. These data point to the strengthening of the weakened proteins quality control system as being responsible for the observed increases in longevity.

5. That one or both of the proteostasis systems are either overwhelmed or are malfunctioning in neurodegeneration is supported by multiple lines of evidence. For example, a decline in the proteostasis network is observed in *C. elegans*. One of the most revealing pieces of evidence is the growing series of discoveries that mutations in key members of this system either directly cause neurodegenerative disorders or are a major risk factor for them. Prominent examples are mutated dyneins and dynactins, the Hsp22 and Hsp27 small heat shock proteins, the p97/VCP AAA-ATPase and molecular chaperone, the p62 cargo protein, and the PINK1 and parkin mitophagy effectors. These connections, only briefly touched upon in this chapter, will be explored in some detail in the disease-specific chapters to follow.

Bibliography

Deter, R. L., & de Duve, C. (1967). Influence of glucagon, an inducer of cellular autophagy, on some physical properties of rat liver lysosomes. *The Journal of Cell Biology, 33*, 437–449.

Droz, B., & Leblond, C. P. (1962). Migration of proteins along the axons of the sciatic nerve. *Science, 137*, 1047–1048.

Grafstein, B., & Foreman, D. S. (1980). Intracellular transport in neurons. *Physiological Reviews, 60*, 1167–1283.

Tytell, M., Black, M. M., Garner, J. A., & Lasek, R. J. (1981). Axonal transport: Each major rate component reflects the movement of distinct macromolecular complexes. *Science, 214*, 179–181.

Weiss, P., & Hiscoe, H. B. (1948). Experiments in the mechanism of nerve growth. *The Journal of Experimental Zoology, 107*, 315–395.

Motor Proteins: Discovery and Structure

Brady, S. T. (1985). A novel brain ATPase with properties expected for the fast axonal transport motor. *Nature, 317*, 73–75.

Carter, A. P., Cho, C., Jin, L., & Vale, R. D. (2011). Crystal structure of the dynein motor domain. *Science, 331*, 1159–1165.

Gibbons, I. R., & Rowe, A. J. (1965). Dynein: A protein with adenosine triphosphate activity from cilia. *Science, 149*, 424–426.

Huxley, A. F., & Simmons, R. M. (1971). Proposed mechanism of force generation in striated muscle. *Nature, 233*, 533–538.

Kull, F. J. (1996). Crystal structure of the kinesin motor domain reveals a structural similarity to myosin. *Nature, 380*, 550–555.

Pollard, T. D., & Korn, E. D. (1973). *Acanthamoeba* myosin II. Interaction with actin and with a new cofactor protein required for actin activation of Mg^{2+} adenosine triphosphate activity. *The Journal of Biological Chemistry, 248*, 4691–4697.

Rayment, I., Rypniewski, W. R., Schmidt-Base, K., Smith, R., Tomchick, D. R., Benning, M. M., et al. (1993). Three-dimensional structure of myosin subfragment-1: A molecular motor. *Science, 261*, 50–58.

Szent-Györgyi, A. (1945). Studies on muscle. *Acta Physiologica Scandinavica, 9*(Suppl. 25), 1–115.

Vale, R. D., Reese, T. S., & Sheetz, M. P. (1985). Identification of a novel force-generating protein, kinesin, involved in microtubule-based motility. *Cell, 42*, 455–462.

Motor Proteins: Nature's Tiniest Bipeds

Finer, J. T., Simmons, R. M., & Spudich, J. A. (1994). Single myosin molecule mechanics-piconewton forces and nanometer steps. *Nature, 368*, 113–119.

Funatsu, T., Harada, Y., Tokunaga, M., Saito, K., & Yanagida, T. (1995). Imaging of single fluorescent molecules and individual ATP turnovers by single myosin molecules in aqueous solution. *Nature, 374*, 555–559.

Hirokawa, N., Noda, Y., Tanaka, Y., & Niwa, S. (2009). Kinesin superfamily motor proteins and intracellular transport. *Nature Reviews. Molecular Cell Biology, 10*, 682–696.

Reck-Peterson, S. L., Yildiz, A., Carter, A. P., Gennerich, A., Zhang, N., & Vale, R. D. (2006). Single-molecule analysis of dynein processive and stepping behavior. *Cell, 126*, 335–348.

Spudich, J. A., & Sivaramakrishnan, S. (2010). Myosin VI: An innovative motor that challenged the swinging lever arm hypothesis. *Nature Reviews. Molecular Cell Biology, 11*, 128–137.

Svoboda, R., Schmidt, C. F., Schnapp, B. J., & Block, S. M. (1993). Direct observation of kinesin stepping by optical trapping interferometry. *Nature, 365*, 721–727.

Veigel, C., & Schmidt, C. F. (2011). Moving into the cell: Single-molecule studies of molecular motors in complex environments. *Nature Reviews. Molecular Cell Biology, 12*, 163–176.

Yildiz, A., Forkey, J. N., McKinney, S. A., Ha, T., Goldman, Y. E., & Selvin, P. R. (2003). Myosin V walks hand-over-hand: Single fluoropore imaging with 1.5-nm localization. *Science, 300,* 2061–2065.

Zala, D., Hinckelmann, M. V., Yu, H., Lyra da Cunha, M. M., Liot, G., Cordelière, F. P., et al. (2013). Vesicular glycolysis provides on-board energy for fast axonal transport. *Cell, 152,* 479–491.

Autophagy

Axe, E. L., Walker, S. A., Manifava, M., Chandra, P., Roderick, H. L., Habermann, A., et al. (2008). Autophagosome formation from membrane compartments enriched in phosphatidylinositol 3-phosphate and dynamically connected to the endoplasmic reticulum. *The Journal of Cell Biology, 182,* 685–701.

Fujita, N., Itoh, T., Omori, H., Fukuda, M., Noda, T., & Yoshimori, T. (2008). The Atg16L complex specifies the site of LC3 lipidation for membrane biogenesis in autophagy. *Molecular Biology of the Cell, 19,* 2092–2100.

Hailey, D. W., Rambold, A. S., Satpute-Krishnan, R., Mitra, K., Sougrat, R., Kim, P. K., et al. (2010). Mitochondria supply membranes for autophagosome biogenesis during starvation. *Cell, 141,* 656–667.

Ichimura, Y., Kirisako, T., Takao, T., Satomi, Y., Shimonishi, Y., Ishihara, N., et al. (2000). A ubiquitin-like system mediates protein lipidation. *Nature, 408,* 488–492.

Johansen, T., & Lamark, T. (2011). Selective autophagy mediated by autophagic adapter proteins. *Autophagy, 7,* 279–296.

Johnston, J. A., Ward, C. L., & Kopito, R. R. (1998). Aggresomes: A cellular response to misfolded proteins. *The Journal of Cell Biology, 143,* 1883–1898.

Kabeya, Y., Mizushima, N., Ueno, T., Yamamoto, A., Kirisako, T., Noda, T., et al. (2000). LC3, a mammalian homologue of yeast Apg8p, is localized in autophagosome membranes after processing. *The EMBO Journal, 19,* 5720–5728.

Kang, R., Zeh, H. J., Lotze, M. T., & Tang, T. (2011). The Beclin 1 network regulates autophagy and apoptosis. *Cell Death and Differentiation, 18,* 571–580.

Kirkin, V., McEwan, D. G., Novak, I., & Dikic, I. (2009). A role for ubiquitin in selective autophagy. *Molecular Cell, 34,* 259–269.

Komatsu, M., Waguri, S., Koike, M., Sou, Y. S., Ueno, T., Hara, T., et al. (2007). Homeostatic levels of p62 control cytoplasmic inclusion body formation in autophagy-deficient mice. *Cell, 131,* 1149–1163.

Kroemer, G., Mariño, G., & Levine, B. (2010). Autophagy and the integrated stress response. *Molecular Cell, 40,* 280–293.

Levine, B., & Klionsky, D. J. (2004). Development of self-digestion: Molecular mechanisms and biological functions of autophagy. *Developmental Cell, 6,* 463–477.

Mizushima, N., & Komatsu, M. (2011). Autophagy: Renovation of cells and tissues. *Cell, 147,* 728–741.

Mizushima, N., Noda, T., Yoshimori, T., Tanaka, Y., Ishii, T., George, M. D., et al. (1998). A protein conjugation system essential for autophagy. *Nature, 395,* 395–398.

Nishida, Y., Arakawa, S., Fujitani, K., Yamaguchi, H., Mizuta, T., Kanaseki, T., et al. (2009). Discovery of Atg5/Atg7-independent alternative macroautophagy. *Nature, 461,* 654–658.

Ravikumar, B., Moreau, K., Jahreiss, L., Puri, C., & Rubinzstein, D. C. (2010). Plasma membrane contributes to the formation of pre-autophagosomal structures. *Nature Cell Biology, 12,* 747–757.

Rogov, V., Dötsch, V., Johansen, T., & Kirkin, V. (2014). Interactions between autophagy receptors and ubiquitin-like proteins for the molecular basis for selective autophagy. *Molecular Cell, 53,* 167–178.

Tsukada, M., & Ohsumi, Y. (1993). Isolation and characterization of autophagy-defective mutants of *Saccharomyces cerevisiae*. *FEBS Letters, 333,* 169–174.

Mitochondria Dynamics and Mitophagy

Chen, H., Vermulst, M., & Chan, D. C. (2010). Mitochondrial fusion is required for mtDNA stability in skeletal muscle and tolerance of mtDNA mutations. *Cell, 141*, 280–289.

Li, Z., Okamoto, K., Hayashi, Y., & Sheng, M. (2004). The importance of dendritic mitochondria in the morphogenesis and plasticity of spines and synapses. *Cell, 119*, 873–887.

López-Lluch, G., Irusta, P. M., Navas, P., & de Cabo, R. (2008a). Mitochondrial biogenesis and healthy aging. *Experimental Gerontology, 43*, 813–819.

López-Lluch, G., Irusta, P. M., Navas, P., & de Cabo, R. (2008b). Mitochondrial biogenesis and healthy aging. *Experimental Gerontology, 43*, 813–819.

Navarro, A., & Boveris, A. (2007a). The mitochondrial energy transduction system and the aging process. *The American Journal of Physiology, 292*, C670–C686.

Navarro, A., & Boveris, A. (2007b). The mitochondrial energy transduction system and the aging process. *The American Journal of Physiology, 292*, C670–C686.

Palay, S. L. (1956). Synapses in the central nervous system. *The Journal of Biophysical and Biochemical Cytology, 2*, 193–202.

Szabadkai, G., & Duchen, M. R. (2008). Mitochondria: The hub of cellular Ca^{2+} signaling. *Physiology (Bethesda), 23*, 84–94.

Twig, G., Elorza, A., Molina, A. J. A., Mohamed, H., Wikstrom, J. D., et al. (2008). Fission and selective fusion govern mitochondrial segregation and elimination by autophagy. *The EMBO Journal, 27*, 433–446.

Verstreken, P., Ly, C. V., Venken, J. T., Koh, T. W., Zhou, Y., & Bellen, H. J. (2005). Synaptic mitochondria are critical for mobilization of reserve pool vesicles at *Drosophila* neuromuscular junctions. *Neuron, 47*, 365–378.

Youle, R. J., & Narendra, D. P. (2011). Mechanism of mitophagy. *Nature Reviews. Molecular Cell Biology, 12*, 9–14.

ERAD and the UPR

Hampton, R. Y., & Sommer, T. (2012). Finding the will and the way of ERAD substrate retranslocation. *Current Opinion in Cell Biology, 24*, 460–466.

Hessa, T., Sharma, A., Mariappan, M., Eshleman, H. D., Gutierrez, E., & Hedge, R. S. (2011). Protein targeting and degradation are coupled for elimination of mislocalized proteins. *Nature, 475*, 394–397.

Hetz, C. (2012). The unfolded proteins response: Controlling cell fate decisions under ER stress and beyond. *Nature Reviews. Molecular Cell Biology, 13*, 89–102.

Høyer-Hansen, M., & Jäättelä, M. (2007). Connecting endoplasmic reticulum stress to autophagy by unfolded protein response and calcium. *Cell Death and Differentiation, 14*, 1576–1582.

Kang, S. W., Rane, N. S., Kim, S. J., Garrison, J. L., Taunton, J., & Hedge, R. S. (2006). Substrate-specific translocational attenuation during ER stress defines a pre-emptive quality control pathway. *Cell, 127*, 999–1013.

Matus, S., Lisbona, F., Torres, M., León, C., Thielen, P., & Hetz, C. (2008). The stress rheostat: An interplay between the unfolded protein response (UPR) and autophagy in neurodegeneration. *Current Molecular Medicine, 8*, 157–172.

Oyadomari, S., Yun, C., Fisher, E. A., Kreglinger, N., Kreibich, G., Oyadomari, M., et al. (2006). Cotranslocational degradation protects the stressed endoplasmic reticulum from protein overload. *Cell, 126*, 727–739.

Pellegrino, M. W., Nargund, A. M., & Haynes, C. M. (2013). Signaling the mitochondrial unfolded proteins response. *Biochimica et Biophysica Acta, 1833*, 410–416.

Smith, M. H., Ploegh, H. L., & Weissman, J. S. (2011). Road to ruin: Targeting proteins for degradation in the endoplasmic reticulum. *Science, 334*, 1086–1090.

Walter, P., & Ron, D. (2011). The unfolded protein response: From stress pathway to homeostatic regulation. *Science, 334*, 1081–1086.

Aging and Caloric Restriction

Hayflick, L., & Moorehead, P. S. (1961). The serial cultivation of human diploid cell strains. *Experimental Cell Research, 25*, 585–621.

Kirkwood, T. B. L. (1977). Evolution of ageing. *Nature, 270*, 301–304.

McCay, C. M., Crowell, M. F., & Maynard, L. A. (1935). The effect of retarded growth upon the length of life span and upon the ultimate body size. *The Journal of Nutrition, 10*, 63–79.

Medawar, P. B. (1952). *An unsolved problem of biology*. London: HK Lewis and Co.

Mortimore, G. E., & Pösö, A. R. (1987). Intracellular protein catabolism and its control during nutrient deprivation and supply. *Annual Review of Nutrition, 7*, 539–564.

Weindruch, R., Walford, R. L., Fligiel, S., & Guthrie, D. (1986). The retardation of aging in mice by dietary restriction: Longevity, cancer, immunity and lifetime energy intake. *The Journal of Nutrition, 116*, 641–654.

Williams, G. C. (1957). Pleiotropy, natural selection and the evolution of senescence. *Evolution, 11*, 398–411.

Signaling Pathways Activated in Response to CR

Brown-Borg, H. M., Borg, K. E., Meliska, C. J., & Bartke, A. (1996). Dwarf mice and the ageing process. *Nature, 384*, 33.

Donmez, G., Arun, A., Chung, C. Y., McLean, P. J., Lindquist, S., & Guarente, L. (2012). SIRT1 protects against α-synuclein aggregation by activating molecular chaperones. *The Journal of Neuroscience, 32*, 124–132.

Haigis, M. C., & Yankner, B. A. (2010). The aging stress response. *Molecular Cell, 40*, 333–344.

Kimura, K. D., Tissenbaum, H. A., Liu, Y., & Ruvkun, G. (1997). Daf-2, an insulin receptor-like gene that regulates longevity and diapause in *Caenorhabditis elegans*. *Science, 277*, 942–946.

Price, N. L., Gomes, A. P., Ling, A. J. Y., Duarte, F. V., Martin-Montalvo, A., North, B. J., et al. (2012). SIRT1 is required for AMPK activation and the beneficial effects of resveratrol on mitochondrial function. *Cell Metabolism, 15*, 675–690.

Ruderman, N. B., Xu, X. J., Nelson, L., Cacicedo, J. M., Saha, A. K., Lan, F., et al. (2010). AMPK and SIRT1: A long-standing partnership ? *American Journal of Physiology, Endocrinology and Metabolism, 298*, E751–E760.

Salminen, A., & Kaarniranta, K. (2012). AMP-activated protein kinase (AMPK) controls the aging process via an integrated signaling network. *Ageing Research Reviews, 11*, 230–241.

Suirtins

Cantó, C., Gerhart-Hines, Z., Feige, J. N., Lagouge, M., Noriega, L., Milne, J. C., et al. (2009). AMPK regulates energy expenditure by modulating NAD metabolism and SIRT1 activity. *Nature, 458*, 1056–1060.

Fulco, M., Cen, Y., Zhao, P., Hoffman, E. P., McBurney, M. W., Sauve, A. A., et al. (2008). Glucose restriction inhibits skeletal myoblast differentiation by activating SIRT1 through AMPK-mediated regulation of Nampt. *Developmental Cell, 14*, 661–673.

Gomes, A. P., Price, N. L., Ling, A. J. Y., Moslehi, J. J., Montgomery, M. K., Rajman, L., et al. (2013). Declining NAD induces a pseudohypoxic state disrupting nuclear-mitochondrial communication during aging. *Cell, 155*, 1624–1638.

Herskovits, A. Z., & Guarente, L. (2013). Sirtuin deacetylases in neurodegenerative diseases of aging. *Cell Research, 23*, 746–758.

Howitz, K. T., Bitterman, K. J., Cohen, H. Y., Lamming, D. W., Lavu, S., Wood, J. G., et al. (2003). Small molecular activators of sirtuins extend Saccharomyces cerevisiae lifespan. *Nature, 425*, 191–196.

Hubbard, B. P., Gomes, A. P., Dai, H., Li, J., Case, A. W., Considine, T., et al. (2013). Evidence for a common mechanism of SIRT1 regulation by allosteric activators. *Science, 339*, 1216–1219.

Lee, I. H., Cao, L., Mostoslavsky, R., Lombard, D. B., Liu, J., Bruns, N. E., et al. (2008). A role for the NAD-dependent deacetylase Sirt1 in the regulation of autophagy. *Proceedings of the National Academy of Sciences of the United States of America, 105*, 3374–3379.

Lin, S. J., Defossez, P. A., & Guarente, L. (2000). Requirement of NAD and *SIR2* for life-span extension by caloric restriction in *Saccharomyces cerevisiae*. *Science, 289*, 2126–2128.

Milne, J., Lambert, P. D., Schenk, S., Carney, D. P., Smith, J. J., Gagne, D. J., et al. (2007). Small molecular activators of SIRT1 as therapeutics for the treatment of type 2 diabetes. *Nature, 450*, 712–716.

Autophagy in Neurodegeneration

Bjørkøy, G., Lamark, T., Brech, A., Outzen, H., Perander, M., Øvervatn, A., et al. (2005a). p62/SQSTM1 forms protein aggregates degraded by autophagy and has a protective effect on huntingtin-induced cell death. *The Journal of Cell Biology, 171*, 603–614.

Hara, T., Nakamura, K., Matsui, M., Yamamoto, A., Nakahara, Y., Suzuki-Migishima, R., et al. (2006). Suppression of basal autophagy in neural cells causes neurodegenerative disease in mice. *Nature, 441*, 885–889.

Komatsu, M., Waguri, S., Chiba, T., Murata, S., Iwata, J. I., Tanida, I., et al. (2006). Loss of autophagy in the central nervous system causes neurodegeneration in mice. *Nature, 441*, 880–884.

Ravikumar, B., Duden, R., & Rubinsztein, D. C. (2002). Aggregate-prone proteins with polyglutamine and polyalanine expansions are degraded by autophagy. *Human Molecular Genetics, 11*, 1107–1117.

Autophagy in Lifespan Extension

Meléndez, A., Tallóczy, Z., Seaman, M., Eskelinen, E. L., Hall, D. H., & Levine, B. (2003). Autophagy genes are essential for dauer development and life-span extension in *C. elegans*. *Science, 301*, 1387–1391.

Pyo, J. O., Yoo, S. M., Ahn, H. H., Nak, J., Hong, S. H., Kam, T. I., et al. (2013). Overexpression of Atg5 in mice activates autophagy and extends lifespan. *Nature Communications 4*, Art. No. 2300.

Rubinsztein, D. C., Mariño, G., & Kroemer, G. (2011). Autophagy and aging. *Cell, 146*, 682–695.

Salminen, A., & Kaarniranta, K. (2009). Regulation of the aging process by autophagy. *Trends in Molecular Medicine, 15*, 217–224.

Aging, Breakdown in Protein Quality Control, and Neurodegeneration

Ben-Zvi, A., Miller, E. A., & Morimoto, R. I. (2008). Collapse of proteostasis represents an early molecular event in *Caenorhabditis elegans* aging. *Proceedings of the National Academy of Sciences of the United States of America, 106*, 14914–14919.

Ciryam, P., Tartaglia, G. G., Morimoto, R. I., Dobson, C. M., & Vendruscolo, M. (2013). Widespread aggregation and neurodegenerative diseases are associated with supersaturated proteins. *Cell Reports, 5*, 1–10.

Crippa, V., Sau, V., Rusmini, P., Boncoraglio, A., Onesto, E., Bolzoni, E., et al. (2010). The small heat shock protein B8 (HspB8) promotes autophagic removal od misfolded proteins involved in amyotrophic lateral sclerosis (ALS). *Human Molecular Genetics, 19*, 3440–3456.

David, D. C., Ollikainen, N., Trinidad, J. C., Cary, M. P., Burlingame, A. L., & Kenyon, C. (2010). Widespread protein aggregation as an inherent part of aging in *C. elegans. PLoS Biology, 8*, 1000450.

Gamerdinger, M., Hajieva, P., Kaya, A. M., Wolfrum, U., Hartl, F. U., & Behl, C. (2009). Protein quality control during aging involves recruitment of the macroautophagy pathway by BAG3. *The EMBO Journal, 28*, 889–901.

Geisler, S., Holmström, K. M., Skujat, D., Fiesel, F. C., Rothfuss, O. C., Kahle, P. J., et al. (2010). PINK1/Parkin-mediated mitophagy is dependent on VDAC1 and p62/SQSTM1. *Nature Cell Biology, 12*, 119–131.

Hafezparast, M., Klocke, R., Ruhrberg, C., Marquardt, A., Ahmad-Annuar, A., Bowen, S., et al. (2003). Mutations in dynein link motor neuron degeneration to defects in retrograde transport. *Science, 300*, 808–812.

Johnson, J. O., Mandrioli, J., Benatar, M., Abramzon, Y., Van Deerlin, V. M., Trojanowski, J. Q., et al. (2010a). Exome sequencing reveals VCP mutations as a case of familial ALS. *Neuron, 68*, 857–864.

Komatsu, M., & Ichimura, Y. (2010). Physiological significance of selective degradation of p62 by autophagy. *FEBS Letters, 584*, 1374–1378.

Laird, F. M. (2008). Motor neuron disease occurring in a mutant dynactin mouse model is characterized by defects in vesicular trafficking. *The Journal of Neuroscience, 28*, 1997–2005.

Meyer, H., Bug, M., & Bremer, S. (2012). Emerging functions of the VCP/p97 AAA-ATPase in the ubiquitin system. *Nature Cell Biology, 14*, 117–123.

Narendra, D. P., Jin, S. M., Tanaka, A., Suen, D. F., Gautier, C. A., Shen, J., et al. (2010). PINK1 is selectively stabilized on impaired mitochondria to activate Parkin. *PLoS Biology, 8*, e1000298.

Olzscha, H., Schermann, S. M., Woerner, A. C., Pinkert, S., Hecht, M. H., Tartaglia, G. G., et al. (2010b). Amyloid-like aggregation sequester numerous metastable proteins with essential cellular functions. *Cell, 144*, 67–78.

Ravikumar, B., Acevedo-Arozena, A., Imarisio, S., Berger, Z., Vacher, C., O'Kane, C. J., et al. (2005). Dynein mutations impair autophagic clearance of aggregate-prone proteins. *Nature Genetics, 37*, 771–776.

Rubino, E., Rainero, I., Chiò, A., Rogaeva, E., Galimberti, D., Fenoglio, P., et al. (2012). SQSTM1 mutations in frontotemporal lobar degeneration and amyotrophic lateral sclerosis. *Neurology, 79*, 1556–1562.

Tresse, E., Salomons, F. A., Vesa, J., Bott, L. C., Kimonis, V., Yao, T. P., et al. (2010). VCP/p97 is essential for maturation of ubiquitin-containing autophagosomes and this function is impaired in mutations that case IBMPFD. *Autophagy, 6*, 217–227.

Vives-Bauza, C., Zhou, C., Huang, Y., Cui, M., de Vries, R. L. A., Kim, J., et al. (2010a). PINK1-dependent recruitment of Parkin to mitochondria in mitophagy. *Proceedings of the National Academy of Sciences of the United States of America, 107*, 378–383.

Chapter 7
Prion Diseases

Recall from Chap. 1 that it took a visit to Papua New Guinea in the mid-1950s. At that time a rare but fatal disease, kuru, had broken out among the primitive *Fore* people. In response to what was clearly an unusual happening, the epidemiologist Carleton Gajdusek (1923–2008) traveled to the region at the invitation of the local medical officer, Vincent Zigas. Once he was in the region Gajdusek discovered that the disease was being transmitted by means of funerary cannibalistic feasts involving ingestion of brain tissue.

In the meantime, in the UK and on the Continent, scrapie outbreaks affecting sheep and goats had been occurring with some regularity. This particular illness has been known for at least 250 years, and was given the descriptive name "scrapie" because of the propensity of affected animals to rub against fences in order to remain upright. It been studied with varying results and with no clear idea as to its nature except it was observed to have a genetic component. Early efforts to establish that it was transmissible failed due to the use of too-short incubation periods. A major breakthrough occurred in the late 1930s when Cuille and Chelle finally established its transmissibility using inoculations of brain and spinal cord tissue and utilizing incubation periods of up 2 years. They reported their results in a series of publications that appeared from 1936 to 1939, and hypothesized that a virus might be the causative agent.

Those finding took on added significance in 1959 when William Hadlow noted the similarity between scrapie and kuru. Those observations led to a landmark experiment in 1966 by Gadjusek in which kuru-infected brain extracts taken from humans were shown to induce the disease in chimpanzees. Similarities of these diseases to another even more obscure disease, Creutzfeld–Jakob disease (CJD), were noted and from these studies there arose a new concept—that of a transmissible dementia—a remarkable discovery for which the Nobel Prize in Physiology and Medicine was awarded to Gajdusek and again later to Stanley Prusiner.

So what is the causative agent of the transmissible spongiform encephalopathies (TSEs) as they are known? That question was addressed by Alper in a series of experiments in which he subjected the isolated scrapie agent to ionizing radiation

© Springer International Publishing Switzerland 2015

191

M. Beckerman, *Fundamentals of Neurodegeneration and Protein Misfolding Disorders*,
Biological and Medical Physics, Biomedical Engineering,
DOI 10.1007/978-3-319-22117-5_7

and UV in order to estimate its size. The result was another surprise—the agent appeared to be too small to be a virus or to involve a nucleic acid. This left proteins acting without any nucleic acid carrier of genetic information as the most likely agent. That study led to a theoretical investigation in 1967 by Griffith who proposed three possible scenarios in which a conformational change from a normal protein to a disease-causing form might take place. That is, he postulated that the disease might be caused by a self-replicating protein. Given the striking nature of that concept it is perhaps not surprising that it was simply rejected by the vast majority of the researchers in the field.

Progress on resolving the conundrum of the causal agent(s) was made by Stanley Prusiner with the development of the means of preparing greatly purified samples of the scrapie agent. Analysis of the resulting material further reduced the potential molecular mass of the molecules and made it less likely that any genetic material was present in the samples. The resulting proteinaceous, rod-like material was given the name "prion" in 1982 by Prusiner. As explained by him, that term stands *protein-aceous infectious particle*, symbolized as PrP. The scrapie-causing particles, designated as PrP^{Sc}, were shown to be resistant to inactivation by procedures that modify nucleic acids further reducing the possibility that a nucleic acid was involved. That effort was followed a few years later in 1985 with the cloning of the gene that encoded a 33–35 kDa, soluble and protease-sensitive protein, PrP^{C} derived from normal and scrapie-infected brains. That effort was accompanied by yet another study that identified a 27–30 kDa protease-resistant core (amino acid residues 90–231) belonging to the 33–35 kDa disease-causing PrP^{Sc} protein. A year later the genes responsible for the normal PrP^{C} and the disease-causing PrP^{Sc} were shown to be the same.

The goal of this chapter is to explore what has been discovered about this remarkable protein, the diseases it causes, and the revisions in thinking about the fundamental biophysical and biochemical capabilities of proteins. Although it is expressed in multiple tissues in the body, the focus is on its behavior in the brain. The chapter begins with a quick overview of how prions reach the brain upon injection in foodstuffs. That introduction is followed by a review of the panoply of human prion diseases and followed by one of the most remarkable features that characterize the prions—the existence of prions strains encoded not in differences in primary sequence but rather in differences in protein conformation. That phenomenon is still under active investigation with several ideas advanced as to how this can actually occur and how strain differences relate to the various human prion diseases.

The central biophysical feature of the prions diseases is the conversion of the normal, or cellular, form of PrP^{C} to the misfolded, beta-sheet enriched PrP^{Sc} through contact with the latter. It is by now well accepted that the disease-causing form of the protein can replicate itself and, consequently, act as an infectious agent. This process is a robust one that survives transit through the body and from one host to another. The precise mechanism underlying this remarkable conversion, and the possible need for cofactors, is still being explored.

An additional dimension to the prion world appeared in 1994. In that year, Wickner reported the discovery of two fungal prions—[URE3] and [PSI$^+$]—that were altered forms of the Ure2p and Sup35p proteins, respectively. Shortly thereafter a third fungal prion—[Het-s]—was discovered. Rather than causing a disease the fungal prions seem to function as non-chromosomal (epigenetic) units of inheritance. These prions have been of great utility to the research community and many of the insights into prion phenomena have been derived from their study. These too will be examined. This short chapter then concludes with a large thought—cell-to-cell transmission of a disorder by self-templating misfolded proteins may not be limited to PrP; it may well extend to other causal agents of systemic amyloidoses and neurodegenerative disorders.

7.1   Prions Take an Invasion Route from the Gut to the Brain

Full length, normal PrP is a cell-surface glycoprotein attached to the extracytoplasmic surface of neurons and other cell types by means of a C-terminal glycosylphosphatidylinositol (GPI) anchor. One of the first questions that may be asked is how do disease-causing forms of this protein get to the brain from the stomach? This type of passage occurs following ingestion of contaminated brain tissue in the case of kuru and contaminated cattle feed in the case of bovine spongiform encephalopathy (BSE), or mad cow disease, in the recent outbreak in Great Britain. As illustrated in Fig. 7.1 several different kinds of immune cells facilitate the uptake from the intestinal lumen. That initial step is followed by entry into the lymphatic system and from there into the nervous system and on to the brain.

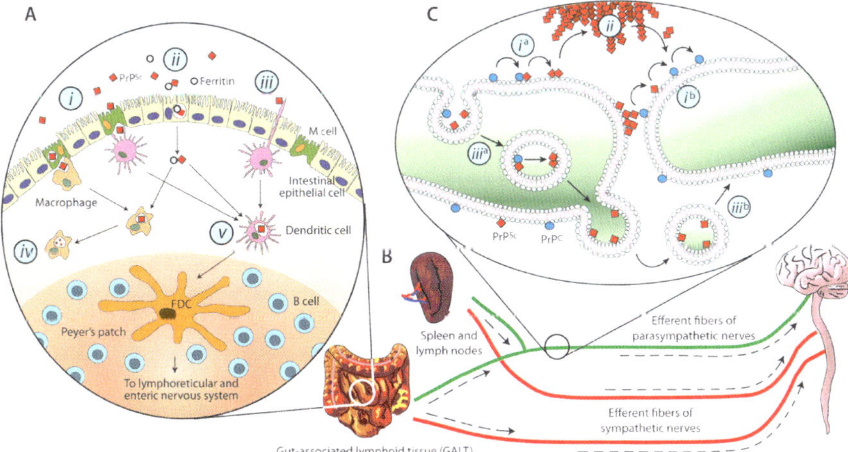

Fig. 7.1 Schematic depictions of the possible invasion routes leading from the gut to the brain taken by prions. (**a**) Prion uptake in the intestinal lumen by various means including: M-cell transcytosis (*i*), intestinal epithelial cell transcytosis (*ii*), or direct capture by dendritic cells (*iii*). In addition, while phagocytic cells such as macrophages may degrade PrP^{Sc} (*iv*) dendritic cells may deliver them to follicular dendritic cells where early accumulation takes place (*v*). (**b**) Amplification in the lymphoid tissues is followed by invasion of the nervous system via peripheral nerves; (**c**) PrP^{Sc} retrograde transport and propagation along neuronal processes either along the cell surface (*i*), through extracellular deposits (*ii*), or by means of vesicle-mediated mechanisms (*iii*) (from Cobb *Biochemistry* 48: 2574 © 2009 American Chemical Society and reprinted with their permission)

Accumulation in the lymphoid tissue (e.g., spleen, lymph nodes, tonsils, appendix, Peyer's patches) is crucial. Several routes from the periphery to the brain are possible all involving passage from lymphoid tissue by means of the efferent fibers of the parasympathetic nerves (e.g., the vagus nerve) or sympathetic nerves (e.g., the splanchnic nerve), which regulate intestinal secretions and motility. Of particular interest are the follicular dendritic cells (FDCs), which possess extremely large surface areas. This property enables them to readily capture prion proteins. FDCs are long-lived cells, and prions can remain there for considerable periods of time, months and years.

7.2   Human Transmissible Spongiform Encephalopathy

The panoply of human prion diseases is presented in Table 7.1. One of the key features of the illnesses is the typical age of onset; another is the incubation period, and still another is the duration of the illness. The age of onset of nvCJD, 29 years, is

Table 7.1 Transmissible spongiform encephalopathy (prion disease) of humans

Human TSE	Clinical signs	Neuropathological signs
Sporadic forms		
Sporadic CJD (sCJD)	Dementia, cerebellar dysfunction	Spongiform changes, gliosis, neuronal loss, infrequent amyloid plaques
Acquired forms		
New variant CJD (nvCJD)	Psychiatric symptoms	Spongiform changes in basal ganglia and thalamus, numerous amyloid plaques surrounded by vacuoles
Kuru	Cerebellar ataxia	Spongiform changes, gliosis, neuronal loss, amyloid plaques
Iatrogenic CJD (iCJD)		
– hGH associated	Gait abnormalities and ataxia	Similar to sCJD but with more numerous amyloid plaques
– Dura associated	Resembles sCJD	Similar to sCJD
Familial forms		
Familial CJD (fCJD)		
– Mutation and polymorphism-dependent	Personality alterations, dementia, Parkinsonism; resembles sCJD	Spongiform changes, mild gliosis, neuronal loss; similar to sCJD
Gerstmann–Sträussler–Scheinker syndrome (GSS)	Varies with affected codon from gait abnormalities to ataxia to dementia to Parkinsonism	Numerous amyloid plaques,neurofibrillary tangles, gliosis, and neuronal loss
Fatal familial insomnia (FFI)	Untreatable sleep disturbances	Profound neuronal loss and mild thalamic gliosis

Early clinical signs are listed in *column 2*; key neuropathological features are listed in *column 3*

notably shorter than that of other forms of CJD which range up to 60 years. Incubation periods also vary considerable, even within the same form of TSE. Some forms such as kuru have incubation periods that can span decades and because of that aspect prion diseases may have gone unidentified for many years if not for Gajdusek's efforts in investigating an outbreak of an obscure illness in remote New Guinea.

Creutzfeldt–Jakob disease (*CJD*) is the most common form of human prion disease with a worldwide incidence of one case per million of population per year. CJD was first described in the early 1920s. At that time Creutzfeldt described a 22-year-old female patient with dementia; his report was followed shortly thereafter by Jakob's, which described several patients with symptoms similar to those of Creutzfeldt's case. The *PRNP* prion gene is located on chromosome 20. As is the case for the other prion diseases, the specific set of clinical symptoms varies with the particular mutations, and with the presence of either methionine or valine codon at position 129. Octapeptide repeat insertions are sometimes encountered and there is another prominent polymorphism at codon 219 (glutamine or lysine) in both sCJD and fCJD. Familial forms of CJD are associated most frequently with mutations at codon 200 and less often with mutations at codons 178, 208, or 210. These mutations do not alter the native structure of PrP^C as much as they do the interactions between PrP^C with PrP^{Sc} and the latter's ability to polymerize.

Gerstmann–Sträussler–Scheinker (*GSS*) *syndrome* is an autosomal dominant disease first described in 1936 by the three authors after which the disease is named. This is a heterogeneous group of disorders with a familial origin. It is most frequently associated with mutations at codon 102 and less often with mutations in several other codons of the prion gene. The codon 102 mutation leads to cerebellar dysfunction, gait abnormalities, and dementia. More generally, the specific manifestations of the illness vary depending on the particular mutation.

Fatal familial insomnia (*FFI*) is the least well known of the human prion diseases. It is an autosomal dominant disorder first described in 1986. It is associated with the D178N mutation of the prion gene along with presence of the methionine codon at position 129. Its most prominent symptoms are untreatable insomnia and hallucinations.

The sporadic form of CJD (sCJD) is the most common variant of TSE. It accounts for 85 % of the cases. There are two main hypotheses as to the cause of sCJD. The first is that the disease arises as a consequence of an age-dependent somatic mutation in the prion gene resulting in the formation of PrP^{Sc}. The second hypothesis is that the disease arises from a spontaneous conversion of PrP^C to PrP^{Sc} in one or more neurons. These changes triggers further conversions leading to the onset of the disease. As the general term spongiform encephalopathies indicates, neuropathological indications of the disease include spongiform changes and also neuronal loss, gliosis (proliferation of glia in response to brain injury), and in some cases amyloid plaques (Table 7.1). In many instances, the amyloid plaques are accompanied by even larger deposits of prionic aggregates that do not satisfy all the biophysical conditions for amyloids.

One of the most troubling aspects of TSE is that of iatrogenic transmission from one person to another. The first instance of this type of acquired CJD was in 1974 when a patient receiving a corneal transplant developed CJD. Other instances of iCJD involve transmission of CJD to persons through contaminated medical instruments, through the receipt of cadaveric pituitary-derived gonadotropin and human growth hormone (hGH), and to persons receiving dura matter grafts.

7.3   Animal Prion Diseases Include Scrapie and Mad Cow Disease

The classical example of an animal prion disease is scrapie, a fatal neurological disorder of sheep and goats. This disease is associated with polymorphisms at three codons in the prion protein gene—136, 154, and 171, with two combinations, VRQ and ARQ—rendering the sheep particularly susceptible to developing scrapie. As discussed earlier, this disease together with the discovery of kuru in Papua-New Guinea led to the establishment of transmissible spongiform encephalopathy as incurable neurodegenerative disorder caused by self-replicating, transmissible prions.

These diseases came to public attention in the later 1980s when two major events occurred. The first was an outbreak of a cattle prion disease, termed bovine spongiform encephalopathy (BSE), or mad cow disease, in Great Britain. By the time the epidemic ended about 5 years later 4.5 million cattle had been destroyed. The second event began in 1996/1997 with the appearance of several reports in *Nature* describing the emergence of not-seen-before form of CJD. Termed new variant Creutzfeldt-Jakob disease (nvCJD), this disease outbreak also took place in Great Britain. It was characterized by an earlier age of onset and with a time-course and neuropathology that differed from the classical forms of CJD. Most alarmingly, the strain responsible for nvCJD was found to be identical to that one causing bovine spongiform encephalopathy. These cases appeared followed several prominent outbreaks of BSE in the UK and elsewhere in Europe. BSE-infected meat products had apparently entered the food-chain and crossed the species barrier, thereby affecting humans. That disturbing finding along with the equally disturbing recognition of mad cow disease's exceptionally long incubation period (up to 50 years) firmly established the disorder in the public mind. It also established that interspecies transmission can occur naturally and not just in the laboratory.

Another dimension to the prion story comes into play with studies of chronic wasting disease (CWD) of cervids—mule deer, white tailed deer, elk, and moose. Improper disposal through burial of carcasses of these animals, and also of infected cattle and sheep, have led to the introduction of prions into the natural environment. Prions released from decomposing carcasses adhere to minerals in the soil and remain bioactive for long periods of time. It is believed the improper handling of carcasses contributed to the severity of the BSE outbreak in Great Britain. In the case of mule deer, decomposing carcasses provide extra nutrients for grasses and the resulting growth attracts deer to the site. In addition to the carcass route there is a fecal-oral route of transmission. While CWD can be transmitted horizontally among cervids, it cannot be readily transmitted to humans, cattle, or sheep. That is, there is a fairly high transmission barrier in the non-cervid direction.

Transmission barriers: In general, prions pass from one species to another far less readily than transit within the same species. This aspect gives rise to the notion of a *species barrier* that inhibits cross-species PrP^{Sc} transmission. The PrP^C primary sequence and three-dimensional structure are both highly conserved among mammalian species. The primary impediment to cross-species transmission is thought to be a mismatch between scrapie prion and host PrP^C conformations that prevent conversion of the latter to the misfolded disease-causing former. The compatibility of the two is sensitive to amino acid sequence in several critical regions so that a single point mutation in one of these regions can either promote or prevent successful seeding by the invasive prions species. Thus, primary sequence issues are a key contributor to the species barrier. In addition, glycoform properties and any other modifications may contribute to the species barrier.

Transmissible mink encephalopathy (TME) is yet another form of prion disease. This rare prion disease affects ranched mink. There are two distinct strains of TME—drowsy (DY) in which the subjects become lethargic and hyper (HY) in which they become hyperexcitable. Studies of TME carried out in the mid-1990s helped establish the strain concept, a central tenet of prion biology. Scrapie prions taken from these animals (1) exhibited distinct biochemical properties, and (2) faithfully converted normal PrP^C form to new DY and HY prions.

One of the frustrations facing researchers in prion diseases is the lack of atomic level PrP^{Sc} structures. However, other pieces of information when put together point to conformational differences as being the critical factor. One of these is the variation in fragment size produced by exposure to proteinase K, a broad-spectrum serine protease that digests proteins. One of the findings from studies of TME was that DY and HY prions are cleaved by proteinase K (PK) at different amino-terminal sites. Later studied carried out using scrapie prions taken from the brains of FFI and CJD patients reinforced the emerging connections between prions, strains, conformations and PK-derived fragment lengths. In FFI, PrP^{Sc} was cleaved at residue 97 to produce 19 kDa fragments, whereas in CJD and sCJD, scrapie prions were cleaved at reside 82 to produce 21 kDa fragments. The implications from these studies were twofold—that prions did not require genetic material to convey disease, and that strains arose from differences in conformation as reflected by their varying PK cleavage sites.

7.4   There Are Multiple Prion Strains Incubation Periods and Pathology

That a single protein can give rise to multiple strains just like a conventional, nucleic acid bearing pathogen is quite remarkable, and has been the subject of considerable interest. This aspect has been studied for over 25 years in mice, in other mammalian species, and in yeasts. By the early 1990s over 20 different strains of mouse prions had been found, and strain differences were found to underlie the various forms of human TSE, as well.

Strains can be distinguished from one another by differences in disease characteristics. These include:

- Clinical signs of the disease;
- Differences in disease progression (e.g., incubation period, survival time);
- Biochemical properties (e.g., responses to proteinase K digestion and heat);
- Distribution in the brain, and
- Histological features.

Differences in strains can arise in several ways. They can appear as a consequence of slight differences in primary sequence. These include tiny differences in primary amino acid sequence from individual to individual due to mutations and polymorphisms. Strains may also arise from differences in posttranslational modifications. They may arise, perhaps even more fundamentally from the intrinsic dynamics of the prion protein itself, which continually samples a host of different conformational states and substates. For example, using FRET single molecule biophysics, yeast Sup35 was found to adopt an ensemble of collapsed and rapidly fluctuating structures on a 20–300 ns timescale. This prion possesses a fluctuating amyloidogenic N-terminal region joined to an extended charged and solubilizing middle region. Putting these and similar findings together gives rise to the widely accepted hypothesis that strains are encoded in the prion's tertiary structure, i.e., in its conformations. *Importantly, strain-associated prion conformations are stabilized through formation of oligomeric seeds.* They can then pass stably from one individual to another where they convert resident PrP proteins to the transmitted prion conformation, thereby giving rise to a stable phenotype.

7.5 The Protein-Only Hypothesis and Its Corollaries

The overriding understanding in the field at present is that:

1. PrP^{Sc} is a beta-sheet rich, amyloidogenic alternate conformation of the PrP protein. In this alternative conformation, the protein is resistant to proteolysis, is detergent insoluble and, as is the situation for other misfolded disease-causing proteins, forms aggregates and amyloid-like structures.
2. Small oligomers, either amorphous in fibrillar in organization, operate as minimal nuclei that stabilize alternative scrapie conformations.
3. These nuclei form seeds that catalyze the conversion of normal form of the protein in the cell to the same alternative conformation.
4. These seeds can be transmitted to other cells where they again can convert normal PrP^{C} to an alternative PrP^{Sc} conformation.
5. Several different alternative three-dimensional conformations can be stabilized through oligomerization. These give rise different strains, each with its distinct set of clinical manifestations and neuropathological signs.

6. Transmission barriers exist between individuals of the same species and between members of different mammalian species; these are determined by how well transmitted scrapie and resident normal prions overlap in primary sequence, posttranslational modifications, and tertiary structure.

Following the discovery that PrP^C and PrP^{Sc} shared the same primary sequence efforts were made to firmly establish the main prion tenets, known as the *protein-only hypothesis*, described above. Those efforts spanning the past 20 years have largely succeeded in establishing that the above-described chain of events is the correct one. That is, prions by themselves are transmissible and infectious, and are the causal agent of the transmissible spongiform encephalopathies such as kuru in humans, scrapie in sheep, and bovine spongiform encephalopathy in cattle.

There are a number of milestones in the efforts to establish the protein-only hypothesis. One of these occurred in 1994 with a demonstration of the cell-free conversion of PrP^C to protease-resistant form similar to PrP^{Sc}. The demonstration highlighted the necessary presence of PrP^{Sc} in the process, thereby providing evidence for direct PrP^C–PrP^{Sc} interactions in the conversion.

A few years later, in the 2001–2005 time period, Soto and colleagues introduced a new process that greatly increased prion concentrations. In this technique called protein misfolding cyclic amplification (PMCA), a mixture containing PrP^{Sc} and a far-larger amount of PrP^C is incubated to grow PrP^{Sc} aggregates. That step is followed by sonication in which the sample is exposed to ultrasound that fragments the aggregates into smaller pieces that serve as new PrP^{Sc} seeds. The two-step, grow/break process is then repeated over many cycles to produce large quantities of PrP^{Sc} aggregates. Inoculation of laboratory animals with the in vitro PMCA-derived scrapie proteins produced scrapie illness identical to that produced by brain-derived material. Additional studies focusing on prion generation and transmissibility provided further evidence for the correctness of the proposed causal chain. It was also was used a few years later to demonstrate the generation of scrapie prions from PrP^C together with co-purifying lipids and synthetic polyanions without requiring the presence of PrP^{Sc} seeds. Those dramatic results served to focus attention of the role(s) of non-genetic cofactors both in cell-free systems and in the brain.

Yet another piece of data supporting the protein-only hypothesis, and linking strain effects to conformation, was provided in the late 1990s when conformation-dependent antibodies were devised. Using a conformation-dependent immunoassay, scrapie prions were first subjected to denaturation and then the fraction of antibodies that bound exposed epitopes on the prions was measured. When this technique was applied to each of eight strains of mouse scrapie, differences in binding fractions were observed, thereby providing further support for the underlying chain of reasoning.

Requirement forcofactors: As noted above cofactors may contribute to the conversion process. A variety of potential cofactors have been identified. These are primarily large biomolecules—polysaccharides, nucleic acids, and proteins. Among the more notable examples are glycosaminoglycans (GAGs) such as heparin, the nucleic acid RNA, and the membrane lipid phosphatidylethanolamine. Studies of these possible contributors to the conversion of PrP^C to PrP^{Sc} have produced several tentative conclusions in spite of variations in results arising from strain and host dependencies. One of these is that most, if not all, of these cofactors are dispensable for the conversion, itself. Templated conversion can occur in their absence, but at exceptionally low rates. The cofactors interact with the PrPs; they influence their conformation and stability, and by that means accelerate the conversion.

7.6   Biogenesis of the PrP Protein

As discussed in the preceding two chapters, newly synthesized proteins are processed, subjected to quality control, and shipped to their cellular destinations. Prosthetic groups—sugars and lipids—are added to proteins destined for insertion in membranes, thereby enabling their attachment. These modifications are made subsequent to translation, in several stages. The overall process resembles an assembly line that builds up the proteins, folds them, inserts them into membranes, sorts them, labels them with targeting sequences, and ships them out.

The Golgi apparatus consists of a stacked system of membrane-enclosed sacs called cisternae. Some of the polysaccharide modifications needed to make glycoproteins are either made or started in the rough ER. They are then sent from the rough ER to the smooth ER where they are encapsulated into transport vesicles pinched off from the smooth ER. The transport vesicles are then sent to the Golgi. The Golgi apparatus takes carbohydrates and attaches then as oligosaccharide side chains to form glycoproteins and complete modifications started in the rough ER. The cellular prion protein, PrP^C, is a glycosylphosphatidylinositol (GPI)-anchored protein. Like other proteins in this class, it is synthesized in the endoplasmic reticulum, processed and finished in the Golgi apparatus, and transported to the cell surface.

Most proteins destined for insertion in the plasma membrane contain covalently linked oligosaccharides that extend out from their extracellular side. These proteins are referred to as glycoproteins. There are two forms of modification, N-linked and O-linked. In N-linked glycoproteins, a carbohydrate, or glycan, is added to a side chain NH$_2$ group of an asparagine amino acid residue. In O-linked glycoproteins, the oligosaccharide chain is appended to a side chain hydroxyl group of a serine or threonine amino acid residue. Carbohydrates are hydrophilic; addition of the N-linked glycans increases solubility of the proteins, and influences protein folding and stability. Their addition also facilitates intracellular transport and targeting, and assists in the signaling roles of the mature proteins. The fully dressed PrP^C with its attached N-linked glycans at Asn-181 and Asn-197, attached copper ions, and GPI anchor is depicted in Fig. 7.2.

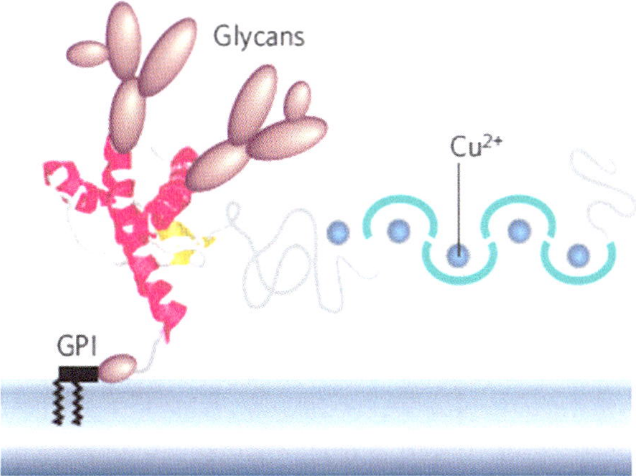

Fig. 7.2 Schematic depiction of a mouse prion protein PrP (residues 23–231). Shown are two domains—a compact C-terminal domain terminated with a GPI anchor and a natively unstructured N-terminal domain that binds copper ions. The compact domain contains three α-helices shown in *red* and a pair of short anti-parallel β-strands depicted in *yellow* (from Caughey *Nature* 443: 603 © 2006 Reprinted by permission from Macmillan Publishers Ltd)

7.7   Structure of the PrP^C Protein

A more detailed look at the PrP's N- and C-terminal domain structure is presented in Fig. 7.3. As can be seen in the figure, PrP contains a 22-amino-acid-residue signal peptide in its N-terminal region that is cleaved. The truncated protein containing residues 23–231 contains an N-terminal disordered region plus a C-terminal compact, globular folding core spanning amino acid residues 121–231. The folding core is predominantly alpha helical in it secondary structure and misfolds into an alternative PrP^{Sc} conformation under the influence disease-causing mutations. The four fatal prion diseases affecting humans (Table 7.1) are rare diseases brought on by missense mutations. These mutations are primarily located within the compact folding core and concentrated in the region about the H2 and H3 helices, while a few mutations causing GSS are found in the stretch from CC2 to H1. In addition, there are two polymorphisms that play a key role in the transformation from benign to lethal conformations in CJD. As remarked upon in Section 7.2, there is a codon 129 polymorphism—methionine or valine, and a codon 219 polymorphism—glutamine or lysine. These do not affect the native prion fold but when partially denatured the polymorphisms (especially the one at codon 129), strongly affect the propensity to form amyloids and the ability to propagate.

A growing number of ligands have been identified. Some of these bind to the octapeptide repeats in N-terminal region, while others bind to specific sequences in the C-terminal core. (These are examined shortly.) The two regions are functionally

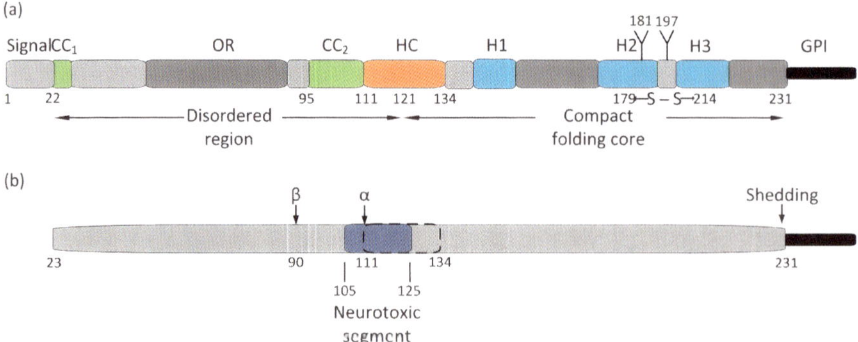

Fig. 7.3 (**a**) Structure of the cellular prion protein. The first 22 amino acid residues comprise a signal peptide that is cleaved leaving a mature 208 residue prion protein. Residues 23–110 constitute a disordered (random coil) region followed by a globular, compact folding core consisting of residues 121–231 and a C-terminal GPI anchor. The N-terminal region contains a pair of positively charged acidic amino acid clusters (CC_1 and CC_2), and an octarepeat region (OR). The protein possesses a hydrophobic core (HC), a single disulfide bond (between residues 179 and 214), a pair of glycosylation sites (at residues 181 and 197), and three alpha helices (H1, H2, and H3). (**b**) Proteolytic processing of the prion protein. There are two sites where proteolytic cleavage occurs— an α-cleavage site at residues 111–112, and a β-cleavage sites in the vicinity of residue 90. In addition, the GPI anchor may be shedded to produce an anchorless prion protein, Also shown (in *dark blue*) is a neurotoxic segment spanning residues 105–125 that overlaps with the hydrophobic core (*dashed lines*)

coupled. They both serve in a neuroprotective capacity under normal conditions and contain segments involved in the generation of neurotoxic signals in disease scenarios.

Full-length PrP is proteolytically processed. These cleavage actions parallel those occurring in the processing of the Alzheimer's disease-associated amyloid beta precursor protein (APP), to be studied in detail in the next chapter. Although they are not yet as well understood as the APP processing, there is accumulating evidence that these similar actions, and the fragments generated by them, have important roles in prion physiology. The reason why becomes clear in an examination of Fig. 7.3b. Two cleavage sites have been identified. The first, the α-cleavage site is located between residues 111 and 112, which lies within the segment of the protein believed to cause its neurotoxic effects. Cleavage at that location eliminates that possibility and imparts a neuroprotective role to the breakage, which generates a detached soluble N1 fragment (containing an important PrP^C–PrP^{Sc} interaction site) and a remaining membrane-bound C1 fragment. Cleavage at the β-site located near the end of the octarepeat region results in the generation of a soluble N2 fragment and a membrane-bound C2 fragment containing an intact neurotoxic segment. Finally, the prion protein may undergo removal of its GPI anchor by a sheddase resulting in the release into the intracellular spaces of a largely intact prion protein.

7.8   PrP Resides in Caveolae and Lipid Rafts

Membrane lipids form gels and liquid states through the cooperative effects of multiple weak noncovalent interactions such as van der Waals forces and hydrogen bonds. As a result there is considerable fluidity of movement within the membrane, and the lipid and protein constituents are free to diffuse laterally. The overall structure is that of a fluid of lipids and membrane-associated proteins undergoing thermally driven Brownian motion. These properties were codified in the *Singer–Nicolson fluid mosaic model* of biological membranes introduced in 1972.

That model was challenged by biophysical studies that posited that lipids in biological membranes might exist in several phases and as a result would sequester membrane lipids and proteins in distinct microdomains. In this depiction, lateral movement may not be quite as free as presented in the Singer–Nicolson model due to mechanical constrains imposed by the cytoskeleton and through the actions of large scaffolding proteins. Instead, there is a mosaic of membrane compartments each with its own distinct population of lipids and proteins. This later model was introduced by Karnovsky and Klausner in 1982, and extended by Simons, van Meer, Ikonen, and others in the years since then.

In more detail, compartments enriched in cholesterol and sphingolipids contain high concentrations of signaling molecules—GPI anchored proteins in their exoplasmic leaflet, multiple transmembrane receptors, and numerous signaling proteins recruited to its cytoplasmic surface. The best characterized of these signaling compartments are the *caveolae* (little caves) and *lipid rafts*. The former are tiny flask-shaped invaginations in the plasma membrane. They are detergent insoluble and are enriched in coat-like materials, caveolins, which bind to cholesterol. The latter, the lipid rafts, differ from caveolae in that they are flat and do not contain caveolins. They are formally defines as small (10–200 nm), heterogeneous, highly dynamic sterol- and sphingolipid-enriched domains that compartmentalize cellular processes. These structures play an important role in signaling at synapses where they enable proteins that have to work together to be localized near one another. Prions preferentially localize to lipid rafts and to a smaller extent to caveolae. These structures and their biophysical properties are discussed further in the Appendix to this chapter.

7.9   Protective Functions of PrP

The prion protein is a synaptic protein expressed at high levels in neurons and found in both presynaptic and postsynaptic compartments. It localizes to detergent-resistant lipid rafts, and is internalized in clathrin-coated vesicles and endosomal organelles. Although its exact physiological functions are not completely characterized, a considerable body of data has been collected supporting the idea that the prion protein normally functions as a signaling platform. The protein does not penetrate to the cytoplasmic leaflet of the cell surface membrane, but is believed to partner with one or more co-receptors that initiate downstream signaling.

Under normal cellular conditions PrP^C acts in a neuroprotective capacity related to its role in synaptic transmission and cell adhesion. A number of cell adhesion and extracellular matrix protein have been identified that bind PrP^C. Prominent among these binding partners are heparan sulfate proteoglycans (HSPGs), vitronectin, and low-density lipoprotein receptor-related protein 1 (LRP1). These bind sequences in the N-terminal region of PrP^C. Similarly, laminins and NCAM among others bind motifs in the C-terminal domain. These molecules act as neurotrophic factors that confer signals essential for growth and preservation of synapses and neurons. Binding of these ligands initiate a series of events including recruitment the tyrosine kinases Fyn or Src, which act as signaling hubs, to initiate a broad spectrum of downstream signaling actions in different neuronal populations (Fig. 7.4a).

A way that cellular PrP acts in a neuroprotective capacity is through its regulation of glutamatergic synaptic transmission. One of its co-receptors is metabotropic glutamate receptor 5 (mGlu5). In response to ligand binding, PrP^C-mGlu5 recruits the tyrosine kinase Fyn to the cytoplasmic face where it initiates a signaling cascade that includes regulatory phosphorylation of nearby NMDA receptor subunits. These actions protect again excessive entry of Ca^{2+} into neurons through the NMDA receptors. Further examples of neurotrophic actions are:

- Maintenance of axonal myelination while PrP removal results in chronic demyelinating polyneuropathy (CDP),
- Preservation of embryonic cell adhesion in zebrafish through its localization of E-cadherin adhesion complexes, and
- Support of sensory information processing in the olfactory system.

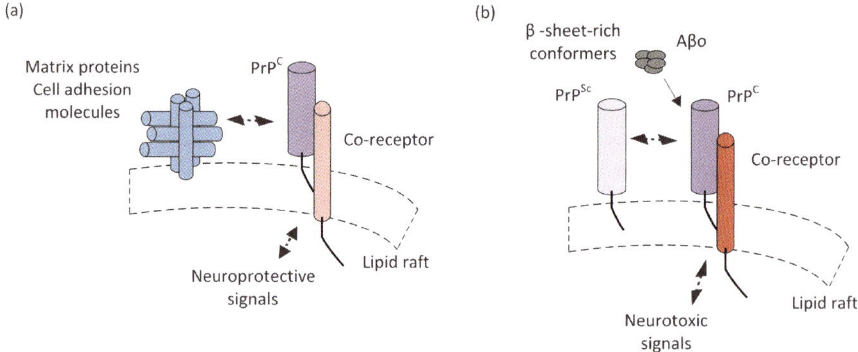

Fig. 7.4 Simplified picture of prion signaling under (**a**) normal and (**b**) disease-causing conditions

7.10   Toxic Actions of PrP

When disease-causing mutations or other abnormal conditions are present PrP^C has an increased propensity to misfold into the scrapie form PrP^{Sc}. In this alternative conformation, it interacts with the normal cellular prions to initiate abnormal signaling. This interaction seems to be a necessary one. In the absence of the cellular form, scrapie PrP does not cause neurodegeneration. As discussed earlier, PrP^{Sc} is a β-sheet-rich conformer. Interestingly, it is not the only β-sheet-rich conformer that interacts with PrP^C. Another set of prominent interactors are the amyloid-β oligomers (Aβo) that have a central place in the Alzheimer's disease pantheon. These aberrant interactions compete with the normal ones leading to shift in signaling from protective to harmful. The altered, neurotoxic signaling occurring in the presence of β-sheet-rich conformer is depicted in Fig. 7.4b. That subject is revisited in the next chapter.

Several regions within PrP have been identified as playing crucial roles in the protein's normal neuroprotective and abnormal neurotoxic activities. The most prominent of these is the central region (CR) spanning residues 105–125. This region, highlighted in Fig. 7.3b, contains a segment enriched in charged residues followed by a sequence supplied with several hydrophobic ones. It is one of the sites bound by PrP^{Sc} and, significantly, deletion of this region renders the resulting protein, ΔCR PrP, toxic to neurons. A second PrP sequence, this one spanning residues 23–31 is also involved in binding PrP^{Sc} and its loss also enhances PrP neurotoxicity.

These results support the notion that competitive interactions promote PrPC-PrPSc toxicity. That is, the matrix proteins and cell adhesion molecules that normally bind to these regions are prevented from doing so when PrP^{Sc} binds to the same sites. Toxicity may result from failure of the blocked PrP^C to convey its normal signals and/or from its redirection into a toxic signaling pathway (Fig. 7.4b).

The above mechanism is not the only contributor to prion toxicity. As is the case for most, if not all, neurological disorders, there is an accumulation of misfolded proteins, synaptic transmission fails, protein quality control declines, and neurons are lost. There is evidence that, in response to a buildup of misfolded PrP oligomers, ER stress increases, protein translation is reduced, and the UPS is inhibited. These conditions when combined with an increased sensitivity to synaptic excitotoxicity and misdirected neurotrophic signaling may be sufficient to send neurons along an irreversible apoptotic trajectory.

7.11   Fungal Prions Do Not Cause Disease

The discovery in 1994 of fungal prions launched a major thrust in prion research. Unlike the mammalian, disease-causing scrapie prion, fungal prions are non-disease-causing, alternatively folded forms of cellular proteins. An ever-growing number of fungal prions have been identified; some of the most prominent of which are listed in Table 7.2. As can be seen in the table the "standard" forms of the

Table 7.2 Fungal prions

Protein	Function	Prion
Het-s	Heterokaryon incompatibility regulator	[Het-s]
Sup35	Translation termination factor	[PSI$^+$]
Ure2	Nitrogen catabolite repressor	[URE3]
Rnq1	Protein template factor	[PIN$^+$], [RNQ$^+$]
Swi1	Chromatin remodeling factor	[SWI$^+$]
Cyc8	Transcription co-repressor	[OCT$^+$]
Mot3	Transcription regulator	[MOT3$^+$]
Spf1	Transcription regulator	[ISP$^+$]

Het-s: *Podospora anserina*; all others: *Saccharomyces cerevisiae*

proteins carry out a diverse set of cellular functions. In contrast to the mammalian PrP^{Sc} prion, the alternatively folded fungal prions function as non-chromosomal elements that confer differences in cellular phenotype to daughter cells.

An informative example is [PSI$^+$], the prion form of the yeast Sup35 protein. As noted in Table 7.2, Sup35 is a translation termination factor. It is needed for termination of mRNA translation and release of the nascent polypeptide chain from the ribosome. This activity is suppressed when Sup35 takes the [PSI$^+$] prion form. In this conformation, the protein collects in aggregates and no longer acts as a translation termination factor resulting in the read-through of nonsense codons. These actions lead to changes in cellular phenotype, which can be beneficial to survival under stressful environmental conditions. In support of that idea, it has been found that the dwell time in the [PSI$^+$] state increases, and consequently there is a higher frequency of prion state induction, in the presence of environmental stresses such as heat and toxic chemicals. Several of the prion-forming proteins listed in Table 7.2 have roles in transcription regulation. In these cases, a loss-of-function resulting from association of the alternatively folded proteins in aggregates can result in useful changes in the cellular transcriptional program. This has been observed with respect to nitrogen utilization, drug resistance, and the presence of oxidative and heavy-metal induced stresses.

One of the most prominent features of the fungal prions is the presence of a modular and transferable prion domain. These domains are usually at least 60 amino acid residues in length. They are enriched in uncharged polar residues such as asparagine (N), glutamine (Q), and tyrosine (Y), and depleted in hydrophobic and charged residues. These chains are intrinsically unstructured but can readily switch to β-sheet-rich forms. With the exception of the [Het-s] prion all the proteins listed in Table 7.2 possess prion (Q/N) domains or small variations thereof.

There are a number of examples where the prion state and its attendant aggregates function as the active (gain-of-function) state. Even more interestingly, there are examples of positive non-disease-related roles played by prion-like proteins in higher organisms. One of these is the prion form of the cytoplasmic polyadenylation element-binding protein (CPEB), which facilitates the maintenance and persistence of memory. One of the key tenets in memory formation is the requirement for local protein synthesis. In its monomeric, non-prion form CPEB inhibits translation of the mRNAs. That situation changes when neurotransmitter serotonin stimulates the stabilization of CPEB in an oligomeric/amyloid prion state. In that active state,

CPEB stimulates the translation of its target mRNAs resulting in the strengthening and stabilization of the synapses. A similar theme is at work in how the mitochondrial adapter protein MAVS stimulates antiviral responses. In response to viral infection, MAVS form large fibrous structures with prion-like self-templating abilities. These structures are the active form of the protein; in that state, they activate and promote the translocation of transcription factors IRF3 and NF-κB to the nucleus where they stimulate transcription of interferons and other antiviral genes. The MAVS, unlike CPEB, does not rely on Q/N sequences. Instead, fibril formation is mediated by caspase activation and recruitment domains (CARDs).

7.12   Structure of the Prion Amyloid-Like Fibrils

Recall that several amyloid structures were presented in Chap. 4. These included the microcrystal structure formed by a short seven-residue Q/N-rich segment of the yeast Sup35 prion (Fig. 4.1), the hydrophobic segment from the amyloid-β_{1-42} (Fig. 4.2), and the portion encompassing residues 218–289 of the yeast [Het-s] prion (Fig. 4.3). All of these structures were built from beta strands oriented perpendicular to the fibril axis but differing from one another in their details. One of these, the Het-s structure, was that of a left-handed β-solenoid or β-helix. This type of fibril was built from short β-strands that alternated with bends resulting in a coiled structure.

Three leading models of how misfolded prions might assemble into fibril-like structures such as these are depicted in Fig. 7.5. The first of these models is the *β-helix model*. It was introduced in 2004 and was based on electron crystallographic analysis of 2D crystals of PrP 27–30 and PrP^{Sc}106 (in which residues in the N- and C-terminals have been deleted leaving 106 residues). The results of the analysis by Govaerts are presented in Fig. 7.5a. As shown in the figure the modeling prediction is for a partial conversion from the predominantly alpha helical PrP^{C} form one with both α-helical and β-strand content. Part of the protein retains its alpha helical organization, while the remainder refolds into a β-helical form similar to the one solved a few years later for the [Het-s] prion. The left-handed alpha helices can easily form trimers, and a fibrillar structure built from these elements is presented in the figure.

The second model of the conversion from normal to scrapie form is the *β-spiral model*. This model was developed by Daggett using molecular dynamics to study the conformational fluctuations in the presence of scrapie-inducing mutations and making use of 2D electron crystallography data. In this model (Fig. 7.5b), a three-stranded β-sheet forms the amyloidogenic core, and there is an additional isolated strand, while the three α-helices are preserved. Both this model and the β-helical model contain a roughly 50:50 α-helical–β-sheet mix consistent with results of biophysical measurements.

The third model, the *parallel, in-register β-sheet model*, differs considerably from other two. It posits the presence in-register β-sheets stacked one on top of the other. In this model, from Surewicz, an extensive unfolding and rearrangements of the protein structure takes place resulting in a canonical cross-β amyloid structure.

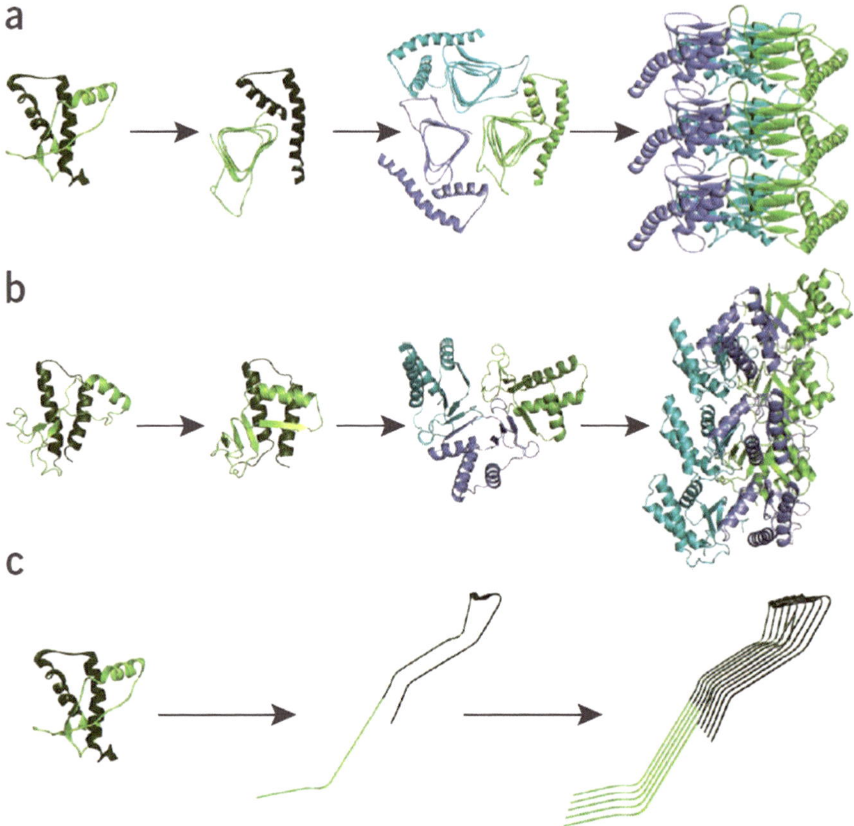

Fig. 7.5 Three-dimensional PrP^{Sc} structures. (**a**) The *β-helix model* in which residues 90–177 in the N-terminal region of PrP 27–30 refolds into a β-helix. (**b**) The *β-spiral model* in which the three alpha helices are preserved while short β-strands form a fibril core. (**c**) The *parallel, in-register β-sheet model* in which a major refolding occurs (from Diaz-Espinoza *Nat. Struct. Mol. Biol.* 19: 170 © 2012 Reprinted by permission from Macmillan Publishers Ltd)

It is supported by data for several yeast prions including Sup35, Rnq1, and Ure2. Further data in support of this model was obtained for recombinant huPrP 90–231 in which residues 160–220 forms the β-sheet core of the stack depicted in Fig. 7.5c.

7.13   Nucleated Conformational Conversion and Fragmentation

The yeast prions have been the subject of a number of experimental studies of amyloid fiber formation. These studies have led to a nucleated conformation conversion model of their assembly. In this model, PrP^{C} and PrP^{Sc} are in reversible

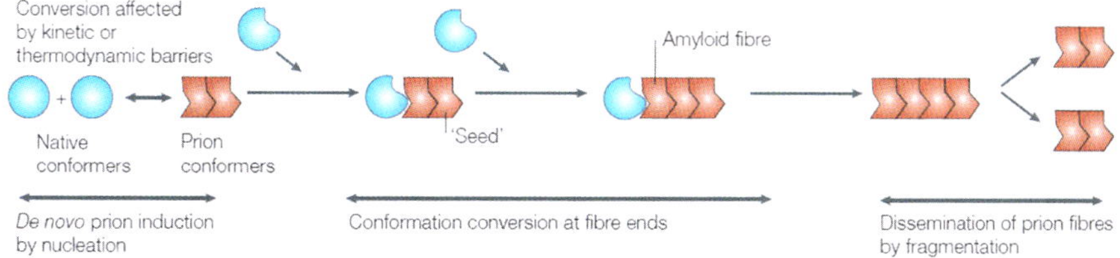

Fig. 7.6 Nucleated conformational conversion. The prion state is stabilized upon formation of a nucleus. This oligomeric assemblage becomes a seed for the accretion and conversion of PrP^C to the PrP^{Sc} form. Fibril growth then takes place followed by fragmentation that makes new seeds (from Shorter *Nat. Rev. Genet.* 6: 435 © 2005 Reprinted by permission from Macmillan Publishers Ltd)

thermodynamic equilibrium with the PrP^C state being favored and the PrP^{Sc} state only transiently populated. Infrequently, several molecules of PrP^{Sc} will assemble and when this occurs the scrapie/amyloidogenic state is stabilized. The stabilized PrP^{Sc} oligomers then form a seed for the further accretion of PrP^{Sc} monomers or oligomers, or for the more typical addition of PrP^C monomers/oligomers. The latter then undergo a nucleated conformational conversion to the amyloidogenic form. The kinetics of this process consists of a lag phase describing the slow formation of a nucleating seed followed by an exponential growth phase of monomer addition. As was the case for other aggregation-prone misfolded proteins, a variety of oligomeric compositions and morphologies can be generated. These processes are depicted in Figs. 7.6 and 7.7.

Molecular chaperones play a central role in fungal prion propagation. Recall from Chap. 5 that members of the Hsp100 family of molecular chaperones prevent protein aggregation and disaggregates protein assemblages. These chaperones form networks with other chaperone family members most notably with Hsp70 chaperones and their Hsp40 co-chaperones. In the case of amorphous protein aggregates, they are able to pull them apart and disassemble them. Their effects on highly ordered amyloid fibrils are slightly different. In these situations, they break apart the fibrils into several pieces; that is, they fragment the fibrils, thereby creating new seeds for prion propagation.

The yeast Hsp100 family member is Hsp104. This protein like other members of that family is an AAA+ hexamer. Substrate amyloid fibers that are threaded through the Hsp104 pore are broken into several pieces. The fibers are chaperoned to the Hsp104 entry sites by the yeast Hsp70/Hsp40 proteins. However, when Hsp104 is overproduced it may block the chaperoning activities of Hsp70/Hsp40 and fragmentation is blocked. This can lead to "prion curing"—the name given to one or more mechanisms whereby the necessary pool of seeds fed by fragmentation is no longer maintained and the prion phenotype disappears.

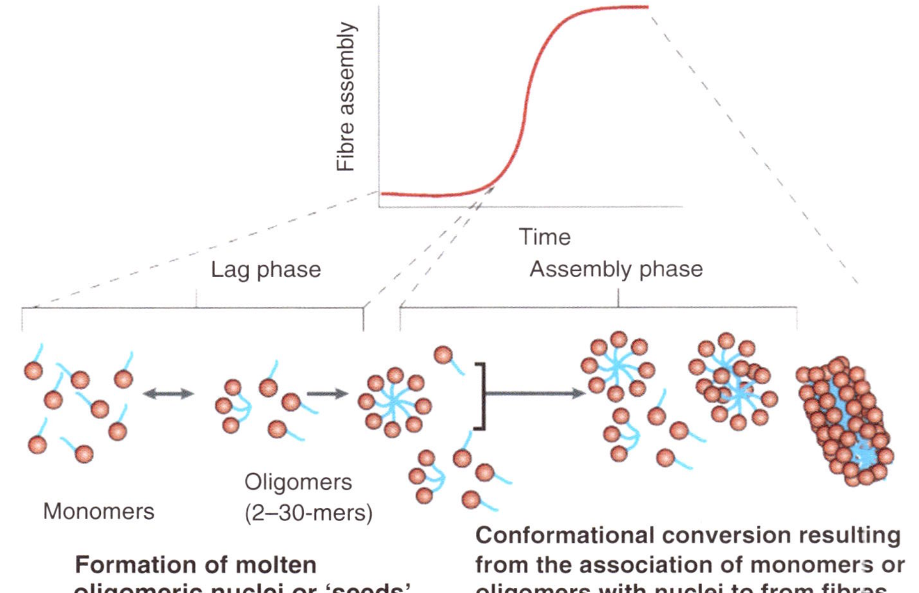

Fig. 7.7 Kinetics of oligomer and fibril formation. The plot shown above the assembly cartoon depicts the two main stages in oligomer/fibril formation—Monomers first assemble into oligomeric nuclei that serve as the seeds for subsequent aggregation and fibril growth. That step is a slow one and is referred to as the lag phase. It is followed by a rapid, exponential-like assembly phase in which most of the conformational conversion and fibril growth occurs (from Shorter *Nat. Rev. Genet.* 6: 435 © 2005 Reprinted by permission from Macmillan Publishers Ltd)

7.14   Cell-to-Cell Spread of Misfolded-Protein Seeds

Neurodegenerative disorders such as Alzheimer's disease and Parkinson's disease spread in a stereotypic manner from one region of the brain to another. These spreading patterns correspond to the degree of clinical pathology. They are widely referred to as *Braak stages*, named after the German anatomist Heiko Braak who pioneered their patho-anatomical study and developed their classification. In the case of AD, the density of amyloid plaques is poorly correlated with the degree of pathology. Instead, Braak relied on the neurofibrillary tangles and neuropil threads to define six stages of disease progression. In the case of PD, Braak used the presence of Lewy bodies enriched in α-synuclein and ubiquitin, and dystrophic neurites similarly enriched in α-synuclein, ubiquitin, and neurofilament to analyze the disease progression and delineate their various stages. Similarly, stereotypic patterns of cortical degeneration have been observed for Huntington's disease.

The natural question is: what is (are) the mechanism(s) underlying the spread of the diseases? The leading hypothesis until recently has been that the disease independently appears in each of these regions as conditions (e.g., age-related decline in protein quality control) progressively arise conducive to their establishment. That theory has been challenged in the last few years by the striking notion and supporting evidence that many of the misfolded proteins implicated in these diseases may behave like prions. That is, seeds may be transmitted from cell-to-cell, and once in a recipient cell the seeds catalyze the conversion of the resident normally folded isoform to a disease-causing conformation.

One of the central ideas emerging from the staging studies is that the spreading pattern of a neurodegenerative disease is not random. Rather, the disease spreads along well-defined anatomical routes. It may spread from one neuron to neighboring neurons, and also move down axons and across to their synaptically connected neighbors. These data can be interpreted as supporting a prion-like spread mechanism, thereby giving rise to efforts to uncover the possible spread mechanisms. *Tunneling nanotubes* are a recently discovered (2004) means of cell-to-cell communication. These are thin membrane channels that establish continuity between cells. They are typically 50–200 nm in diameter and up to several cell diameters in length, and facilitate the direct cell-to-cell transfer of organelles and proteins. These have been found to be used by prions for intercellular spread. Their use by prions is depicted in Fig. 7.8, which illustrates tunneling nanotubular and synaptic connectivity.

Multivesicular bodies (MVBs) transport cellular components to the lysosomal compartment for degradation. They also transport materials to the plasma membrane, with which they fuse and release their contents into the extracellular spaces. Intraluminal vesicles are small vesicles contained within the MVBs that sequester degradation-bound material. These small vesicles are referred to as *exosomes* when destined for the plasma membrane and extracellular release. The exosomes and their contents can then be captured by other cells through internalization or by means of receptor binding. This means of transport serves a useful communication role in the immune system and other cell types.

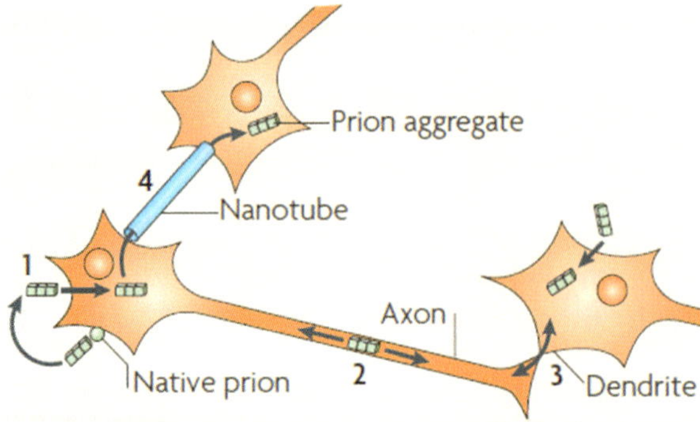

Fig. 7.8 Cell-to-cell spread via direct contacts. Prions are internalized by means of endocytosis (*step 1*). They can move bidirectionally through an axon (*step 2*) and across the synaptic cleft to dendrites on a recipient cell (*step 3*). Alternatively, they can move from cell to cell through nanotubes (*step 4*) (from Brundin *Nat. Rev. Mol. Cell Biol.* 11: 301 © 2010 Reprinted by permission from Macmillan Publishers Ltd)

Recently, exosomes have been found to be utilized by neurons to get rid of unwanted PrP^C and PrP^{Sc} prions presumably functioning as yet another backup protein quality control mechanism. However, any exosome bearing disease-causing proteins becomes a Trojan horse. This mechanism of spread is not limited to PrP prions. There is increasing evidence for a prion-like, exosome-mediated spread of tau and α-synuclein, and of other key players in neurodegeneration. Figure 7.9 depicts the encapsulation and release of tau, α-synuclein and huntingtin in exosomes, microvesicles, and through membrane pores, and their subsequent uptake by nearby or distant neighbors.

7.15   Summary

1. Prions are self-propagating alternatively folded forms of the cellular PrP protein. They are responsible for causing transmissible spongiform encephalopathies (TSEs) among different mammalian species. A prominent example of an animal TSE is bovine spongiform encephalopathy that infects cattle; another is scrapie that attacks sheep and goats. There are four prion disorders affecting humans. These are:

   - Creutzfeldt–Jakob disease
   - Gerstmann–Sträussler–Scheinker syndrome
   - Fatal familial insomnia
   - Kuru

   The name "prion" was coined by Stanley Prusiner in 1982 as a contraction of the term "proteinaceous infectious particle". The normal (cellular) prion protein is designated PrP^C. The misfolded, disease-causing forms are referred

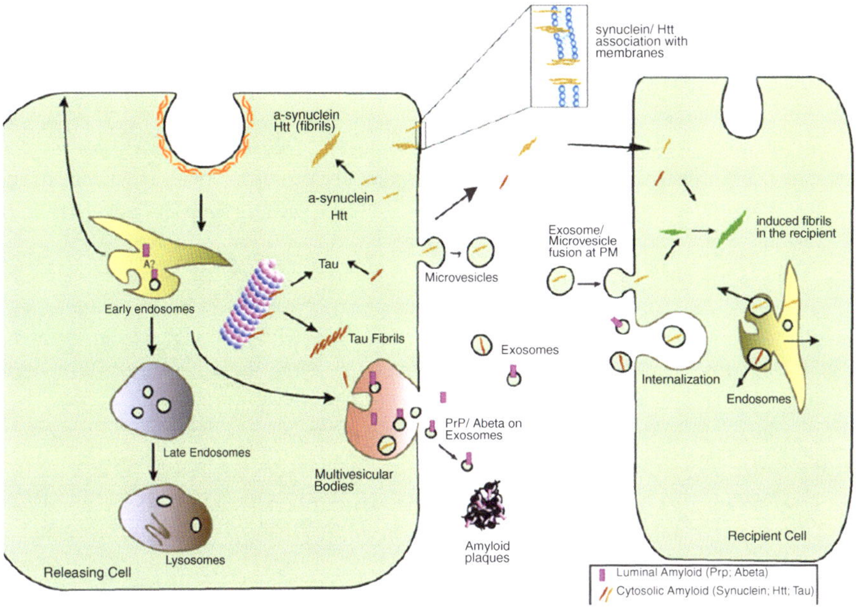

Fig. 7.9 Cell-to-cell routes of prion-like spread of neurodegenerative diseases. Several possible routes of release are depicted. One of these is direct translocation across the plasma membrane either through pores or by some other means. Another release mechanism is through formation of membrane blebs and microvesicles. Still another, and one for which considerable evidence has been acquired, is by means of exosomes. Many but not all of these utilize the ER-Golgi machinery. These vesicles undergo fusion and internalization by recipient cells resulting with the release of their contents (from Aguzzi *Nat. Struct. Mol. Biol.* 6: 598 © 2009 Reprinted by permission from Macmillan Publishers Ltd)

to in one of several ways, depending in part on how they were produced. The most common designation is PrP^{Sc} (scrapie form); others are PrP^{Res} (protein-ase K resistant form), and PrP 27–30 (independent folding core form).

2. Prion diseases are transmissible dementias. Prions by themselves without the involvement of gene-encoding nucleic acids are able to transmit the disease (1) from cell-to-cell within an individual, (2) from individual to individual and in some instances across mammalian species, and (3) from generation to generation in the case of fungi. There are several ways that a prion can transit from one cell to another in the brain. Prions may ride inside exosomes and microvesicles, exit through pores in the plasma membrane, transit through nanotubes, and move across the synaptic cleft. They may also be taken up in contaminated foodstuffs by a recipient individual, and subsequently move from gut through the lymphatic system and eventually (perhaps taking years) end up in the brain. Incubation periods can take decades making study of these fatal illnesses particularly challenging. In general, transmission across species lines is more difficult than that occurring within a given species. Transmission effectiveness depends on how well seed and host PrPs match. The resistance to cross-species transfer is referred to as the species, or transmission, barrier.

3. The prion protein in its scrapie form has the property that it can stimulate the refolding of the cellular PrP^C into the scrapie PrP^{Sc} form. This occurs when a nucleating seed is formed that stabilizes the protein in its alternative conformation. Monomers can then be added to the seed to produce large amyloidogenic aggregates. Fragmentation may play an important role by breaking apart fibrillar structures into smaller pieces that become new seeds for fibril growth. In fungi, this process is aided crucially by the heat shock protein Hsp104 and is partner chaperones, Hsp70 and Hsp40.

   As is the case for viruses and bacteria, prion strains exist. These are encoded in the specific three-dimensional conformation of the prion, in the detailed primary sequence, and in modifications to the three-dimensional structure of the protein. The resulting conformational variations impart significant difference in disease characteristics to the subject—onset and duration of the illness can vary considerable as do its clinical signs and neuropathology.

4. PrP is a GPI-anchored glycoprotein that attaches to the outer surface of the plasma membrane in neurons and other cell types. Like other proteins in that category it is imported into the ER and from there passes through the Golgi apparatus. In these compartments, further finishing and quality control actions takes place. PrP ligands have been identified. These consist primarily of cell adhesion and matrix proteins, and cellular PrP functions in a neuroprotective capacity. If that is so, then co-receptors are needed that transduce the signals since PrP^C by itself cannot convey a signal across the plasma membrane. Several possible candidate co-receptors have been identified.

   The PrP protein is localized to lipid rafts. Interactions between PrP^C and PrP^{Sc} at the cell surface are essential in disease generation, and the specific lipid composition of the membrane has an influence on these interactions. Deleting PrP^C abolishes the disease even when PrP^{Sc} aggregates are present. And removing the GPI anchor also alleviates the disease even through plaques may form. The implications are that PrP^{Sc} oligomers bind to PrP^C at the rafts resulting in the transduction into the cell of disease-causing signals. Competition for binding between normal and abnormal ligands, with the latter including Aβ oligomers, may well determine disease susceptibility.

5. Fungal prions have been studied extensively. In fungi, prion strains do not cause illness but instead enable the yeasts to rapidly adapt to changing environmental conditions. Strains may well provide a means of epigenetic inheritance from mother to daughter cells. Useful mammalian prions have been found as well, and in these instances it is the prion-like form that is active.

6. Prion-like behaviors may occur in several of the systemic amyloidoses and in many of the neurological disorders. Evidence has been uncovered for prion-like cell-to-cell transmission of the AA amyloid protein in the case of the AA systemic amyloidosis, and tau, α-synuclein, huntingtin and other primary causal agents of neurodegeneration. These findings are potentially of great significance and are explored in the remaining chapters of the book.

Appendix. Lipid Membrane Composition and Biophysical Properties

The mobility properties of the constituents of biological membranes depend both on the amount of cholesterol and the type of lipids present. Membrane lipids consist of a head group, a backbone, and a tail region (Fig. 7.10). Phosphoglycerides contain a glycerol backbone, whereas sphingolipids utilize sphingosine. Lipids found in biological membranes vary in acyl chain length and degree of saturation. Chains possess an even numbers of carbons typically between 14 and 24 with 16, 18, and 20. Chains with one or more doubly bonded carbons are unsaturated. These bonds are rigid and introduce kinks in the chain. These kinks cause irregularities or voids to appear in the array and, consequently, these molecules cannot be packed tightly. In contrast, in a fully saturated acyl chain the carbon–carbon atoms are covalently linked by single bonds. These carbon atoms are able to maximize the number of bonds with hydrogen atoms. As a result the chains can freely rotate about their carbon–carbon bonds and can pack tightly.

Cholesterol plays an important role in determining the fluidity of the membrane compartments. It is smaller than the phospholipids and sphingolipids. As the concentration of cholesterol increases the lipid membrane becomes less like a disordered gel and more like an ordered liquid in which the lipids are more tightly packed together. Three main types of lipid membranes are illustrated in Fig. 7.11. The

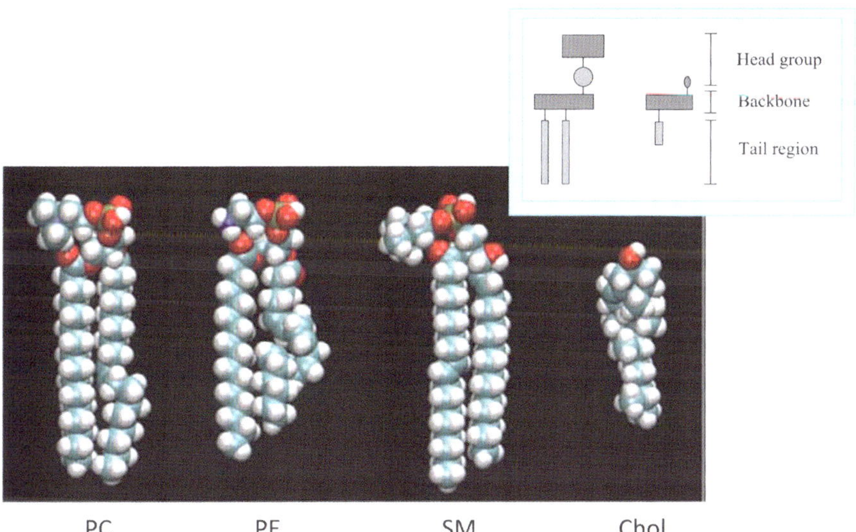

Fig. 7.10 Membrane lipid and cholesterol structure. *PC* phosphatidylcholine (also abbreviated PtdCho), *PE* phosphatidylethanolamine (PtdEtn), *SM* sphingomyelin, *Chol* cholesterol. The inset depicts, in block diagram form, the skeletal structure of the molecules with a typical lipid shown on the *left* and a smaller cholesterol molecule pictured on the *right*. (Main figure from van Meer *EMBO J*. 24: 3159 © 2005 Reprinted with permission from John Wiley and Sons)

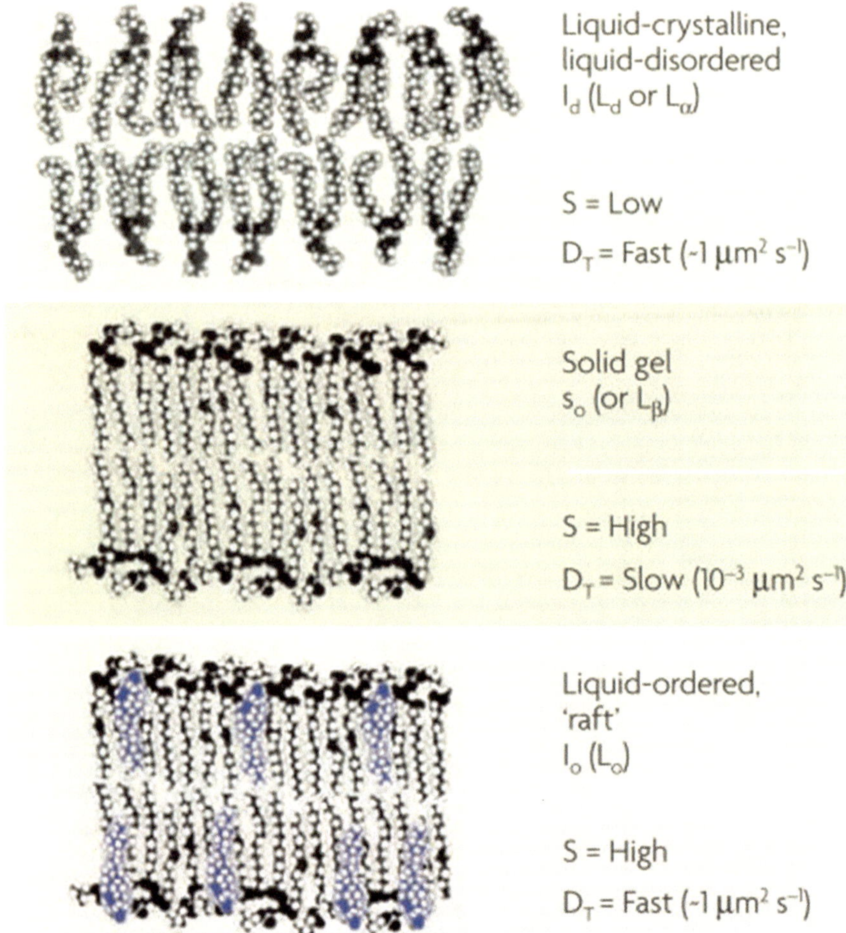

Fig. 7.11 Depiction of the lipid phases typical of biological membranes. From *top* to *bottom*: liquid-disordered (l_d), solid gel (s_o), and liquid-ordered (l_o) in which there is a high cholesterol content (*blue*). The chain segment order (*S*) and diffusion constant (D_T) for each of the phases are shown (from van Meer *Nat. Rev. Mol. Cell Biol.* 9: 112 © 2008 Reprinted by permission from Macmillan Publishers Ltd)

upper panel depicts a liquid disordered membrane composed of unsaturated and fairly short phosphatidylcholine (or phosphatidylethanolamine) lipid molecules. At the other extreme the bottom panel illustrates the organization of a membrane composed of longer, saturated and fairly straight sphingomyelin molecules plus a considerable amount of cholesterol. This membrane structure is typical of a lipid raft.

The biophysical properties of these membrane compartments are captured by phase diagrams such as the one shown in Fig. 7.12. This diagram summarizes the possible phases exhibited by mixtures containing various proportions of an unsaturated lipid, a saturated lipid, and cholesterol. In this figure, it can be seen that as the

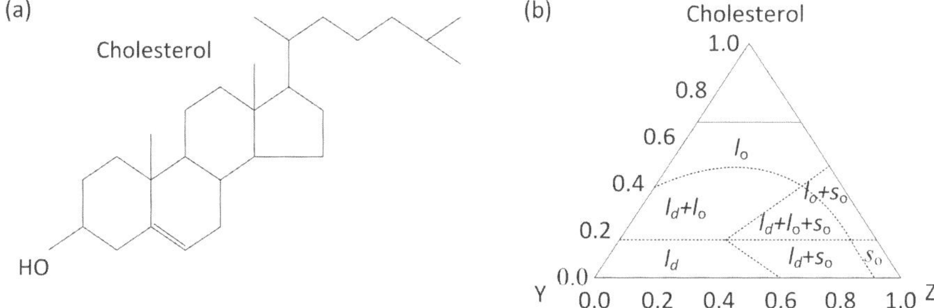

Fig. 7.12 Structure of cholesterol and a lipid membrane phase diagram. (**a**) Rigid structure of the cholesterol molecule. (**b**) The phases of a hypothetical mixture of cholesterol, an unsaturated lipid Y (e.g., PC) and a saturated lipid Z (e.g., SM) at temperatures near or at the physiological temperature, 37 °C

cholesterol content increases the liquid-ordered (l_o) phase begins to dominate. Membrane compartments lacking cholesterol but with a high unsaturated lipid content are disordered (l_d), while saturated lipids generate gel-like (s_o) structures.

Bibliography

Alper, T., Cramp, W. A., Haig, D. A., & Clarke, M. C. (1967). Does the agent of scrapie replicate without nucleic acid? *Nature, 214,* 764–766.

Alper, T., Haig, D. A., & Clarke, M. C. (1966). The exceptionally small size of the scrapie agent. *Biochemical and Biophysical Research Communications, 22,* 278–284.

Baron, G. S., Wehrly, K., Dorward, D. W., Chesebro, B., & Caughey, B. (2002). Conversion of raft-associated prion protein to the protease-resistant state requires insertion of PrP-res (PrP[Sc]) into contiguous membranes. *The EMBO Journal, 21,* 1031–1040.

Basler, K., Oesch, B., Scott, M., Westaway, D., Wälchli, M., Groth, D. F., et al. (1986). Scrapie and cellular PrP isoforms are encoded by the same chromosomal gene. *Cell, 46,* 417–428.

Brandner, S., Isenmann, S., Raeber, A., Fischer, M., Sailer, A., Kobayashi, Y., et al. (1996). Normal host prion protein necessary for scrapie-induced neurotoxicity. *Nature, 379,* 339–343.

Castilla, J., Saá, P., Hetz, C., & Soto, C. (2005). *In vitro* generation of infectious scrapie prions. *Cell, 121,* 195–206.

Chesebro, B., Race, R., Wehrly, K., Nishio, J., Bloom, M., Lechner, D., et al. (1985). Identification of scrapie prion protein-specific mRNA in scrapie-infected and uninfected brain. *Nature, 315,* 331–333.

Chesebro, B., Trifilo, M., Race, R., Meade-White, K., Teng, C., LaCasse, R., et al. (2005). Anchorless prion protein results in infectious amyloid disease without clinical scrapie. *Science, 308,* 1435–1439.

Deleault, N. R., Harris, B. T., Reese, J. R., & Supattapone, S. (2007). Formation of native prions from minimal components *in vitro. Proceedings of the National Academy of Sciences of the United States of America, 104,* 9741–9746.

Deleault, N. R., Piro, J. R., Walsh, D. J., Wang, F., Ma, J., Geoghegan, J. C., & Supattapone, S. (2012). Isolation of phosphatidylethanolamine as a solitary cofactor for prion formation in the absence of nucleic acids. *Proceedings of the National Academy of Sciences of the United States of America, 109,* 8546–8551.

Gajdusek, D. C., Gibbs, C. J., Jr., & Alpers, M. (1966). Experimental transmission of a kuru-like syndrome to chimpanzees. *Nature, 209,* 794–796.

Gajdusek, D. C., & Zigas, V. (1957). Degenerative disease of the central nervous system in New Guinea: Epidemic occurrence of "kuru" in the native population. *New England Journal of Medicine, 257*, 974–978.

Gibbs, C. J., Jr., Gajdusek, D. C., Asher, D. M., Alpers, M. P., Beck, E., Daniel, P. M., et al. (1968). Creutzfeldt-Jakob disease (spongiform encephalopathy): Transmission to the chimpanzee. *Science, 161*, 388–389.

Griffith, J. S. (1967). Self-replication and scrapie. *Nature, 215*, 1043–1044.

Hadlow, W. J. (1959). Scrapie and kuru. *Lancet, 2*, 289–290.

Kocisko, D. A., Come, J. H., Priola, S. A., Chesebro, B., Raymond, G. J., Lansbury, P. T., et al. (1994). Cell-free formation of protease-resistant prion protein. *Nature, 370*, 471–474.

Mallucci, G., Dickinson, A., Linehan, J., Klöhn, P. C., Brandner, S., & Collinge, J. (2003). Depleting neuronal PrP in prion infection prevents disease and reverses spongiosis. *Science, 302*, 871–874.

Makarava, N., Kovacs, G. G., Bocharova, O., Savtchenko, R., Alexeeva, I., Budka, H., et al. (2010). Recombinant prion protein induces a new transmissible prion disease in wild-type animals. *Acta Neuropathologica, 119*, 177–187.

Oesch, B., Westaway, D., Wälchli, M., McKinley, M. P., Kent, B. H., Aebersold, R., et al. (1985). A cellular gene encoding scrapie PrP 27–30 proteins. *Cell, 40*, 735–746.

Pan, K. M., Baldwin, M., Nguyen, J., Gasset, M., Serban, A., Groth, D., et al. (1993). Conversion of α-helices into β-sheets features in the formation of the scrapie prion proteins. *Proceedings of the National Academy of Sciences of the United States of America, 90*, 10962–10966.

Prusiner, S. B. (1982). Novel proteinaceous infectious particles cause scrapie. *Science, 216*, 136–144.

Prusiner, S. B., McKinley, M. P., Bowman, K. A., Bolton, D. C., Bendheim, P. E., Groth, D. F., et al. (1983). Scrapie prions aggregate to form amyloid-like birefringent rods. *Cell, 35*, 349–358.

Riek, R., Hornemann, S., Wider, G., Billeter, M., Glockshuber, R., & Wüthrich, K. (1996). NMR structure of the mouse prion protein domain PrP(121–231). *Nature, 382*, 180–182.

Sandberg, M. K., Al-Doujaily, H., Sharps, B., Clarke, A. R., & Collinge, J. (2011). Prion propagation and toxicity *in vivo* occur in two distinct mechanistic phases. *Nature, 470*, 540–542.

Silveira, J. R., Raymond, G. J., Hughson, A. G., Race, R. E., Sim, V. L., Hayes, S. F., et al. (2005). The most infectious prion protein particles. *Nature, 437*, 257–261.

Wong, C., Xiong, L. W., Horiuchi, M., Raymond, L., Wehrly, K., Chesebro, B., et al. (2001). Sulfated glycans and elevated temperature stimulate PrPSc-dependent cell-free formation of protease-resistant prion protein. *The EMBO Journal, 20*, 377–386.

Prion Passage from Gut to Brain

Mabbott, N. A., & MacPherson, G. G. (2006). Prions and their lethal journey to the brain. *Nature Reviews. Microbiology, 4*, 201–211.

Prion Diseases in Animals and Humans

Bruce, M. E., Will, R. G., Ironside, J. W., McConnell, I., Drummond, D., Suttie, A., et al. (1997). Transmissions to mice indicate that 'new variant' CJD is caused by the BSE agent. *Nature, 389*, 498–501.

Castilla, J., Gonzalez-Romero, D., Saá, P., Morales, R., de Castro, J., & Soto, C. (2008). Crossing the species barrier by PrPSc replication *in vitro* generates unique infectious prions. *Cell, 134*, 757–765.

Chesebro, B. (2003). Introduction to the transmissible spongiform encephalopathies or prion diseases. *British Medical Bulletin, 66*, 1–20.

Collins, S., McLean, C. A., & Masters, C. L. (2001). Gerstmann-Sträussler-Scheinker syndrome, fatal familial insomnia, and kuru: A review of three less common human transmissible spongiform encephalopathies. *Journal of Clinical Neuroscience, 8*, 387–397.

Hill, A. F., Desbruslais, M., Joiner, S., Sidle, K. C. L., Gowland, I., Collinge, J., et al. (1997). The same prion strain causes vCJD and BSE. *Nature, 389*, 448–450.

Johnson, C. J., Phillips, K. E., Schramm, P. T., McKenzie, D., Aiken, J. M., & Pedersen, J. A. (2006). Prions adhere to soil minerals and remain infectious. *PLoS Pathogens, 2*, e32.

Miller, M. W., Williams, E. S., Hobbs, N. T., & Wolfe, L. L. (2004). Environmental sources of prion transmission in mule deer. *Emerging Infectious Diseases, 10*, 1003–1006.

Raymond, G. J., Bossers, A., Raymond, L. D., O'Rourke, K. I., McHolland, L. E., Bryant, P. K. III, et al. (2000). Evidence of a molecular barrier limiting susceptibility of humans, cattle and sheep to chronic wasting disease. *The EMBO Journal, 19*, 4425–4430.

Tamgüney, G., Miller, M. W., Wolfe, L. L., Sirochman, T. M., Glidden, D. V., Palmer, C., et al. (2009). Asymptomatic deer excrete infectious prions in feces. *Nature, 461*, 529–532.

Vanik, D. L., Surewicz, K. A., & Surewicz, W. K. (2004). Molecular basis of barriers for interspecies transmissibility of mammalian prions. *Molecular Cell, 14*, 139–145.

Wadsworth, J. D. F., Hill, A., Beck, J. A., & Collinge, J. (2003). Molecular and clinical classification of human prion diseases. *British Medical Bulletin, 66*, 241–254.

Prion Strains

Beassen, R. A., Kocisko, D. A., Raymond, G. J., Nandan, S., Lansbury, P. T., & Caughey, B. (1995). Non-genetic propagation of strain-specific properties of scrapie prion. *Nature, 375*, 698–700.

Bessen, R. A., & Marsh, R. F. (1994). Distinct PrP properties suggest the molecular basis of strain variation in transmissible mink encephalopathy. *Journal of General Virology, 68*, 7859–7868.

Collinge, J., & Clarke, A. R. (2007). A general model of prion strains and their pathogenicity. *Science, 318*, 930–936.

Jones, E. M., & Surewicz, W. K. (2005). Fibril conformation as the basis of species- and strain-dependent seeding specificity of mammalian prion amyloids. *Cell, 121*, 163–172.

Legname, G., Baskakov, I. V., Nguyen, H. O. B., Riesner, D., Cohen, F. E., DeArmond, S. J., et al. (2004). Synthetic mammalian prions. *Science, 305*, 673–676.

Saborio, G. P., Permanne, B., & Soto, C. (2001). Sensitive detection of pathological prion protein by cyclic amplification of protein misfolding. *Nature, 411*, 810–813.

Safar, J., Wille, H., Itri, V., Groth, D., Serban, H., Torchia, M., et al. (1998). Eight prion strains have PrP molecules with different conformations. *Nature Medicine, 4*, 1157–1165.

Tanaka, M., Chien, P., Naber, N., Cooke, R., & Weissman, J. S. (2004). Conformational variations in an infectious protein determine prion strain differences. *Nature, 428*, 323–328.

Telling, G. C., Parchi, P., DeArmond, S. J., Cortelli, P., Montagna, P., Gabizon, R., et al. (1996). Evidence for the conformation of the pathogenic isoform of the prion protein enciphering and propagating prion diversity. *Science, 274*, 2079–2082.

Toyama, B. H., Kelly, M. J. S., Gross, J. D., & Weissman, J. S. (2007). The structural basis of yeast prion strain variants. *Nature, 449*, 233–237.

Prion Biogenesis, Structure and Function

Bremer, J., Baumann, F., Tiberi, C., Wessig, C., Fischer, H., Schwarz, P., et al. (2010). Axonal prion protein is required for peripheral myelin maintenance. *Nature Neuroscience, 13*, 310–318.

Collinge, J. (1994). Prion protein is necessary for normal synaptic function. *Nature, 370*, 295–297.

Khosravani, H., Zhang, Y., Tsutsui, S., Hameed, S., Altier, C., Hamid, J., et al. (2008). Prion protein attenuates excitotoxicity by inhibiting NMDA receptors. *Journal of Cell Biology, 181,* 551–565.

Le Pichon, C. E., Valley, M. T., Polymenidou, M., Chesler, A. T., Sagdullaev, B. T., Aguzzi, A., et al. (2009). Olfactory behavior and physiology are disrupted in prion protein knockout mice. *Nature Neuroscience, 12,* 60–69.

Málaga-Trillo, E., Solis, G. P., Schrock, Y., Geiss, C., Luncz, L., Thomanetz, V., et al. (2009). Regulation of embryonic cell adhesion by the prion protein. *PLoS Biology, 7,* e1000055.

Mouillet-Richard, S., Ermonval, M., Chebassier, C., Laplanche, J. L., Lehrmann, S., Launay, J. M., et al. (2000). Signal transduction through prion protein. *Science, 289,* 1925–1928.

Santuccione, A., Sytnyk, V., Leshchyns'ka, I., & Schachner, M. (2005). Prion protein recruits its neuronal receptor NCAM to lipid rafts to activate p59[fyn] and to enhance neurite outgrowth. *Journal of Cell Biology, 169,* 341–354.

Stahl, N., Baldwin, M. A., Teplow, D. B., Hood, L., Gibson, B. W., Burlingame, A. L., et al. (1993). Structural studies of the scrapie prion protein using mass spectroscopy and amino acid sequencing. *Biochemistry, 32,* 1991–2002.

Taylor, D. R., & Hooper, N. M. (2006). The prion protein and lipid rafts. *Molecular Membrane Biology, 23,* 89–99.

Prion Toxicity

Hetz, C., Russelakis-Carneiro, M., Maundrell, K., Castilla, J., & Soto, C. (2003). Caspase-12 and endoplasmic reticulum stress mediate neurotoxicity of pathological prion protein. *The EMBO Journal, 22,* 5435–5445.

Kristiansen, M., Deriziotis, P., Dimcheff, D. E., Jachson, G. S., Ovaa, H., Naumann, H., et al. (2007). Disease-associated prion protein oligomers inhibit the 26S proteasome. *Molecular Cell, 26,* 175–188.

Laurén, J., Gimbel, D. A., Nygaard, H. B., Gilbert, J. W., & Strittmatter, S. M. (2009). Cellular prion protein mediates impairment of synaptic plasticity by amyloid-β oligomers. *Nature, 457,* 1128–1132.

Moreno, J. A., Radford, H., Peretti, D., Steinert, J. R., Verity, N., Martin, M. G., et al. (2012). Sustained translational repression by eIF2α-P mediates prion neurodegeneration. *Nature, 485,* 507–511.

Rane, N. S., Kang, S. W., Chakrabarti, O., Feigenbaum, L., & Hedge, R. S. (2008). Reduced translocation of nascent prion protein during ER stress contributes to neurodegeneration. *Developmental Cell, 15,* 359–370.

Resenberger, U. K., Harmeier, A., Woerner, A. C., Goodman, J. L., Müller, V., Krichnan, R., et al. (2011). The cellular prion protein mediates neurotoxic signaling of β-sheet-rich conformers independent of prion replication. *The EMBO Journal, 30,* 2057–2070.

Sonati, T., Reimann, R. R., Falsig, J., Baral, P. K., O'Connor, T., Hornemann, S., et al. (2013). The toxicity of antiprion antibodies is mediated by the flexible tail of the prion protein. *Nature, 501,* 102–106.

Um, J. W., Nygaard, H. B., Heiss, J. K., Kostylev, M. A., Stagi, M., Vortmeyer, A., et al. (2012). Alzheimer amyloid-β oligomer bound to postsynaptic prion protein activates Fyn to impair neurons. *Nature Neuroscience, 15,* 1227–1235.

Structure of the Prion Amyloid Fibrils

Cobb, N. J., Sönnichsen, F. D., Mchaourab, H., & Surewicz, W. K. (2007). Molecular architecture of human prion protein amyloid: A parallel, in-register β-structure. *Proceedings of the National Academy of Sciences of the United States of America, 104*, 18946.

DeMarco, M. L., & Daggett, V. (2004). From conversion to aggregation: Protofibril formation of the prion protein. *Proceedings of the National Academy of Sciences of the United States of America, 101*, 2293–2298.

Govaerts, C., Wille, H., Prusiner, S. B., & Cohen, F. E. (2004). Evidence for assembly of prions with left-handed β-helices into trimers. *Proceedings of the National Academy of Sciences of the United States of America, 101*, 8342–8347.

Krishnan, R., & Lindquist, S. L. (2005). Structural insights into a yeast prion illuminate nucleation and strain diversity. *Nature, 435*, 765–772.

Shewmaker, F., Wichner, R. B., & Tycko, R. (2006). Amyloid of the prion domain of Sup35p has an in-register parallel β-sheet structure. *Proceedings of the National Academy of Sciences of the United States of America, 103*, 19754–19759.

Wickner, R. B., Dyda, F., & Tycko, R. (2008). Amyloid of Rnq1p, the basis of the [PIN+] prion, has a parallel in-register β-sheet structure. *Proceedings of the National Academy of Sciences of the United States of America, 105*, 2403–2408.

Fungal Prions

Chernoff, Y. O., Lindquist, S. L., Ono, B., Inge-Vechtomov, S. G., & Liebman, S. W. (1995). Role of the chaperone protein Hsp104 in propagation of the yeast prion-like factor [psi+]. *Science, 268*, 880–884.

Halfmann, R., Jarosz, D. F., Jones, S. K., Chang, A., Lancaster, A. K., & Lindquist, S. (2012). Prions are a common mechanism for phenotypic inheritance in wild yeasts. *Nature, 482*, 363–368.

Liebman, S. W., & Chernoff, Y. O. (2012). Prions in yeast. *Genetics, 191*, 1041–1072.

Mukhopadhyay, S., Krishnan, R., Lemke, E. A., Lindquist, S., & Deniz, A. A. (2007). A natively unfolded yeast prion monomer adopts an ensemble of collapsed and rapidly fluctuating structures. *Proceedings of the National Academy of Sciences of the United States of America, 104*, 2649–2654.

Serio, T. R. (2000). Nucleated conformational conversion and the replication of conformational information by a prion determinant. *Science, 289*, 1317–1321.

Shorter, J., & Lindquist, S. (2004). Hsp104 catalyzes formation and elimination of self-replicating Sup35 prion conformers. *Science, 304*, 1793–1797.

True, H. L., Berlin, I., & Lindquist, S. L. (2004). Epigenetic regulation of translation reveals hidden genetic variation to produce complex traits. *Nature, 431*, 184–187.

Tuite, M. F., & Serio, T. R. (2010). The prion hypothesis: From biological anomaly to basic regulatory mechanism. *Nature Reviews. Molecular Cell Biology, 11*, 823–833.

Tyedmers, J., Madariaga, M. L., & Lindquist, S. (2008). Prion switching in response to environmental stress. *PLoS Biology, 6*, e294.

Wickner, R. B. (1994). [URE3] is an altered URE2 protein: Evidence for a prion analog in Saccharomyces cerevisiae. *Science, 264*, 566–569.

Beneficial Prions in Higher Organisms

Hou, F., Sun, L., Zheng, H., Skaug, B., Jiang, Q. X., & Chen, Z. J. (2011). MAVS forms functional prion-like aggregates to activate and propagate antiviral innate immune response. *Cell, 146*, 448–461.

Si, K., Choi, Y. B., White-Grindley, E., Majumdar, A., & Kandel, E. R. (2010). *Aplysia* CPEB can form prion-like multimers in sensory neurons that contribute to long-term facilitation. *Cell, 140*, 421–435.

Si, K., Giustetto, M., Etkin, A., Hsu, R., Janisiewicz, A. M., Miniaci, M. C., et al. (2003). A neuronal isoform of CPEB regulates local protein synthesis and stabilizes synapse-specific long-term facilitation in Aplysia. *Cell, 115*, 893–904.

Cell-to-Cell Spread

Braak, H., & Braak, E. (1991). Neuropathological staging of Alzheimer-related changes. *Acta Neuropathologica, 82*, 239–259.

Braak, H., del Tredici, K., Rüb, U., de Vos, R. A., Jansen Steur, E. N., & Braak, E. (2003). Staging of brain pathology related to sporadic Parkinson's disease. *Neurobiology of Aging, 24*, 197–211.

Fevrier, B., Vilette, D., Archer, F., Loew, D., Faigle, W., Vidal, M., et al. (2004). Cells release prions in association with exosomes. *Proceedings of the National Academy of Sciences of the United States of America, 101*, 9683–9688.

Gousset, K., Schiff, E., Langevin, C., Marijanovic, Z., Caputo, A., Browman, D. T., et al. (2009). Prions hijack tunneling nanotubes for intercellular spread. *Nature Cell Biology, 11*, 328–336.

Lundmark, K., Westermark, G. T., Nyström, S., Murphy, C. L., Solomon, A., & Westermark, P. (2002). Transmissibility of systemic amyloidosis by a prion-like mechanism. *Proceedings of the National Academy of Sciences of the United States of America, 99*, 6979–6984.

Rustom, A., Saffrich, R., Markovic, I., Walther, P., & Gerdes, H. H. (2004). Nanotubular highways for intercellular organelle transport. *Science, 303*, 1007–1010.

Simons, M., & Raposo, G. (2009). Exosomes—Vesicular carriers for intercellular communication. *Current Opinion in Cell Biology, 21*, 575–581.

Théry, C., Ostrowski, M., & Segura, E. (2009). Membrane vesicles as conveyors of immune responses. *Nature Reviews. Immunology, 9*, 581–593.

Lipid Membranes

Karnovsky, M. J., Kleinfeld, A. M., Hoover, R. L., & Klausner, R. D. (1982). The concept of lipid domains in membranes. *The Journal of Cell Biology, 94*, 1–6.

Simons, K., & Ikonen, E. (1997). Functional rafts in cell membranes. *Nature, 387*, 569–572.

Simons, K., & van Meer, G. (1988). Lipid sorting in epithelial cells. *Biochemistry, 27*, 6197–6202.

Singer, S. J., & Nicolson, G. L. (1972). The fluid mosaic model of the structure of cell membranes. *Science, 175*, 720–731.

Chapter 8
Alzheimer's Disease

The search for a specific biochemical cause of Alzheimer's disease began in the 1960s and 1970s, and was inspired by the burgeoning success of levodopa treatment for Parkinson's disease. In the case of Alzheimer's disease, brain regions associated with higher brain functions, primarily the hippocampus and neocortex, are affected. Investigations of those regions in the presence of senile dementia led to the discovery that (1) there are deficits in presynaptic terminals in the enzyme that catalyzes the synthesis of the neurotransmitter acetylcholine (ACh), namely, choline acetyltransferase (ChAT); (2) acetylcholine has a role in learning and memory, and (3) blocking its release leads to memory impairment.

This collective set of findings gave rise to the *cholinergic hypothesis* of Alzheimer's disease. That hypothesis was articulated in an influential 1982 paper by Bartus and led to the development of a number of drugs that act as cholinesterase inhibitors, that is, as agents that prevent acetylcholinesterase (AChE) from hydrolyzing acetylcholine in the synaptic cleft. Since that time it has become increasingly apparent that the primary effects of cholinergic depletion are on attention and only small improvements in symptoms of AD are seen in patients receiving e ACh-directed drugs. Following the discoveries of the involvement of amyloid-β in AD the main focus shifted away from the early groundbreaking observations on the cholinergic system to the glutamatergic system, the other main central excitatory neurotransmitter, and to pathologies derived from misfolded amyloid β.

8.1 The Amyloid Cascade Hypothesis

By the early 1990s the amyloid cascade hypothesis had replaced the cholinergic hypothesis as the leading idea as to the cause of Alzheimer's disease. This articulation has been revised and expanded since its initial formulation in 1992 by Hardy and Higgins. In its initial form, the amyloid cascade hypothesis had as its

© Springer International Publishing Switzerland 2015

M. Beckerman, *Fundamentals of Neurodegeneration and Protein Misfolding Disorders*,
Biological and Medical Physics, Biomedical Engineering,
DOI 10.1007/978-3-319-22117-5_8

main focus the buildup of amyloid plaques that collect in the extracellular spaces. These deposits were thought to initiate a cascade of events leading to the development of the neurofibrillary tangles and the fatal loss of neurons in the affected regions of the brain resulting in dementia and death.

Several key developments guided and supported this hypothesis. The first of these, as discussed in Chap. 1, was the identification in 1984 by Glenner and Wong of the 39–43 amino acid long amyloid β peptides as the main component of the extracellular plaques seen in Alzheimer's disease. That finding was followed by Masters' confirmation in 1985, and by the discovery in 1987 of the gene that encodes the amyloid precursor protein on chromosome 21. The discovery was followed by that of the presenilin genes, PSEN1 and PSEN2, on chromosome 14 and 1, respectively, that encode proteins involved in processing the amyloid precursor protein, further strengthened the case for Aβ. In these genetic studies, the focus was on individuals suffering from early onset (ages less than 60 years) autosomal dominant familial forms of the disease (EOFAD, or EOAD). It is customary to distinguish between the early familial forms and later sporadic and familial forms of Alzheimer's disease, or LOAD. In the latter case, the main risk factor is the presence of the apolipoprotein E4 (ApoE4) allele. The ApoE protein has a role in Aβ clearance, thereby supplying yet another connection between Aβ and Alzheimer's disease.

However, it was soon noted that the correlations between the presence and severity of the extracellular amyloid plaques and the clinical progression of the illness were weak, at best. The amyloid cascade hypothesis was then modified to reflect these findings along with the growing belief that the main toxic species was not the large and highly visible fibrillar deposits, but rather the smaller, more mobile and highly reactive Aβ oligomers. That shift in focus was promoted by a convergence of results of studies of several different neurodegenerative disorders including AD. The revised and current cascade hypothesis has as its main theme an inappropriate buildup of oligomeric species as the key initiator of dysfunction. That leaves as the paramount challenge identifying the exact manner or manners in which the Aβ oligomers are toxic with the most likely sites being the synapses and the blood–brain barrier, leading to progressive loss of neurons and failing neural circuitry.

This chapter presents the amyloid cascade hypothesis of Alzheimer's disease as it currently stands. It begins with an examination of the two types of sequential processing of the amyloid precursor protein by proteolytic enzymes, and with an overview its possible regulation by trafficking, transcription, translation, and epigenetic elements. This is followed by a brief introduction to the different oligomeric species produced when there is an imbalance between production and clearance. This issue is particularly relevant given that a unique toxic species has not been identified and may not exist. This introduction is then followed by an overview of the three main clearance mechanisms. As already noted, the single most important genetic risk factor for developing LOAD is the presence of the ApoE4 allele. How expression of this factor influences Aβ aggregation and toxicity is not yet apparent but there are a number of plausible scenarios each supported by a growing body of data. These are explored in the content of Aβ clearance.

The blood–brain barrier and neurovascular unit are examined next. During the last few years the blood–brain barrier and the associated neurovascular unit have drawn increasing attention. Alzheimer's disease is a multifaceted disorder. The "old" idea of a vascular dementia has been reborn within a broader framework whereby neural and vascular damage coexist and impact one another. A central tenet of the amyloid cascade hypothesis is that of synaptic damage. Synaptic dysfunction is explored along with findings that many of the key molecular players are lipid raft residents. Those discussions are accompanied by overviews of how inflammation, vascular dysfunction, and the cholinergic system impact the disorder, and then the second key observation of Alzheimer—the presence of neurofibrillary tangles—is examined. Tau is the principal component of the neurofibrillary tangles, and is examined in the context of how it may act as a downstream effector of Aβ toxicity. Its broader set of actions awaits the chapter on tauopathies.

8.2   APP Processing and Generation of the Amyloid-β Protein

The Aβ peptide is derived from a larger β-amyloid precursor protein (APP) through a series of proteolytic cleavage operations. These are carried out by enzymatic proteins and protein complexes referred to as α-secretases, β-secretases, and γ-secretases. The presenilins, key risk factors in EOAD, are part of the γ-secretase complex, which has three other members—Nicastrin, Aph-1, and Pen-2. As illustrated in Fig. 8.1 the presenilins are 8-pass transmembrane proteins proteolytically cleaved into two chains. A pair of aspartate residues positioned in opposition to one another is crucial for presenilin's APP cleavage actions. APP passes through the gap formed by the segments containing the aspartates, and is cleaved. The presenilins carry out their catalytic activities together with the single-pass Nicrastrin protein, the 7-pass Aph-1 protein, and the 2-pass Pen-2 protein, which help assemble and stabilize the complex.

8.2.1   APP Processing

The processing of type I single pass transmembrane APP by the secretases involves making two cuts to the protein—the first cut is made outside the membrane and the second one within the lipid bilayer. The second type of processing is a fairly recent discovery and is known as regulated intramembrane proteolysis (RIP) (see Appendix 1). In the case of APP, there are two distinct ways of making the two cuts. The first way, the non-amyloidogenic pathway, uses the α-secretase to make the first cut. The second way, the amyloidogenic pathway, does not utilize the

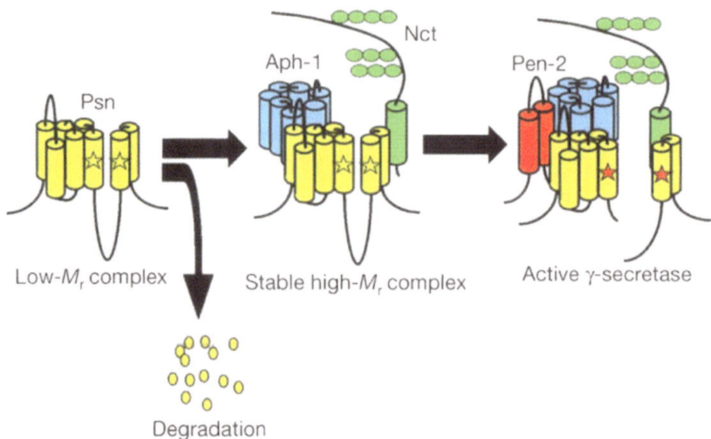

Fig. 8.1 Model of assembly and activation of the catalytically active γ-secretase complex. *Psn* Presenilin, *Aph-1* anterior pharynx defective-1, *Nct* Nicastrin, *Pen-2* presenilin enhancer-2. The positions of the critical aspartate residues are indicated by the pair of opposed stars (from Takasugi *Nature* 422: 438 © 2003 Reprinted by permission from Macmillan Publishers Ltd)

α-secretase but instead employs the β-secretase and that cut is made in a different location. The γ-secretase and its presenilins carry out the second cut in both routes.

In more detail, APP is cleaved at one of two alternative extracellular locations termed the α- and β-sites. This operation is commonly referred to as ectodomain shedding because an ectodomain stub is shed into the luminal space. The γ-secretases that carry out the subsequent cleavages create two fragments—an APP intracellular domain (AICD) and either the P3 or Aβ peptide depending on which enzyme carries out the first cut. As illustrated in the upper part of Fig. 8.2 the ectodomain generated by the α-secretase is designated as APPsα (or alternatively as sAPPα). In the second step, the γ-secretase generates the AICD and a P83 fragment. These operations may be contrasted with the amyloidogenic processing by the β- and γ-secretases shown in the lower portion of Fig. 8.2. In this situation, an sAPPβ stub is created and in the following cut the AICD and Aβ peptide are produced. In the case of cleavage at β sites, but not the α-sites, the fragments include the Alzheimer's disease-causing 39–43-amino-acid-residue forms of the Aβ amyloid protein. The 42 residue form is favored by the mutated Presenilins, and is especially prone to aggregation.

8.2.2 The Aβ Peptide

Aβ is a metalloprotein and, not surprisingly, elevated levels of Fe, Cu and Zn are present in amyloid plaques. Increased amounts of these metals within the synaptic environment contribute to increases in oxidative stress and in the rates of aggregation and fibril growth. Support for these observations comes from the following.

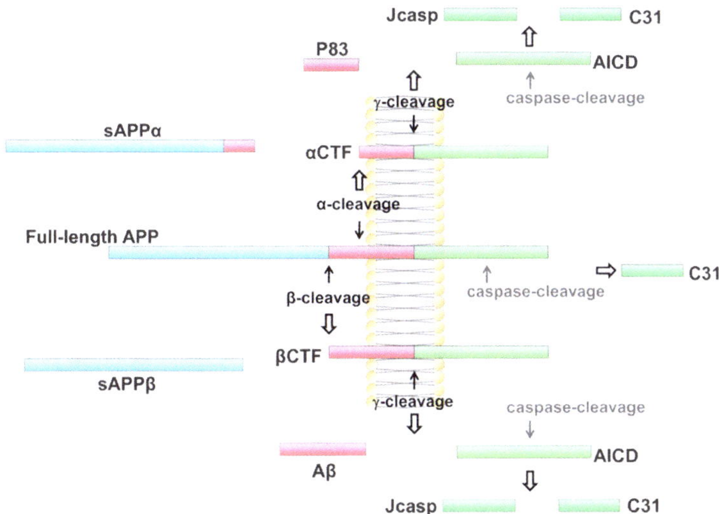

Fig. 8.2 Nonamyloidogenic and amyloidogenic processing of the amyloid precursor protein by α-, β-, and γ-secretases. Full-length APP undergoes non-amyloidogenic cleavage by the α- and γ-secretases, and amyloidogenic cleavage by the β- and γ-secretases (from Zhang 2011 *Mol. Brain* 4: art. 3; reprinted with permission through a Creative Commons Attribution License)

First, Aβ has high affinity binding sites for Cu and Zn, and Fenton chemistry is facilitated by these binding events resulting in an increased production of hydroxyl radicals. Secondly, α-secretases require Zn, β-secretase interacts with Cu and, in those situations where there is a lack of this metal, APP processing is affected. Thirdly, the presenilins facilitate Cu and Zn uptake and turnover. Lastly, the aggregation process itself is accelerated when Aβ coordinates Fe, Cu, or Zn.

Multiple isoforms of Aβ exist. Some of the isoforms are generated by proteolytic cleavage of the most N-terminal residues, while others are subject to a variety of posttranslational modifications. One particularly noteworthy modification is to the glutamine residue at position 3 (by glutaminyl cyclase) to form pyroglutamate, pE3. Modifications such as isomerization, metal-induced oxidation and phosphorylation, especially those involving the N-terminal-most residues, can accelerate the formation of aggregates and fibrils. The most prominent isoforms in the hippocampus and cortex of AD sufferers as determined by means of mass spectroscopy appear to be:

- $A\beta_{1-42}$
- $A\beta_{pE3-42}$
- $A\beta_{4-42}$
- $A\beta_{1-40}$

This bullet list is an ordered one; that is, the $A\beta_{x-42}$ isoforms are thought to be more toxic than the $A\beta_{x-40}$ ones. The reason for this difference is easy to understand upon noting that residues 41 and 42 are isoleucine and alanine, respectively (Fig. 8.3). Both of these residues are hydrophobic (Table 2.1) and their presence renders the

Fig. 8.3 The amyloid-beta (Aβ) peptide. The transmembrane portion of Aβ is shown in *blue*. The locations of the three secretase cleavage sites are shown above the sequence

peptide far more likely to form aggregates. In addition to these isoforms, a number of proteolytic enzymes may be active resulting in isoforms with x values ranging up to 10, $A\beta_{1-y}$ isoforms may be present, with y assuming values from 37 to 43, and various combinations of the two types of action may occur. The most prominent of the resulting isoforms are included in the bullet list.

8.2.3 Identity of the α- and β-Secretases

The α-secretases and β-secretases responsible for cleaving the APPs at the α- and β-sites are members of two families of proteases. One or more members of the "a disintegrin and metalloprotease" (ADAM) family function as α-secretases. The ADAM proteases are single pass transmembrane proteins with a N-terminal signal peptide, followed by a pro-domain. Two members of this family have roles in APP metabolism—ADAM17, also called TACE and ADAM10. ADAM10 has been identified as the main family members with ADAM17 perhaps acting as a backup.

The β-site APP cleavage enzyme 1 (BACE1) protein is the only beta secretase found to date. This transmembrane protein is a membrane-associated aspartyl protease belonging to the pepsin family. It possesses pre and pro domains followed by a catalytic domain, transmembrane helix, and cytoplasmic segment. BACE1 is the rate limiting enzyme in the formation of Aβ peptides.

8.3 Regulation of APP and Secretase Expression and Activity Occurs at Multiple Levels

APP and the secretases are tightly regulated at the trafficking, transcription, translation, and posttranslational levels. Full length APP is synthesized in the endoplasmic reticulum (ER). It is then transported through the trans-Golgi network (TGN), one of the main sites of APP residency in neurons. There it undergoes posttranslational modifications such as glycosylation and phosphorylation. It is transported to the cell surface in TGN-produced secretory vesicles where it may be cleaved by the α-secretases or reenter the endosomal pathway enclosed in clathrin-coated vesicles. Then, it may be sent back to the plasma membrane or transit to lysosomes for degradation. These trafficking steps are sketched in Fig. 8.4.

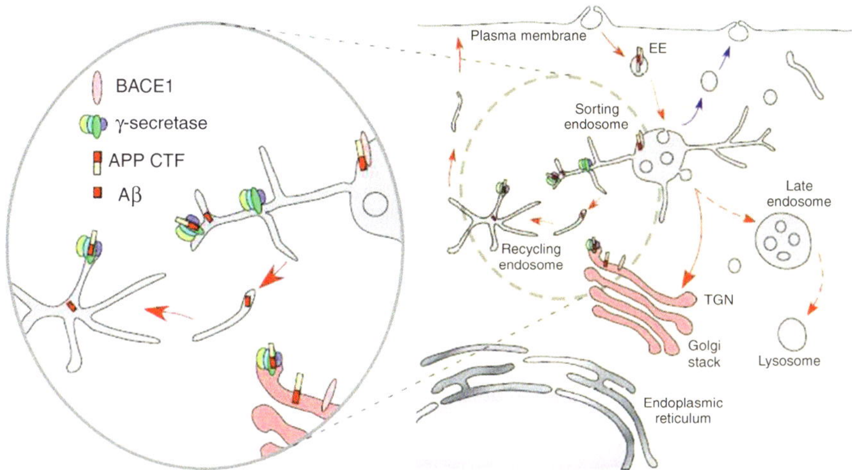

Fig. 8.4 Cellular trafficking of APP, Aβ, and BACE1. [The full-length protein (depicted as *short bars*) is synthesized in the ER, passes through the TGN to the cell surface (1) where is subject to recycling through the endosomal network (2) either back to the plasma membrane (3) or to the lysosomal compartment for degradation] (from Vetrivel *Biochim. Biophys. Acta* 1801: 860 © 2010 Reprinted by permission from Elsevier)

The β-secretase BACE1 follows a similar trafficking pattern, and in the Golgi its pro-domain is removed. It too cycles between the endosomes and plasma membrane. Its dwelling in the early endosomes permits it to come into contact with APP and cleave it, while residency in late endosomes leads to degradation in lysosomes. The four components that comprise the γ-secretase are synthesized and assembled in the ER. They then cycle between the ER and Golgi with a small number exiting to the plasma membrane and endosomal vesicles where they are able to carry out the second, intramembrane cleavage of the α- or β-CTF. The TGN and early endosomes function as sorting hubs where proteins are placed into cargo vesicles and sent to various locations. These processes are regulated. *GGAs* (Golgi-localized γ-ear-containing ARF binding proteins) mediate the trafficking between TGN and endosomes. Other complexes known as *retromers* regulate cargo selection and budding activities. All of these activities can influence the interactions between BACE1 and APP.

The promoter for BACE1 is a complex one with putative binding sites for a number of transcription factors. Its promoter is TATA-less with high CG content and with a Sp1 binding element (as is the case for APP and PrP), characteristics of a housekeeping gene. Other prominent binding sites are present, as well, such as those for NF-κB and PPARγ. The former regulates APP, β-secretase, and γ-secretase gene transcription. As has been found in many settings, the actions of NF-κB can be either repressive or stimulatory depending on the presence or absence of cellular stresses. Under healthy, non-stressed conditions NF-κB represses transcription of BACE1 but when the cells are continually exposed to stressful conditions such as those present in AD (e.g., high Aβ levels) NF-κB promotes BACE1 gene expression. PPARγ activity is thought to repress BACE1. However, activated microglia and astrocytes secrete pro-inflammatory cytokines and these signaling molecules suppress PPARγ leading to an increased production of BACE1. Overall, a positive feedback loop may well occur in which a resulting increased Aβ plaque buildup stimulates further inflammatory response leading to a greater production of BACE1 and hence more Aβ plaque buildup.

The expression levels of BACE1 are also regulated translationally and posttranslationally. One of the properties of the AD brain (and also of the non-AD brains of young/middle-aged individuals carrying the ApoE4 allele) is reduced glucose utilization. A link between reduced glucose utilization and BACE1 activity is provided by the translation initiation factor eIF2α. This control element increases BACE1, Aβ and APP levels when it is phosphorylated in response to energy deprivation stress. A second route whereby cellular stresses contribute to BACE1 activation is through actions by noncoding RNAs, most notably by the microRNAs miR-29a/b-1 and miR-107. MicroRNAs act as negative regulators of translation of substrate mRNAs. In AD brains, the activities of these microRNAs are reduced leading to an increased translation of BACE1 and a corresponding increase in Aβ levels.

8.4 Apolipoprotein E4 and Cholesterol Transport

Cholesterol is essential for proper synaptic and neuronal function. About 25 % of the total cholesterol in the body resides in the brain. Most of it resides in myelin sheaths with the remainder embedded in the plasma membranes of neurons and astrocytes. In short, cholesterol is:

- A major component of the myelin sheath and **needed for axonal growth and repair**.
- Used in synapse formation and remodeling, actions **essential for learning and memory**.
- The key ingredient in lipid rafts, **critical for synaptic transmission**.

Cholesterol was first discovered in gallstones by François Poulletiere de la Salle in 1769. It was subsequently rediscovered and named cholesterine by Eugene Chevreul in 1815, and identified in blood by Boudet in 1833. As noted by Brown and Goldstein in their Nobel lecture, cholesterol has fascinated scientists ever since its discovery with 13 Nobel prizes having been awarded to scientists who studied this small molecule. Cholesterol cannot circulate by itself. Instead, it has to be transported by lipid transport vesicles—the *lipoproteins* (see Appendix 2). Uncovering the nature of this fat transport system has spanned three centuries— from its initial surmise by Robert Boyle in 1665 to the discovery of the low-density lipoprotein (LDL) receptor by Brown and Goldstein in 1974. Since then this transport system has been linked to a number of disorders, the most prominent being atherosclerosis and Alzheimer's disease.

Apolipoproteins are amphipathic, lipid-binding protein constituents of the lipoprotein particles. There are several different kinds of apolipoproteins. These are designated by letters—A, B, C, etc. and sometimes numerically to distinguish different isoforms. For example, Apo B has two main isoforms, designed as B-48 and B-100. The former, a truncated version of Apo B-100, is synthesized in the gut and the latter is produced in the liver.

There are several different classes of lipoprotein particles. These, the lipids that they chaperone, and the responsible apolipoproteins are listed in Table 8.2 in Appendix 2. The apolipoproteins are large, and they wrap around and enable the transport of the hydrophobic lipids from one place to another in the body. For example, a typical LDL particle would contain one large apolipoprotein B particle plus about 2500 lipid molecules. These would be arranged in the following way. There would be core consisting of cholesteryl esters and triglycerides; plus an outer layer consisting of a smaller number of free cholesterol molecules plus a phospholipid monolayer. The entire assembly would be enveloped by the large apoB protein, which makes a pocket for the lipid ball.

Apolipoprotein E has three common isoforms, encoded by the APOE gene E2, E3, and E4 alleles. These differ from one another through the presence of either Cys or Arg at amino acid positions 112 and 158. The frequency of each of these alleles in the general population and among AD patients is presented in Table 8.1. As shown in the table the E3 allele is the most common and the E2 allele the least common. Possession of the E4 allele is the major risk factor for development of late onset familial and sporadic Alzheimer's disease. Individual heterozygous for this allele have a 47 % chance of developing AD with a mean age of onset of 76 years. Individuals homozygous for this gene have an even greater chance of developing AD. For them the average age of onset decreases to 68 years with a frequency of 91 %.

Apolipoprotein E (ApoE) is the main lipoprotein in the brain and cerebrospinal fluid (CSF). The abundance of ApoE in the brain is high with levels second only to that of the liver. The liver cannot supply the brain with ApoE since these particles are too large to pass through the blood–brain barrier (BBB). Instead, the main suppliers of ApoE are the astrocytes with microglia and neurons providing the remainder. These cells maintain cholesterol and lipid homeostasis in the brain and ensure that the essential functions performed by cholesterol described above can be carried out.

Table 8.1 Apolipoprotein E2, E3, and E4 alleles and their frequencies in the general population and among AD patients (modified from Liu 2013 *Nat. Rev. Neurol.* 9: 106)

ApoE allele	Position 112	Position 158	Freq. (%) General	Freq. (%) AD
E2	Cys	Cys	8	4
E3	Cys	Arg	78	59
E4	Arg	Arg	14	37

8.5   Aβ Clearance and ApoE

Amyloid-β generation and clearance are complementary actions. Under healthy conditions, clearance is sufficiently robust and there is a balance between the two that prevents large deposits of amyloid-β from developing. Clearance of amyloid-β occurs in one of three ways (Fig. 8.5):

- By means of enzymatic degradation;
- Through uptake and lysosomal degradation by microglia and astrocytes, and
- From export through the blood–brain barrier and clearance by liver and kidneys.

Several enzymes, most notably neprilysin (NEP), insulin-degrading enzyme (IDE), tissue plasminogen activator (tPA), and matrix metalloproteinases (MMPs), are responsible for degrading Aβ peptides. These enzymes reside in several places within multiple cell types and in the extracellular spaces either free or attached to

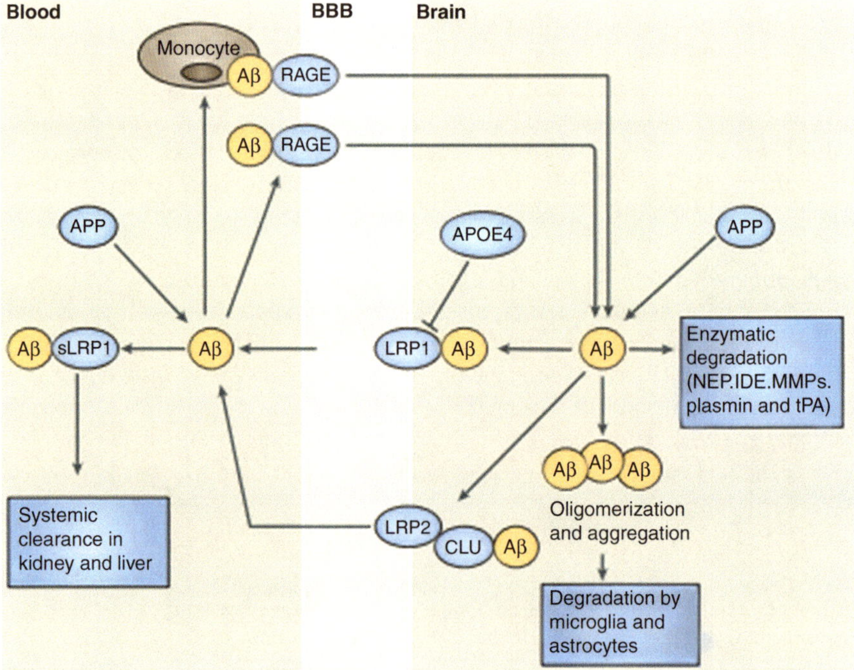

Fig. 8.5 Three methods of clearance of Aβ oligomers. *LRP1, 2* low-density lipoprotein receptor-related protein 1, 2, *CLU* clusterin (also known as apolipoprotein J or ApoJ), *RAGE* receptor for advanced glycation end-products. See text for discussion (from Zlokovic *Nat. Rev. Neurosci.* 12: 723 © 2011 Reprinted by permission from Macmillan Publishers Ltd)

the outer surface of the plasma membranes of those cells. Misfolded $A\beta_{1-40}$ and especially $A\beta_{1-42}$ resist degradation by aggregating, and this activity is enhanced by associations with heavy metals such as zinc and copper. These cations coordinate pairs of Aβ peptides, thereby facilitating aggregation into protease-resistant amyloid plaques.

Astrocytes and microglia accumulate in senile plaques. They not only secrete Aβ-peptide degrading enzymes but also express receptors that enable their uptake and phagocytosis. Prominent among the receptors are members of the low-density lipoprotein receptor-related protein receptor (LRP) family, scavenger receptors, toll-like receptors (TLRs) and other danger receptors, complement receptors, and the receptor for advanced glycation end-products (RAGE). Aβ oligomers are difficult to deal with especially when present at high concentrations. As a result, phagocytosis is frequently impaired and an inflammatory response induced by the now activated glial cells. These cells release a variety of pro-inflammatory enzymes and reactive oxygen and nitrogen species. Interestingly, ApoE and especially the E4 allele compete with Aβ for binding to LRP1 and by that means ApoE can influence Aβ clearance.

APP is expressed throughout the body. Import of Aβ into the brain and export from the brain are mediated by RAGE and LRP1/2, respectively. RAGE is a multi-ligand receptor belonging to the Ig (immunoglobulin) superfamily. It is responsible for transporting unbound circulating Aβ peptides across the blood–brain barrier. Under normal conditions clearance by LRP is far more rapid than influx by RAGE, and the concentration of amyloid-β peptides in the brain is kept at a low level.

8.6   A Variety of Aβ Oligomeric Assemblages Are Formed

The revised and expanded amyloid cascade hypothesis posits that an inappropriate buildup of Aβ oligomers is the key initiating event in AD. Multiple lines of evidence acquired starting in the 1990s support this belief. In particular, the accretion of amyloid fibrils was found to correlate poorly with cognitive impairment, whereas the buildup of soluble, non-fibrillar Aβ oligomers correlated far more strongly with the severity of the disease. The production and secretion of soluble oligomers is a normal cellular process and thus by itself is not an indication of anything amiss. Rather, it is the production of large quantities of non-normal oligomeric assemblages that are the putative causative agents. These arise as a result of missense mutations in the APP proteins and presenilins, as a consequence of the ApoE4 allele, and due to gradually emerging age and stress-related factors including those associated with the metabolic syndrome.

A variety of oligomeric species may be formed from $A\beta_{1-40}$ and the highly toxic $A\beta_{1-42}$ peptides. Each of these species may populate several conformations endowed with specific set of toxicities, and these can interconvert from one oligomeric form to another. This multiplicity of assemblages and conformations creates difficulties

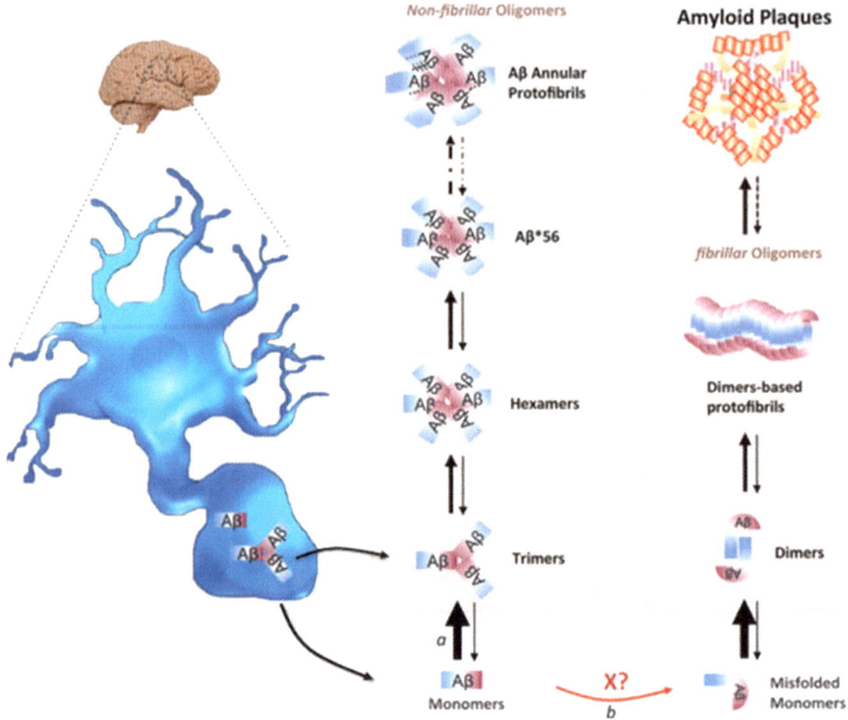

Fig. 8.6 Hypothetical dual pathway model of Aβ-oligomer assembly. Neurons secrete Aβ monomers and trimers. These can assemble and interconvert into larger-n species along the non-fibrillar pathway. The monomers may misfold, stimulated by unknown factor X, and form dimers. The dimers can assemble into higher-n oligomers and into the large fibrillar structures, the amyloid plaques (from Larson *J. Neurochem.* 120: 125 © 2012 John Wiley and Sons, and reprinted with their permission)

in designing drugs aimed at countering their toxic effects. This aspect is illustrated in Fig. 8.6 in which the various oligomeric species that have been discovered are placed in two assembly pathways. One of these leads from monomers through trimers and through a set of non-fibrillar oligomers terminating in annular protofibrils. These structures bear a resemblance to bacterial toxins and for that reason it has been suggested that the Aβ oligomers are toxic through their ability to form pores in membranes. The other pathway has dimers as its basic building block and terminates with formation of amyloid plaques.

Soluble and stable oligomers of various sizes and morphologies have been designated in several ways. One of these is as Aβ-derived diffusible ligands, or ADDLs. This term is intended to highlight their role as ligands for synaptic receptors (to be discussed further shortly). Another term for these is as globular amyloid β-peptide 1–42 oligomers. This naming emphasizes the primacy of the longer length Aβ-peptide in AD. One of the key findings over the years is the presence of low-n

(dimers to octamers) Aβ oligomeric species, especially dimers, in AD tissue samples. As already noted, the model shown in Fig. 8.6 places the dimers in the lead position of the fibrillar (amyloid plaque) pathway and trimers in the non-amyloidogenic pathway.

8.7 Aβ Serves in a Neurotrophic and Neuroprotective Capacity

Amyloid-β is a synaptic protein, and the earliest and strongest signs of AD take the form of synaptic deficits. Aberrations in synaptic structure and plasticity appear long before deposits of Aβ or tau can be detected and independent of plaque buildup. Aβ is intimately connected to synaptic activity. It is synthesized and released into the extracellular (interstitial) spaces in response to synaptic activity, and localizes to presynaptic and postsynaptic compartments. Once synthesized, Aβ regulates cholinergic and glutamatergic receptor function, and synaptic transmission. These findings support the idea that under normal conditions Aβ has a protective role. It functions as a homeostatic regulator of synaptic plasticity operating through one or more feedback loops to maintain synaptic strength in its proper operating range.

Neural activity stimulates increased production of Aβ through presynaptic and postsynaptic mechanisms. First, depolarization at the presynaptic terminal triggers a Ca^{2+} influx resulting in an exocytosis of synaptic vesicles. Synaptic vesicle membranes are internalized via clathrin-coated pits result in an increased concentration of APP in lipid raft-bearing endosomes where they come into contact with BACE1 and γ-secretases. Aβ is then released into the extracellular spaces. Secondly, at postsynaptic terminals, NMDA-induced reduction in α-secretase activity occurs furthering the increased production of Aβ. The overall increased production of Aβ leads to synaptic depression and internalization of AMPA receptors resulting in spine loss and rendering the synapses silent. This serves as a negative feedback to throttle back excitatory synaptic transmission that has become too strong.

The amyloid precursor protein and its cleavage products regulate synaptic structure and function at several levels.

- In the adult, they

  - Maintain proper dendritic spine morphology,
  - Control glutamatergic and cholinergic receptor activity, and
  - Act as de facto antioxidants through their metal sequestration activities.

- During development these proteins and their peptide products

  - Promote cell–cell and cell–substrate adhesion, and
  - Facilitate synaptic formation and maturation.

These processes and how they might go awry in AD are now discussed in greater detail.

8.8 Aβ and Synaptic Dysfunction

8.8.1 Synaptic Plasticity at Glutamatergic Synapses

Memories are encoded in the plastic changes in the strengths of the synaptic connections between neurons in the hippocampus and elsewhere in the brain. Memory loss in these regions is the principal sign of AD, and because of that a great amount of attention has been directed at uncovering the molecular machinery behind memory formation and loss at synapses. This machinery has been explored extensively over the past 40 years using laboratory models of synaptic plasticity in test subjects such as mice. These models, most notably long-term potentiation (LTP) in which the synaptic connections are strengthened and long-term depression (LTD) in which they are weakened, are believed to have a general applicability throughout the brain and give basic insights into what is happening in people.

Action potentials trigger neurotransmitter release from the presynaptic axonal terminals and these bind receptors embedded in the postsynaptic density in dendritic terminals and spines, small dendritic protrusions that receive the signals transmitted from axons. A considerable amount of machinery resides in both terminals to supports these actions. On the presynaptic side there is the machinery that ties the arrival of the action potentials to release of the neurotransmitter-loaded vesicles and the machinery that prepares the vesicles and loads them. The postsynaptic side, commonly referred to as the postsynaptic density (PSD) contains a rich assortment of signal receptors, voltage-gated ion channels, signal transducing protein kinases and phosphatases, anchors, adapters and scaffolding proteins, calcium stores, and mRNAs along with the machinery to translate and regulate them.

Three kinds of glutamatergic receptors work together in the hippocampal neurons—N-methyl-D-aspartate (NMDA) receptors, α-amino-3-hydroxy-5-methyl-4-isoxazolepropionate (AMPA) receptors, and metabotropic glutamate receptors (mGluRs). NMDA receptors are rather special in that they are gated both by ligand binding and membrane depolarization. When these ion channels open they permit the entry of Ca^{2+} ions into the neuron. High levels of NMDA receptor stimulation triggers a signaling cascade resulting in biosynthesis, transport, and insertion of additional AMPA receptors, and to an increase in the number of dendritic spines. The nascent ligand-gated ionotropic (NMDA and AMPA) receptors are transported from the ER to the Golgi and from there travel along the microtubule rail system to dendritic membrane sites. Upon arrival they interact with the actin cytoskeleton and other elements of the PSD. These interactions result in a stronger response to the arrival of neurotransmitter and thus potentiate the signal. Conversely, repeated low levels of NMDA receptor stimulation produce the converse biophysical reaction. The result is depression of the synaptic response to neurotransmitter arrival. Over time, AMPA are internalized and there is a loss of dendritic spines. The former process underlies LTP, while the second kind of modification produces LTD. These processes are modulated by the mGluRs and by nicotinic acetylcholine receptors containing the α7 subunit (α7-nAChR) that are also present and co-localize with the NMDA and AMPA receptors.

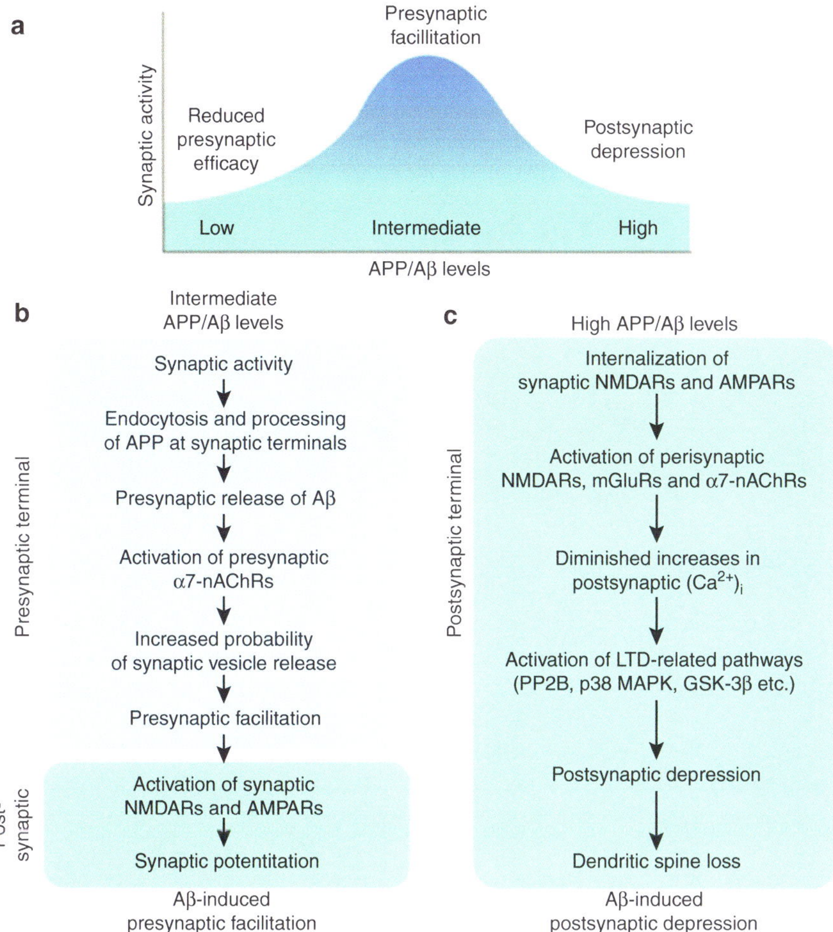

Fig. 8.7 APP acts at pre- and post-synaptic terminals to regulate synaptic transmission Intermediate (normal) levels of Aβ promote proper functioning at synapses whereas levels that are either too high or too low have deleterious effects at the postsynaptic terminal (from Palop *Nat. Neurosci.* 13: 812–818 © 2010 Reprinted by permission from Macmillan Publishers Ltd)

Binding of amyloid-β peptides to NMDA receptors influences synaptic transmission in several ways. Expression levels matter and Fig. 8.7 presents an encapsulation of this aspect. At physiologically normal levels Aβ acting at presynaptic and postsynaptic terminals potentiates synaptic transmission. However, if excessive levels of Aβ peptides are present and persist over time they stimulate continued increases in signaling and calcium entry leading to a dangerous excitotoxicity. This sets the stage the internalization and removal of first the NMDA receptors and then the AMPA receptors followed by loss of dendritic spines. These events derail synaptic plasticity and replace it with a persistent depression of synaptic transmission in what is perhaps the earliest stage in the disease. Additional modulatory influences come from the presence of prion PrP^C receptors. Binding of Aβ oligomers to these receptors stimulates receptor clustering and activation of additional inappropriate signaling leading to a further excessive entry of calcium, excitotoxicity, and other damaging effects leading to further receptor and spine loss and synaptic failure.

8.8.2 The Cholinergic System

The study of the cholinergic system dates back to the earliest days of neuroscience. Acetylcholine was, in fact, the first neurotransmitter to be identified. Its discovery and characterization by Henry Dale and Otto Loewi led to their receiving the Nobel Prize in 1936. As noted earlier in this chapter the cholinergic system was an early focus in Alzheimer's research. Emphasis on this system arose from studies that established that, in Alzheimer's disease, there was a selective deterioration of cholinergic neurons located in the basal forebrain and that project to the hippocampal region and cortex. This loss was accompanied by a decline in cholinergic neurotransmission and led to development of a number of drugs that treat this deficiency and are still in use today.

Nicotinic acetylcholine receptors containing the α7 subunit are highly expressed in the basal forebrain neurons that project to the hippocampus. These receptors are pentameric ion channels that enable the entry of Ca^{2+} and Na^+ cations. The channels have a high permeability to calcium and as a result influence a variety of calcium-dependent cellular processes. These receptors are present in presynaptic terminals and perisynaptically in the somal/dendritic compartments (see Fig. 8.7). At the presynaptic terminals they modulate calcium homeostasis and the release of acetylcholine, while at the dendritic spines they alter the synaptic currents involved in sensory information processing, learning, memory formation, and protective anti-apoptotic actions. Significantly, these channels function as high-affinity receptors for amyloid-β. In the presence of high levels of Aβ peptides, their normal modulatory and protective roles are lost and instead binding contributes to the destructive hyperexcitability.

8.8.3 Calcium Signaling

Calcium is an ideally suited second messenger, more so than monovalent ions such as sodium or potassium or chloride, and divalent cations such as magnesium. Because of its size and coordination chemistry calcium is well matched to the typical sizes of protein binding cavities, readily forms bonds with proteins, and is able to induce conformational changes. Unlike other second messengers, calcium is not synthesized by cells. Instead, there are two reservoirs of calcium—the extracellular spaces outside the cell and the calcium stores located within the cell. The calcium concentration in the extracellular spaces is on the order of 2 mM, some 20,000 times greater than the resting levels within the cell. Calcium is sequestered within the cell in intracellular stores, regions enriched in calcium buffers located in the lumen of the endoplasmic reticulum, the matrix of mitochondria, and in the Golgi.

The duration of a calcium signal is short. Intracellular calcium levels are restored to their base values fairly rapidly. Buffering agents bind calcium ions before they can diffuse appreciably from their entry point. Free calcium path lengths, the distance traveled by calcium ions before being bound, average less than 0.5 μm, which is far smaller than the linear dimensions, 10–30 μm, of typical eukaryotic cells. In addition to being buffered, ATP-driven calcium pumps located in the plasma membrane rapidly remove calcium ions from the cell, and other ATP-driven pumps transport calcium back into the intracellular stores. The take-up of calcium by buffers, along with its rapid pumping out of the cytosol, produces a sharp localization of the calcium signal both in space and time.

8.9 Amyloid-β and Raft Lipids Regulate One Another

Recall from the last chapter that lipid rafts are cholesterol- and sphingolipid-enriched microdomains. These structures play an important role in aging and Alzheimer's disease pathology. Lipid rafts are situated in the trans-Golgi and endosomal membranes, and in the plasma membrane of the somal, axonal, and dendritic compartments. In these locations, they provide needed support for the receptors and associated synaptic machinery described in the previous sections. NMDA, AMPA, mGluRs, and α7-AChRs are raft residents as are the amyloid precursor protein and BACE1. In contrast, the α-secretase resides in the plasma membrane in non-raft locations. Targeting sequences possessed by the proteins facilitate their recruitment to the rafts. There, they can carry out their proteolytic, signaling and control functions and regulate the lipid composition of the rafts through positive and negative feedback loops.

The lipid content of the rafts is altered in aged and AD brains. Sphingomyelin (SM) levels are reduced leading to increases in ceramide content. Ceramide is a pro-apoptotic signaling molecule. It is generated through the action of sphingomyelinase (SMase), a hydrolytic enzyme that breaks down sphingomyelin into phosphocholine and ceramide. SMase is activated by cellular stresses and these include not only inflammation but also Aβ peptides, most notably the long $A\beta_{1-42}$ form. Ceramide stimulates Aβ biogenesis by increasing the half-like of BACE1, thereby setting up a positive feedback loop to drive neurons towards an apoptotic outcome. Under normal operating conditions these actions might protect the brain against defective or excessively stressed neurons, but under aging and AD conditions it establishes ceramide as a potential downstream effector of Aβ toxicity.

The uptake of cholesterol-laden ApoE particles is mediated by LRP1 receptors. The endocytosed lipoproteins are hydrolyzed in lysosomal compartments resulting in the release of their cholesterol content, which then can be incorporated into lipid rafts. The localization of APP to cholesterol-rich rafts is promoted by the presence of a cholesterol-sequestration site on C99, the portion of APP that remains membrane-bound after BACE1 cleavage and release of the ectodomain. Once the second cut is made by the γ-secretase the AICD translocates to the nucleus where it acts at the LRP1/2 promoter to suppress LRP1 expression, thereby throttling back cholesterol uptake. The Aβ peptide, primarily Aβ_{1-40}, acts in concert with these actions by negatively regulating hydroxymethylglutaryl-CoA reductase (HMGR), the main enzyme in cholesterol biosynthesis.

Gangliosides are a prominent family of glycosphingolipid constituents of neuronal lipid rafts. Aβ oligomers bind strongly to gangliosides such as GM1. These interactions sequester Aβ on the rafts, and trigger structural changes. The consequences of these interactions are complex with cholesterol binding an added modulatory factor. Some studies highlight the potential for these interactions to reduce Aβ-oligomer levels in neurons through peripheral clearance, while others emphasize that GM1 may act as a cofactor that seeds Aβ fibrillization. All agree that these interactions along with those involving cholesterol are important.

The cholesterol content of the lipid rafts influences BACE1 and gamma secretase activity. Increases in cholesterol levels promote the association of BACE1 with lipid rafts leading to an increased amyloidogenic processing of APP and a greater production of Aβ. Components of the gamma secretase complex localize to the rafts, as well. These results point to biophysical manipulations of the raft's lipid components as a way of altering the amyloidogenic processing of APP.

8.10 Amyloid-β Oligomers Initiate Aberrant Signaling

Considerable efforts have been made to identify receptors for Aβo. Finding the correct receptors is made difficult by the sheer variety of oligomeric species and peptide chains that can be produced. One of the putative receptors is PrP. Recall from the previous chapter that the cellular prion protein cannot by itself transduce a signal into neurons but, instead, partners with co-receptors to trigger signaling. The receptors then recruit and activate tyrosine kinases such as Fyn to stimulate further downstream signaling. These pathways can be exploited by Aβo to induce aberrant and potentially toxic signaling cascades at the synapse. An example of how this route can lead to excitotoxicity is depicted in Fig. 8.8a. In this situation, the metabotropic glutamate receptor mGlu5 function as a co-receptor that activates Fyn. That kinase, in turn, phosphorylates nearby NMDA receptor subunits leading to excessive entry of Ca^{2+} into the postsynaptic terminal.

Several candidate receptors for Aβo have been identified. Besides PrP[C], ephrinB2 receptors (EphB2) have been found to bind amyloid-β oligomers. The competitive binding of Aβo in place of its normal ligand Ephrin-B2, results in the internalization and degradation of the Eph receptors as shown schematically in Fig. 8.8b. Under normal conditions EphB2 regulates NMDA receptor-mediated calcium entry and by that means regulates synaptic plasticity. The reduction in EphB2 levels impairs that function resulting in inadequate NMDA activity. Another recently

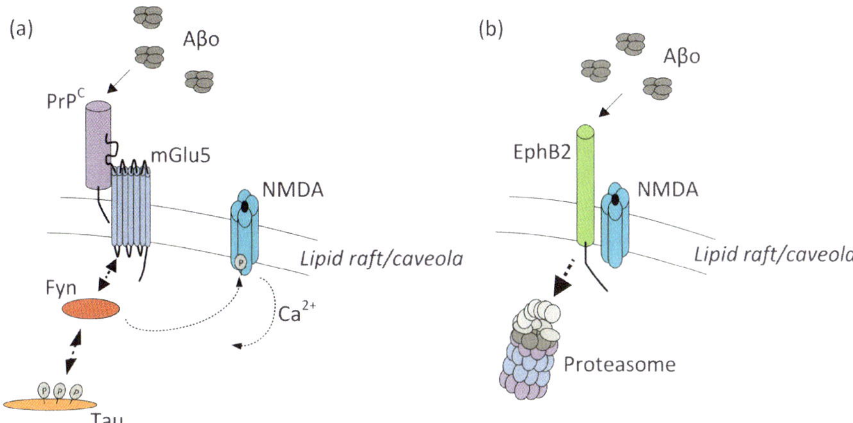

Fig. 8.8 Aberrant signaling by amyloid-β oligomers (Aβo). (**a**) Signaling initiated by binding to the PrP^C/mGlu5/Fyn complex leading to excessive activation of NMDA ion channels and increased excitotoxicity. Mislocalized tau further recruits Fyn to the raft-associated signaling complex resulting in increased phosphorylation and stimulation of NMDA (and other) ion channels. (**b**) Reduced synaptic plasticity resulting from binding of Aβo to the EphB2 receptors. This action displaces Ephrin ligands; the EphB2 receptors are internalized and degraded by the proteasome, and the NMDA ion channels fail to open when needed

identified receptor for Aβ oligomers is the leukocyte immunoglobulin-like receptor B2 (LilrB2). In this situation, Aβo binding initiates a signaling cascade in which the protein phosphatases PP2A and PP2B are stimulated, cofilin is activated, and actin is depolymerized. The result is shrinkage of dendritic spines and loss of synaptic plasticity.

Finally, Aβo can signal through PrP^C in more than one way. Another PrP^C co-receptor is the low-density lipoprotein receptor-related protein-1 (LRP1). When Aβo binds PrP^C/LRP1 it impairs its ability to inhibit the activity of BACE1, one of its normal functions that occur upon entry into the endocytic pathway. This particular activity depends on the presence of a fibrillar/hydrophobic conformation on the part of the Aβ oligomers; soluble/less-hydrophobic oligomeric forms do not bind PrP^C as readily. In sum, aberrant Aβo signaling can produce a number of toxic events, most notably, excitotoxicity, receptor malfunction, and dendritic spine loss, all leading to synaptic failure and network breakdown.

The term "excitotoxicity" was coined by James Olnay in 1969 and arises as a consequence of excessive Ca^{2+} entry through NMDA glutamate receptors, which consequently damages cells rather than protects them. In exploring these diverging health-promoting and cell death effects, it has emerged that there are two distinct sets of NMDA receptors—one set at synapses and the other at extrasynaptic sites. The former set initiates a Ca^{2+}-mediated signaling cascade to the nucleus that acts in a protective, anti-apoptotic capacity, while the latter triggers a different signaling cascade that supports cell death. The differences between the two derive from their activation of opposing cellular signaling pathways and alternate gene expression programs.

Inappropriate Aβ levels can disrupt a careful balance between synaptic and extrasynaptic NMDA receptors. One of the ways it does so is by stimulating glutamate release from glial cells through activation of α7-nAChRs. This production favors extrasynaptic receptor activation. When this activity occurs concurrently with the NMDA-stimulated increases in Aβ production it generates a positive feedback loop that further shifts the balance between survival and death towards the latter.

8.11 The Blood–Brain Barrier and Neurovascular Unit

The term *blood–brain barrier* refers to the selectively permeable interface connecting the central nervous system to the circulation. This interface performs a number of essential functions. It:

- Enables entry into the brain of nutrients;
- Facilitates the removal from the brain of waste products;
- Prevents entry into the brain of harmful chemicals, and
- Regulates fluid and ion movement, thereby providing an environment that is conducive to synaptic transmission and neural function.

While the brain microvasculature possesses a large surface area that helps maintain brain homeostasis, breakdowns in its operation can contribute to disease. The blood–brain barrier is also an impediment to entry into the brain of large molecules and that category including many potentially helpful drugs.

The concept of a blood–brain barrier took over a century to fully emerge. Early hints to its existence were noted by Paul Ehrlich (1854–1915) in the late nineteenth century through studies in which he injected dyes into the circulation. These were observed to penetrate tissues throughout the body except the cerebral spinal fluid and brain. Follow-up studies by his student further demonstrated the existence of some sort of barrier to entry of various dyes into the brain. That led to the concept of a restrictive space or blood–brain barrier. The precise identity of that space was somewhat mysterious given the failure of different probes to find any special material separating blood and brain. Instead, it became apparent in the years that followed that the vascular endothelial cells themselves formed the interface.

A key observation leading to the maturation of the concept was the discovery that the endothelial cells were connected to one another by tight junctions. These junctions prevent direct, or para-cellular, passage of molecules between the cells and forced all passage to occur in a trans-cellular fashion through the cells. This type of passage is mediated by transporters, receptors, and other agents, thereby enabling precise control of what materials can pass and which ones cannot. Key cells in these operations are not only the endothelial cells, but also pericytes and astrocytes that wrap about the blood vessels and assist the endothelial cells and neurons (Fig. 8.9). These cells along with the extracellular matrix all work together and function as an integrated *neurovascular unit*.

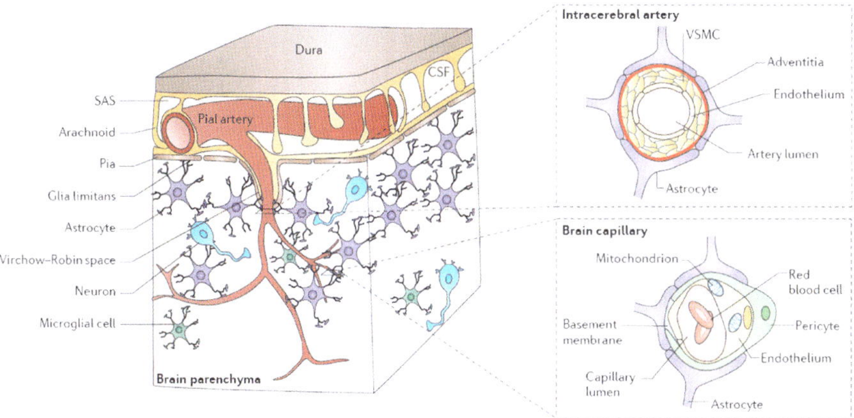

Fig. 8.9 Brain microcirculation and the neurovascular unit. Depicted is the subarachnoid space (SAS) containing the cerebrospinal fluid (CSF). The large pial arteries branch into penetrating intracerebral arteries which further branch into smaller arteries and capillaries, the principal sites of oxygen and nutrient exchange. Several different types of cells make contact and envelop the intracerebral arteries and brain capillaries. Their cellular organizations are shown to the right. *VSMC:* vascular smooth muscle cell (from Zlokovic *Nat. Rev. Neurosci.* 12: 723 © 2011 Reprinted by permission from Macmillan Publishers Ltd)

8.12   Aβ and Vascular Injury

Up until the 1950s and 1960s the prevailing view was that dementia in elderly persons was a consequence of impaired blood flow to the brain. That vascular impairment was connected in some way to dementia had been noted by Alzheimer in the late 1890s and also by Otto Binswanger (1852–1929) in 1894. Attempts were made to separate dementias originating from vascular impairment from senile dementia due to other causes and the term "arteriosclerotic dementia" was applied to the former by Emil Kraepelin (1856–1926) in his widely used 1896 textbook on psychiatry. Other terms used to describe the different types of age-related vascular degeneration were Binswanger disease, vascular cognitive impairment, vascular dementia, and lastly, hardening of the arteries.

Cerebral amyloid angiopathy (CAA) accompanies Alzheimer's disease and is widely seen in the elderly, in persons beyond the age 85. This condition is characterized by a buildup of amyloid-β in the walls of the small arteries, capillaries and venules. It is a typical feature of AD being present in 90 % of the cases. Damage of this type to the BBB can occur either sporadically or in familial forms such as the Dutch (E22Q) and Iowa (D23N) mutant forms of APP. This type of impairment accompanies the imbalances between production and clearance, and mirrors the Aβ deposits and damage to the neurons. Nowadays, there is an increased awareness that risk factors such as obesity, diabetes, hypertension, and atherosclerosis might contribute to neurodegenerative disorders by creating conditions that impair the normal functions of the BBB. Interestingly, this aspect of the disease provides a link to all of the prominent risk factors—to the aged, to people possessing the ApoE4 allele, and to individuals suffering from various forms of the metabolic syndrome (hypertension, type II diabetes, and/or obesity). In many instances, the blood vessels become deformed, localized ischemia occurs, and oxidative stress and inflammation ensue. These conditions induce cerebral hypoperfusion, which, in turn, stimulates CAA, thereby setting up a positive feedback loop that progressively worsens cerebral and neural function over time. Shown in Fig. 8.10 is a depiction of how Aβ oligomers and degraded brain capillary function can jointly set off a cascade of damaging events under these conditions.

8.13   Inflammation in Alzheimer's Disease

Astrocytes are one of the principal components of the neurovascular unit. They coordinate local neural activity with the local blood supply, help supply neurons with nutrients, and provide homeostatic control over the local microenvironment. Astrocyte end feet are critically positioned between neurons and vasculature. This positioning has given rise to the concept of a tripartite synapse consisting of a presynaptic terminal, postsynaptic process, and astrocyte end feet. In their synaptic role, the astrocytes monitor and exert control over the chemical milieu in the synaptic cleft and its surroundings. They sense the extent of the synaptic activity, clear neurotransmitters from the cleft, and regulate vasoconstriction and vasodilation.

Astrocytes, like neurons, are excitable cells. They do not generate action potentials but instead utilize calcium waves and oscillations to stimulate release of gliotransmitters. Astroglia become excited in response to glutamate and other neurotransmitters released by neurons. The amplitude and frequency of the calcium oscillations are determined jointly by the intrinsic properties of the astrocytes and the neuronal signals being received. Neurons, in turn, are sensitive to the release of the gliotransmitters and other molecules and the interactions are bidirectional.

That glia cells surround amyloid plaques and might contribute to the disease progression was first suggested in 1910 by Alzheimer. Earlier, Virchow had discovered the presence of large numbers of non-neural cells in the brain, but thought of them simply as connective tissue. The function of astrocytes was a mystery to Ramon y Cajal and a century later their multiple roles in the tripartite synapse, in the neurovascular unit, and in neurodegenerative disease are still being uncovered. One of the key discoveries is that amyloid-β peptides trigger reactive astrogliosis, a condition whereby programmatic, morphological, and molecular changes turn the astrocytes from supporters of the environment to destroyers of perceived damaged and potentially harmful cells in their vicinity. The reversals generated by these programming changes are depicted in Fig. 8.11.

Microglia are the immune system cells of the brain. They were discovered and named in 1932 by the Spanish neuroscientist Pio del Rio-Hortega (1882–1945). These cells enter the brain during early embryonic development prior to formation of the blood–brain barrier. They disperse throughout the brain regions and transform

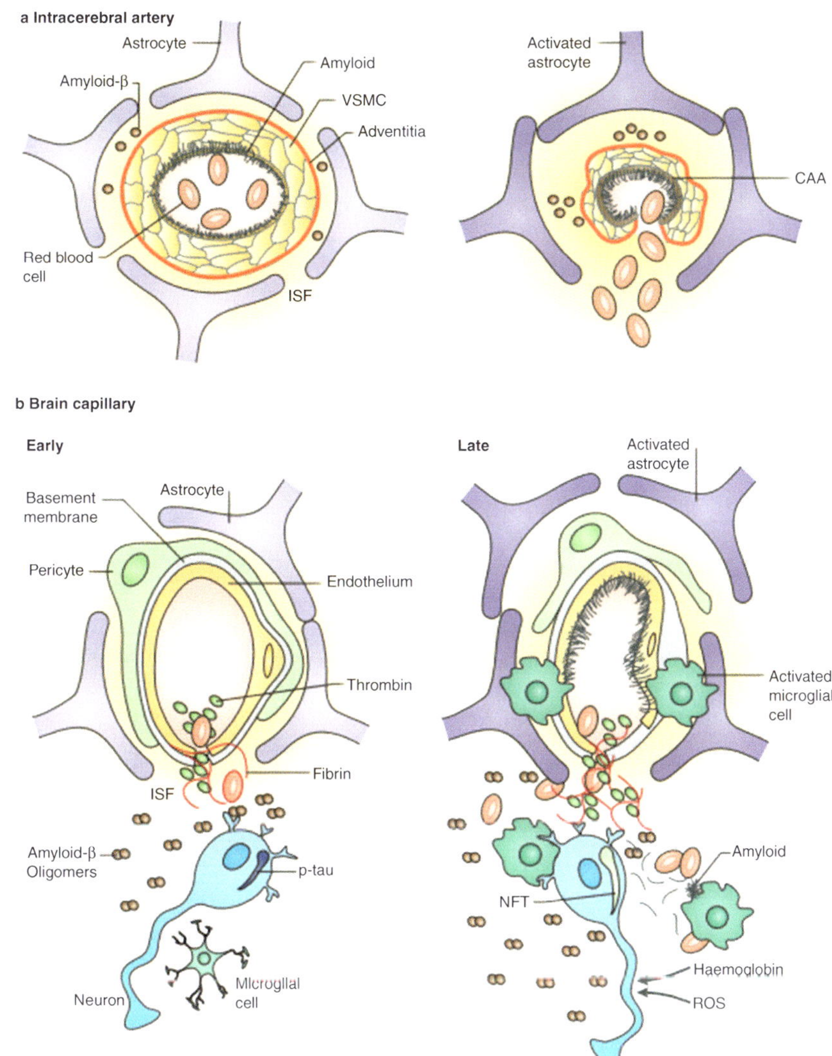

Fig. 8.10 Joint actions of Aβ oligomers and BBB dysfunction. Early: An impaired BBB leads to micro-hemorrhaging and the release of blood-derived toxic molecules such as thrombin and fibrin that negatively impact synaptic function. Faulty clearance through the BBB results in a buildup of Aβ oligomers that dysregulate synaptic function, and hyperphosphorylated tau accumulates. Late: An increased release of molecules from pericytes, activated astrocytes and microglia, and neurons, has occurred. There is an amyloid buildup along the capillary walls. The proper functioning of the BBB in handling influx and efflux is degraded further contributing to increased inflammation, neurofibrillary tangles, and synaptic breakdown (from Zlokovic *Nat. Rev. Neurosci.* 12: 723 © 2011 Reprinted by permission from Macmillan Publishers Ltd)

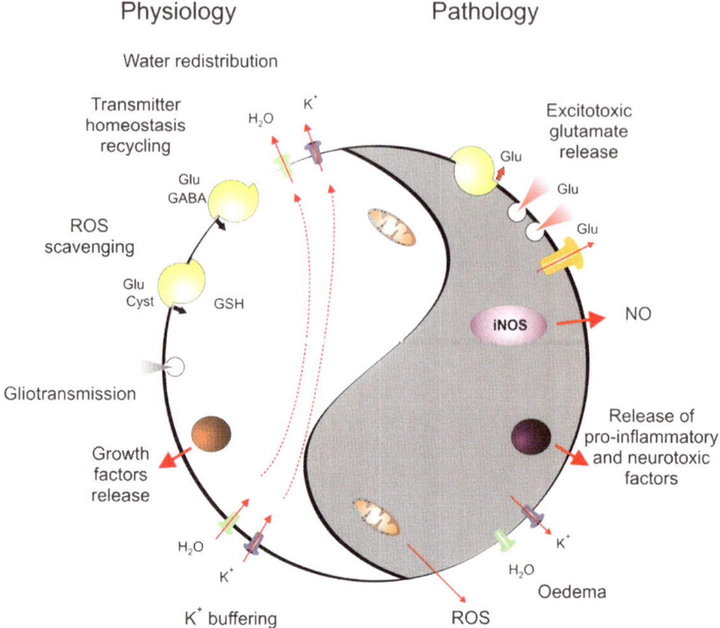

Fig. 8.11 Maintenance of homeostasis by astrocytes under normal and pathological (metabolic stressed) conditions. *Glu:* glutamate, *GABA:* γ-aminobutyric acid (a prominent inhibitory neurotransmitter), *ROS:* reactive oxygen species, *NO:* nitric oxide (from Heneka *Brain Res. Rev.* 63: 189–211 © 2010 Reprinted by permission from Elsevier)

into highly branched and ramified resting microglia. In this form, they continually monitor their immediate environment for signs of injury and infection making use of their motile processes and protrusions. Once signs of damage are detected they become activated and alter their morphology.

Brain microglia like macrophages found elsewhere in the body utilize a variety of intracellular sensing platforms, referred to as inflammasomes, to detect the presence of pathogens and other signs of danger. One of the best characterized sensor platforms is the NLRP3 (NACHT, LRR, and PYD domain-containing protein 3) inflammasome. This detector of danger signals is assembled and activated in macrophages elsewhere in the body and by brain microglia in response to danger signals such as uric acid crystals, silica, aluminum salts, and, as it turns out, Aβ. The presence of Aβ leads to activation of NLRP3 (NALP3) and the generation and launching of a pro-inflammatory cascade of interleukins and other chemicals. Like the danger signs encountered in other diseases such as type 2 diabetes (islet amyloid peptide) and atherosclerosis (cholesterol crystals) the continual presence of stimulatory Aβ in the local environment results in chronic inflammation.

The complement system is active in AD. In a landmark 1992 paper, Rogers reported that aggregates of Aβ such as the senile plaques bind C1q, the pattern-recognizing lead element of the complement system. The complement system, composed of more than 30 proteins, is part of the innate immune system. It is tasked with recognizing, opsonizing (in which identified pathogens are marked for phagocytosis), stimulating inflammation, and killing pathogens (through activation of its membrane attack complex, or MAC). Like the activated microglia, the complement system can become overstimulated and fail to resolve, thereby worsening the inflammatory effects of the Aβ aggregates. In Alzheimer's disease, this process initiated by C1q occurs in the absence of any pathogen and is a prime example of *autotoxicity*.

Proteins of the complement system not only help get rid of pathogens but also help sculpt neural circuits during development. They do so by tagging unwanted synaptic connections with C1q and C3 complement proteins. These are released by microglia and neurons, and the synapses so marked are then removed by the microglia functioning as phagocytes. One provocative surmise is that this process might become reactivated during neurodegeneration resulting in synaptic loss.

8.14 Tau Is a Highly Dynamic Microtubule-Associated Protein

The tau protein was discovered and named by Marc Kirchner and his coworkers in 1975 in their search for binding partners involved in the assembly of tubulin into microtubules. Tau is primarily an axonal protein of between 352 and 441 amino acid residues, encoded by the MAPT gene on chromosome 17q21, and consisting of six major isoforms generated by alternative splicing. That tau comprised the core of the neurofibrillary tangles first described by Alois Alzheimer over a century ago was established by several groups in the 1985–1986 time-period. The term *paired helical filaments (PHFs)* was introduced by Kidd in 1963 to describe the appearance of the tangles but this term is now recognized as being somewhat inaccurate. The tangles are far more polymorphic and are best described simply as tau fibers.

Tau proteins are natively unfolded and this property is fully exploited in their binding to the microtubules. By assuming an extended shape the tau proteins are able to establish contact with multiple tubulin dimers, thereby serving as rail ties. They have few hydrophobic amino acids and an excess of positively charges both properties supporting an unfolded native state. Recall that microtubules are the rails over which motor proteins transport their cargo to distant sites. The ties formed by tau stabilize these rails. However, the rails are constantly being reworked and modified. In order to function effectively the ties are highly dynamic. They are able to grip the microtubules tightly when necessary and weaken and detach whenever the rails need to uncouple and reform. Phosphorylation drives these dynamic processes by adding negative charges that weaken the attachment of tau to the tubulins.

Tau contains multiple (80) phosphorylation sites, about 18 % of the amino acid residues are either serines or threonines. Many of these residues are phosphorylated as part of tau's normal function as a regulator of microtubule stability and growth and as a regulator of fast axonal transport. However, in AD at least 30 serine/threonine resides are modified to produce excessively hyperphosphorylated tau proteins that with the addition of these extra negative charges detach from the microtubules. These aberrantly phosphorylated proteins are found in the NFTs. A transition from the extended native state to a more folded and aggregation-prone shape occurs in the formation of the tau fibers and smaller tau aggregates.

Phosphorylation by itself does not drive this transition. Instead, excessive phosphorylation works together with aggregation-prone hydrophobic patches and proteolytic cleavage to generate the tau fibers and smaller aggregates. Tau possesses several hydrophobic patches and these seed the transition. These alterations are further promoted by proteolytic cleavage of N- and C-terminal segments. These generate shorter, tau fragments that are far-more aggregation prone.

8.15 Tau Mediates Aβ Toxicity

One of the enduring questions in the field has been: are the two major signs of AD—the amyloid plaques and the neurofibrillary tangles—causally linked and, if so, how? A start to answering that question emerged in 2001 with findings from two groups that implied that tau acts downstream from Aβ in some as yet undefined pathway or pathways. Specifically, it was found that Aβ fibrils accelerated the formation of neurofibrillary tangles. Those findings implied a connection between the two pathologies and, furthermore, placed tau pathology downstream from Aβ actions. These surmises were strengthened by subsequent discoveries. In the next few years, it was found that reducing tau levels ameliorated Aβ-induced excitotoxicity and axonal transport defects.

Today, most depictions of how Alzheimer's disease progresses place malfunctioning tau downstream of a toxic buildup of Aβ oligomers. One critical finding is that the binding of soluble Aβ oligomers at the synapse stimulates the hyperphosphorylation of tau over a broad range of sites. These events lead to its dissociation from the microtubules and its mislocation to the dendritic compartment. One of the consequences of the accumulation of tau in dendrites is the recruitment of Fyn to the cytosolic face of the rafts. Once Fyn has been localized there it phosphorylates and stabilizes the glutamatergic receptors leading to excessive Ca^{2+} entry into the cell and excitotoxicity(Fig. 8.8a). By this means a lethal triad consisting of Aβ, tau and Fyn is able to trigger a cascade of deleterious consequences—excitotoxicity, receptor malfunction, dendritic spine loss, synaptic failure, and network breakdown.

Amyloid beta peptides and tau may operate synergistically through positive feedback to drive Alzheimer's disease. Stresses brought on by oligomers and aggregates of misfolded Aβ stimulate activity by stress kinases and proteolytic enzymes such as caspases 3 and 6. These stresses persist; the kinases phosphorylate tau and the caspase cleave their N- and C-terminal segments leading to tau fragments with a far-greater propensity for misfolding and aggregation. Receptor trafficking is impaired and the several degenerative processes are exacerbated by failures in chaperone-mediated autophagy to degrade the tau fragments (yet another recent finding). This occurs because of inefficient translocation across the lysosomal membrane resulting in a buildup of tau oligomers. The resulting malfunctions in axonal transport and lysosomal clearance generate further stresses and Aβ-driven damage. The small effects synergize with the other damaging processes, and they accumulate, until the synapses fail and the neurons die.

8.16 Aβ Can Undergo a Prion-Like Spread from One Cell to Another

Alzheimer's disease begins in the hippocampus where it impairs short-term memory formation. It spreads from the hippocampus to other limbic regions such as the amygdala and entorrhinal cortex, and from these regions propagates to higher associative areas of the brain producing losses in memory, judgment, and reasoning, and changes in behavior.

In Chap. 7, the notion was introduced that many of the misfolded proteins with central roles in neurodegeneration may behave like prions. That is, seeds can be transmitted from cell-to-cell, and once in a recipient cell the seeds catalyze the conversion of the resident normally folded isoform to disease-causing ones. That idea is not a new one but rather had been advanced since the early 1980s when similarities between prion diseases and amyloid-β neurodegeneration were first noted. Accumulating evidence and support for this important idea has been found for most if not all of the leading agents of neurodegeneration including Aβ.

The subject is a complex one and the conditions under which such as transfer can and cannot occur are still being discovered. For example, as is the case for prion diseases, there must be a good matching between the toxic agent and the potential host neurons in terms of receptiveness of the host protein to conversion through templating. The presence or absence of cofactors can influence the outcome, the adequacy of cellular clearance is important, regional vulnerability plays a role, and adequate incubation times are needed. The best studied and established examples of the prion-mechanism are for tau and α-synuclein and this major concept is discussed in greater detail in the next chapters.

8.17 Summary

1. The last 20 years have witnessed tremendous progress in understanding the causes of Alzheimer's disease. These advances are centered about the amyloid cascade hypothesis, which posits an inappropriate buildup of amyloid-β oligomers as the primary causative event in the onset of the disease. The buildup occurs through an imbalance between generation and clearance of the Aβ peptides. The oligomers attach to and degrade synaptic function leading to a breakdown in

network connectivity, neural loss, dementia, and death. Vascular impairment, inflammation, and tau misfolding and aggregation contribute to the disease progression acting either in parallel or as accelerants to the Aβ oligomer buildup and burgeoning synaptic failures.

2. Evidence pointing to the involvement of aberrant amyloid-β processing in AD has been supplied by studies of EOAD in which mutations in APP and the presenilins, the enzymatically active arm of the γ-secretase complex are major risk factors. There are two processing pathways. In the non-amyloidogenic route, α-secretase (ADAM10 and sometimes ADAM17) cleavage is following by γ-secretase cleavage. In the amyloidogenic pathway, β-secretase (BACE1) cleavage is following by γ-secretase cleavage to produce Aβ peptides ranging in length from 39 to 43 amino acids. There is strong evidence that the long, $A\beta_{1-42}$ form is the primary culprit in Alzheimer's disease.

APP and the secretases are tightly regulated at the trafficking, transcription, translation, and posttranslational levels. These proteins are synthesized in the endoplasmic reticulum, transported through the trans-Golgi network, sent in secretory vesicles to the plasma membrane, cycle in the endosomal pathway, and sent to lysosomes. Several protein complexes, especially GGAs and retromers, regulate their trafficking and by that means influence the actions of the secretases. The promoters for APP and the secretases are complex and similar to those of housekeeping genes. Cellular stresses (e.g., metabolic and inflammation-related) influence their transcription programs leading to a greater production of BACE1 and thus to a larger Aβ oligomer and plaque buildup. Additional control elements responsive to stresses are present in the translational and posttranslational machinery. In the latter case, several microRNAs have been identified that increase BACE1 translation.

In its original 1992 formulation the focus of attention was on the extracellular amyloid plaques first seen and described by Alois Alzheimer. Since then the primary emphasis has shifted to the smaller, soluble, and more mobile and reactive oligomers. These correlate far better with the disease progression than do the amyloid plaques. A variety of oligomeric species are produced; these vary in size from dimers and trimers to 56-mers and beyond, and exhibit and broad spectrum of morphologies from linear structures to ring- or pore-shaped ones. Some species are regarded as being more toxic than others. For example, dimers and believed to be more toxic and amyloidogenic than trimers.

3. Additional insights into the causes of AD come from studies of LOAD. Here, attention is directed by the identification of the ApoE4 allele as the predominant genetic risk factor in AD. This risk factor along with aging and the metabolic syndrome drive late-onset familial and the far-more prevalent sporadic forms of the disease. One key result of the LOAD studies is the finding that impaired clearance of Aβ peptides is a major factor in disease onset and progression, and to point towards the need to better understand the possible vascular and inflammatory connections. These are intimately tied together in the neurovascular unit and tripartite synapse. The former describes the tight association of astrocytes, pericytes and endothelial cells in the operation of the blood–brain barrier, while

the latter describes the equally close association of astrocytes and the presynaptic and postsynaptic terminals in regulating synaptic function.

4. Excessive levels and aberrant forms of Aβ peptides lead to synaptic failure. When present at physiologically normal expression levels and forms, Aβ peptides carry out their proper tasks in presynaptic and postsynaptic terminals. However, if large numbers of pathological forms are present a cascading series of damaging events ensues. At physiologically normal levels Aβ potentiates synaptic transmission. However, when excessive levels of Aβ peptides are present for long periods of time they stimulate excessive signaling and calcium entry, and generate an excitotoxic environment. Under these conditions the NMDA and AMPA receptors are removed and dendritic spines are lost. These events derail synaptic plasticity, replacing it with a persistent LTD in what is believed to be the earliest sign in the disease.

 Receptors critically involved in synaptic transmission are situated in lipid rafts, microdomains enriched in cholesterol and sphingolipids such as sphingomyelin. These detergent-resistant liquid-ordered structures facilitate the recruitment of the necessary signaling machinery to the presynaptic and postsynaptic terminals as well as the endosomal positioning of APP and BACE1 in close proximity to one another. Once in the rafts, key lipid components—cholesterol, sphingomyelin, and gangliosides–are subject to Aβ feedback regulation.

5. Inflammation and the deterioration of the neurovasculature act in concert with, and as part of, the amyloid cascade in Alzheimer's disease. Astrocytes change their genetic programming from one that supports synaptic transmission and environmental homeostasis to one that destroys perceived damaged and potentially harmful cells in their vicinity. Microglia, the innate immune cells of the brain, treats amyloid buildups as danger signs and become activated. Working together with an equally activated complement system, a persistent inflammatory response that fails to resolve is generated. Aβ oligomers adhere to the inner walls of the microvasculature, act as irritants, generate additional persistent damage and inflammatory responses, and contribute to the deterioration of the vasculature. These activities generate cerebral amyloid angiopathy, and the impaired supportive neurovasculature worsens the growing damage to synaptic transmission and health of the neurons.

6. The second major observation made in 1907 by Alois Alzheimer was the presence of the intracellular tangles within the diseased neurons. These tangles have been subsequently shown to be composed of misfolded aggregates of the tau protein. The tau protein is natively unfolded and utilizes the addition and subtraction of negative charges supplied through phosphorylation to attach and detach from the microtubules where it functions as rail ties. The tau proteins involved in AD have dissociated from their normal microtubule location and formed fibrous structures by means of their hydrophobic stretches. These misfolded and aggregated tau proteins are hyperphosphorylated and have had their N- and/or C-terminals truncated resulting in fragments that are highly aggregation-prone. In AD, tau acts as yet another downstream effector of Aβ toxicity.

A long-standing mystery has been the failure to clearly identify how Aβ and tau relate to one another in AD. That mystery is slowly being solved. For example, there have been discoveries of synaptic intermediaries such as Fyn, findings that tau becomes mislocated to presynaptic terminals, and observations of impaired lysosomal degradation. There are also emerging insights on how Aβ-generated stresses might promote excessive phosphorylation and cleavages by caspases.

Appendix 1. Regulated Intramembrane Proteolysis (RIP)

The cleavage of transmembrane protein such as APP occurs both outside and within their membrane-spanning portions. This process, first uncovered in the mid to late 1990s, is known as regulated intramembrane proteolysis (RIP). The stubs and other fragments proteolytically generated from the Type I or Type II single-pass proteins can either be discarded or used as signaling molecules. In the latter instance, the signaling pathway to the nucleus is remarkable short and direct, bypassing the typical protein kinase, phosphatase and other signal transducers.

A number of signaling proteins are produced in this manner. These include Notch, a key developmental and repair regulator of neurons, sterol regulatory element-binding protein (SREPB), a major regulator of cholesterol biosynthesis, and tumor necrosis factor α (TNFα) a key mediator of inflammation. In all of these cases, extracellular cleavage occurs first that generates an ectodomain stub and reduces the length of the portion remaining for the second intramembrane cleavage step. ADAM10 and TACE (tumor necrosis factor-α converting enzyme) are responsible in many instances for the first cut and γ-secretase with its presenilins for the second step. A number of BACE1 substrates have been identified, as well. These include LRP1, neuregulins (proteins involved in myelination and synapse formation, and a schizophrenia risk factor), and voltage-gated sodium channels, among others. Lastly, the γ-secretase is remarkably promiscuous. By 2011 at least 90 substrates had been identified. In those instances where the fragments are simply discarded and have no signaling role, the picture is one where the γ-secretase functions as a member of the cellular protein quality control machinery.

Turning to the nuclear transit of APP and Notch signaling elements, the amyloid precursor protein intracellular domain (AICD) is generated through actions of gamma-secretase on the APP at one of a number of sites. Cuts are made at the γ-cleavage sites (at residues 38, 40, 42) C-terminal to the β-cleavage site, or alternatively to an ε-cleavage site (49) or a ζ-cleavage site (46) within the transmembrane portion of the APP. These operations and the resulting actions of the fragments resemble the situation generated by the presenilins on the Notch protein. In the case of Notch, the released C-terminal fragment, referred to as the Notch intracellular domain (NICD) translocates to the nuclear where it functions as a transcription factor. The actions taken in the case of the APP are slightly different. The AICD influences transcription by activating the Fe65 adaptor protein and this dimer interacts with other the histone acetyltransferase Tip60 to produce a transcriptionally active complex. Genes influenced include those promoting developmental and injury related alterations to the actin cytoskeleton structure.

Lastly, designing drugs that treat neurodegenerative disorders is a challenging endeavor. In the case of Alzheimer's disease, the hurdles are at least threefold. First, the drugs must get past the blood–brain barrier. Secondly, drugs that directly target Aβ peptides must deal with the existence of multiple oligomeric forms. Thirdly, drugs aimed at the secretases must take into account the presence of substrates other than APP. BACE1 is arguably the most promising drug target because of its role as the rate-limiting enzyme, but here too the recent discovery of other substrates presents challenges.

Appendix 2. The Five Types of Lipoprotein Particles

There are five kinds of lipoproteins, each differing in size, density and composition. The physical and chemical properties of these cholesterol, triglyceride and phospholipid transport molecules are summarized in Table 8.2. These entries have been ordered from the largest and least dense chylomicrons to the smallest and most dense high-density lipoproteins (HDLs). The alterations in size and density reflect differences in composition. The chylomicrons mostly transport triglycerides, while the LDLs and HDLs primarily convey cholesterol and phospholipids and carry only small amounts of TGs.

Chylomicrons are produced in the small intestine by absorptive cells (enterocytes), and enter the bloodstream via the lymphatic system. They transport TG derived from foodstuffs to the liver, adipose tissue, cardiac and skeletal tissues. Very low density lipoproteins (VLDLs) are synthesized in the liver. They are the main vehicle for transport of triglycerides from the liver to adipocytes and muscle tissue for energy storage and production of energy through oxidation. The fatty acid component of the VLDLs is released to the muscle cells and adipocytes; this loss together with the loss of the ApoCs leaves a VLDL remnant, termed an intermediate

Table 8.2 Physical properties and representative values for the composition of the lipoprotein particles

Lipoprotein particle	Density (g/ml)	Chol (%)	TG (%)	PL (%)	Prot (%)	Apo
Chylomicron	<0.96	4	87	8	1	A,B,C,E
VLDL	0.96–1.006	22	50	20	8	B,C,E
IDL	1.006–1.019	42	30	20	8	B,C,E
LDL	1.019–1.063	50	10	20	20	B
HDL	1.063–1.210	35	5	30	30	A,C,E

Abbreviations: *Chol* cholesterol, *TG* triacylglycerides (triglycerides), *PL* phospholipids, *Prot* protein, *Apo* specific apolipoproteins involved in the transport of cholesterol, triglycerides, and phospholipids

density lipoprotein (IDL). Some VLDLs are taken up by the liver, while others are transformed in the blood and become LDLs. The LDLs transport cholesterol, phospholipids and small amounts of TGs from the liver and intestines to other organs and peripheral tissues in the body; HLDs remove and return them to the liver for eventual removal from the body.

Bibliography

Bartus, R. T., Dean, R. L., III, Beer, B., & Lippa, A. S. (1982). The cholinergic hypothesis of geriatric memory dysfunction. *Science, 217*, 408–414.

Blessed, G., Tomlinson, B. E., & Roth, M. (1968). The association between quantitative measures of dementia and of senile change in the cerebral grey matter of elderly subjects. *The British Journal of Psychiatry, 114*, 797–811.

Davies, P., & Maloney, A. J. (1976). Selective loss of central cholinergic neurons in Alzheimer's disease. *Lancet, 2*, 1403.

Goate, A., Chartier-Harlin, M. C., Mullan, M., Brown, J., Crawford, F., Fidani, L., et al. (1991). Segregation of a missense mutation in the amyloid precursor protein gene with familial Alzheimer's disease. *Nature, 349*, 704–706.

Hardy, J. A., & Higgins, G. A. (1992). Alzheimer's disease: The amyloid cascade hypothesis. *Science, 256*, 184–185.

Hardy, J., & Selkoe, D. J. (2002). The amyloid hypothesis of Alzheimer's disease: Progress and problems on the road to therapeutics. *Science, 297*, 353–356.

Levy-Lahad, E., Wasco, W., Poorkaj, P., Romano, D. M., Oshima, J., Pettingell, W. H., et al. (1995). Candidate gene for the chromosome 1 familial Alzheimer's disease locus. *Science, 269*, 973–977.

Sherrington, R., Rogaev, E. I., Liang, Y., Rogaeva, E. A., Levesque, G., Ikeda, M., et al. (1995). Cloning of a gene bearing missense mutations in early-onset familial Alzheimer's disease. *Nature, 375*, 754–760.

St. George-Hyslop, P. H., Tanzi, R. E., Polinsky, R. J., Haines, J. L., Nee, L., Watkins, P. C., et al. (1987b). The genetic defect causing familial Alzheimer's disease maps on chromosome 21. *Science, 235*, 885–890.

Tanzi, R. E., & Bertram, L. (2005). Twenty years of the Alzheimer's disease amyloid hypothesis: A genetic perspective. *Cell, 120*, 545–555.

APP Processing

Chami, L., Buggia-Prévot, V., Duplan, E., Delprete, D., Chami, M., Peyron, J. F., et al. (2012). Nuclear factor-κB regulates β APP and β - and γ-secretases differentially at physiological and supraphysiological Aβ concentrations. *The Journal of Biological Chemistry, 287*, 24573–24584.

Christensen, M. A., Zhou, W., Qing, H., Lehman, A., Philipsen, S., & Song, W. (2004). Transcriptional regulation of BACE1, the β-amyloid protein β-secretase by Sp1. *Molecular and Cellular Biology, 24*, 865–874.

De Strooper, B., Saftig, P., Craessaerts, K., Vanderstichele, H., Guhde, G., Annaert, W., et al. (1998). Deficiency of prersenilin-1 inhibits the normal cleavage of amyloid precursor protein. *Nature, 391*, 387–390.

Hébert, S. S., Horré, H., Nicolaï, L., Papadopoulou, A. S., Mandemakers, W., Silahtaroglu, A. N., et al. (2008). Loss of microRNA cluster miR-29a/b-1 in sporadic Alzheimer's disease correlates with increased BACE1/β -secretase expression. *Proceedings of the National Academy of Sciences of the United States of America, 105*, 6415–6420.

Kuhn, P. H., Wang, H., Dislich, B., Colombo, A., Zeitschel, U., Ellwart, J. W., et al. (2010). ADAM10 is the physiologically relevant, constitutive α- secretase of the amyloid precursor protein in primary neurons. *The EMBO Journal, 29*, 3020–3032.

Lammich, S., Kojro, E., Postina, R., Gilbert, S., Pfeiffer, R., Jasionowski, M., et al. (1999). Constitutive and regulated α-secretase cleavage of Alzheimer's amyloid precursor protein by a disintegrin metalloprotease. *Proceedings of the National Academy of Sciences of the United States of America, 96*, 3922–3927.

O'Connor, T., Doherty-Sadleir, K. R., Maus, E., Velliquette, R. A., Zhao, J., Cole, S. L., et al. (2008). Phosphorylation of the translation initiation factor eIF2α increases BACE1 levels and promotes amyloidogenesis. *Neuron, 60*, 988–1009.

Roberts, B. R., Ryan, T. M., Bush, A. I., Masters, C. L., & Duce, J. A. (2012). The role of metallobiology and amyloid-β peptides in Alzheimer's disease. *Journal of Neurochemistry, 120*(Suppl. 1), 149–166.

Sastre, M., Dewachter, I., Rossner, S., Bogdanovic, N., Rosen, E., Borghgraef, P., et al. (2006). Nonsteroidal anti-inflammatory drugs repress β - secretase gene promoter activity by the activation of PPARγ. *Proceedings of the National Academy of Sciences of the United States of America, 103*, 443–448.

Vassar, R., Bennett, B. D., Babu-Khan, S., Khan, S., Mendiaz, E. A., Denis, P., et al. (1999). β -Secretase cleavage of Alzheimer's amyloid precursor protein by the transmmembrane aspartic protease BACE. *Science, 286*, 735–741.

Vassar, R., Kuhn, P. H., Haass, C., Kennedy, M. E., Rajendran, L., Wong, P. C., et al. (2014). Function, therapeutic potential and cell biology of BACE proteases: Current status and future prospects. *Journal of Neurochemistry, 130*, 4–28. doi:10.1111/jnc.12715.

Wang, W. X., Rajeev, B. W., Stromberg, A. J., Ren, N., Tang, G., Huang, Q., et al. (2008). The expression of microRNA miR-107 decreases early in Alzheimer's disease and may accelerate disease progression through regulation of β -site amyloid precursor protein-cleaving enzyme 1. *Journal of Neuroscience, 28*, 1213–1223.

Apolipoprotein E, Its Receptors and Cholesterol

Björkhem, I., & Meaney, S. (2004). Brain cholesterol: Long secret life behind a barrier. *Arteriosclerosis, Thrombosis, and Vascular Biology, 24*, 806–815.

Brown, M. S., & Goldstein, J. L. (1986). A receptor-mediated pathway for cholesterol homeostasis. *Science, 232*, 34–47 (Nobel lecture).

Corder, E. H., Saunders, A. M., Strittmatter, W. J., Schmechel, D. E., Gaskell, P. C., Small, G. W., et al. (1993). Gene dosage of apolipoprotein E type 4 allele and the risk of Alzheimer's disease in late onset families. *Science, 261*, 921–923.

Grimm, M. O., Grimm, H. S., Pätzold, A. J., Zinser, E. G., Halonen, R., Duering, M., et al. (2005). Regulation of cholesterol and sphingomyelin metabolism by amyloid-β and presenilin. *Nature Cell Biology, 7*, 1118–1123.

Holtzman, D. M., Herz, J., & Bu, G. (2012). Apolipoprotein E and Apolipoprotein E receptors: Normal biology and roles in Alzheimer's disease. *Cold Spring Harbor Perspectives in Medicine, 2*, a006312.

Liu, Q., Zerbinatti, C. V., Zhang, J., Hoe, H. S., Wang, B., Cole, S. L., et al. (2007). Amyloid precursor protein regulates brain apolipoprotein E and cholesterol metabolism through lipoprotein receptor LRP1. *Neuron, 56*, 66–78.

Liu, C. C., Kanekiyo, T., Xu, H., & Bu, G. (2013). Apolipoprotein E and Alzheimer's disease: Risk, mechanisms and therapy. *Nature Reviews. Neurology, 9*, 106–118.

Mahley, R. W., Weisgraber, K. H., & Huang, Y. (2006). Apolipoprotein E4: A causative factor and therapeutic target in neuropathology, including Alzheimer's disease. *Proceedings of the National Academy of Sciences of the United States of America, 103*, 5644–5651.

Aβ Oligomers

Barghorn, S., Nimmrich, V., Striebinger, A., Krantz, C., Keller, P., Janson, B., et al. (2005). Globular amyloid β -peptide$_{1-42}$ oligomer—A homogeneous and stable neuropathological protein in Alzheimer's disease. *Journal of Neurochemistry, 95*, 834–847.

Kayed, R., Pensalfini, A., Margol, L., Sokolov, Y., Sarsoza, F., Head, E., et al. (2009). Annular protofibrils are a structurally and functionally distinct type of amyloid oligomer. *The Journal of Biological Chemistry, 284*, 4230–4237.

Larson, M. E., & Lesné, S. E. (2012). Soluble Aβ oligomer production and toxicity. *Journal of Neurochemistry, 120*, 125–139.

Walsh, D. M., & Selkoe, D. J. (2007). Aβ oligomers—A decade of discovery. *Journal of Neurochemistry, 101*, 1172–1184.

Synaptic Activity and Aβ-Mediated Neuroprotection

Cirrito, J. R., Yamada, K. A., Finn, M. B., Sloviter, R. S., Bales, K. R., May, P. C., et al. (2005). Synaptic activity regulates interstitial fluid amyloid-β levels in vivo. *Neuron, 48*, 913–922.

Hsia, A. Y., Masliah, E., McConlogue, L., Yu, G. Q., Tatsuno, G., Hu, K., et al. (1999). Plaque-independent disruption of neural circuits in Alzheimer's disease mouse models. *Proceedings of the National Academy of Sciences of the United States of America, 96*, 3228–3233.

Hsieh, H., Boehm, J., Sato, C., Iwatsubo, T., Tomita, T., Sisoda, S., et al. (2006). AMPA-R removal underlies Aβ -induced synaptic depression and dendritic spine loss. *Neuron, 52*, 831–843.

Kamenetz, F., Tomita, T., Hsieh, H., Seabrook, G., Borchelt, D., Iwatsubo, T., et al. (2003). APP processing and synaptic plasticity. *Neuron, 37*, 925–937.

Lesné, S., Koh, M. T., Kotilinek, L., Kayed, R., Glabe, C. G., Yang, A., et al. (2006b). A specific amyloid-β protein assembly in the brain impairs memory. *Nature, 440*, 352–357.

Puzzo, D., Privitera, L., Fa, M., Staniszewski, A., Hashimoto, G., Aziz, F., et al. (2011). Endogenous amyloid-β is necessary for hippocampal synaptic plasticity and memory. *Annals of Neurology, 69*, 819–830.

Terry, R. D., Masliah, E., Salmon, D. P., Butters, N., DeTeresa, R., Hill, R., et al. (1991). Physical basis of cognitive alterations in Alzheimer's disease: Synapse loss is the major correlate in cognitive impairment. *Annals of Neurology, 30*, 572–580.

Tyan, S. H., Shih, A. Y. J., Walsh, J. J., Murayama, H., Sarsoza, F., Ku, L., et al. (2012). Amyloid precursor protein (APP) regulates synaptic structure and function. *Molecular and Cellular Neurosciences, 51*, 43–52.

Amyloid-β and Synaptic Function

Abramov, E., Dolev, I., Fogel, H., Ciccotosto, G. D., Ruff, E., & Slutsky, I. (2009). Amyloid-β as a positive endogenous regulator of release probability at hippocampal synapses. *Nature Neuroscience, 12*, 1567–1576.

Bliss, T. V. P., & Collingridge, G. L. (1993). A synaptic model of memory: Long-term potentiation in the hippocampus. *Nature, 361*, 31–39.

Bliss, T. V. P., & Lømo, T. (1973). Long-lasting potentiation of synaptic transmission in the dentate area of the anaesthetized rabbit following stimulation of the perforant path. *The Journal of Physiology, 232*, 331–356.

Lacor, P. N., Buniel, M. C., Furlow, P. W., Clemente, A. S., Velasco, P. T., Wood, M., et al. (2007). Aβ oligomers induced aberrations in synapse composition, shape and density provide a molecular basis for loss of connectivity in Alzheimer's disease. *Journal of Neuroscience, 27*, 796–807.

Lambert, M. P., Barlow, A. K., Chromy, B. A., Edwards, C., Freed, R., Liosatos, M., et al. (1998b). Diffusible, nonfibrillar ligands derived from Aβ$_{1-42}$ are potent central nervous system neurotoxins. *Proceedings of the National Academy of Sciences of the United States of America, 95*, 6448–6453.

Liu, Q. S., Kawai, H., & Berg, D. K. (2001). β-Amyloid peptide blocks the response of α7-containing nicotinic receptors on hippocampal neurons. *Proceedings of the National Academy of Sciences of the United States of America, 98*, 4734–4739.

Renner, M., Lacor, P. N., Velasco, P. T., Xu, J., Contractor, A., Klein, W. L., et al. (2010). Deleterious effects of amyloid β oligomers acting as an extracellular scaffold for mGluR5. *Neuron, 66*, 739–754.

Shankar, G. M., Li, S., Mehta, T. H., Garcia-Munoz, A., Shepardson, N. E., Smith, I., et al. (2008b). Amyloid-β protein dimers isolated directly from Alzheimer brains impair synaptic plasticity and memory. *Nature Medicine, 14*, 837–842.

Snyder, E. M., Nong, Y., Almeida, C. G., Paul, S., Moran, T., Choi, E. Y., et al. (2005). Regulation of NMDA receptor trafficking by amyloid-β. *Nature Neuroscience, 8*, 1051–1058.

Wang, H. Y., Lee, D. H. S., D'Andrea, M. R., Peterson, P. A., Shank, R. P., & Reitz, A. B. (2000). β-Amyloid$_{1-42}$ binds to α7 nicotinic acetylcholine receptor with high affinity. *The Journal of Biological Chemistry, 275*, 5626–5632.

Aβo Receptors and Aberrant Signaling

Cissé, M., Halabisky, B., Harris, J., Devidze, N., Dubal, D. B., Sun, B., et al. (2011). Reversing EphB2 depletion rescues cognitive function in Alzheimer model. *Nature, 469*, 47–52.

Freier, D. B., Nicoll, A. J., Klyubin, I., Panico, S., Mc Donald, J. M., Risse, E., et al. (2011). Interaction between prion protein and toxic amyloid β assemblies can be therapeutically targeted at multiple sites. *Nature Communications 2*, art. 336.

Kim, T., Vidal, G. S., Djurisic, M., William, C. M., Birnbaum, M. E., Garcia, K. C., et al. (2013). Human LilrB2 is a β amyloid receptor and its murine homolog PirB regulates synaptic plasticity in an Alzheimer's model. *Science, 341*, 1399–1404.

Larson, M., Sherman, M. A., Amar, F., Nuvolone, M., Schneider, J. A., Bennett, D. A., et al. (2012). The complex PrP -Fyn couples human oligomeric Aβ with pathological tau changes in Alzheimer's disease. *Journal of Neuroscience, 32*, 16857–16871.

Roberson, E. D. (2011a). Amyloid-β/Fyn-induced synaptic, network, and cognitive impairments depend on Tau levels in multiple mouse models of Alzheimer's disease. *Journal of Neuroscience, 31*, 700–711.

Rushworth, J. V., Griffiths, H. H., Watt, N. T., & Hooper, N. M. (2013). Prion protein-mediated toxicity of amyloid-β oligomers requires lipid rafts and the transmembrane LRP1. *The Journal of Biological Chemistry, 288*, 8935–8951.

Um, J. W., Kaufman, A. C., Kostylev, M., Heiss, J. K., Stagi, M., Takahashi, H., et al. (2013). Metabotropic glutamate receptor 5 is a coreceptor for Alzheimer Aβ oligomer bound to cellular prion protein. *Neuron, 79*, 887–902.

Calcium Signaling, Excitotoxicity, and Extrasynaptic NMDA Receptors

Berridge, M. J. (1998). Neuronal calcium signaling. *Neuron, 21,* 13–26.
Choi, D. W. (1990). Glutamate neurotoxicity and diseases of the nervous system. *Neuron, 1,* 623–634.
Clapham, D. E. (2007). Calcium signaling. *Cell, 131,* 1047–1058.
Hardingham, G. E., Fukunaga, Y., & Bading, H. (2002). Extrasynaptic NMDARs oppose synaptic NMDARs by triggering CREB shut-off and cell death pathways. *Nature Neuroscience, 5,* 405–414.
Olney, J. W. (1969). Brain lesions, obesity, and other disturbances in mice treated with monosodium glutamate. *Science, 164,* 719–721.
Talantova, M., Sanz-Blasco, S., Zhang, X., Xia, P., Akhtar, M. W., Okamoto, S., et al. (2013). Aβ induces astrocyte glutamate release, extrasynaptic NMDA receptor activation, and synaptic loss. *Proceedings of the National Academy of Sciences of the United States of America, 110,* E2518–E2527.

Aβ at the Rafts

Ariga, T., McDonald, M. P., & Yu, R. K. (2008). Role of ganglioside metabolism in the pathogenesis of Alzheimer's disease—A review. *Journal of Lipid Research, 49,* 1157–1175.
Di Paolo, G., & Kim, T. W. (2011). Linking lipids to Alzheimer's disease: Cholesterol and beyond. *Nature Reviews. Neuroscience, 12,* 284–296.
Kalvodova, L., Kahya, N., Schwille, P., Ehehalt, R., Verkade, P., Drechsel, D., et al. (2005). Lipids as modulators of proteolytic activity of BACE. *The Journal of Biological Chemistry, 280,* 36815–36823.
Matsuoka, Y., Saito, M., LaFrancois, J., Saito, M., Gaynor, K., Olm, V., et al. (2003). Novel therapeutic approach for the treatment of Alzheimer's disease by peripheral administration of agents with an affinity for β -amyloid. *Journal of Neuroscience, 23,* 29–33.
Vetrivel, K. S., & Thinakaran, G. (2010). Membrane rafts in Alzheimer's disease beta-amyloid production. *Biochimica et Biophysica Acta, 1801,* 860–867.

Blood-Brain Barrier and Neurovascular Unit

Abbott, N. J., Rönnbäck, L., & Hansson, E. (2006). Astrocyte-endothelial interactions at the blood-brain barrier. *Nature Reviews. Neuroscience, 7,* 41–53.
Brightman, M. W., & Reese, T. S. (1969). Junctions between intimately opposed cell membranes in the vertebrate brain. *The Journal of Cell Biology, 40,* 648–677.
Reese, T. S., & Karnovsky, M. J. (1967). Fine structural localization of a blood-brain barrier to exogenous peroxidase. *The Journal of Cell Biology, 34,* 207–217.
Suzuki, N., Cheung, T. T., Cai, X. D., Odaka, A., Otvos, L., Jr., Eckman, C., et al. (1994). An increased percentage of long amyloid β protein secreted by familial amyloid β protein precursor (β APP$_{717}$) mutants. *Science, 264,* 1336–1340.
Zlokovic, B. V. (2008). The blood-brain barrier in health and chronic neurodegenerative disorders. *Neuron, 57,* 178–201.
Zlokovic, B. V. (2011). Neurovascular pathways to neurodegeneration in Alzheimer's disease and other disorders. *Nature Reviews. Neuroscience, 12,* 723–738.

Aβ and Vascular Injury

Buée, L., Hof, P. R., Bouras, C., Delacourte, A., Perl, D. P., Morrison, J. H., et al. (1994). Pathological alterations in the cerebral microvasculature in Alzheimer's disease and related dementing disorders. *Acta Neuropathologica, 87*, 469–480.

Iadecola, C. (2010). The overlap between neurodegenerative and vascular factors in the pathogenesis of dementia. *Acta Neuropathologica, 120*, 287–296.

Inflammation in AD

Heneka, M. T., Kummer, M. P., Stutz, A., Delekate, A., Schwartz, S., Vieira-Saecker, A., et al. (2013). NLRP3 is activated in Alzheimer's disease and contributes to pathology in APP/PS1 mice. *Nature, 493*, 674–678.

McGeer, P. L., & McGeer, E. G. (2001). Inflammation, autotoxicity and Alzheimer's disease. *Neurobiology of Aging, 22*, 799–809.

Nimmerjahn, A., Kirchhoff, F., & Helmchen, F. (2005). Resting microglial cells are highly dynamic surveillants of brain parenchyma in vivo. *Science, 308*, 1314–1318.

Oberheim, N. A., Takano, T., Han, X., He, W., Lin, J. H. C., Wang, F., et al. (2009). Uniquely hominid features of adult human astrocytes. *Journal of Neuroscience, 29*, 3276–3287.

Rogers, J., Cooper, N. R., Webster, S., Schultz, J., McGeer, P. L., Styren, S. D., et al. (1992). Complement activation by β-amyloid in Alzheimer's disease. *Proceedings of the National Academy of Sciences of the United States of America, 89*, 10016–10020.

Halle, A., Hornung, V., Petzold, G. C., Stewart, C. R., Monks, B. G., Reinheckel, T., et al. (2008). The NALP3 inflammasome is involved in the innate immune response to amyloid-β. *Nature Immunology, 9*, 857–865.

Schafer, D. P., Lehrman, E. K., Kautzman, A. G., Koyama, R., Mardinly, A. R., Yamasaki, R., et al. (2012). Microglia sculpt postnatal neural circuits in an activity and complement-dependent manner. *Neuron, 74*, 691–705.

Stephan, A. H., Madison, D. V., Mateos, J. M., Fraser, D. A., Lovelett, E. A., Coutellier, L., et al. (2013). A dramatic increase of C1q protein in the CNS during normal aging. *Journal of Neuroscience, 33*, 13460–13474.

Stevens, B., Allen, N. J., Vazquez, L. E., Howell, G. R., Christopherson, K. S., Nouri, N., et al. (2007). The classical complement cascade mediates CNS synapse elimination. *Cell, 131*, 1164–1168.

Volterra, A., & Meldolesi, J. (2005). Astrocytes, from brain glue to communication elements: The revolution continues. *Nature Reviews. Neuroscience, 6*, 626–640.

APP Oligomers and Tau

Augustinack, J. C., Schneider, A., Mandelkow, E. M., & Hyman, B. T. (2002). Specific tau phosphorylation sites correlate with severity of neuronal cytopathology in Alzheimer's disease. *Acta Neuropathologica, 103*, 26–35.

de Calignon, A., Fox, L. M., Pitstick, R., Carlson, G. A., Bacskai, B. J., Spires-Jones, T. L., et al. (2010). Caspase activation precedes and leads to tangles. *Nature, 464*, 1201–1204.

Götz, J., Chen, F., van Dorpe, J., & Nitsch, R. M. (2001). Formation of neurofibrillary tangles in P301L tau transgenic mice induced by Aβ42 fibrils. *Science, 293*, 1491–1495.

Hoover, B. R., Reed, M. N., Su, J., Penrod, R. D., Kotilinek, L. A., Grant, M. K., et al. (2010). Tau mislocation to dendritic spines mediates synaptic dysfunction independently of neurodegeneration. *Neuron, 68*, 1067–1081.

Ittner, L. M., Ke, Y. D., Delerue, F., Bi, M., Gladbach, A., van Eersel, J., et al. (2010). Dendritic function of tau mediates amyloid-β toxicity in Alzheimer's disease mouse models. *Cell, 142*, 387–397.

Jin, M., Shephardson, N., Yang, T., Chen, G., Walsh, D., & Selkoe, D. J. (2011). Soluble amyloid β-protein dimers isolated from Alzheimer cortex directly induce Tau hyperphosphorylation and neuritic degeneration. *Proceedings of the National Academy of Sciences of the United States of America, 108*, 5819–5824.

Lewis, J., Dickson, D. W., Lin, W. L., Chisholm, L., Corral, L., Jones, G., et al. (2001). Enhanced neurofibrillary degeneration in transgenic mice expressing mutant tau and APP. *Science, 293*, 1487–1491.

Oddo, S., Caccamo, A., Shepherd, J. D., Murphy, M. P., Golde, T. E., Kayed, R., et al. (2003). Triple-transgenic model of Alzheimer's disease with plaques and tangles: Intracellular Aβ and synaptic dysfunction. *Neuron, 39*, 409–421.

Roberson, E. D. (2011b). Amyloid-β/Fyn-induced synaptic, network, and cognitive impairments depend on Tau levels in multiple mouse models of Alzheimer's disease. *Journal of Neuroscience, 31*, 700–711.

Shipton, O. A., Leitz, J. R., Dworzak, J., Acton, C. E. J., Tunbridge, E. M., Denk, F., et al. (2011). Tau protein is required for amyloid β -induced impairment of hippocampal long-term potentiation. *Journal of Neuroscience, 31*, 1688–1692.

Vossel, K. A., Zhang, K., Brodbeck, J., Daub, A. C., Sharma, P., Finkbeiner, S., et al. (2010). Tau reduction prevents Aβ -induced defects in axonal transport. *Science, 330*, 198.

Wang, Y., Martinez-Vicente, M., Krüger, U., Kaushik, S., Wong, E., Mandelkow, E. M., et al. (2009). Tau fragmentation, aggregation and clearance: The dual role of lysosomal processing. *Human Molecular Genetics, 18*, 4153–4170.

Zempel, H., Thies, E., Mandelkow, E., & Mandelkow, E. M. (2010). Aβ oligomers cause localized Ca^{2+} elevation, missorting of endogenous tau into dendrites, tau phosphorylation, and destruction of microtubules and spines. *Journal of Neuroscience, 30*, 11938–11950.

Zempel, H., Luedtke, J., Kumar, Y., Biernat, J., Dawson, H., Mandelkow, E., et al. (2013). Amyloid-β oligomers induce damage via tau-dependent microtubule severing by TTLL6 and spastin. *The EMBO Journal, 32*, 2920–2937.

Prion-Like Spread

Bero, A. W., Yan, P., Roh, J. H., Cirrito, J. R., Stewart, F. R., Raichle, M. E., et al. (2011). Neuronal activity regulates the regional vulnerability to amyloid-β deposition. *Nature Neuroscience, 14*, 750–756.

Domert, J., Rao, S. B., Agholme, L., Brorsson, A. C., Marcusson, J., Hallbeck, M., et al. (2014). Spreading of amyloid-β peptides via neuritic cell-to-cell transfer is dependent on insufficient cellular clearance. *Neurobiology of Disease, 65*, 82–92.

Eisele, Y. S., Obermüller, U., Heilbronner, G., Baumann, F., Kaeser, S. A., Wolburg, H., et al. (2010). Peripherally applied Aβ -containing inoculates induce cerebral β -amyloidosis. *Science, 330*, 980–982.

Meyer-Luehmann, M., Coomaraswamy, J., Bolmont, T., Kaeser, S., Schaefer, C., Kilger, E., et al. (2006). Exogenous induction of cerebral β - amyloidogenesis is governed by agent and host. *Science, 313*, 1781–1784.

Nath, S., Agholme, L., Kurudenkandy, F. R., Granseth, B., Marcusson, J., & Hallbeck, M. (2012). Spreading of neurodegenerative pathology via neuron-to-neuron transmission of β -amyloid. *Journal of Neuroscience, 32,* 8767–8777.

Stöhr, J., Watts, J. C., Mensinger, Z. L., Oehler, A., Grillo, S. K., DeArmond, S. J., et al. (2012). Purified and synthetic Alzheimer's amyloid beta (Aβ) prions. *Proceedings of the National Academy of Sciences of the United States of America, 109,* 11025–11030.

Regulated Intramembrane Proteolysis

Brown, M. S., Ye, J., Rawson, R. B., & Goldstein, J. L. (2000). Regulated intramembrane proteolysis: A control mechanism conserved from bacteria to humans. *Cell, 100,* 391–398.

Cao, X., & Südhof, T. C. (2001). A transcriptionally [correction of transcriptively] active complex of APP with Fe56 and the histone acetyltransferase Tip60. *Science, 293,* 115–120.

De Strooper, B., Vassar, R., & Golde, T. (2010). The secretases: Enzymes with therapeutic potential in Alzheimer disease. *Nature Reviews. Neurology, 6,* 99–107.

Kopan, R., & Ilagen, M. X. (2009). The canonical Notch signaling pathway: Unfolding the activation mechanism. *Cell, 137,* 216–233.

Lichtenthaler, S. F., Haase, C., & Steiner, G. H. (2011). Regulated intramembrane proteolysis— Lessons from amyloid precursor protein processing. *Journal of Neurochemistry, 117,* 779–796.

Lipoproteins

Mahley, R. W. (1988). Apolipoprotein E: Cholesterol transport protein with expanding role in cell biology. *Science, 240,* 622–630.

Olson, R. E. (1998). Discovery of the lipoproteins, their role in fat transport and their significance as risk factors. *The Journal of Nutrition, 128,* 439S–443S.

Chapter 9
Parkinson's Disease

The year 2017 will mark the 200th anniversary of the 1817 publication of James Parkinson's monograph "An Essay on the Shaking Palsy." In his monograph, Parkinson described six individuals suffering from a condition involving a bent posture, involuntary tremulous motion and weakened muscular action. Today, Parkinson's disease (PD) in its various forms is recognized as the second most prevalent neurodegenerative disorder. It has an incidence rate of 1 % for individuals age 65 and older and that rate increases to 5 % for people past the age of 85.

There are six major clinical symptoms of the syndrome known as *parkinsonism.* Three of these are early signs of the "shaking palsy." These are: tremor at rest, rigidity, and slowness of movement (bradykinesia). The remaining three signs appear later as the condition worsens. These signs are: postural abnormalities (bent posture), postural instabilities, and freezing of gait often resulting in falls. These are the key clinical indications of the syndrome, and as such were described in the classical textbooks on neurology by Gowers (1893), Oppenheim (1911), and Wechsler (1932).

Parkinson's disease, the chief form of parkinsonism, begins in a specific region of the midbrain known as the substantia nigra *pars compacta* (SNc) and from there spreads to other brain region. The association of the disorder with specific anatomical regions took many years to establish. First, researchers in the late 1800s had found that in the basal ganglia and substantia nigra were affected in the disease. The latter region is enriched in the dark pigment neuromelanin, hence its name, and undergoes depigmentation when neurons are lost. The connection to the SNc was strengthened by Lewy and by Konstantin Tretiakoff (1892–1958), who in his 1919 doctoral dissertation identified the SNc as the primary starting point of the disease and named the Lewy bodies, the chief histopathological sign of PD, after Lewy.

SNc neurons use dopamine as their neurotransmitter, and project their axons to neurons in the striatum responsible for initiating and controlling movement. The lack of adequate dopamine signaling results in failed motor control, and leads in the later stages of the disease to cell death and dementia. A landmark event in treating

© Springer International Publishing Switzerland 2015
M. Beckerman, *Fundamentals of Neurodegeneration and Protein Misfolding Disorders,*
Biological and Medical Physics, Biomedical Engineering,
DOI 10.1007/978-3-319-22117-5_9

Parkinson's disease was the discovery in 1957 by Arvid Carlsson that L-dopa could reverse the symptoms of parkinsonism. Up until that time, depletion of serotonin was thought to be responsible for the observed symptoms, but that idea was shown to be incorrect by Carlsson. His discovery was followed shortly thereafter in 1960 by findings by Herbert Ehringer and Oleh Hornykiewicz that dopamine levels were depleted in the striatum of humans suffering from parkinsonism. That discovery established for the first time the association of an abnormality in a particular chemical occurring in a specific region of the brain with a neurological disorder. That discovery led to the initiation by Hornykiewicz of clinical trials with levodopa in 1961 and by the mid-1970s that treatment had become the accepted approach for reducing the symptoms of parkinsonism.

For many years it was felt that PD lacked a genetic component. Instead, environmental factors leading to oxidative stress were thought to be responsible for the disease. Support for this belief was provided by the discovery in 1983 that the chemical agent 1-methyl-4-phenyl-1,2,3,6-tetrahydropyridine (MPTP) can produce signs in neurons closely resembling those of parkinsonism. This chemical agent can readily pass through the blood–brain barrier (BBB); it is oxidized to a highly toxic MPP^+ form by the enzyme monoamine oxidase B (MAO-B), and is taken up by the dopamine transporter. Once in the neurons it (1) inhibits complex I of the mitochondrial electron transport chain (ETC), (2) impairs respiration, (3) increases production of superoxide, and (4) disrupts calcium homeostasis and associated cellular functions. In a further development of an environmental factor theme, the widely used pesticide rotenone was found to be an even better trigger of parkinsonism-like symptoms in mammalian neurons. Here too the main site of action of the agent was mitochondrial complex I. And a few years later dopaminergic neurons were found to be vulnerable to the herbicide paraquat.

Evidence keeps accumulating pointing to the involvement of environmental toxins including heavy metals in the etiology of Parkinson's disease. One finding that is consistent across almost all studies is the generation of oxidative stress, that is, the toxins damage components of the mitochondrial ETC resulting in increased reactive oxygen species (ROS) and reactive nitrogen species (RNS) production (Table 9.1). These toxins are widely used in the laboratory today to reproduce features of PD in test subjects. Using these experimental models, researchers can dissect the steps in the pathogenesis of the disease and test possible therapeutic drugs.

There are over one million Parkinson's disease sufferers in the United States and this neurological disorder is second only to Alzheimer's disease in its frequency of occurrence. The focus on environmental factors for its cause began to shift in 1997 with two sets of findings. First, Polymeropoulos discovered that the appearance of PD in certain individuals was associated with mutations in α-synuclein, a small, 140-amino-acid-residue protein enriched in the presynaptic terminals of neurons. Second, Spillantini established that the same protein, α-synuclein, was the main component of the Lewy bodies, the postmortem defining feature of Parkinson's disease and dementia with Lewy bodies.

Three PD-causing mutations in SNCA, the gene that encodes α-synuclein, were found. The first, the 1997 discovery, involved the substitution of threonine for alanine at position 53 (A53T). The second, found within the next year, was that of a

Table 9.1 Environmental chemicals and other toxins widely used to model Parkinson's disease

Toxins	Description
MPTP	1-Methyl-4-phenyl-1,2,3,6-tetrahydropyridine generates parkinsonism through the actions of its metabolite MPP$^+$. It is lipophilic and readily passes through the BBB. It damages mitochondrial complex I
CCCP	Carbonyl cyanide m-chlorophenylhydrazone is a mitochondrial membrane depolarizer. It inhibits oxidative phosphorylation and triggers mitophagy
Rotenone	Lipophilic insecticide that readily passes through the BBB. Damages mitochondrial complex I. Activates microglia. Generates α-synuclein-positive Lewy bodies
Paraquat	Herbicide that undergoes channel-mediated passage through the BBB. Damages mitochondrial complex I. When combined with maneb generates a form of PD comparable in magnitude to MPTP-induced forms
Maneb	Maneb, a fungicide, utilizes manganese ethylene-bis-dithiocarbamate (Mn-EBDC) as its main active ingredient. Damages mitochondrial complex III
6-OHDA	6-Hydroxydopanine is used in the laboratory to create a PD model that can be used in drug development. It cannot pass the BBB but can be administered directly through injection. It impairs mitochondrial complex I

proline substitution for alanine at position 30 (A30P) and the third in 2004 was that of E46K. In just the last few years, three more PD-associated mutations in the SNCA gene have been identified. These mutations, namely H50Q, G51D and A53E, are clustered in a narrow region spanned by the previously identified E46K and A53T mutations. These mutations occur in families with a history of developing PD in an autosomal dominant fashion. By altering α-synuclein's conformation the mutations influence its ability to oligomerize and fibrillize. They also produce changes in its subcellular localization and alter its function. Further reinforcement of the connection between PD and α-synuclein came from multiple experimental studies showing that loss of α-synuclein results in failures in the nigrostriatal dopamine system and leads to neurodegeneration.

α-Synuclein expression levels are important. The connection between concentration and disease was firmly established by the discovery in 2003 of a triplication in the SNCA gene of Parkinson's disease sufferers. It was reinforced the following year when a duplication in SNCA was shown to cause PD. In another, large-scale genetic study, polymorphisms in the SNCA promoter were uncovered. These three studies firmly establish that as the expression level of α-synuclein increases so does the risk for developing PD. This theme is a familiar one having been encountered and discussed at length with regard to amyloid-β generation and clearance in the last chapter.

9.1   Multiple Pathways to Parkinson's Disease

This chapter begins with an examination of what makes the substantia nigra neurons so special that they and they alone are the locus of the disease onset. These neurons possess a number of striking features centered on their unusual morphology and

Table 9.2 Key genetic factors implicated in the pathogenesis of Parkinson's disease

Locus	Protein	Description
PARK1—4q21	α-Synuclein	Presynaptic proteins; AD
PARK2—6q26-27	Parkin	E3 ubiquitin ligase; AR
PARK6—1p35-36	PINK1	Mitochondrial serine/threonine kinase; AR
PARK7—1p36	DJ-1	Protects against oxidative stress; AR
PARK8—12p11.2	LRRK2	Mixed lineage kinase; AD

AD: autosomal dominant inheritance, *AR:* autosomal recessive inheritance. In addition to the PARK1 SNCA locus, a PARK4 locus was initially assigned. The latter was subsequently found to correspond to a triplication in the SNCA gene

unique biophysical properties. These properties are examined along with an overview of the circuitry involved in PD. Alpha-synuclein is then examined. Since its discovery a growing list of PD-associated mutated genes has been uncovered. The most prominent and best studied of these genes and the proteins they encode are listed in Table 9.2. As noted in the table leucine-rich repeat kinase 2 (LRRK2), like α-synuclein, is inherited in an autosomal dominant fashion, while the remaining risk factors are inherited in an autosomal recessive manner. Significantly, three of these risk factors—phosphatase and tensin homolog (PTEN)-induced putative kinase 1 (PINK1), parkin, and DJ-1 function in protein quality control pathways. Leucine-rich repeat kinase 2 (LRRK2) operates as a lipid-raft/caveolae-associated signaling platform, and it too has a prominent role in protein quality control.

Lewy bodies are spherically shaped, 5–25 μm in diameter, structures. They contain a dense core of filamentous and granular material surrounded by a filamentous halo. They are composed primarily of neurofilaments, ubiquitin, heat shock proteins, and α-synuclein. These structures are present in almost all sporadic or idiopathic cases of PD that comprise the vast majority of the cases of PD. They are present in most instances of LRRK2-associated PD and in all SNCA-associated PD instances. However, they are largely absent in PINK1 and parkin-associated PD, and their presence or absence in DJ-1 associated forms is unknown.

In Alzheimer's disease, the various risk factors all seemed to involve the β-amyloid peptide in one manner or another. They either had a role in its generation, or alternatively, in its clearance. Thus, a single pathway appears at the causal center of the disease. It is already clear from the selective presence or absence of Lewy bodies that PD is a more complex phenomenon. PINK1 and parkin do define a clear pathway, one that involves mitochondrial quality control. That pathway is examined in this chapter along with DJ-1, which resides in a parallel mitochondrial-protective pathway to that of PINK1/parkin. The placement of α-synuclein and LRRK2 in a single pathway is a more challenging endeavor. One place where they do come together is in their mutual impairment of chaperone-mediated autophagy, and that locus of commonality is looked before the mitochondrial pathways are discussed.

Lastly, the risk factors listed in Table 9.2 are not the only ones that have been found. Instead, there is a growing list of recently discovered genetic risk factors. Some of these point to yet another pathway to PD, one that centrally involves the lysosome and its dysfunction due to mutations in its enzymes and its non-enzymatic transport proteins. These emerging risk factors are discussed towards the end of the chapter. The chapter then ends with an overview of a way that cells dispose of potentially dangerous material when all other means are blocked—they release the materials into the extracellular spaces.

9.2 Nigrostriatal Neurons Have a Number of Special Properties

Recall that in the pioneering 1952 study by Hodgkin and Huxley, two types of voltage-dependent ion channels were identified. First, the opening of sodium channel allowed Na^+ ions to enter the neuron resulting in a decrease in the net negative charge of the neural interior, thereby depolarizing the plasma membrane. That was followed after a short delay by the opening of potassium ion channels resulting in an efflux of K^+ ions and the repolarization of the membrane. The dual actions involving the opening and closing of sodium and potassium channels produced an action potential that propagated down the squid giant axon.

Central nervous system (CNS) neurons have far more complex signaling, communication and control tasks to perform than that of the squid giant axon and their action potentials reflect the varying requirements. Shown in Fig. 9.1 are three examples of different action potentials and accompanying firing patterns—bursting, tonic, and rhythmic/pacemaking—exhibited by CNS neurons. The repertoire of voltage-gated ion channels possessed by these neurons is quite impressive and underlies their ability to produce the different firing patterns. In particular, there are several dozen different kinds of voltage-dependent potassium channels including some that are dependent on the intracellular calcium concentration in their immediate vicinity. Also, there are also a large number of voltage-dependent calcium channels that allow Ca^{2+} ions to enter the neurons.

The elevation of calcium levels in neurons has multiple effects and its concentration is tightly regulated. Under normal and quiescent conditions, its intracellular concentration is kept some three orders of magnitude lower than its extracellular concentration. Any transient increase in concentration is handled rapidly by an assortment of channels, pumps, and other calcium transporters. These either send the calcium ions back out to the extracellular spaces or into intracellular storage locations. As discussed in the last chapter, the calcium handling system localizes the Ca^{2+} increases both in space and time, thereby enabling the ion to function as a signaling intermediary, referred to as a second messenger. For example, in response to the arrival of an action potential at the presynaptic terminal Ca^{2+} enters through voltage-gated calcium channels. The Ca^{2+} ions diffuse to calcium sensors located on the outer surface of the neurotransmitter-loaded vesicles. This arrival activates effector proteins that trigger vesicle membrane fusion and release of the neurotransmitters into the synaptic cleft.

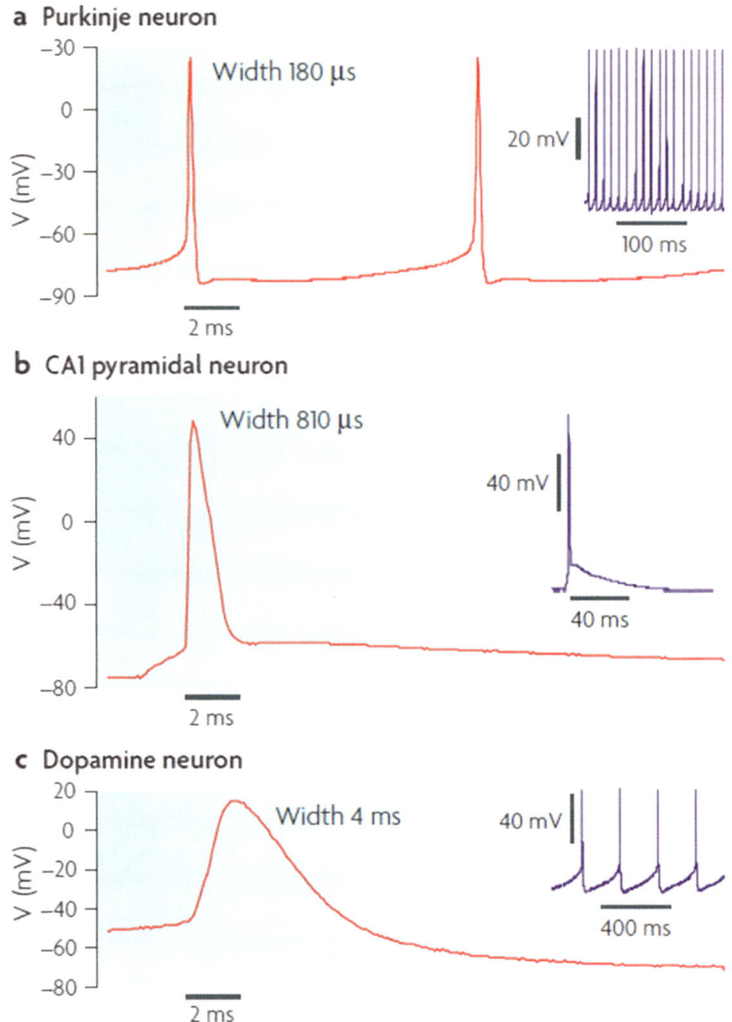

Fig. 9.1 Action potential shapes and firing patterns. (**a**) Burst firing by cerebellar Purkinje neurons; (**b**) Tonic firing of hippocampal CA1 pyramidal neurons; (**c**) Pacemaker firing by SNc dopaminergic neurons (from Bean *Nat. Rev. Neurosci.* 8: 451 © 2007 Reprinted by permission from Macmillan Publishers Ltd)

Substantia nigra *pars compacta* neurons have a number of special properties that create stressed conditions. These properties render the neurons vulnerable to age-related decline of proteostasis and to buildups in misfolded proteins especially those identified in Table 9.1. SNc neurons:

- *Possess highly ramified, poorly myelinated axons.* Unlike most other neurons their axons form dense axonal bushes. This architecture provides every DA-secreting neuron with a large number of DA releasing sites that can influences tens of thousands of striatal neurons.

- *Function as autonomous pacemakers.* SNc neurons spike on their own in the absence of synaptic input. This enables the neurons to continuously supply dopamine to the recipient striatal neurons. In order to spike continuously in that manner, the neurons exhibit rather depolarized membrane conditions typical of pacemakers. They have a high metabolic burden and rely heavily on oxidative phosphorylation to meet their metabolic needs.
- *Have weakcalcium handlingcapabilities*: Calcium concentrations are tightly regulated in most cells. Calcium is sequestered in intracellular stores and rapidly captured and pumped back out of the cells subsequent to spiking. SNc neurons utilize L-type Ca^{2+} channels that assist in their pacemaking. These channels permit the entry of large amounts of calcium, but the SNc neurons have poor calcium buffering and handling capabilities. Thus, they are more susceptible to calcium overload and calcium-induced excitotoxicity than most other neurons.

9.3 Dopamine Modulates Striatal Synaptic Transmission

The circuitry responsible for Parkinson's disease is centered in the striatum, which functions as the gateway to the basal ganglia. Striatal medium spiny projection neurons (MSNs) receive glutamatergic inputs from neurons in the cortex and thalamus. These centrally positioned neurons, in turn, project to a variety of cortical targets involved in body movement. Additional inputs are supplied to the MSN dendritic spines, and to the glutamate-releasing axon terminals providing the corticostriatal inputs, by dopamine-releasing axons from neurons located in the substantia nigra pars compacta (SNc). The loss of the dopamine releasing SNc neurons results in dysregulated synaptic transmission at the corticostriatal synapses. In particular, the motor control neurons receive malformed and excessively stimulatory signals from the striatal projection neurons that result in movement deficits and Parkinson's disease.

In more detail, dopamine is a neuromodulator—molecules that produce changes in the electrical excitability of neurons and neural circuits. These molecules (serotonin and histamine are other examples of neuromodulators) bind members of the seven-pass, G-protein-coupled receptor (GPCR) family such as the metabotropic glutamate receptors (mGluRs) introduced in the previous chapter. Ligand binding to these GPCRs initiates a series of signaling events that regulate neurotransmitter detection by the NMDA and other ligand-gated ion channels on the postsynaptic side and neurotransmitter release properties on the presynaptic side. Calcium, sodium and potassium fluxes through the ion channels are altered and membrane electrical excitability is updated and adjusted according to the additional needs (Fig. 9.2). By these methods, dopaminergic inputs acting at presynaptic and postsynaptic sites regulate glutamatergic synaptic transmission.

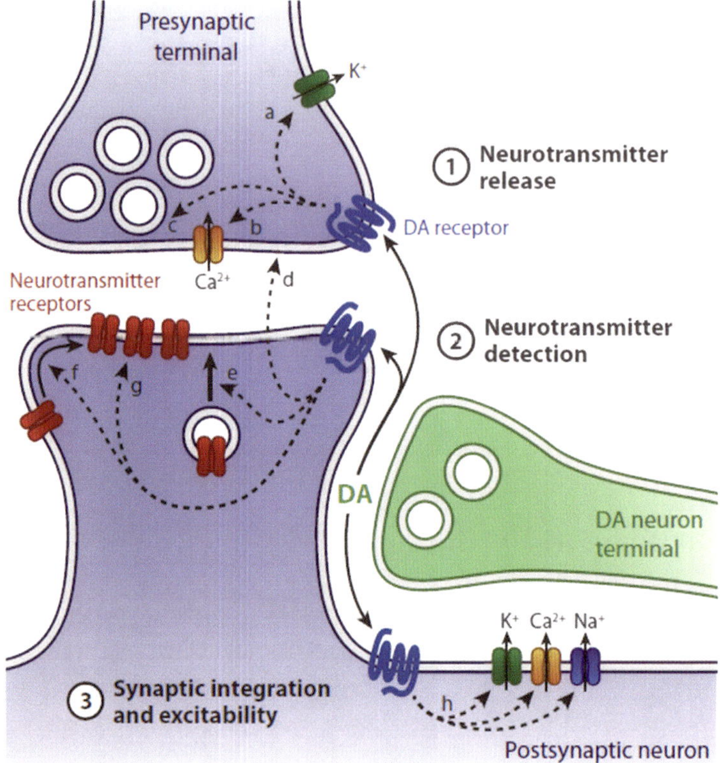

Fig. 9.2 Dopaminergic modulation of corticostriatal synaptic transmission. Release of dopamine from the SNc axon terminal binds dopaminergic, seven-pass G-protein-coupled receptors on pre- and postsynaptic terminals. These actions modulate neurotransmitter release from the presynaptic corticothalamic axon terminal and neurotransmitter detection and synaptic integration at the postsynaptic, striatal dendritic terminal (from Tritsch *Neuron* 76: 33 © 2012 Reprinted by permission from Elsevier)

9.4 Mutations and Overexpression of α-Synuclein Cause Autosomal Dominant Parkinson's Disease

Alpha-synuclein has been discovered several times. It was first discovered in the neurons of the *Torpedo californica* electric ray by Maroteaux in 1988. Using an antiserum against cholinergic synaptic vesicles, the protein was identified in presynaptic terminals of neurons and portions of the nuclear envelop, hence its name *synuclein*. It was next uncovered in 1993 by Uéda as the precursor protein NACP, the NAC being a 35-amino acid fragment found to co-localize with Aβ in Alzheimer's disease plaques but not derived from Aβ or its precursor. Hence it bears the appellation "non-Aβ component of the AD amyloid (or NAC)." Those findings were followed by the Spillantini and Polymeropoulos discoveries.

α-Synuclein is a vertebrate-specific protein highly expressed in the brain, especially in the neocortex, hippocampus and substantia nigra. Nowadays, it sits on top of the pantheon of high agents of Parkinson's disease listed in Table 9.2. Parkinson's disease is not the only neurodegenerative disorder characterized by α-synuclein-positive filamentous inclusions in neurons or glia. Rather, several neurological disorders, each with its own distinct set of clinical and pathological features, have α-synuclein-positive inclusions. The most prominent of the *alphasynucleinopathies* are:

- *Dementia withLewy bodies(DLB)*: Other names for this disorder are Lewy body dementia (LBD) and LB variant of AD (LBV/AD). Prominent among the clinical features of this disorder are extrapyramidal signs such as bradykinesia (slowness of movement), cognitive decline, visual hallucinations or delusions, and sensitivity to antipsychotic (neuroleptic) medications. Described originally by Okazaki in 1961 this has emerged as one of the most common forms of dementia although it is less prevalent than Alzheimer's.
- *Multiple system atrophy (MSA)*: clinical features include parkinsonism along with autonomic and gait abnormalities, cerebellar signs, and cognitive decline. A major pathological feature is the presence of tubulofilamentous inclusion in oligodendrocytes referred to as glial cytoplasmic inclusions (GCIs). These are found in multiple regions of the brain.
- *Neurodegeneration with brain iron accumulation, type I (NBIA)*: This disorder is characterized clinically by parkinsonism accompanied by cognitive decline, cerebellar signs and bulbar symptoms. Its chief pathological feature is the buildup of iron and axonal swellings termed spheroids. GCIs and LBs are commonly encountered, as well.

9.5 Structure and Function of α-Synuclein

α-Synuclein possesses three domains. It has a positively charged N-terminal amphipathic region in which are embedded a series of KTKEGV repeats that mediate binding to negatively charged membranes. It possesses a central non amyloid component (Uéda's NAC) domain containing a stretch of hydrophobic residues responsible for its tendency to aggregate and form β-sheets, and it has a C-terminal acidic region. These domains and the locations of the six known PD-causing mutations are presented in Fig. 9.3.

One of its most important properties is its ability to develop amphipathic α-helical secondary structure in its N-terminal region upon association with (negatively charged) lipid rafts. When doing so, the C-terminal region remains free and unstructured. Not only does α-synuclein associate with these membrane microdomains but also operates as a sensor of membrane curvature. This second key property enables the protein to preferentially associate with the more highly curved vesicles that package neurotransmitters in presynaptic terminals and other curved membranes such as those that bound mitochondria.

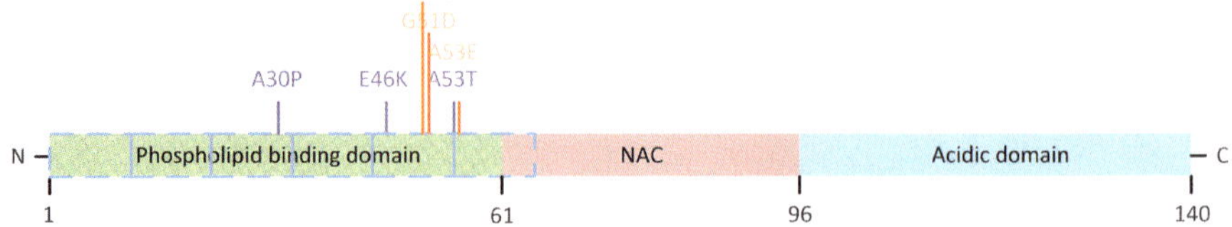

Fig. 9.3 α-Synuclein domain structure and disease-causing mutations. Shown are the N-terminal phospholipid/lipid raft binding domain, the central hydrophobic and aggregation-prone NAC, and the C-terminal negatively charged acidic domain. A series of repeats (depicted by the *blue dashed boxes*) in the N-terminal each contains an imperfect, KTKEGV-like consensus motif. The original three disease-causing mutations are depicted in *purple* and the recently found trio in *orange*

Like other intrinsically disordered proteins (discussed in Section 4.10) it is highly flexible and sensitive to environmental conditions. It has been called a "chameleon" for its ability to assume any of a number of alternative conformations depending on its immediate surroundings. α-Synuclein has been explored using a variety of techniques ranging from low resolution ensemble studies to high resolution single-molecule biophysics. In recent explorations using optical tweezers, the monomeric protein (cytosolic form) was found fluctuate between large numbers of marginally stable structures, and possessed an energy landscape that closely resembled the schematic depicted presented by Uversky in Fig. 4.12b. That is, its energy landscape was flat and rugged with energy barriers on the order of 2–$3\,k_BT$. Those topographical features were in contrast to the funnel-shaped and smooth landscapes typical of well-folded globular proteins (Fig. 4.12a).

A new generation of optical microscopic methods have been developed that provide nanoscale images of living cells. These methods fill in the gap between low resolution live imaging and high resolution electron microscopy in nonliving tissue samples. The diffraction ($\lambda/2$) resolution limit of lens-based optical microscopes was put forth by the German physicist Ernst Abbe (1840–1905) in 1873. This limit has held sway until recently when new fluorescence methods such as stimulated emission depletion (STED) microscopy and fluorescence recovery after photobleaching (FRAP) were developed. These new developments overcome the diffraction barrier and make possible nanoscale imaging in living systems.

One of the areas being explored with these new probes is synaptic vesicle dynamics. Synaptic vesicles are approximately 40 nm in size and these reside in synaptic boutons on the order of 1 μm in diameter. The new methods reveal that while some vesicles are localized to a single synaptic bouton others form a pool that circulates between multiple boutons. By this means synapses can significantly alter their vesicle composition in a matter of minutes. Interestingly, it turns out that α-synuclein exerts considerable influence over vesicle dynamics and interacts with a variety of presynaptic vesicle-associated proteins. A number of these critical interactions had been identified leading to the emergence of a consistent picture of impaired vesicle dynamics and synaptic dysfunction arising from overexpressed (by a factor of 2 or 3) and mutated α-synuclein.

That α-synuclein interacts with proteins involved in synaptic vesicle formation and dynamics is supported by multiple studies. Neurotransmitter release occurs when neurotransmitter-laden synaptic vesicles fuse with the plasma membrane. This step is regulated by soluble N-ethylmaleimide-sensitive factor (NSF) attachment protein receptor (SNARE) and Sec1/Munc18-like proteins. Alpha-synuclein has been shown to promote assembly and disassembly of these essential complexes. Another key protein is the presynaptic vesicle chaperone, cysteine string protein-α (CSPα), which assists in the localization and exocytosis secretory vesicles containing neurotransmitters and peptides such as dopamine. CSPα helps direct chaperone complexes to the SNARE machinery and helps preserve the correct folding of proteins. α-Synuclein operates not only in conjunction with SNARE complexes but also cooperatively with CSPα. A third locus of α-synuclein activity is in its ability to inhibit vesicle trafficking from the endoplasmic reticulum to the Golgi. One of the earliest signs of Parkinson's disease is the accumulation of clusters of vesicles indicative of emerging dysregulation and malfunctions in vesicle trafficking and dopamine handling. Sitting at the center of a cascade of events excessive alpha-synuclein helps to bring this about.

When mutated or overexpressed the protective activities of α-synuclein are lost and in their place the protein misfolds, aggregates, and promotes neurodegeneration. One of the main goals in the field has been and remains to identify the precise toxic forms of the molecule so that effective pharmacological treatments can be devised. According to the widely held toxic oligomerhypothesis the filamentous deposits found in the Lewy bodies are not the primary causal agents. Instead, the smaller oligomeric species are responsible. This hypothesis is supported by an impressive body of evidence. First, there is the poor correlation between the presence of Lewy bodies and Parkinson's disease. Lewy bodies are commonly found at autopsy in the brains of elderly but otherwise neurologically normal brains. Secondly, oligomers generated through laboratory-induced genetic manipulations of SNCA reproduce many of the PD-associated presynaptic deficits. Thirdly, progress has been made in uncovering the biophysics of the transformations. The E46K and A53T mutations, in particular, produce deficits in presynaptic protection; they encode proteins that expose an increased fraction of hydrophobic surface area, and possess considerable β-sheet structure.

Mutated and overexpressed alpha-synuclein aggregates into a variety of oligomeric forms ranging in size from dimers, trimers and tetramers up to 100-mers or larger. These can assume a wide range of morphologies from linear and spherical structures to annular rings to large fibrillar assemblages. The latter, generated through nucleated-polymerization, can subsequently fragment and generate more oligomers. Each oligomeric species exists as an ensemble of conformations residing in a dynamic equilibrium and interconverting into one another, and into other oligomeric forms. Some of these conformers have a higher beta-sheet content and hydrophobicity than the others. Significantly, low and high β-sheet classes of conformers are separated by energy barriers that slow down the transition from harmless conformations to the toxic ones and may provide a pharmacological point of entry.

Excess, misfolded, damaged, aggregated and otherwise unwanted proteins are degraded by the ubiquitin-proteasome system, by macroautophagy, and in the case of selected (long-lived) proteins bearing a KFERQ-like motif by chaperone-mediated autophagy (CMA). The latter is a form of autophagy in which substrates are not packaged in vesicles, but instead, are sent one-by-one directly to and into lysosomes for degradation. These systems complement one another. For example, if a protein is too large to be handled by the proteasome it can be degraded by one of the autophagic pathways. If the protein has formed an aggregate it can still be encapsulated in an autophagosome and shipped for lysosomal degradation via macroautophagic transport. If there are circumstances where when one pathway becomes blocked, the other pathways can compensate for the loss by increasing their participation.

In a healthy cell, α-synuclein can be cleared both by macroautophagy and CMA. However, these systems fail in PD leading to a buildup of toxic oligomers, large aggregates, and fibrils. While there is information for how PD forms of α-synuclein can impair macroautophagy through its disruption of the autophagosome assembly process, it is the breakdown of CMA at the lysosome that has elicited the most interest. That is because not only α-synuclein but also several other parkinsonism-associated mutated proteins negatively impact autophagy and lysosomal function. One of these is LRRK2. That protein and its blocking effect on CMA are examined next.

9.6 Mutations in LRRK2 Cause Autosomal Dominant Parkinson's Disease

In 2002, Funayama identified a new locus for an autosomal dominant form of Parkinson's disease. That locus carrying the PARK8 designation mapped to chromosome 12p11.2-q13.1. Two years later, in 2004, independent studies by Zinprich and Paisán-Ruíz identified the encoded protein, LRRK2, and several of its disease-causing mutations. Since many of the families examined by Paisán-Ruíz were of Basque descent, he named the protein "dardarin," from the Basque word *dardara*, meaning trembling. A series of papers appearing shortly thereafter identified the G2019S mutation in LRRK2 as the single most commonly encountered mutation in PD sufferers. In some ethnic population groups, 30–40 % of all PD cases can be traced to that one mutation.

LRRK2 and alpha-synuclein both generate autosomal dominant forms of PD with similar features. The proteins, however, are strikingly different. While alpha-synuclein is small and contains large unstructured stretches, LRRK2 is huge and highly structured. The LRRK2 gene contains 51 exons, and encodes a protein containing 2527 amino acid residues organized into six well-defined domains (Fig. 9.4). This protein belongs to the so-called ROCO family of proteins containing not one but two distinct enzymatic activities—a guanosine triphosphatase (GTPase)

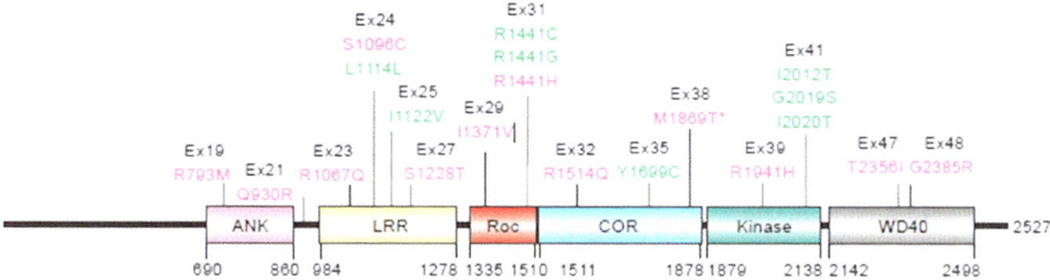

Fig. 9.4 LRRK2 domain structure and disease-causing mutations. Shown are the ankyrin (ANK), leucine-rich repeat (LRR), Ras-of-complex (Roc), C-terminal of Ras (COR), kinase, and WD40 domains. Not depicted: an armadillo domain N-terminal to the ankyrin domain. Disease-causing amino acid substitutions are denoted in *green*; other putative pathogenic sequences in *magenta*, and the corresponding exon numbers are indicated in *black* (from Mata *Trends Neurosci.* 29: 286 © 2006 Reprinted by permission from Elsevier)

activity (Ras-of-complex, Roc) and a kinase activity with a C-terminal of Ras (COR) domain in between. In addition to its enzymatic activities, the armadillo, ankyrin, leucine-rich repeat and WD40 domains confer on LRRK2 the ability to carry out protein–protein interactions and serve as a signaling scaffold.

The presence of two enzymatic activities is highly unusual and there is considerable controversy regarding their relationships and functioning. The most widely accepted model is one where the GTPase activates the kinase and the latter functions as the output unit. In that model, the GTPase must be in its ON state for the kinase domain to be catalytically active. Another requirement in this model is autophosphorylation. The functional form of LRRK2 is that of a dimer (oriented head-to-head), and this arrangement facilitates autophosphorylation at a number of sites including some located in the catalytically active site of the kinase. Most of the key disease-causing mutations are located in the central part of the protein. The two mutations occurring at sites in the Roc GTPase domain and the one located in the COR inhibit the GTPase but do not prevent the downstream kinase activity in a clear-cut manner, while the G2019S mutation located in the kinase domain actually enhances the kinase activity by a factor of three.

A further complexity arises from the presence of the protein–protein interaction domains. Their presence implies that LRRK2 functions as part of a signaling network with both upstream and downstream signaling partners. A number of these have now been identified among which are members of the 14-3-3 family of signaling modules. The importance of this interaction is derived from 14-3-3 ability to influence both the cellular location of LRRK2 and its propensity to aggregate. It turns out that phosphorylation at S910 and S939 sites located in between the ankyrin and LRR domains of LRRK2 is a crucial event that enables 14-3-3 binding. While either the kinase domain of LRRK2 or an upstream kinase can, in principle, carry out this task recent evidence points to one or more external kinases as the primary agents. Mutations such as R1441C/G/H and I2020T that abolish phosphorylation and 14-3-3 binding promote LRRK2 aggregation.

One major clue as to the functions of LRRK2 was the finding that the protein localizes to lipid rafts and caveolae. Recall that these specialized microdomains can be found embedded in the plasma membrane and intracellularly in organelle and vesicle membranes. LRRK2 can be detected in all of these locales. Recent studies emphasized possible locations in autophagic vacuoles (AVs) and in intraluminal vesicles (ILVs) of multivesicular bodies (MVBs), while earlier studies hinted at localization in lysosomes, endosomes, the mitochondrial outer membrane, the Golgi, the plasma membrane, and synaptic vesicles. At these sites, LRRK2 participates in the biogenesis and/or regulation of the intracellular membranous structures. Of special note is its role in autophagy; it has been found to regulate autophagic vacuole activity, clearance of trans-Golgi-derived vesicles through the autophagic-lysosomal pathway, and CMA (to be discussed in the next section). In addition, it plays a role in presynaptic vesicle storage and mobilization.

9.7 Chaperone-Mediated Autophagy Is Blocked by Mutated α-Synuclein and LRRK2

Recall from Chap. 6 that in autophagy unwanted proteins, protein aggregates, and organelles are sent to lysosomes for hydrolytic degradation. CMA differs from macroautophagy (autophagy) and microautophagy in two ways. First, rather than packaging the substrates in transport vesicles the proteins targeted for degradation in CMA are sent one-by-one directly through the lysosomal membrane and into the hydrolytic compartment for degradation. Secondly, in autophagy and microautophagy the targeting of misfolded and damaged proteins is nonspecific, while CMA targets only a specific subset of proteins. The main elements of CMA were uncovered in the late 1980s and early to mid1990s and motivated by a suggestion by J. Fred Dice that there might be a form of autophagy that targeted specific proteins for degradation in lysosomes. Experiments carried out in that time period in several groups and especially by Dice, Ana Maria Cuervo, and their coworkers led to the uncovering of the main elements of this pathway and confirmed Dice's prescient surmise.

The main elements of CMA consist of a recognition sequence, a chaperone, and a receptor. Proteins destined for degradation by CMA contain a KFERQ-like sequence. Motifs of this type are found on about 30 % of all the proteins. These sequences are recognized by heat shock cognate protein of 70 kDa (Hsc70), a constitutively expressed member of the Hsp70 family of molecular chaperones. The chaperone-bound substrates are then conveyed to the lysosome where they attach to the LAMP-2A receptors. As is the situation discussed in earlier chapters, proteins are translocation across the membrane in an unfolded form. In these operations, depicted in Fig. 9.5, newly synthesized LAMP-2A receptors reside in lipid rafts where (1) their mobility is restricted and where after a short time they are degraded, or (2) CMA is activated and the LAMP-2A receptors exit the rafts to carry out their translocation tasks.

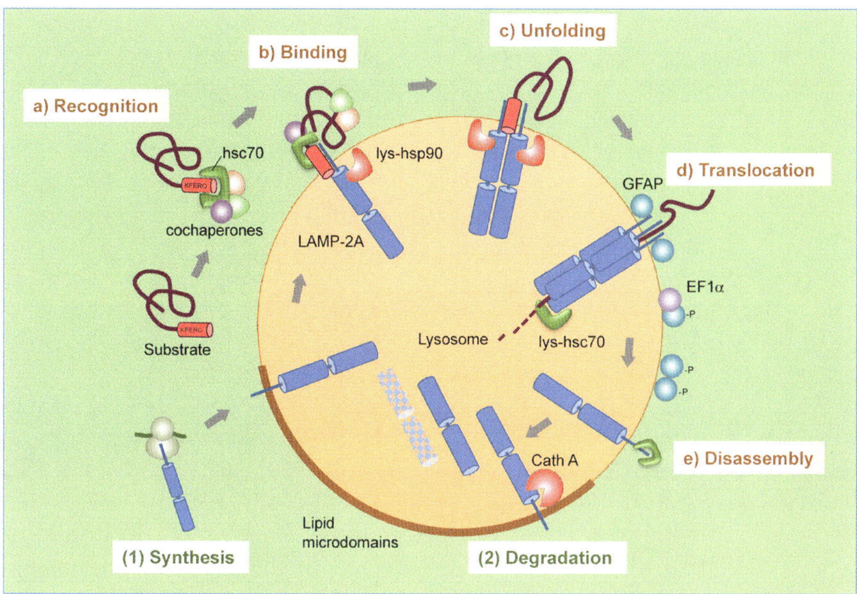

Fig. 9.5 Steps in chaperone-mediated autophagy: (**a**) Recognition of substrate proteins by hsc70/cochaperones; (**b**) binding of substrate-chaperone complex to LAMP-2A; (**c**) unfolding of the substrate; (**d**) LAMP-2A multimerization and substrate translocation followed by lysosomal degradation; (**e**) disassembly of LAMP-2A complex. Regulation of LAMP-2A at the lysosomal membrane occurs through (*1*) de novo synthesis and (*2*) sequestration and degradation in lipid rafts (from Kaushik *Trends Cell Biol.* 22: 407 © 2012 Reprinted by permission from Elsevier)

One of the ways that PD-associated α-synuclein and LRRK2 exerts their toxic effects is through their jamming of the LAMP-2A-apparatus. Not only do these blocking actions impair the removal of other damaged monomers and aggregates of the same species, and of each other, so they build up to toxic levels, but they also impair removal of other damaged and unwanted proteins. A particularly revealing example of the negative effects of blocking CMA by mutated α-synuclein or LRRK2 is provided by the blocking of neuronal survival factor MEF2D (myocyte enhancer factor 2D). This transcription factor is required for neuronal survival. It continuously shuttles to the cytoplasm where it binds to Hsc70 and is then transported to lysosomes for degradation. Under normal conditions CMA removes inactive MEF2D from the active pool and by that means contributes to survival. The blocking by α-synuclein/LRRK2 prevents this clearing operation.

9.8   The PINK1/Parkin Pathway Regulates Mitophagy

Badly damaged and unrepairable mitochondria undergo mitophagy, a specialized form of macroautophagy in which mitochondria are encapsulated in autophagosomes and shipped to lysosomes for degradation. That disposal route is part of a mitochondrial quality control system that includes mitochondrial chaperones that refold misfolded proteins, and proteases that degrade unrecoverable ones. It includes the machinery to ship outer membrane proteins to the ubiquitin-proteasome system for disposal, and the machinery to initiate mitophagy that removes the remainder.

In order for mitophagy to take place, a damaged mitochondrion must be isolated from the remaining healthy mitochondrial network. That step involves activating the mitochondrial fission machinery and inactivating the fusion apparatus. The protein Drp1 is a critical mediator of fission and has to be activated. It organizes a multiprotein complex that encircles the outer mitochondrial membrane and exerts mechanical scission forces. Three GTPases—OPA1 functioning at the inner membrane and Mfn1 and Mfn2 at the outer membrane—regulate fusion. These need to be inactivated and the inner membrane protease OMA1 activated.

Mutations in the PARK2 that encodes parkin and the PARK6 gene that encodes PINK1 head up the list of risk factors for developing autosomal recessive forms of Parkinson's disease (ARPD). These proteins are essential components of the mitochondrial quality control system. That parkin was a cause of ARPD was discovered by Kitano in 1998. The discovery that PINK1 was also a cause of ARPD followed a few years later by Valente in 2004. A landmark in understanding the causes of ARPD was the realization shortly thereafter that both parkin and PINK1 operate in a single pathway. In that pathway, PINK1 functions as a sensor of mitochondrial damage and parkin as its effector, facilitating removal of oxidized and damaged proteins, or alternatively, mediating removal of the damaged organelle.

This pathway operates in the following way. PINK1 is a 581-amino-acid-residue serine/threonine kinase. It possesses an N-terminal mitochondrial targeting sequence and a serine/threonine kinase domain. This protein, like other proteins bearing a mitochondrial targeting sequence, is transported into mitochondria by transporter of the outer membrane (TOM) and transporter of the inner membrane (TIM) complexes. Under normal conditions, PINK1 is acted upon by one or more mitochondrial proteases such as presenilin-associated rhomboid-like protease (PARL) or mitochondrial processing peptidase (MPP) and degraded. Under abnormal membrane potential ($\Delta\Psi_{m}$) conditions PINK1 is not degraded but instead accumulates and is stabilized on the outer mitochondrial membrane with its kinase domain facing the cytoplasm primed to recruit parkin. This situation is depicted in Fig. 9.6.

Parkin is a 465-amino-acid-residue E3 ubiquitin ligase. It contains an ubiquitin-like Ubl domain in its N-terminal, and a series of motifs in its C-terminal that provides an E3 ubiquitin ligase capability to the protein. A long-held mystery was how PINK1 recruits parkin to the mitochondrial outer membrane. That mystery is being solved. Among the recent findings is that stabilized and activated PINK1 phosphorylates mitofusin 2 (Mfn2). This OMM protein is subsequently ubiquitinated by parkin and these operations are consistent with Mfn2 participating in the recruitment and serving as a parkin receptor. That finding was followed in early 2014 by reports that ubiquitin is phosphorylated by PINK1. This action may well be a central one as it not only appears to facilitate the recruitment of parkin to the mitochondrial outer membrane but also activates the ubiquitin ligase. That finding followed

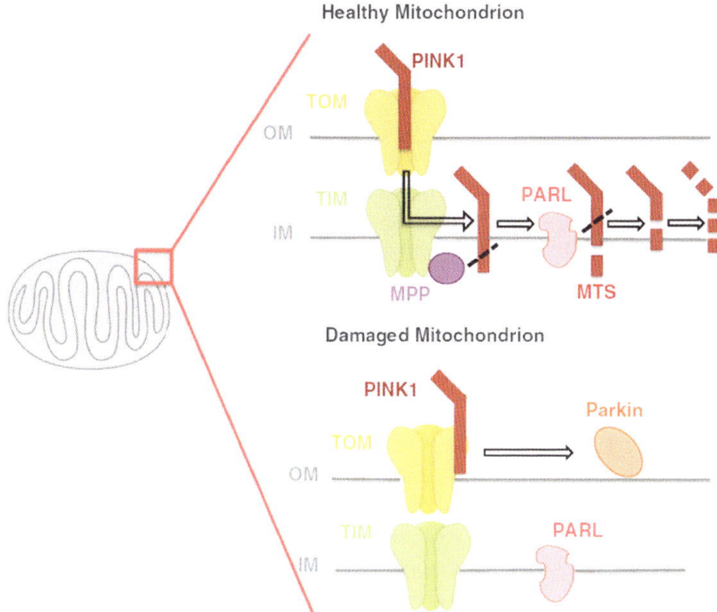

Fig. 9.6   PINK1 is imported into the inner mitochondrial membrane via the TOM/TIM complexes. The TIM complex-associated protease, mitochondrial MPP, cleaves the PINK1 mitochondrial targeting sequence (MTS). PINK1 is also chopped into pieces by the inner membrane presenilin-associated rhomboid-like (PARL) protease and proteolytically degraded. Loss of membrane potential prevents the import of PINK1 leading to the accumulation of unprocessed PINK1 on the outer membrane surface where it associates with the TOM complex and recruits cytosolic parkin to the damaged mitochondria (from Ashrafi *Cell Death Diff.* 20: 31 © 2013 Reprinted by permission from Macmillan Publishers Ltd)

earlier studies in which PINK1 was found to phosphorylate parkin at the corresponding site, Ser65, in its amino-terminal Ubl domain. That step primes the E3 ligase for subsequent activation by phospho-ubiquitin.

Once parkin relocates to the mitochondrial outer membrane from the cytosol and is activated it ubiquitinates its substrates. Progress has been made on identifying parkin's targets and understanding how their ubiquitination triggers mitophagy and other neuroprotective actions. Consistent with the needed fission/fusion steps, parkin ubiquitinates OPA1 and Drp1, and both PINK1 and parkin tag a protein called Miro for degradation. That protein tethers the OMM to kinesin motor proteins and its degradation arrests the movement of the mitochondrion. These are first steps in removing damaged mitochondria from the remaining healthy parts of the mitochondrial network, and begin the process of taking apart and removing the damaged mitochondria. The outer membrane proteins are then shipped to the 26S proteasome for degradation. That step is followed by mitophagy of the inner membrane and matrix proteins.

9.9 Neuroprotective Actions Take Place Independent of Mitophagy

PINK1 and especially parkin serve in a broad neuroprotective capacity. While many of their protective activities involve mitophagy, others occur independently of autophagy. These additional activities are launched in instances where they damage to the mitochondria is modest and not irrevocable. Some of these additional protective steps involve the regulation of gene expression.

One of the first parkin substrates to be identified was a protein named PARIS. That protein accumulates in the brains of PD patients suffering from parkin-inactivating mutations, and also in sporadic PD when parkin is inactivated through oxidative stresses (Fig. 9.7a). PARIS acts as a transcription repressor and its primary substrate is PGC-1α, a master regulator of mitochondrial function and ROS responses. Repression of PGC-1α by PARIS prevents these actions and dopaminergic neurons die off. Another of parkin's substrates is NEMO, a regulatory subunit of the NF-κB's upstream activating complex. NF-κB is a central stress-activated transcription factor. One of the proteins whose transcription is increased by parkin-mediated ubiquitination of NEMO is the mitochondrial fusion regulator OPA1. That action protects the mitochondria from apoptotic cell death as depicted in Fig. 9.7b.

A third neuroprotective parkin substrate is aminoacyl-tRNA synthase complex interacting multifunctional protein-2 (AIMP2, also known as p38/JTV-1). This protein is found in Lewy body inclusions of PD sufferers, and consistent with it being

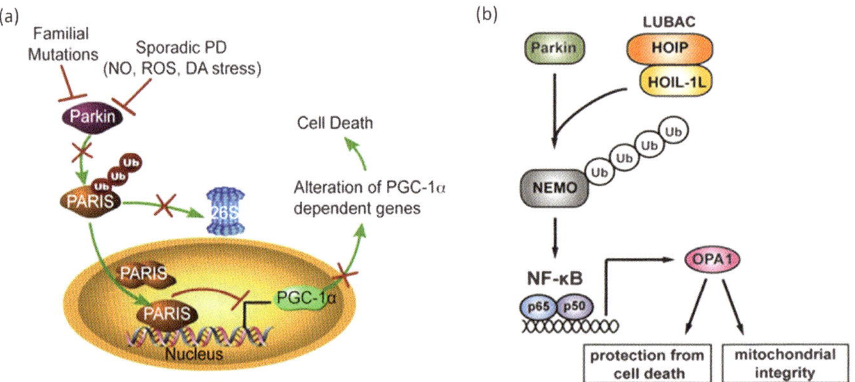

Fig. 9.7 (**a**) Parkin-PARIS-PGC-1α pathways in which PARIS maintains the balance of PGC-1a levels. In PD, parkin is inactivated by mutations, reactive oxygen species (ROS), nitrosative (NO) or dopaminergic (DA) stress. PARIS accumulates as a result and continuously inhibits PGC-1a transcription leading to reduction in PGC-1α-dependent gene expression (from Shin *Cell* 144: 689 © 2011 Reprinted by permission from Elsevier). (**b**) Parkin-NEMO/NF-κB-OPA1 pathway in which activation of NF-κB gene expression serves in a neuroprotective capacity by reducing the likelihood of apoptosis (from Müller-Rischart *Mol. Cell* 49: 908 © 2013 Reprinted by permission from Elsevier)

a parkin substrate its levels are elevated when parkin is inactivated either through mutations or MPTP treatment. One of AIMP2's interaction partners is PARP1 and its concentration builds up when AIMP2 is overexpressed. Too much PARP1 triggers a form of cell death known as *parthanatos*. This is a caspase-independent form of cell death distinct from apoptosis and necrosis. In this form of cell death, PARP1 overexpression results in a buildup of poly(ADP-ribose) (PAR), which translocates to mitochondria and facilitates the release of apoptosis-inducing factor (AIF). This factor then translocates to the nucleus where it triggers DNA fragmentation, nuclear condensation, and cell death.

There is also evidence for neuroprotective role of the PINK1-parkin pathway in mitochondrial quality control. This arises in part because signs of mitochondrial damage can be present prior to drastic loss in mitochondrial membrane potential. One of these is the nascent buildup of oxidized, misfolded and damaged proteins. Instead of overseeing the autophagic removal of these partially damaged mitochondria, PINK1 and parkin can direct the removal of the oxidized and misfolded proteins from the mitochondria, thereby promoting its recovery. These proteins are packaged in mitochondrial-derived vesicles (MDVs) and sent on to lysosomes for degradation.

9.10 DJ-1 Protects Against Oxidative Stress

DJ-1 is a 189-amino-acid-residue protein that folds into a compact 3D shape. In 2003, mutations in PARK7, the DJ-1 gene, were found to be associated with autosomal recessive early-onset Parkinson's disease. Like autosomal recessive genes PARK2 and PARK6, loss of DJ-1 results in parkinsonism. The most prominent of these mutations, L166P, abolishes the ability of the protein to form dimers, its physiologically active form. When mutated in this way, the protein loses stability and unfolds.

The protein is found in several locations in the cell—in the cytosol, in the nucleus, and in the mitochondria where it carries out its protective actions. There is good evidence that DJ-1 acts in a protective capacity against mitochondrial damage in a pathway parallel to that of PINK1 and parkin. Like PINK1 and parkin it helps regulate mitochondrial dynamics and autophagy. Deletion of this protein results in mito-chondrial fragmentation, loss of mitochondrial membrane potential, increased production of ROS and induction of autophagy markers. These defects can be rescued through PINK1/parkin overexpression. Although the exact mechanisms whereby DJ-1 exerts its protective actions have long been a mystery, progress has been made to uncover the possible mechanisms.

DJ-1 has a trio of exposed cysteine residues—Cys46, Cys53, and Cys106. Of the three Cys106 has attracted considerable attention because of its critical location near or within the protein's catalytic site. Recall that cysteine possesses a highly reaction thiol (SH) group on its side chain. Cys106 is especially susceptible to oxidation in which SOH, SO_2H, and SO_3H may be formed. Its localization to mitochondria plus its propensity to undergo these modifications enables DJ-1 to serve as a ROS sensor. There is also a reciprocal relationship between DJ-1 and PINK1/ parkin brought about by DJ-1's ability to restrict ROS production.

One focus of attention has been on calcium handling given the extreme sensitivity of SNc neurons to this aspect. Mitochondria buffer the increased calcium load present in SNc neurons. DJ-1 localizes to the interface between the ER and mitochondria where it regulates calcium trafficking and ER-mitochondrial tethering. Interestingly, there is evidence for localization of an α-synuclein subpopulation at these sites, formally referred to as mitochondria-associated ER membranes (MAMs). An emerging theme in these investigations is the onset of calcium mishandling brought on by oxidative stresses and vice versa, thereby creating a yet another vicious cycle.

More generally, DJ-1 is believed to be pleiotropic in its actions. There is support for the idea that when it localizes to the nucleus it regulates transcription of genes that confer protection against oxidative stresses. In particular, it can mediates the induction of Nrf2 (nuclear factor erythroid 2-related factor 2), a master regulator of electrophilic and oxidative stresses in the cell. There is also evidence that when outside the nucleus it can bind RNA in an oxidative stress-dependent manner and modulate translational activities. Finally, there is evidence for the participation of DJ-1 in several signaling pathways. Prominent among these additional protective roles are (1) its shifting of the signaling balance away from apoptosis through its inhibition of ASK1 (apoptosis signal-regulating kinase 1) signaling and (2) its activation of neuroprotective Akt signaling.

9.11 Other Risk Factors for Developing Parkinsonism Have Been Identified

The risk factors listed in Table 9.1 are not the only ones found so far for developing PD. Several others have been identified as having possible causal links to PD and other forms of parkinsonism. The most prominent of these emerging risk factors are listed in Table 9.3. These risk factors fit in with the pathways and themes

Table 9.3 Additional risk factors for developing PD and other forms of parkinsonism

Locus	Protein	Description
PARK5—4p14	UCH-L1	Deubiquitinating enzyme, AD
PARK9—1p36	ATP13A2	Neuronal P-type ATPase; AR, AT
PARK13—2p13	Omi/HtrA2	Mitochondrial protease, AD
PARK14—22q13.1	PLA2G6	Phospholipase A$_2$; AR, AT
PARK15—22q12-13	Fbxo7	E3 ubiquitin ligase; AR, AT
------------1q21	Glucocerebrosidase	Lysosomal enzyme, AR

AD: autosomal dominant, *AR:* autosomal recessive. *AT:* Atypical (non-PD) forms of parkinsonism, which includes progressive supranuclear palsy (PSP), corticobasal degeneration (CBD), multiple system atrophy (MSA), and dementia with Lewy bodies (DLB)

encompassed by the others and broaden their sweep. Most notably they highlight the growing importance of lysosomal quality control in the etiology of Parkinson's disease.

9.11.1 Lysosomal Storage Disorders

More than 50 different kinds of lysosomal storage disorders have been identified to date. The first of these Tay–Sachs disease was discovered in 1881 and Gaucher disease one year later in 1882, while Niemann–Pick disease emerged in 1914. Most of these diseases arise from mutations in lysosomal transport proteins and to a lesser extent in enzymes needed to degrade their cellular substrates. Niemann–Pick disease types A and B arises from a deficiency in the enzyme sphingomyelinase; in Tay–Sachs disease hexosaminidase A malfunctions result in its inability to degrade GM2 gangliosides, and in Gaucher disease the enzyme glucocerebrosidase is defective because of mutations in GBA, the gene that encodes it.

Mutations in these lysosomal proteins result in toxic buildups of undegraded metabolites and other potentially harmful molecules. Lysosomal acidification is impaired as are lysosomal functions resulting in failed autophagy. These diseases have a number of features in common. For instance, enlargements in the liver and spleen are widely encountered. In the neural forms of diseases as in Gaucher disease, substrates such as cathepsin D (an aspartic protease) are released into the cytosol, Ca^{2+} homeostasis is disrupted, and that molecule is released, as well. And because lysosomes are the downstream endpoint of the autophagic removal pathways, their impairment allows toxic oligomers of α-synuclein and other misfolded proteins to accumulate. Worse, it is speculated that glucocerebrosidase and α-synuclein may interact in ways that further contribute to the developing ataxias and cognitive decline.

Autosomal recessive mutations in PARK9, the gene that encodes ATP13A2, a large 1180-amino-acid-residue P-type ATPase, are associated with parkinsonism. These are ion and lipid cation pumps that catalyze the decomposition of ATP into ADP and use energy derived from ATP to carry out their cation pumping activities. Mutations in this protein were found in 2006 to give rise to Kufor-Rakeb syndrome, a rare form of juvenile, early-onset parkinsonism. Highly expressed in the substantia nigra *pars compacta* and in pyramidal neurons throughout the cerebral cortex, this cation transporter localizes to the acidic membrane compartments such as lysosomes (pH 4.6), late endosomes (pH 5.5), and early endosomes (pH 6.2). To date the best candidate cation substrates are manganese and zinc. There is recent evidence that impaired sequestration of Zn^{2+} in acidic vesicles leads to lysosomal deficiencies and to excessive cytosolic concentrations that can negatively impact mitochondria by stimulating ROS buildup.

9.11.2   Ubiquitin Carboxyl-Terminal Hydrolase L1 (UCH-L1)

Enzymes that perform deubiquitinating activities are important contributors to protein quality control. UCH-L1 is a 223 amino acid residue deubiquitinating enzyme. It is abundant, comprising more than 1 % of the total soluble protein in neurons, where it maintains a pool of available ubiquitin molecules. A mutation to UCH-L1, I93M, is associated with a rare familial form of PD. Both the mutated form and a carbonylated one (modified through the addition of a carbon monoxide group) become insoluble and exhibit inappropriate interactions with other proteins. Aberrant forms of UCH-L1 have been detected in Lewy bodies and co-localize with α-synuclein with whom it it may act as an ubiquitin ligase. There is evidence for interactions with LAMP-2A, Hsc70 and Hsp90, all prominently involved in chaperone-mediated autophagy.

9.11.3   Other Members of the PINK1/Parkin Pathway

HtrA2/Omi is a mitochondrial protease that removes misfolded proteins by chopping them up into small pieces. It possesses an N-terminal mitochondrial targeting sequence, a serine protease domain, and a C-terminal PDZ protein–protein interaction domain that recognizes exposed hydrophobic patches. This protein is phosphorylated by mitochondrial PINK1, but apparently functions in a pathway branch independent of parkin. This second branch would preserve the health of the mitochondrion by degrading misfolded and damaged proteins, whereas the other, parkin branch, would remove damaged mitochondria through mitophagy.

Fbxo7 is another component of the PINK1/parkin pathway. F-box proteins (FBPs) such as Fbxo7 are substrate-recruiting subunits of SCF-type multi-subunit E3 ubiquitin ligase complexes. However, the FBPs are atypical proteins and perform other tasks besides their E3 ligase recruitment ones. Fbxo7 functions as part of the PINK1/parkin pathway that initiates mitophagy.

9.11.4   Maintenance of Lipid Homeostasis

Phospholipase A2 group VI (PLA2G6) encodes a protein critical for the maintenance of lipid metabolism and cellular membrane homeostasis. It does so by regulating the levels of phosphatidylcholine present in the membranes. Its failures to do so because of mutations lead to the brain iron and axonal spheroid pathologies seen in some cases of NBIA and in a second disorder called infantile neuroaxonal dystrophy (INAD).

9.12   Nitrosative stress Can Harm SNc Neurons and Contribute to PD

It is surprising that a small highly diffusible molecule such as nitric oxide (NO) can function in a signaling capacity, but it does. Two sets of discoveries established that nitric oxide is indeed a legitimate signaling molecule. The first of these was the discovery that nitric oxide generated by endothelial cells relaxes vascular smooth muscle cells. That discovery took place in several phases in the 1980s. Robert Furchgott made the discovery that some sort of signal was being sent from vascular endothelial cells to smooth muscle cells instructing them to relax. That somewhat mysterious signal was named endothelium-derived relaxing factor, or EDRF. A few years earlier, Ferid Murad had found that nitric oxide can signal smooth muscle cells to relax, and finally Louis Ignarro and Salvador Moncada reported that the signaling molecule found by Furchgott was, in fact, nitric oxide. The second key discovery, made in 1988 by John Garthwaite, was that the same molecule released from vascular endothelial cells was produced by cerebellar neurons when their NMDA receptors were stimulated by glutamate.

In recent years, S-nitrosylation has emerged as a major route of NO bioactivity. In S-nitrosylation, an NO group is covalently appended to a thiol (sulfhydryl, SH) group on a functionally critical cysteine side chain. This posttranslational modification alters the functional properties of the recipient, either positively or negatively. A prime example of how NO may act through S-nitrosylation in a neuroprotective capacity is its effect on NMDA receptors. Nitric oxide synthase enzymes are tethered in close proximity to NMDA receptors. These enzymes are sensitive to intracellular calcium levels; the calcium-binding protein calmodulin (CaM) and Ca^{2+} are among the cofactors needed for its activation. Entry of Ca^{2+} through glutamate-bound NMDA receptors activates nNOS which releases NO leading to S-nitrosylation of the regulatory subunits of the receptor. This alteration forces the closing of the ion channel, thereby preventing the potential excitotoxic entry of Ca^{2+} into the neuron (Fig. 9.8).

While low levels of nitric oxide serve in a signaling capacity, excessive production of NO can be harmful to the neurons. Two notable examples of excessive S-nitrosylation are its effects on protein disulfide isomerase (PDI) and parkin. Both actions are believed to contribute to Parkinson's disease and other neurodegenerative disorders. Protein disulfide isomerase is an endoplasmic reticulum enzyme that catalyzes the formation and breakage of disulfide bonds between cysteine residues, and by that means chaperones the recovery of misfolded proteins. The ability of PDI to carry out these tasks is impaired when critical cysteines are S-nitrosylated. Turning now to parkin, the effects of mutations on parkin's E3 ligase activity have been difficult to assess with results depending in large part on the particular mutation. Parkin contains an unusually large number of cysteine residues, 35 to be exact. S-nitrosylation of exposed cysteines has been found to reduce the protein's solubility and increase its tendency to aggregate. A buildup of S-nitrosylated parkin aggregates has been detected in cases of sporadic PD.

There is another way that excessive nitric oxide levels can be dangerous. While neither superoxide nor NO is particularly harmful by itself, the two may combine to

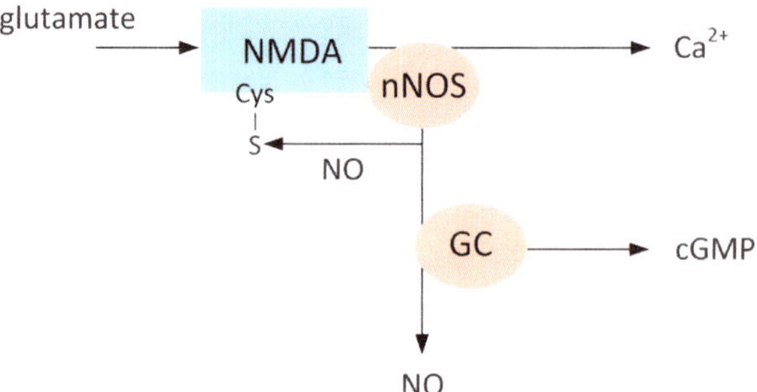

Fig. 9.8 Entry of Ca²⁺ through dual voltage-gated and glutamate-activated NMDA receptors leads to generation of nitric oxide and cyclic GMP second messengers. In this signaling process, guanylyl cyclase (GC) functions as an NO receptor. This is a haem-containing enzyme that catalyzes the conversion of guanosine triphosphate (GTP) into 3′5′-cyclic guanosine monophosphate (cGMP). cGMP regulates neurotransmission through its effects on cyclic nucleotide gated ion channels and cGMP-activated kinases. Nitric oxide also serves in a protective capacity by S-nitrosylating exposed thiols on the NMDA receptor's regulatory subunits

form the exceptionally dangerous free radical peroxynitrite (ONOO⁻). The reaction of the two to form peroxynitrite is highly favorable and will occur almost 100 % of the time without the assistance of an enzyme. Under normal cellular conditions, superoxide is rapidly removed by superoxide dismutase (SOD) and other scavenging enzymes, and NO diffuses out of the neurons and into red blood cells, where it is removed by reacting with oxyhemoglobin and forming nitrite.

Returning now to the discussion of dopamine and pacemaking SNc neurons presented in the beginning of the chapter, increased levels of dopamine are present in the cytosol of SNc neurons and exacerbate these effects. The increases in cytosolic dopamine are substantial, roughly by a factor or 2–3. Cytosolic dopamine promotes oxidative stress and the buildup of toxic metabolites. It is metabolized by monoamine oxidase (MOA) to produce hydrogen peroxide and 3,4-dihydroxyphenylacetaldehyde (DOPAL) and through auto-oxidation to generate reactive oxygen species and dopamine quinones. Both DOPAL and dopamine quinones are highly toxic. This situation is worsened by elevated entry of calcium into the cell through the L-type calcium channels coupled to poor calcium handling within the cell. Its presence places an increased burden on the mitochondria, and interferes with the signaling networks dependent upon Ca²⁺ second messengering.

9.13 A Chronic Inflammatory Milieu Contributes to Parkinson's Disease

Inflammation in the form of reactive microglia contributes to Parkinson's disease. The role of microglia in Parkinson's disease was posited by Patrick and Edith McGeer in 1988 when they observed the presence of activated microglia in the brains of PD patients at autopsy. Dopaminergic neurons are particularly vulnerable to chronic inflammation due to their limited antioxidant capabilities and the presence of large numbers of microglia in the substantia nigra. The contribution of reactive microglia to the disease can be observed not only at autopsy of PD patients, but also in experimental mouse models of PD induced by the application of lipopolysaccharide (LPS), an endotoxin derived from gram-negative bacteria, and pesticides such as paraquat and rotenone. At low concentrations, the pesticides act through microglia to damage neurons, while LPS is a potent activator of microglia.

Once activated the microglia together with dead and dying neurons create a vicious circle that exacerbates the disease. This comes about in the following way. Dying neurons secrete a variety of molecules into the extracellular spaces. Some of these act as neurotoxins and activate resident microglia. One of these is the dark pigment neuromelanin (NM). Composed of melanin, peptides, and lipids, it forms insoluble clumps, thereby endowing it with a long resident time in the extracellular spaces. Another factor released into the substantial nigra by the neurons is aggregated α-synuclein. The participation of α-synuclein is noteworthy as it was for some time regarded as residing exclusively within neurons. That picture was overturned with the discovery that α-synuclein is secreted into the extracellular spaces where it is taken up not only by microglia but also by astrocytes. One activated, the microglia and astrocytes release a host of chemical agents. The resulting chronic inflammatory conditions do not resolve, and further damage the dopaminergic neurons, thereby closing the vicious circle.

One of the agents released by the reactive microglia is nitric oxide. There are three nitric oxide synthases. Neuronal NOS (nNOS) and endothelial NOS (eNOS) are constitutive while the third, inducible NOS (iNOS) is upregulated in response to inflammatory stimuli such as lipopolysaccharides and tumor necrosis factor α (TNFα). Once activated iNOS enables microglia to produce large quantities of nitric oxide. When combined with superoxide this leads to the production of peroxynitrite which is highly effective in killing invading bacteria and tumor cells. When released into the substantia nigra under chronic inflammatory conditions, nitric oxide interacts with α-synuclein within the neurons leading to stabilization of a more β-sheet-rich conformation and an increased propensity to aggregate. It also damages mitochondria and sends neurons further along an apoptotic pathway.

9.14   Cell to Cell Spread by α-Synuclein

Cells communicate with cells near and far in several ways. First discovered in 1983, exosomes are export vesicles laden with proteins, lipids, mRNAs and microRNAs. They facilitate the transfer of genetic information and mediate cell-to-cell communication. These vesicles are generated in the endocytic pathway. Intraluminal vesicles (ILVs), 40–100 nm in size, are first formed through budding of endosomal membranes. These are contained within larger multivesicular bodies (MVBs). Rather than being sent to lysosomes or undergoing recycling back to the Golgi and ER, the MVBs fuse with the plasma membrane and expel their exosomes into the extracellular spaces (Fig. 9.9). They are then taken up by recipient cells in a number of different ways ranging from receptor-mediated uptake to fusion with the plasma membrane.

This communication mechanism is exploited by a variety of cell types for good or ill. For example, exosomes facilitate the exchange of genetic information, and assist in lymphocyte activation and development of immunological tolerance. On the other side of the ledger, exosomes are also exploited by tumor cells and by viruses. But, returning to the positive side, exosomes offer a promising way of carrying out gene therapy and vaccine development. For example, in a recent study, short interfering RNAs were delivered to the brain via exosomes and used to knock-down BACE1 a key therapeutic target in Alzheimer's disease treatments.

If the autophagic-lysosmal pathways are all blocked due to failures in, for example, acidification or membrane transport then cells may get rid of unwanted, misfolded and possibly toxic protein oligomers using exosomal export. If these "Trojan horses" are taken up by other cells then the misfolded oligomers might induce otherwise normal counterparts in the recipient cells to misfold and aggregate. As noted briefly in earlier chapters, there is increasing evidence that prions and β-amyloid oligomers can spread from one region of the brain to another by this means. There is even stronger evidence for PD-associated α-synuclein spread in this manner. Support for this mechanism of prion-like spread was provided by researchers working independently in several laboratories beginning in 2009 and continuing to the present.

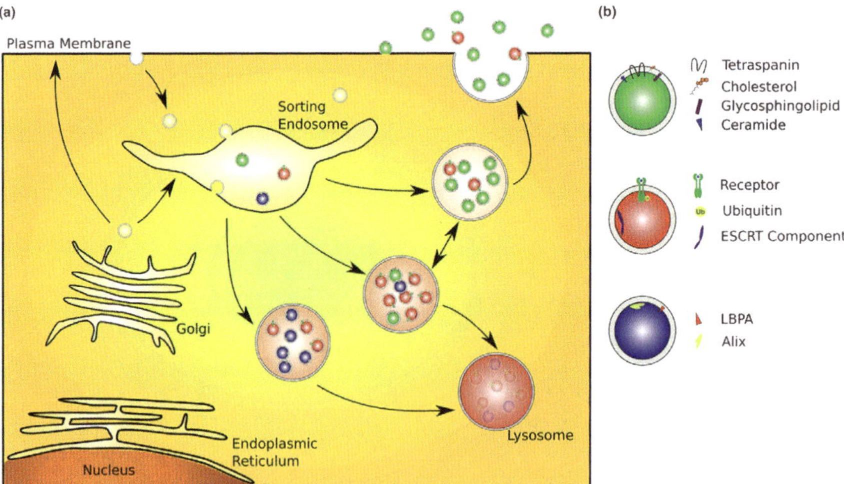

Fig. 9.9 Sorting of cargo into different ILVs (*red*, *green* and *blue*) based on their possible contents shown to the right of the ILV subtypes. *LBPA* lyso-bisphosphatidic acid, a phospholipid, *Alix* an adapter/scaffold protein; Tetraspanin a membrane protein (from Simons *Curr. Opin. Cell Biol.* 21: 575 © 2009 Reprinted by permission from Elsevier)

9.15 Summary

Neurons in the striatum coordinate body movement. They receive input from the cortex and thalamus and send output to regions in the central nervous system that control movement. Dopamine-secreting neurons in the substantia nigra *pars compacta* modulate and regulate these signals. In the absence of dopaminergic input to both presynaptic and postsynaptic corticostriatal terminals, the striatal output signals go awry resulting in movement disorders clinically characterized as parkinsonism, chief among which is Parkinson's disease. People with movement disorders suffer from

- Ataxias: a broad term that encompasses clumsiness, instability and imbalance, lack of coordination, disjointed movements and falling, and tremors;
- Tremors: refers to shaking or trembling of the hands or of arms, head, etc., and
- Dystonias: meaning involuntary muscle contractions/spasms.

In more detail, there are six specific signs of parkinsonism. These are tremor at rest, rigidity, slowness of movement, a bent aspect, postural instabilities, and freezing of gait/falling. Parkinson's disease, the most commonly encountered movement disorder, is characterized by many of these symptoms along with other signs. Atypical, or non-PD, forms of parkinsonism are: progressive supranuclear palsy (PSP), corticobasal degeneration (CBD), multiple system atrophy (MSA), and dementia with Lewy bodies (DLB). These differ from classical PD in their symptoms and responses to medications.

1. Neurons in the SNc are pacemakers, continually spiking and supplying dopamine to the striatum. This functionality places a special metabolic burden on the neurons forcing them to rely heavily on oxidative phosphorylation for ATP generation. The neurons also possess highly ramified, poorly myelinated axons, and utilize L-type Ca^{2+} channels for pacemaking. The L-type Ca^{2+} channels allow large amounts of calcium to enter the neurons, yet the cells have weak calcium handling capabilities. For all of these reasons, SNc neurons especially sensitive to age-related decline in protein quality control.

   Calcium is a second messenger and to enable this function intracellular calcium levels are maintained at a low level so that all increases in intracellular calcium levels are transient and localized in space. Once it enters the cell, calcium is rapidly pumped back out, and sequestered and buffered internally. The main loci of storage within the cell are (1) intracellular calcium stores situated in the endoplasmic reticulum, (2) mitochondria, and (3) acidic vesicles and organelles including lysosomes.

2. A number of environmental toxins especially agents used in agriculture to control pests, and heavy metals, are thought to give rise to parkinsonism. This association is supported by a growing number of epidemiological studies, but a direct causal chain connecting these agents to the disease is lacking. In more recent years, genetic mutations have risen to primacy in the study of the disease. That shift began with the discovery of α-synuclein genetic mutations and duplications

Table 9.4 Risk factors for developing Parkinson's disease

Locus	Protein	Locus	Protein
PARK1/4	α-Synuclein	PARK9	ATP13A2
PARK2	Parkin	PARK13	HtrA2/Omi
PARK5	UCH-L1	PARK14	PLA2G6
PARK6	PINK1	PARK15	Fbxo7
PARK7	DJ-1	–	Glucocerebrosidase
PARK8	LRRK2		

in familial forms of PD. That link was strengthened with the accompanying discovery that α-synuclein was the main component of Lewy bodies, the postmortem defining feature of PD.

3. α-Synuclein is a small, highly flexible protein. It associates preferentially with lipid rafts especially those present on curved surfaces such as those encapsulating synaptic vesicles. It localizes to presynaptic terminals where it appears to regulate dopamine-laden vesicle trafficking and dynamics. Mutated, overexpressed, and dopamine-modified forms of α-synuclein misfold and aggregate into a number of structures that vary in size and morphology. Because of its extreme flexibility and sensitivity to its local environment, identification of the main toxic species has been challenging endeavor and a consensus has not been achieved. Parkinson's disease is not the only disorder associated with α-synuclein misfolding and aggregation. Rather, there is a spectrum of disorders collectively referred to as α-synucleinopathies. These are:

- Parkinson's disease
- Dementia with Lewy bodies
- Multiple system atrophy
- Neurodegeneration with brain iron accumulation, type I

4. A growing number of mutated proteins have been found to either directly produce or confer risk to autosomal dominant and autosomal recessive forms of Parkinson's disease. Today, the list of risk factors (Table 9.4) includes:

Rather than defining a single and simple pathway to Parkinson's disease the causal and risk factors listed above paint a far broader (and more complex) picture. In this expanded worldview, protein quality control includes not only direct prevention of protein misfolding and aggregation but also mitochondrial and lysosomal quality control. Maintaining mitochondrial and lysosomal quality control is essential for maintaining proteostasis. Dysfunctional mitochondria fail to generate ATP; if they are not removed by mitophagy they degrade the healthy part of the mitochondrial network, generate and release dangerous quantities of ROS and apoptotic factors, and fail to maintain calcium homeostasis. Similarly, if dysfunctional lysosomes are present, degradation of misfolded and aggregated proteins does not occur in a timely manner. Instead, potentially toxic entities collect while dysfunctional organelles release acidic, disruptive, and other types of toxic components into the cytosol. These situations generate toxic intracellular environments that over time result in death of the neurons. With PD the sweep encompassed by the term protein misfolding disorder is broadened to include mitochondrial and lysosomal quality control.

5. LRRK2 and alpha-synuclein both generate autosomal dominant forms of Parkinson's disease. One of the ways that α-synuclein and LRRK2 contribute to PD is through their jamming of the LAMP-2A apparatus that transports selected CMA substrates into the lysosome. These blocking actions prevent the removal of other monomers and aggregates of the same species, of each other's, and of other damaged and unwanted proteins. LRRK2 is a complex protein. It possesses two distinct enzymatic activities and several protein–protein interaction domains. Like α-synuclein it associates with lipid rafts, and can be found in multiple cellular raft locations where it has a role in the behavior of intracellular membranous structures. It regulates autophagic vacuole activity, clearance of trans-Golgi-derived vesicles through the autophagic-lysosomal pathway, and presynaptic vesicle storage and mobilization.

6. Unlike α-synuclein and LRRK2, PINK1 and parkin belong to a well-defined pathway, one that maintains mitochondrial quality control. In this pathway, PINK1, a mitochondrial kinase, serves as a damage sensor and parkin, an E3 ubiquitin ligase, acts as a downstream effector. These proteins maintain mitochondrial health in two distinct ways: (1) they regulate the autophagic removal and disposal of badly damaged mitochondria, and (2) they mediate the clearance of potentially harmful oxidized and misfolded mitochondrial proteins before they irrevocably damage the mitochondria. This pathway appears to be quite distinct from the one loosely defined by α-synuclein and LRRK2; for instance, Lewy bodies are largely absent. Another PD-associated protein, DJ-1, helps maintain mitochondrial quality control, as well. It detects the presence of ROS and seems to operate in a parallel mitochondrial-protective pathway to that of PINK1/parkin.

7. The proper maintenance of lysosomal homeostasis is important. That aspect is highlighted by ATP13A2 and glucocerebrosidase, two recently identified PD risk factors that negatively impact lysosomal quality control. All three of the autophagic pathways converge on the lysosome, and if the lysosomal endpoint fails for any reason so does the disposal of misfolded proteins, dangerous oligomers, large aggregates, and dysfunctional organelles. Furthermore, like the mitochondria, these organelles harbor a variety of harmful chemicals that if released into the cytosol can create a toxic environment and initiate cell death.

8. Lastly, evidence has been acquired showing that α-synuclein is released into the extracellular spaces along with neuromelanin by damaged SNc neurons. These agents stimulate responses by resident microglia and astrocytes, creating chronic inflammatory conditions that further damage the stressed neurons. There is also accumulating evidence for a prion-like spread of α-synuclein (and others) to other neurons. This possibility has generated considerable excitement in the field. The prevailing belief had been that outbreaks of neurodegenerative illnesses such as PD happened more or less independently in different brain regions. That viewpoint has undergone a radical change with the discovery of the

direct spread from one region to another by misfolded proteins. The misfolded proteins subsequently trigger the misfolding and aggregation of resident wild-type counterparts in the recipient regions, thereby propagating the disease in a prion-like manner. The discovery of these routes of spread opens up new avenues for therapeutic intervention in the fatal disorders.

Bibliography

Environmental Toxins

Betarbet, R., Canet-Aviles, R. M., Sherer, T. B., Mastroberardino, P. G., McLendon, C., Kim, J. H., et al. (2006). Intersecting pathways to neurodegeneration in Parkinson's disease: Effects of the pesticide rotenone on DJ-1, α-synuclein, and the ubiquitin-proteasome system. *Neurobiology of Disease, 22*, 404–420.

Costello, S., Cockburn, M., Bronstein, J., Zhang, X., & Ritz, B. (2009). Parkinson's disease and residential exposure to maneb and paraquat from agricultural applications in the Central Valley of California. *American Journal of Epidemiology, 169*, 919–926.

McCormack, A. L., Thiruchelvam, M., Manning-Bog, A. B., Thiffault, C., Langston, J. W., Cory-Slechta, D. A., et al. (2002). Environmental risk factors and Parkinson's disease: Selective degeneration of nigral dopaminergic neurons caused by the herbicide paraquat. *Neurobiology of Disease, 10*, 119–127.

Tanner, C. M., Kamel, F., Ross, G. W., Hoppin, J. A., Goldman, S. M., Korell, M., et al. (2011). Rotenone, paraquat, and Parkinson's disease. *Environmental Health Perspectives, 119*, 866–872.

Discovery of α-Synuclein and Its Connection to PD

Chartier-Harlin, M. C., Kachergus, J., Roumier, C., Mouroux, V., Douay, X., Lincoln, S., et al. (2004). α-Synuclein locus duplication as a cause of familial Parkinson's disease. *Lancet, 364*, 1167–1169.

Maraganore, D. M., de Andrade, M., Elbaz, A., Farrer, M. J., Ionnidis, J. P., Krüger, R., et al. (2006). Collaborative analysis of α-synuclein gene promoter variability and Parkinson disease. *JAMA, 296*, 661–670.

Polymeropoulos, M. H., Lavedan, C., Leroy, E., Ide, S. E., Dehejia, A., Dutra, A., et al. (1997). Mutation in the α-synuclein gene identified in families with Parkinson's disease. *Science, 276*, 2045–2047.

Singleton, A. B., Farrer, M., Johnson, J., Singleton, A., Hague, S., Kachergus, J., et al. (2003). α-Synuclein triplication causes Parkinson's disease. *Science, 302*, 841.

Spillantini, M. G., Crowther, R. A., Jakes, R., Hasegawa, M., & Goedert, M. (1998). α-Synuclein in filamentous inclusions of Lewy bodies from Parkinson's disease and dementia with Lewy bodies. *Proceedings of the National Academy of Sciences of the United States of America, 95*, 6469–6473.

Spillantini, M. G., Schmidt, M. L., Lee, V. M. Y., Trojanowski, J. Q., Jakes, R., & Goedert, M. (1997). α-Synuclein in Lewy bodies. *Nature, 388*, 839–840.

Physiology of SNc Neurons

Matsuda, W., Furuta, T., Nakamura, K. C., Hioki, H., Fujiyama, F., Arai, R., et al. (2009). Single nigrostriatal dopaminergic neurons form widely spread and highly dense axonal arborizations in the neostriatum. *The Journal of Neuroscience, 29*, 444–453.

Puopolo, M., Ravoila, E., & Bean, B. P. (2007). Roles of subthreshold calcium current and sodium current in spontaneous firing of mouse midbrain dopamine neurons. *The Journal of Neuroscience, 27*, 645–656.

Surmeier, D. J., Guzman, J. N., & Sánchez-Padilla, J. (2010). Calcium, cellular aging, and selective neuronal vulnerability in Parkinson's disease. *Cell Calcium, 47*, 175–182.

Dopaminergic Modulation of Synaptic Transmission

Kreitzer, A. C., & Malenka, R. C. (2008). Striatal plasticity and basal ganglia circuit function. *Neuron, 60*, 543–554.

Shen, W., Flajolet, M., Greengard, P., & Surmeier, D. J. (2008). Dichotomous dopaminergic control of striatal synaptic plasticity. *Science, 321*, 848–851.

Tritsch, N. X., & Sabatini, B. L. (2012). Dopamergic modulation of synaptic transmission in cortex and striatum. *Neuron, 76*, 33–50.

Biophysical Properties of α-Synuclein

Davidson, W. S., Jonas, A., Clayton, D. F., & George, J. M. (1998). Stabilization of α-synuclein secondary structure upon binding to synthetic membranes. *The Journal of Biological Chemistry, 273*, 9443–9449.

Eliezer, D., Kutluay, E., Bussell, R., Jr., & Browne, G. (2001). Conformational properties of α-synuclein in its free and lipid-associated states. *Journal of Molecular Biology, 307*, 1061–1073.

Ferreon, A. C. M., Gambin, Y., Lemke, E. A., & Deniz, A. A. (2009). Interplay of α-synuclein binding and conformational switching probed by single-molecule fluorescence. *Proceedings of the National Academy of Sciences of the United States of America, 106*, 5645–5650.

Fortin, D. L., Troyer, M. D., Nakamura, K., Kubo, S. I., Anthony, M. D., & Edwards, R. H. (2004). Lipid rafts mediate the synaptic localization of α-synuclein. *The Journal of Neuroscience, 24*, 6715–6723.

Jo, E., McLaurin, J. A., Yip, C. M., St. George-Hyslop, P., & Fraser, P. E. (2000). α-Synuclein membrane interactions and lipid specificity. *The Journal of Biological Chemistry, 275*, 34328–34334.

Middleton, E. R., & Rhoades, E. (2010). Effect of curvature and composition on α-synuclein binding to lipid vesicles. *Biophysical Journal, 99*, 2279–2288.

Solanki, A., Neupane, K., & Woodside, M. T. (2014). Single-molecule force spectroscopy of rapidly fluctuating, marginally stable structures in the intrinsically disordered protein α-synuclein. *Physical Review Letters, 112*, 158103.

Westphal, V., Rizzoli, S. O., Lauterbach, M. A., Kamin, D., Jahn, R., & Hell, S. W. (2008). Video-rate far-field optical nanoscopy dissects synaptic vesicle movement. *Science, 320*, 246–249.

Presynaptic Functions of α-Synuclein

Burré, J., Sharma, M., Tsetsenis, T., Buchman, V., Etherton, M. R., & Südhof, T. C. (2010). α-Synuclein promotes SNARE complex assembly in vivo and in vitro. *Science, 329,* 1663–1667.

Chandra, S., Gallardo, G., Fernánsez-Chacón, R., Schlüter, O. M., & Südhof, T. C. (2005). α-Synuclein cooperates with CSPα in preventing neurodegeneration. *Cell, 123,* 383–396.

Cooper, A. A., Gitler, A. D., Cashikar, A., Haynes, C. M., Hill, K. J., Bhullar, B., et al. (2006). α-Synuclein blocks ER-Golgi traffic and Rab1 rescues neuron loss in Parkinson's models. *Science, 313,* 324–328.

Gitler, A. D., Bevis, B. J., Shorter, J., Strathearn, K. E., Hamamichi, S., Su, L. J., et al. (2008). The Parkinson's disease protein α-synuclein disrupts cellular Rab homeostasis. *Proceedings of the National Academy of Sciences of the United States of America, 105,* 145–150.

Nemani, V. M., Lu, W., Berge, V., Nakamura, K., Onoa, B., Lee, M. K., et al. (2010). Increased expression of α-synuclein reduces neurotransmitter release by inhibiting synaptic vesicle reclustering after endocytosis. *Neuron, 65,* 66–79.

Toxic Oligomers

Cremades, N., Cohen, S. I. A., Deas, E., Abramov, A. Y., Chen, A. Y., Orte, A., et al. (2012). Direct observation of the interconversion of normal and toxic forms of α-synuclein. *Cell, 149,* 1048–1059.

Winner, B., Jappelli, R., Maji, S. K., Desplats, P. A., Boyer, L., Aigner, S., et al. (2011). In vivo demonstration that α-synuclein oligomers are toxic. *Proceedings of the National Academy of Sciences of the United States of America, 108,* 4194–4199.

LRRK2 Structure, Function and Localization

Alegre-Abarrategui, J., Christian, H., Lufino, M. M. P., Mutihac, R., Venda, L. L., Ansorge, O., et al. (2009). LRRK2 regulates autophagic activity and localizes to specific membrane microdomains in a novel human genomic reporter cellular model. *Human Molecular Genetics, 18,* 4022–4034.

Beilina, A., Rudenko, I. N., Kaganovich, A., Civiero, L., Chau, H., Kalia, S. K., et al. (2014). Unbiased screen for interactors of leucine-rich repeat kinase 2 supports a common pathway for sporadic and familial Parkinson's disease. *Proceedings of the National Academy of Sciences of the United States of America, 111,* 2626–2631.

Biskup, S., Moore, D. J., Celsi, F., Higashi, S., West, A. B., Andrabi, S. A., et al. (2006). Localization of LRRK2 to membranes and vesicular structures in mammalian brain. *Annals of Neurology, 60,* 557–569.

Mata, I. F., Wedemeyer, W. J., Farrer, M. J., Taylor, J. P., & Gallo, K. A. (2006). LRRK2 in Parkinson's disease: Protein domains and functional insights. *Trends in Neurosciences, 29,* 286–293.

Paisán-Ruiz, C., Jain, S., Evans, E. W., Gilks, W. P., Simón, J., van der Brug, M., et al. (2004). Cloning of the gene containing mutations that cause PARK8-linked Parkinson's disease. *Neuron, 44,* 595–600.

Piccoli, G., Condliffe, S. B., Bauer, M., Giesert, F., Boldt, K., De Astis, S., et al. (2011). LRRK2 controls synaptic vesicle storage and mobilization within the recycling pool. *The Journal of Neuroscience, 31,* 2225–2237.

Zinprich, A., Biskup, S., Leitner, P., Lichtner, P., Farrer, M., Lincoln, S., et al. (2004). Mutations in LRRK2 cause autosomal-dominant Parkinsonism with pleomorphic pathology. *Neuron, 44,* 601–607.

LRRK2 and Chaperone-Mediated Autophagy

Cuervo, A. M., & Dice, J. F. (1996). A receptor for the selective uptake and degradation of protein by lysosomes. *Science, 273*, 501–503.

Nichols, R. J., Dzamko, N., Morrice, N. A., Campbell, D. G., Deak, M., Ordureau, A., et al. (2010). 14-3-3 binding to LRRK2 is disrupted by multiple Parkinson's disease-associated mutations and regulates cytoplasmic localization. *The Biochemical Journal, 430*, 393–404.

Orenstein, S. J., Kuo, S. H., Tasset, I., Arias, E., Koga, H., Fernandez-Carasa, I., et al. (2013). Interplay of LRRK2 with chaperone-mediated autophagy. *Nature Neuroscience, 16*, 394–406.

Tong, Y., Yamaguchi, H., Giaime, E., Boyle, S., Kopan, R., Kelleher, R. J., III, et al. (2010). Loss of leucine-rich repeat kinase 2 causes impairment of protein degradation pathways, accumulation of α-synuclein, and apoptotic cell death in aged mice. *Proceedings of the National Academy of Sciences of the United States of America, 107*, 9879–9884.

Yang, Q., She, H., Gearing, M., Colla, E., Lee, M., Shacka, J. J., et al. (2009). Regulation of neuronal survival factor MEF2D by chaperonemediated autophagy. *Science, 323*, 124–127.

The PINK1-Parkin Pathway

Chen, Y., & Dorn, G. W., II. (2013). PINK1-phosphorylated mitofusin 2 is a parkin receptor for culling damaged mitochondria. *Science, 340*, 471–475.

Clark, I. E., Dodson, M. W., Jiang, C., Cao, J. H., Huh, J. R., Seol, J. H., et al. (2006). Drosophila pink1 is required for mitochondrial function and interacts genetically with parkin. *Nature, 441*, 1162–1166.

Greene, A. W., Grenier, K., Aguileta, M. A., Muise, S., Farazifard, R., Haque, M. E., et al. (2012). Mitochondrial processing peptidase regulates PINK1 processing, import and parkin recruitment. *EMBO Reports, 13*, 378–385.

Jin, S. M., Lazarou, M., Wang, C., Kane, L. A., Narendra, D. P., & Youle, R. J. (2010). Mitochondrial membrane potential regulates PINK1 import and proteolytic destabilization by PARL. *The Journal of Cell Biology, 191*, 933–942.

Kazlauskaite, A., Kondapalli, C., Gourlay, R., Campbell, D. G., Ritorto, M. S., Hoffman, K., et al. (2014). Parkin is activated by PINK1-dependent phosphorylation of ubiquitin at Ser. *The Biochemical Journal, 460*, 127–139.

Kitada, T., Asakawa, S., Hattori, N., Matsumine, H., Yamamura, Y., Minoshima, S., et al. (1998). Mutations in the parkin gene cause autosomal recessive juvenile parkinsonism. *Nature, 392*, 605–608.

Lazarou, M., Jin, S. M., Kane, L. A., & Youle, R. J. (2012). Role of PINK1 binding to the TOM complex and alternate intracellular membranes in recruitment and activation of the E3 ligase parkin. *Developmental Cell, 22*, 320–333.

Narendra, D., Tanaka, A., Suen, D. F., & Youle, R. J. (2008). Parkin is recruited selectively to impaired mitochondria and promotes their autophagy. *The Journal of Cell Biology, 183*, 795–803.

Park, J., Lee, S. B., Lee, S., Kim, Y., Song, S., Kim, S., et al. (2006). Mitochondrial dysfunction in Drosophila PINK1 mutants is complemented by parkin. *Nature, 441*, 1157–1161.

Poole, A. C., Thomas, R. E., Andrews, L. A., McBride, H. M., Whitworth, A. J., & Pallanck, L. J. (2008). The PINK1/parkin pathway regulates mitochondrial morphology. *Proceedings of the National Academy of Sciences of the United States of America, 105*, 1638–1643.

Valente, E. M., Abou-Sleiman, P. M., Caputo, V., Muqit, M. M., Harvey, K., Gispert, S., et al. (2004). Hereditary early-onset Parkinson's disease caused by mutations in PINK1. *Science, 304*, 1158–1160.

Vives-Bauza, C., Zhou, C., Huang, Y., Cui, M., de Vries, R. L. A., Kim, J., et al. (2010). PINK1-dependent recruitment of Parkin to mitochondria in mitophagy. *Proceedings of the National Academy of Sciences of the United States of America, 107*, 378–383.

Parkin's UPS/Mitophagic Actions

Chan, N. C., Salazar, A. M., Pham, A. H., Sweredoski, M. J., Kolawa, N. J., Graham, R. L. J., et al. (2011). Broad activation of the ubiquitin-proteasome system by Parkin is critical for mitophagy. *Human Molecular Genetics, 20*, 1726–1737.

Misko, A., Jiang, S., Wegorzewska, I., Milbrandt, J., & Baloh, R. H. (2010). Mitofusion 2 is necessary for transport of axonal mitochondria and interacts with the Miro/Milton complex. *The Journal of Neuroscience, 30*, 4232–4240.

Wang, X., Winter, D., Ashrafi, G., Schlehe, J., Wong, Y. L., Selkoe, D., et al. (2011). PINK1 and Parkin target Miro for phosphorylation and degradation to arrest mitochondrial motility. *Cell, 147*, 893–906.

Yoshii, S. R., Kishi, C., Ishihara, N., & Mizushima, N. (2011). Parkin mediates proteasome-dependent protein degradation and rupture of the outer mitochondrial membrane. *The Journal of Biological Chemistry, 286*, 19630–19640.

Ziviani, E., Tao, R. N., & Whitworth, A. J. (2010). *Drosophila* parkin requires PINK1 for mitochondrial translocation and ubiquitinates mitofusions. *Proceedings of the National Academy of Sciences of the United States of America, 107*, 5018–5023.

Neuroprotective Actions Independent of Mitophagy

Lee, Y., Karuppagounder, S. S., Shin, J. H., Lee, Y. I., Ko, H. S., Swing, D., et al. (2013). Parthanatos mediates AIMP2 activated age dependent dopaminergic neuronal loss. *Nature Neuroscience, 16*, 1392–1400.

McLelland, G. L., Soubannier, V., Chen, C. X., McBride, H. M., & Fon, E. A. (2014). Parkin and PINK1 function in a vesicular trafficking pathway regulating mitochondrial quality control. *The EMBO Journal, 33*, 282–295.

Müller-Rischart, A. K., Pilsl, A., Beaudette, P., Patra, M., Hadian, K., Funke, M., et al. (2013). The E3 ligase parkin maintains mitochondrial integrity by increasing linear ubiquitination of NEMO. *Molecular Cell, 49*, 908–921.

Shin, J. H., Ko, H. S., Kang, H., Lee, Y., Lee, Y. I., Pletinkova, O., et al. (2011). PARIS (ZNF746) repression of PGC-1α contributes to neurodegeneration in Parkinson's disease. *Cell, 144*, 689–702.

DJ-1 Mutations and Oxidation

Blackinton, J., Lakshminarasimhan, M., Thomas, K. J., Ahmad, R., Greggio, E., Raza, A. S., et al. (2009). Formation of a stabilized cysteine sulfinic acid is critical for the mitochondrial function of the parkinsonism protein DJ-1. *The Journal of Biological Chemistry, 284*, 6476–6485.

Bonifati, V., Rizzu, P., van Baren, M. J., Schaap, O., Breedveld, G. J., Krieger, E., et al. (2003). Mutations in the DJ-1 gene associated with autosomal recessive early-onset parkinsonism. *Science, 299*, 256–259.

Canet-Avilés, R. M., Wilson, M. A., Miller, D. W., Ahmad, R., McLendon, C., Bandyopadhyay, S., et al. (2004). The Parkinson's disease protein DJ-1 is neuroprotective due to cysteine-sulfinic acid-driven mitochondrial localization. *Proceedings of the National Academy of Sciences of the United States of America, 101*, 9103–9108.

Irrcher, I., Aleyasin, H., Seifert, E. L., Hewitt, S. J., Chhabra, S., Phillips, M., et al. (2010). Loss of the Parkinson's disease-linked gene DJ-1 perturbs mitochondrial dynamics. *Human Molecular Genetics, 19*, 3734–3746.

Krebiehl, G., Ruckerbauer, S., Burbulla, L. F., Kieper, N., Maurer, B., Waak, J., et al. (2010). Reduced basal autophagy and impaired mitochondrial dynamics due to loss of Parkinson's disease-associated protein DJ-1. *PLoS ONE, 5*, e9367.

Raturi, A., & Simmen, T. (2013). Where the endoplasmic reticulum and the mitochondrion tie the knot: The mitochondrial-associated membrane. *Biochimica et Biophysica Acta, 1833*, 213–224.

Thomas, K. J., McCoy, M. K., Blackinton, J., Beilina, A., van der Brug, M., Sandebring, A., et al. (2011). DJ-1 acts in parallel to the PINK1/parkin pathway to control mitochondrial function and autophagy. *Human Molecular Genetics, 20*, 40–50.

Emerging Genetic Risk Factors

Burchell, V. S., Nelson, D. E., Sanchez-Martinez, A., Delgado-Camprubi, M., Ivatt, R. M., Pogson, J. H., et al. (2013). The Parkinson's disease genes Fbxo7 and parkin interact to mediate mitophagy. *Nature Neuroscience, 16*, 1257–1265.

Cullen, V., Sardi, S. P., Ng, J., Xu, Y. H., Sun, Y., Tomlinson, J. J., et al. (2011). Acid β-glucosidase mutants linked to Gaucher disease, Parkinson's disease, and Lewy body dementia alter α-synuclein processing. *Annals of Neurology, 69*, 940–953.

Dehay, B., Ramirez, A., Martinez-Vicente, M., Perier, C., Canron, M. H., Doudnikoff, E., et al. (2012). Loss of P-type ATPase ATP13A2/PARK9 function induces general lysosomal deficiency and leads to Parkinson disease neurodegeneration. *Proceedings of the National Academy of Sciences of the United States of America, 109*, 9611–9616.

Gitler, A. D., Chesi, A., Geddie, M. L., Strathearn, K. E., Hamamichi, S., Hill, K. J., et al. (2009). α-Synuclein is part of a diverse and highly conserved interaction network that includes PARK9 and manganese toxicity. *Nature Genetics, 41*, 308–315.

Gregory, A., Westaway, S. K., Holm, I. E., Kotzbauer, P. T., Hogarth, P., Sonek, S., et al. (2008). Neurodegeneration associated with genetic defects in phospholipase A. *Neurology, 71*, 1402–1409.

Kabuta, T., Furuta, A., Aoki, S., Furuta, K., & Wada, K. (2008). Aberrant interaction between Parkinson disease-associated mutant UCH-L1 and the lysosomal receptor for chaperone-mediated autophagy. *The Journal of Biological Chemistry, 283*, 23731–23738.

Kong, S. M. Y., Chan, B. K. K., Park, J. S., Hill, K. J., Aitken, J. B., Cottle, L., et al. (2014). Parkinson's disease-linked human PARK9/ATP13A2 maintains zinc homeostasis and promotes α-synuclein externalization via exosomes. *Human Molecular Genetics, 23*, 2816–2833.

Mazzulli, J. R., Xu, Y. H., Sun, Y., Knight, A. L., Mclean, P. J., Caldwell, G. A., et al. (2011). Gaucher disease glucocerebrosidase and α-synuclein form a bidirectional pathogenic loop in synucleinopathies. *Cell, 146*, 37–52.

Morgan, N. V., Westaway, S. K., Morton, J. E. V., Gregory, A., Gissen, P., Sonek, S., et al. (2006). PLA2G6, encoding a phospholipase A, is mutated in neurodegenerative disorders with high brain iron. *Nature Genetics, 7*, 752–754.

Park, J. S., Koentjoro, B., Velvers, D., Mackay-Sim, A., & Sue, C. M. (2014). Parkinson's disease-associated human ATP13A2 (PARK9) deficiency causes zinc dyshomeostasis and mitochondrial dysfunction. *Human Molecular Genetics, 23*, 2802–2815.

Plum-Favreau, H., Klupsch, K., Moisoi, N., Gandhi, S., Kjaer, S., Frith, D., et al. (2007). The mitochondrial protease HtrA2 is regulated by Parkinson's disease-associated kinase PINK1. *Nature Cell Biology, 9*, 1243–1252.

Whitworth, A. J., Lee, J. R., Ho, V. M. W., Flick, R., Chowdhury, R., & McQuibban, G. A. (2008).
Rhomboid-7 and HtrA2/Omi act in a common pathway with the Parkinson's disease factors
PINK1 and parkin. *Disease Models & Mechanisms, 1*, 168–174.

Nitric Oxide Signaling Discovered

Arnold, W. P., Mittal, C. K., Katsuki, S., & Murad, F. (1977). Nitric oxide activates guanylate
cyclase and increases guanosine 3':5'-cyclic monophosphate levels in various tissue prepara-
tions. *Proceedings of the National Academy of Sciences of the United States of America, 74*,
3203–3207.
Furchgott, R. F., & Zawadzki, J. V. (1980). The obligatory role of endothelial cells in the relaxation
of arterial smooth muscle by acetylcholine. *Nature, 288*, 373–376.
Garthwaite, J., Chalres, S. L., & Chess-Williams, R. (1988). Endothelium-derived relaxing factor
release on activation of NMDA receptors suggests role as intercellular messenger in the brain.
Nature, 336, 385–388.
Ignarro, L. J., Buga, G. M., Wood, K. S., Byrns, R. E., & Chaudhuri, G. (1987). Endothelium-
derived relaxing factor produced and released from artery and vein is nitric oxide. *Proceedings
of the National Academy of Sciences of the United States of America, 84*, 9265–9269.
Palmer, R. M. J., Ashton, D. S., & Moncada, S. (1988). Vascular endothelial cells synthesize nitric
oxide from L-arginine. *Nature, 333*, 664–666.
Palmer, R. M., Ferrige, A. G., & Moncada, S. (1987). Nitric oxide release accounts for the biological activity of endothelium-derived relaxing factor. *Nature, 327*, 524–526.

Neuroprotective and Neurodestructive Actions of Nitric Oxide and Other Stress Factors

Chung, K. K. K., Thomas, B., Li, X., Pletnikova, O., Troncoso, J. C., Marsh, L., et al. (2004).
S-nitrosylation of parkin regulates ubiquitination and compromises parkin's protective function.
Science, 304, 1328–1331.
LaVoie, M. J., Ostaszewski, B. L., Weihofen, A., Schlossmacher, M. G., & Selkoe, D. J. (2005).
Dopamine covalently modifies and functionally inactivates parkin. *Nature Medicine, 11*,
1214–1221.
Lipton, S. A., Choi, Y. B., Pan, Z. H., Lei, S. Z., Chen, H. S. V., Sucher, N. J., et al. (1993). A
redox-based mechanism for the neuroprotective and neurodestructive effects of nitric oxide and
related nitroso-compounds. *Nature, 364*, 626–632.
Mosharov, E. V., Larsen, K. E., Kanter, E., Phillips, K. A., Wilson, K., Schmitz, Y., et al. (2009).
Interplay between cytosolic dopamine, calcium, and α-synuclein causes selective death of sub-
stantia nigra neurons. *Neuron, 62*, 218–229.
Palacino, J. J., Sagi, D., Goldberg, M. S., Krauss, S., Motz, C., Wacker, M., et al. (2004).
Mitochondrial dysfunction and oxidative damage in parkindeficient mice. *The Journal of
Biological Chemistry, 279*, 18614–18622.
Uehara, T., Nakamura, T., Yao, D., Shi, Z. Q., Gu, Z., Ma, Y., et al. (2006). S-nitrosylated protein-
disulphide isomerase links protein misfolding to neurodegeneration. *Nature, 441*, 513–517.
Yao, D., Gu, Z., Nakamura, T., Shi, Z. Q., Ma, Y., Gaston, B., et al. (2004). Nitrosative stress linked
to sporadic Parkinson's disease: S-nitrosylation of parkin regulates its E3 ubiquitin ligase activity.
Proceedings of the National Academy of Sciences of the United States of America, 101,
10810–10814.

Chronic Inflammation and iNOS

Gao, H. M., Jiang, J., Wilson, B., Zhang, W., Hong, J. S., & Liu, B. (2002). Microglial activation-mediated delayed and progressive degeneration of rat nigral dopaminergic neurons: Relevance to Parkinson's disease. *Journal of Neurochemistry, 81*, 1285–1297.

Lee, H. J., Suk, J. E., Patrick, C., Bae, E. J., Cho, J. H., Rho, S., et al. (2010). Direct transfer of α-synuclein from neuron to astroglia causes inflammatory responses in synucleinopathies. *The Journal of Biological Chemistry, 285*, 9262–9272.

Liberatore, G. T., Jackson-Lewis, V., Vukosavic, S., Mandir, A. S., Vila, M., McAuliffe, W. G., et al. (1999). Inducible nitric oxide synthase stimulates dopaminergic neurodegeneration in the MPTP model of Parkinson disease. *Nature Medicine, 5*, 1403–1409.

McGeer, P. L., Itagaki, S., Boyes, B. E., & McGeer, E. G. (1988). Reactive microglia and positive for HLA-DR in the substantia nigra of Parkinson's and Alzheimer's disease brains. *Neurology, 38*, 1285–1291.

Zhang, W., Wang, T., Pei, Z., Miller, D. S., Wu, Z., Block, M. L., et al. (2005). Aggregated α-synuclein activated microglia: A process leading to disease progression in Parkinson's disease. *The FASEB Journal, 19*, 533–542.

Exosomes and Cell-to-Cell Spread

Desplats, P., Lee, H. J., Bae, E. J., Patrick, C., Rockenstein, E., Crews, L., et al. (2009). Inclusion formation and neuronal cell death through neuronto-neuron transmission of α-synuclein. *Proceedings of the National Academy of Sciences of the United States of America, 106*, 13010–13015.

Emmanouilidou, E., Melachroinou, K., Roumeliotis, T., Garbis, S. D., Ntzouni, M., Marqaritis, L. H., et al. (2010). Cell-produced α-synuclein is secreted in a calcium-dependent manner by exosomes and impacts neuronal survival. *The Journal of Neuroscience, 30*, 6838–6851.

Hansen, C., Angot, E., Bergström, A. L., Steiner, J. A., Pieri, L., Paul, G., et al. (2011). α-Synuclein propagates from mouse brain to grafted dopaminergic neurons and seeds aggregation in cultured human cells. *The Journal of Clinical Investigation, 121*, 715–725.

Harding, C., Hauser, J., & Stahl, P. (1983). Receptor-mediated endosytosis of transferrin and recycling of the transferrin receptor in rat reticulocytes. *The Journal of Cell Biology, 97*, 329–339.

Luk, K. C., Kehm, V. M., Zhang, B., O'Brien, P., Trojanowski, J. Q., & Lee, V. M. Y. (2011). Intracerebral inoculation of pathological α-synuclein initiates a rapidly progressive neurodegenerative α-synucleinopathy in mice. *The Journal of Experimental Medicine, 209*, 975–986.

Pan, B. T., & Johnstone, R. M. (1983). Fate of the transferrin receptor during maturation of sheep reticulocytes in vitro: Selective externalization of the receptor. *Cell, 33*, 967–978.

Chapter 10
Huntington's Disease and Other Unstable Repeat Disorders

The year 2001 witnessed the publication of the human genome sequence by two consortia, one headed by Eric Lander (in *Nature*) and the other led by Crag Venter (in *Science*). The number of protein coding sequences, in the 20,000–30,000 range, and their chief characteristics, were reported. The coding sequences accounted for no more than about 1.5 % of the total genome. Both reports discussed this dramatic finding in detail and presented an overview of the entirety of the genome. In Lander's report, repeat sequences were noted as spanning 50 % of the genome. That such sequences are present was known for some time and made famous under the rubric "selfish DNA" or "selfish genes." These sequences were divided into five classes by the Lander-led team—(1) transposon-derived repeats; (2) processed pseudogenes; (3) simple sequence repeats; (4) segmental duplications—blocks of DNA copied from one place to another, and (5) blocks of tandem-repeated sequences such as those encountered at centromeres and telomeres. The first group, transposon-derived repeats, constituted the bulk of the repeat content comprising 45 % of the genome. The third group, simple sequence repeats, consisted of repeated doublets such as $(CA)_n$, triplets such as $(CGG)_n$, repeated quartets, and so on. These repeats, also known *microsatellites*, accounted for 3 % of the genome.

That the vast bulk of repeat-laden DNA is not useless, inert, or entirely benign has become increasingly apparent during the past few years. One of the chief examples of this aspect is the role that unstable simple sequence repeats have in human disease. At least 30 neurodegenerative disorders are due to them. These diseases include Huntington's disease, a variety of ataxias, monotonic dystrophy, the most common form of autosomal dominant muscular dystrophy, and fragile X syndrome, a leading form of inherited mental retardation. Up to now the main type of disease-causing mutations resulting in protein misfolding was the point mutation. Here, another kind of mutation is considered, one in which the number of simple sequence repeats increases from generation to generation from a normal number of repeats which may number up to 50, to disease-associated ever larger

© Springer International Publishing Switzerland 2015   301
M. Beckerman, *Fundamentals of Neurodegeneration and Protein Misfolding Disorders*,
Biological and Medical Physics, Biomedical Engineering,
DOI 10.1007/978-3-319-22117-5_10

numbers ranging into the thousands. These increases correlate with an increasing severity of the illness, a phenomenon known as *genetic anticipation.*

Nine neurodegenerative disorders are known to be caused by unstable CAG repeats situated in protein coding regions. These repeats encode the amino acid glutamine, hence the name polyQ diseases. These neurodegenerative disorders are listed in Table 10.1. Huntington's disease heads this list. It is caused by unstable repeats in the huntingtin gene (HTT) situated on the short arm of chromosome 4 at location 4p16.3. This gene encodes the huntingtin (Htt) protein. Individuals possessing expanded numbers of these repeats develop Huntington's disease, a fatal disorder that affects neurons in the brain resulting in motor dysfunction and progressing to cognitive decline, dementia, and death 15–20 years after onset of the illness. The primary targets are the medium spiny neurons (MSNs) situated in the striatum that deteriorate and eventually die off. These neurons receive cortical and thalamic input and release the inhibitory neurotransmitter γ-aminobutyric acid (GABA) to multiple regions involved in motor control. Like the illnesses discussed earlier, this disease spreads to other regions of the brain, in this case, to neurons in the hippocampus and cortex. Incidence of this disease among white populations is roughly 5–7 individuals per 100,000.

The first of the CAG repeat disorders to be discovered was that of the androgen receptor which gives rise to spinal and bulbar muscular atrophy (SBMA), also known as Kennedy disease. The disease was first described by William Kennedy and coworkers in 1968 but it was not until 1991 that La Spada and colleagues isolated the cause of the disease as the presence of expanded CAG repeats in the gene encoding the androgen receptor. The gene is located on the long arm of the X-chromosome, and the mutation is inherited in an autosomal recessive manner. Later in the same year, 1991, a pair of papers appeared reporting that Fragile X syndrome arises as a consequence of an unstable CGG repeat in the FMR1 gene, also located on chromosome X. In this case, the repeat expansion occurs in the 5′ UTR (untranslated region). That finding answered what had been called the *Sherman paradox*—the until-then unexplained increases in the disease susceptibility among successive generations observed in the mid-1980s by Stephanie Sherman.

Neurons in several areas of the brain are affected in the CAG repeat disorders. Prominent among these regions are spinal motor nuclei, brainstem nuclei, basal ganglia, and cerebellum. The spinocerebellar ataxias are the most common form of ataxia, a class of disorders that affect more than 150,000 people in the USA alone. Recall that the cerebellum coordinates muscle movement and regulates motor learning and control. Six of the nine entries in Table 10.1 are of spinocerebellar ataxias. Each of these illnesses affects a different portion of the cerebellum and each one is due to unstable expanded CAG repeats in a specific protein. The clinical features of each of the polyQ diseases are distinct, and depend crucially on repeat length and upon the biophysical properties and normal cellular functions of the affected protein. They do, however, share a number of notable features. One of these is the aforementioned increases from generation to generation in the severity of the illnesses due to the expanding number of unstable repeats. Another is the widespread appearance of neuronal intranuclear inclusions (NIIs).

Table 10.1 Polyglutamine (polyQ) repeat disorders

Neurological disease	Protein	Normal repeat length	Disease repeat length	Protein function
Huntington's disease (HD)	Huntingtin (Htt)	6–34	36–121	Transcription, transport
Spinal and bulbar muscular atrophy (SBMA)	Androgen receptor	9–36	38–62	Nuclear receptor
DentatoRubral and PallidoLuysian atrophy (DRPLA)	Atrophin-1	7–34	49–88	Transcription
Spinocerebellar ataxia 1 (SCA1)	Ataxin-1	6–39	40–82	Transcription
Spinocerebellar ataxia 2 (SCA2)	Ataxin-2	15–24	34–200	RNA metabolism
Spinocerebellar ataxia 3 (SCA3)	Ataxin-3	13–36	61–84	Deubiquitination
Spinocerebellar ataxia 6 (SCA6)	α1A-VDCC subunit, and α1ACT	4–18	19–33	P/Q type VDCC subunit
Spinocerebellar ataxia 7 (SCA7)	Ataxin-7	4–35	36–460	Transcription
Spinocerebellar ataxia 17 (SCA17)	TATA-box binding protein (TBP)	25–42	47–63	Transcription

Unlike the situation encountered in Parkinson's disease where an impressive number of risk factors and disease pathways have been uncovered, and where there are both familial and sporadic forms, HD arises from a single causative factor—mutations in the huntingtin protein encoded by the HTT gene. For that reason it might have been thought that HD and the other polyQ disorders are straightforward, and easy to understand and treat. However, the huntingtin protein is large and remarkably complex. Like Alzheimer's and Parkinson's diseases, the underlying disease mechanisms and progression are neither simple nor easy to uncover. Huntingtin contains large stretches of unstructured sequences, is cleaved into different-sized fragments, is encountered with different repeat lengths, and is subject to a variety of posttranslational modifications.

Significantly, HD shares a number of biophysical and cellular features with many if not most of the other polyQ disorders, and with AD and PD. Chief among them are transcriptional and axonal transport failures, loss of calcium homeostasis, dysregulated mitochondria, all leading to the loss of protein quality control, synaptic failure, and neuron death. Their exploration in this chapter begins with a look at the unusual, problem-causing secondary structures generated by the disease-causing expanded repeats. The biogenesis, membrane localization, and clearance of mutant htt follow with special attention to the central role exon 1 in causing Huntington's disease. The exploration of Huntington's disease concludes with an examination of huntingtin's normal functions and disease-causing roles in intracellular vesicular trafficking, synaptic function, and mitochondrial biogenesis.

Those discussions are followed by an examination of the ataxias and the second major class of unstable repeat disorders, namely, those involving expanded repeats in non-protein-coding regions. This second group of disorders includes, besides fragile X syndrome, monotonic dystrophy 1 and 2, Friedrich's ataxia, and several prominent forms of mental retardation. Study of these diseases will recapitulate one of the major themes of the polyQ class, the sequestration of proteins leading to their loss of function, and introduce a new mechanism—splicing errors—that makes a bridge to the RNA-centric disorders that will be encountered in the next and final chapter of the textbook.

10.1   Unusual Secondary Structures Underlie Trinucleotide Repeat Disorders

Instabilities in repeat lengths have their origin in the unusual secondary structures they generate. Several different kinds of exceptional secondary structures are formed by the expanded repeats. Depending on the composition of the repeats and the process being undertaken these structures may involve ssDNA, dsDNA, or triplex DNA; others are established by DNA–RNA hybrid complexes and still others by RNA. Some will form hairpin-like protuberances, while others form loops, out-of-register, and slipped-strand realignments. These structural elements produce kinetic traps, become "sticky" DNA, and cause numerous difficulties in replication, repair, and recombination. Once the number of repeats passes a certain threshold, they lead to cancers and neurodegenerative diseases. The different kinds of aberrant structures and the problems they cause are summarized in Fig. 10.1.

One of the first theories on how these secondary structure elements might produce generation-to-generation increases in repeat length was the replication model. The central idea behind this model was that the unusual secondary structures formed by the repeats act as impediments to replication and cause replication forks to slip and stall at those locations. Cells are equipped with a number of repair systems that fix damaged DNA and restart stalled replication forks. These factors, most notably the MSH2/MSH3 proteins that comprise the mismatch repair (MMR) recognition complex, and the base-excision repair (BER) 8-oxoguanine glycosylase (OGG1), are implicated in the repeat expansions. In addition, there is evidence for the involvement of recombination factors, *cis* acting elements such as the multifunctional transcription regulator CTCF, and epigenetics. The relative contributions of each of these factors to the disease etiology depend critically on the biophysical and biochemical properties of the repeats and cell-type under consideration. In carrying out the repair, additional repeat elements become inserted. In Fig. 10.1 a variety of consequences during replication, repair, recombination, and transcription have been listed.

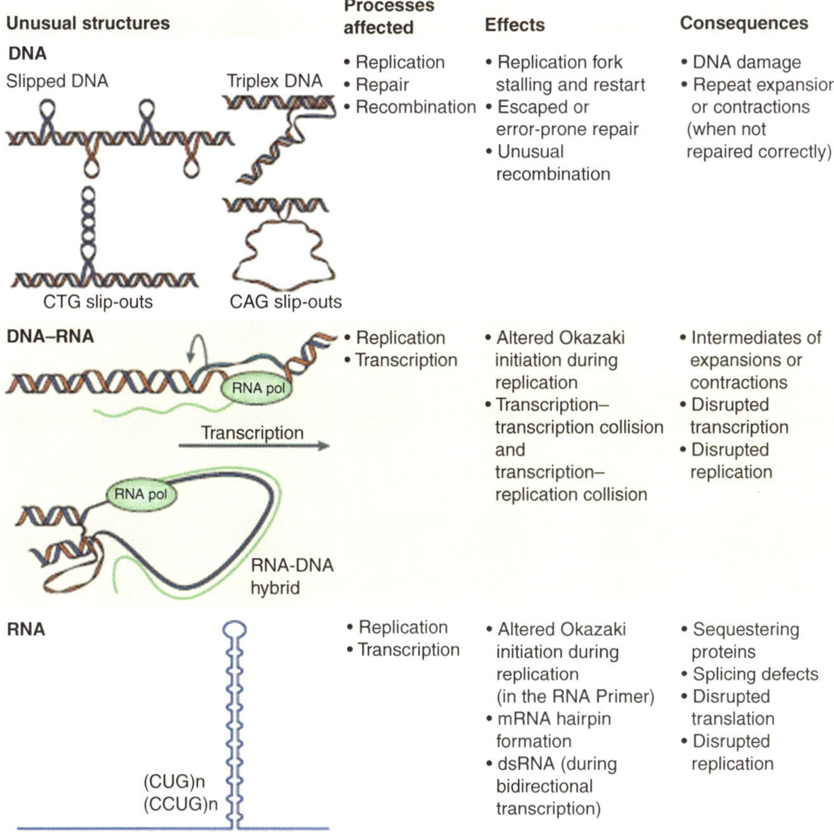

Fig. 10.1 Unusual problem-causing secondary structures generated by expanded repeats. *Upper panel*: slipped DNA, triplex DNA, and CTG and CAG slip-outs. *Middle panel*: DNA-RNA hybrid looped structures. *Lower panel*: RNA hairpins. *Columns 2, 3*, and *4* list the most commonly affected processes, mechanistic effects, and ultimate consequences associated with these aberrant structures (from Castel *Nat. Rev. Mol. Cell Biol.* 11. 165 © 2010 Reprinted by permission from Macmillan Publishers Ltd)

10.2   Structure of the Huntingtin Protein

The 3144-amino-acid-residue huntingtin protein is organized into N- and C-terminal domains (Fig. 10.2). As noted earlier the protein is both large and complex. It has, in fact, defied attempts at full-length crystallization. However, efforts have been successful at biophysical characterizations of exon 1. Exon 1 encodes a 17 residue Htt^{NT} sequence followed by the variable number of polyQ repeats and a proline-rich region (polyP). The importance of exon 1 in Huntington's disease was brought out in early studies showing that expression of exon 1 alone was sufficient to recapitulate most of the features of the disease. For that reason exon 1 is its most significant

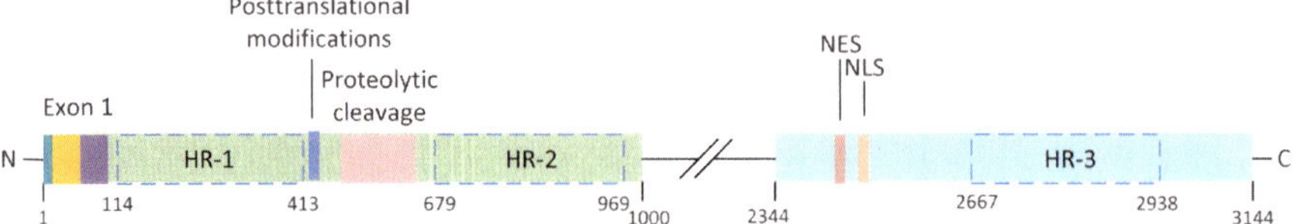

Fig. 10.2 Huntingtin structure. Exon 1 contains an N-terminal fragment (*blue*), PolyQ repeats (*yellow*), and proline-rich sequence (*purple*). The central portion of the N-terminal domain contains amino acid residues subject to posttranslational modifications, and a region where proteolytic cleavage by caspases and calpains occurs. The C-terminal region contains a nuclear export sequence (NES) and a nuclear localization (import) sequence (NLS). Huntingtin also has three large regions each containing several HEAT repeats (HRs); these are outlined by *dashed boxes*

feature. In addition to exon 1, the N-terminal Htt domain contains (1) a largely unstructured region subject to cleavage by proteolytic enzymes, and (2) multiple residues subject to posttranslational modifications. The C-terminal region does not appear to be as interesting as the N-terminal domain, but does contain nuclear import and export sequences. Lastly, Huntingtin possesses several HEAT repeat regions. Two of these are situated in the N-terminal domain and one in the C-terminal domain. All of these structural features are now discussed in greater detail.

10.2.1   The PolyQ Repeats

A landmark in understanding the significance of the glutamine repeats in HD and other polyQ disorders was the introduction in 1994 by Max Perutz of the *polar zipper* concept. In Perutz's polar zipper, the polyQ chains form β-sheets held together by hydrogen bonds between main-chain and side-chain amides. These bonds form the links of the zipper. In introducing his model, Perutz had observed that (1) polyQ tracts readily bind other polyQ tracts and form insoluble clumps, and (2) many transcription factors contain polyQ tracts and these sequences may serve a normal role in transcription. Htt with its polyQ tract of regularly spaced glutamines is able to bind not only to itself but also to these other polyQ containing proteins. In HD, the Htt polyQ stretches misfold into β-sheets that oligomerize and form fibrillar structures highly resistant to degradation and clearance by the ubiquitin–proteasome system. This basic idea has been built upon and expanded in the intervening years.

10.2.2   The PolyQ Flanking Sequences

As shown in Fig. 10.2 the polyQ repeats located in exon-1 are accompanied by N- and C-terminal flanking sequences. There is a 17-residue sequence N-terminal to the polyQ repeats (Htt^{NT}) and a 38-residue proline-rich sequence C-terminal to the polyQ's. The latter consists of three components—two strings of proline residues separated by a proline-rich stretch. In general, both Htt^{NT} and polyP influence the folding, aggregation and toxicity of the polyQ repeats. The extent of their influences varies depending on several factors including the polyQ tract length, modifications to the sequences, and cellular context; in some cases, the influences are profound, in others they are negligible. The flanking sequences and their varying influences are discussed in greater detail in the next section.

10.2.3   Proteolytic Cleavage and Aberrant Splicing

One of the most consistent features of the polyQ disorders is the presence of fragments of the mutated proteins in the neuronal intranuclear inclusions. These fragments are generated by cleavage of the full-length proteins by members of the caspase and calpain families of proteases. Evidence for this type of processing has been found for huntingtin, the androgen receptor, atrophin-1, ataxin-3, and ataxin-7. The huntingtin protein possesses a number of caspase and calpain cleavage sites. These cleavage operations generate N-terminal fragments that are toxic to the striatal neurons. As shown in Fig. 10.2 these sites are clustered together in a region in between the two N-terminal heat repeat regions (from residues 469 to 586).

Cleavage by calpains is significant for yet another reason. Calpains are Ca^{2+}-dependent cysteine proteases. When activated these proteases target critical signaling elements, thereby facilitating synaptic remodeling and memory formation. Under normal operating conditions calpain activity is maintained at a low, controlled level, but Ca^{2+} overload triggers calpain hyperactivation as part of a protective feedback response aimed at substrates such as NMDA receptors that enable calcium entry.

Lastly, aberrant splicing of exon 1 has been found to lead to the formation of small Htt fragments. Small exon-1/intron-1 polyadenylated mRNA transcripts have been detected in brains of mice expressing mtHtt. Translation of these transcripts produces the exon-1 Htt fragments. It was surmised in that study that associations of splicing factors such as SFSR6 with the expanded polyQ tracts precipitate the aberrant splicing events.

10.2.4   Posttranslational Modifications

Huntingtin is subject to different kinds of posttranslational modification—phosphorylation, acetylation, palmitoylation, ubiquitination, and SUMOylation. These actions couple cellular signaling and regulatory pathways to the protein so that huntingtin's subcellular localization, clearance, and processing are responsive to cellular needs. One cluster of sites where phosphorylation and acetylation occurs is located just outside the first HEAT region. Those sites are shown in Fig. 10.2. Other sites are situated further C-terminal to that region in between the HEAT repeat sequences.

A particularly interesting locus for posttranslational modifications is situated in the Htt^{NT}. That short stretch contains phosphorylation sites at serine 13 and serine 16 and competitive ubiquitination—SUMOylation sites at lysine 6, lysine 9, and lysine 15. Ubiquitination usually targets the protein or protein fragments for proteasomal clearance, while SUMOylation stabilizes the protein and reduces aggregation. Concerted actions by the enzymes responsible for phosphorylation and ubiquitination/SUMOylation take place with phosphorylation preparing the adjacent sites for subsequent Ub/SUMO modifications. By these and similar actions, posttranslational modifications influence huntingtin's stability, clearance, subcellular localization, aggregation, fibrillization, and toxicity.

10.2.5   The HEAT Repeats

These structures serve as platforms or scaffolds that enable multiple proteins to come into close contact with one another, thereby mediating their interaction and signaling roles. The acronym HEAT stands for huntingtin, elongation factor 3, protein phosphatase 2A, and TOR1 protein kinase. These are the first four proteins found to contain these stacked structures. HEAT repeats are amino acid sequences approximately 47 or so residues in length. Each repeat is composed of a helix-turn-helix motif; these are stacked one after the other to form curved rod-like structures, hence the name α-RODs. These structures expose large amounts of solvent-exposed surface for binding of large molecules hence their suitability as mediators of protein–protein and protein–DNA interactions.

10.3   Role of Exon 1 in Huntingtin Aggregation

The biophysical properties of the expanded polyQ repeats depend on all three components of exon 1. That is, they depend on Htt^{NT}, the number of polyQ repeats, and the polyP. Those dependences are illustrated in the next two figures. The first of these figures (Fig. 10.3a) illustrates the kinetic acceleration capability of the Htt^{NT}, while the polyP sequence slows the progression and serves in a protective capacity. The second figure (Fig. 10.3b) shows the dramatic speedup in aggregation as the polyQ length increases. The data included in these figures was obtained at several different micromolar concentrations and those values are indicated in the plots.

The Htt^{NT}, by itself, tends to adopt an α-helical conformation and aggregate into α-helical oligomers. Adding the polyQ repeats to the Htt^{NT} generates a Htt^{NT}-polyQ peptide. This structure first assembles into oligomers via N-terminal helix-helix interactions and then converts to a form in which polyQ beta-hairpins or other beta-stranded structures occupy a nascent core and the N-terminal alpha helices reside on the periphery. The importance of the N-terminal segment in physiological settings is enhanced by its membrane-binding ability. The N-terminal α-helical segment anchors the Htt fragments to lipid membranes and by that means facilitates the

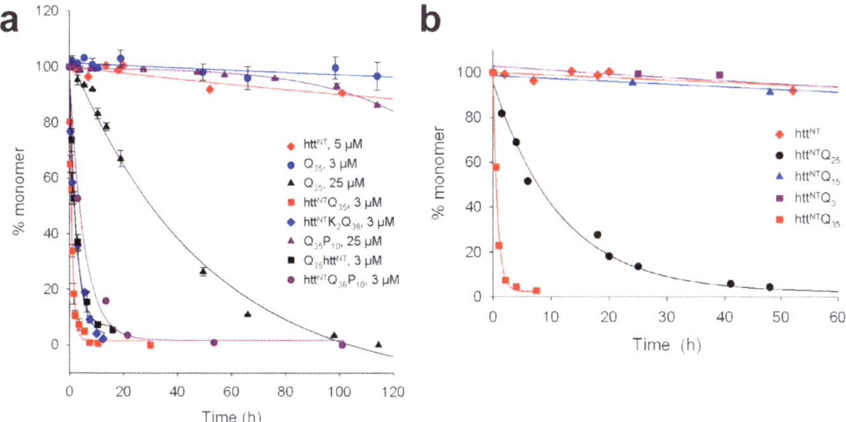

Fig. 10.3 Huntingtin exon 1 aggregation kinetics. (**a**) Data showing the effects of appending Htt^{NT} and P_{10} peptides to Q_{35} and Q_{36} repeats. (**b**) Illustration of the role of polyQ repeat length on the aggregation kinetics (from Thakur *Nat. Struct. Mol. Biol.* 16: 380 © 2009 Reprinted by permission from Macmillan Publishers Ltd)

formation of close associations between fragments. These aggregation-promoting activities are enhanced by the presence of acidic phospholipids in the membranes.

The proline-rich region situated C-terminal to the polyQ repeats has a role, too. As can be seen in Fig. 10.3a the addition of a P_{10} stretch reduces the rate of monomer depletion and inhibits aggregation. It does so by shifting the monomer equilibrium towards aggregation-incompetent conformations. It and other, longer stretches interfere with the formation of the beta sheets and shift the ensemble equilibrium towards formation of more compact, helical structures. They destabilize non-fibrillar aggregates and oligomers and by that means reduce their toxicity.

Overall, the Htt^{NT} and polyP segment act synergistically to:

- Decrease the overall solubility of polyQ-bearing fragments,
- Destabilize non-fibrillar aggregates and intermediates, and
- Accelerate fibril formation.

10.4 Htt Misfolding, Oligomerization, and Aggregation

10.4.1 Aggregates and Intranuclear Inclusions

Like other disease-causing proteins such as Aβ and α-synuclein, Htt oligomers of different sizes and morphologies may be produced. The ensemble of forms includes besides monomers, native and misfolded, small annular structures and large ring-shaped ones, and still larger amorphous aggregates and well-developed fibrillar structures. The intranuclear inclusions that sit at the top of the hierarchy of forms consist of granules, filaments, and randomly and parallel-oriented fibrils.

Proteins other than Htt and fragments thereof are often present in these inclusions. One prominent group consists of proteins associated with the maintenance of proteostasis in the cell—molecular chaperones and components of the ubiquitin–proteasome system. Another important group of proteins present in the NIIs are transcription factors, especially those possessing polyQ stretches. Lastly, the amyloid-like fibrils are composed primarily of N-terminal fragments of varying sizes.

In huntingtin, there is evidence in support of both models of fibril formation. First, there is good evidence for a classical nucleated polymerization mechanism of fibril formation. In this pathway, homogeneous monomers or tetramers misfold into conformations with a high beta-sheet content. The monomers and small oligomeric species serve as seeds for the rapid, polymerization phase of fibril growth that follow the slow, lag-phase dominated nucleation step. Secondly, studies have been made of heterogeneous mixes of conformers with varying propensities to form fibrils. Results of theoretical studies of these entities support the alternative picture in which oligomers first form and these then undergo a nucleated conformational conversion of their constituents into fibril-forming beta-sheet conformers.

10.4.2 Action of Molecular Chaperones

Recall from Chap. 5 that some chaperones refold partially unfolded and misfolded proteins; others maintain solubility and prevent aggregation, or re-solubilize clumped proteins (Fig. 5.7). Still other molecular chaperones, especially those activated under conditions of chronic stress, direct protein aggregates to the UPS and/or into autophagic pathways for disposal (Fig. 5.10). PolyQ-containing aggregates, and especially those containing mutated, cleaved and modified htt, are difficult to handle. These are either targeted for degradation by the UPS or autophagy, or alternatively, sequestered in intracellular inclusions.

Interestingly, many of the neurodegenerative-disease-causing mutated proteins including α-synuclein (PD), the androgen receptor (SMBA), and tau (FTLD) are Hsp90 "clients"; that is, they are bound and stabilized by the molecular chaperone. That has led to the development of Hsp90 inhibitors. These are intended to prevent the stabilization of aberrant proteins by Hsp90, thereby promoting their clearance by the UPS, autophagy, and sequestration.

10.5 Clearance of mHtt by the UPS, Autophagy and Sequestration in Inclusions

10.5.1 The UPS and Autophagy

Mutant Htt and its proteolytic fragments are cleared by the proteasome, but imperfectly. The proteasome cannot entirely digest the chains. Instead, it releases a variety of fragments for further hydrolysis. The fragments, if not cleared, can aggregate and further promote the onset of disease. In addition, there is a jamming issue in which excessively long polyQ peptides impair the UPS resulting in failed degradation of other misfolded and aggregated proteins. These actions precede inclusion body formation, and give rise to the idea that intracellular inclusion body formation is, in fact, a protective cellular response, and indicative of stressed protein quality control.

Autophagy takes up much of the burden of clearing full-length and fragmented mHtt. Recall from Chap. 6 that Ulk1, a key initiator of autophagy, is bound to the mTOR signaling complex. Htt aggregates interact with the mTOR complex and with Ulk1. These interactions free Ulk1 from mTOR inhibition and enable it to initiate autophagy. Htt also interacts with p62. It seems that unlike other aggregation- and neurodegeneration-prone proteins, huntingtin has a normal role as a regulator of autophagy. As a first step, members of the HSPB and DNAJ families of molecular chaperones bind hydrophobic patches in the Htt^{NT} region of huntingtin fragments that have been released from the proteasome. This action disrupts their further aggregation and fibrillization. However, the beneficial interactions between Htt and the autophagic machinery are either too weak or go awry in mutant-driven disease-situations. Empty autophagosomes have been observed in which cargo engulfment has failed to occur, and defects in autophagosome transport have been uncovered in which lysosomal membrane fusion and degradation have not taken place.

10.5.2   Sequestration in Inclusions

The TCP1 ring complex/chaperonin containing TCP1 (TRiC/CCT) chaperonins were briefly discussed in Chap. 5. Recall from that discussion that these are Type II chaperonins belonging to the Hsp60 family. These molecular chaperones complete the folding of difficult-to-handle and kinetically trapped proteins began by Hsp70 family members. TCP1, a subunit of CCT, interacts with mHtt. Through this interaction CCT promotes the incorporation of mHtt into large inclusion bodies. Once there, they are bound by Hsp70, and that action prevents cell death pathways from being activated. These sequential actions serve to detoxify neurons expressing mHtt by sequestering them in the largely inert inclusions.

A number of proteins are sequestered in cytoplasmic and nuclear aggregates by mutant huntingtin. Heat shock proteins and transcription factors were already mentioned. One of the proteins involved in gene transcription that is bound by mHtt is the CREB binding protein, CBP, a transcription co-activator. This protein, consistent with Perutz's seminal 1994 observation, contains a polyglutamine stretch. However, the sequestration of proteins by mHtt extends beyond interactions with polyQ-bearing proteins belonging to the transcription machinery. For example, the polyP sequence in exon 1 is capable of mediating the sequestration of proteins containing WW and SH3 domains. Those interactions are implicated in the sequestration of several vesicle associated proteins by truncated wtHtt and by mHtt fragments. Thus, protection by sequestration has its limits, reached when too many proteins that are needed are tied up in these structures.

It is widely believed that the proteostasis network consisting of molecular chaperones, the ubiquitin–proteasome system, endoplasmic reticulum associated protein degradation, and the three kinds of autophagy decline with age. That decline, coupled to the continued and increasing presence of misfolded proteins that require clearance, eventually tips the scale in favor of a buildup of toxic species. Mutant huntingtin and other neurodegenerative, disorder-associated misfolded proteins exacerbate this shifting balance by tying up declining supplies of key proteostasis components and sequestering them in intracellular inclusions. Consistent with this picture, ubiquitin, Hsp40 family members, and other components of the proteostasis network are found attached to NIIs containing N-terminal fragments of mHtt. This scenario provides a plausible explanation for the late-onset age-dependence of HD, PD, AD, and other neurodegenerative disorders.

10.6 Huntingtin Membrane Localization and Trafficking Roles

Huntingtin full-length and N-terminal fragments associate with lipid membranes. The protein–surface interactions are important for several reasons. They (1) influence the subcellular localization of normal and mutant forms of these proteins, (2) promote aggregation, and (3) help determine the toxicity of the oligomers. It is a partnership. Among the physicochemical factor contributed by the oligomers are its flexibility, hydrophobic surface exposure and stability, while the surface provides lipid composition, especially cholesterol and acidic ganglioside GM1 content, along with physical properties such as charge, membrane curvature and lipid phase. Together, these physicochemical factors determine which lipid membranes are used by Htt in carrying out its normal functions, and the relative toxicity of their abnormal offspring at those locations or at alternative ones.

Huntingtin is found in multiple membrane-associated locations. It resides most often in the cytosol but sometimes in the nucleus. Huntingtin is seen at the plasma membrane, found connected to synaptic vesicles, and tied to microtubules. It localizes to mitochondria, the endoplasmic reticulum, and the Golgi complex. Huntingtin has a role in endocytic and exocytic vesicular trafficking, and, as discussed above, is a component of the autophagic machinery.

Recall from Chap. 6 that microtubule-mediated transport is used to move cargo over long distances, while actin filaments mediate transport locally over short distances. In fast axonal transport (FAT), cargo is transported along microtubule tracks at rates up to 1 μ/s. Slow axonal transport takes place at speeds 100 times lower, on the order of 1 mm/day. Mitochondria are transported by FAT, but at reduced rates punctuated by stops and reverses designed to deliver mitochondria where they are needed.

Huntingtin mediates long-distance transport. As depicted in Fig. 10.4 Htt interacts with Huntingtin-associated proteins HAP-1 and Hap-40, with dynactin, and with regulators of vesicle endocytosis such as Rab5. By those means Htt serves as a scaffold for the attachment of cargo to the microtubule-based kinesin and myosin motor proteins. However, mutant huntingtin and especially mutant N-terminal fragments perturb the scaffolding operation resulting in stalled movement and impaired axonal transport of not only mitochondria and other organelles but also signaling proteins such as brain-derived neurotrophic factor (BDNF). It does so, at least in part, by binding and titrating HAP-1, dynamin (needed for endocytosis) and other vesicle-associated proteins in inclusion bodies. In addition, mutant huntingtin

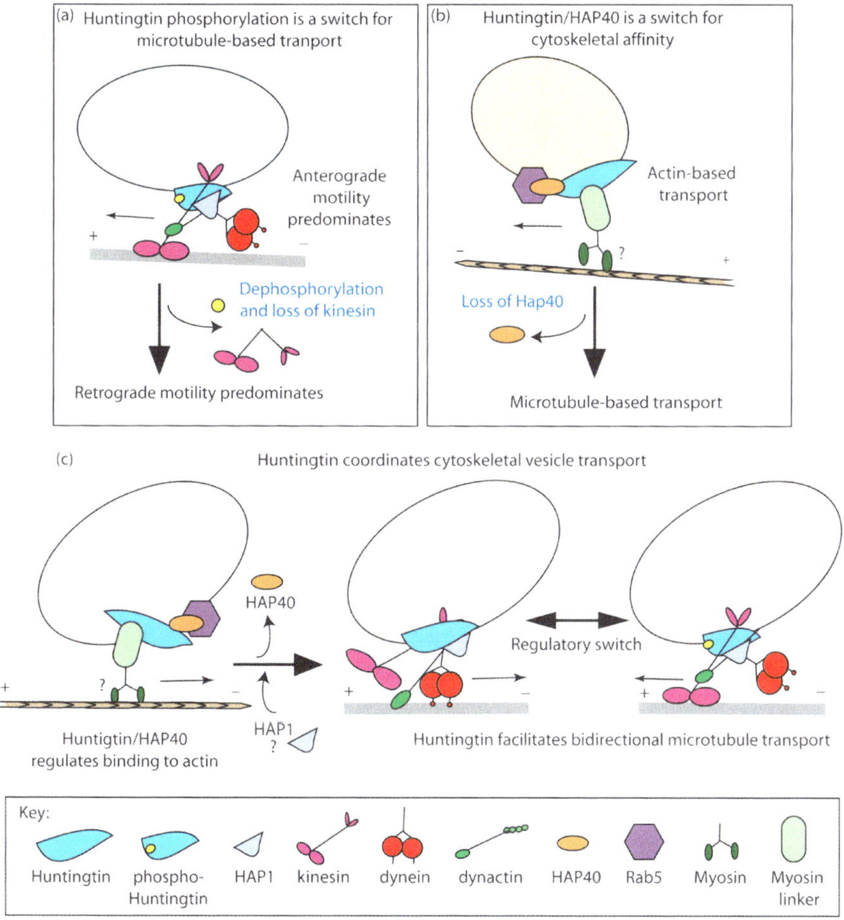

Fig. 10.4 Facilitation of vesicular transport by huntingtin. (**a**) Phosphorylation at serine 421 serves as a molecular switch between anterograde (kinesin) transport and retrograde (dynein/dynactin) motility. (**b**) Htt/Hap40/Rab5 complex mediates endosome mobility between microtubules and actin filaments. (**c**) Coordination of cytoskeletal vesicle transport by huntingtin, which mediates attachment to actin and bidirectional transport along microtubules. In all these models, huntingtin functions as a scaffolding protein (from Caviston *Trends Cell Biol.* 19: 147 © 2009 Reprinted by permission from Elsevier)

affects mitochondrial fusion–fission dynamics. It binds the mitochondrial fission GTPase DRP1, thereby disrupting fusion–fission dynamics and further preventing the normal anterograde and retrograde movement to and from synapses. There is also evidence that mHtt cause excessive S-nitrosylation of DRP1 leading to mitochondrial fragmentation and synaptic damage. The importance of S-nitrosylation in neurological disorders was discussed in the context of PD in Chap. 9.

10.7 Intracellular Vesicular Trafficking: BDNF and TrkB

Growth factor signaling pathways are utilized not only during development but also in adult situations. Brain derived neurotrophic factor (BDNF) is synthesized in cortical cells. It is then transported down the axons and secreted out across the corticostriatal synaptic cleft where it binds to TrkB and $p75^{NTR}$ receptors on striatal dendrites. The ligand–receptor complexes are subsequently internalized, undergoing retrograde vesicular transport to the cell body where they trigger pro-survival responses that maintain the health of the striatal neurons. Lack of proper BDNF-stimulated signaling has been implicated in several psychiatric disorders including depression, addiction, schizophrenia, and the childhood disorder known as Rhett syndrome.

The TrkB and $p75^{NTR}$ receptors act in opposition to one another. TrkB, a high-affinity BDNF receptor, activates a number of transcription factors that support neural survival and continued synaptic plasticity. In contrast, signaling via $p75^{NTR}$, a low-affinity BDNF receptor, negatively regulates synaptic plasticity and dendritic spike morphology. Under normal conditions there is a careful balance between these actions. That balance is disturbed by mutant huntingtin. This is accomplished in two major ways. First, wild-type huntingtin acts at the transcriptional level to promote transcription of BDNF. It accomplishes this by sequestering an inhibitory element called REST/NRSF (repressor element-1 transcription factor/neuron restrictive silencer factor) in the cytoplasm, thereby freeing the BDNF promoter from this inhibitory complex. Mutant huntingtin is defective in this action resulting in a BDNF deficit.

Secondly, mutant huntingtin acts at the transport level in both presynaptic and postsynaptic neurons to disrupt the TrkB/$p75^{NTR}$ balance. It carries out this second set of actions by inhibiting BDNF and TrkB transport. In its role as scaffolding protein wild-type huntingtin facilitates transport of BDNF along microtubules. As noted in the preceding section, mutant huntingtin, in contrast, impedes proper movement along the rail system. In addition to its facilitation of anterograde transport of BDNF ligands, huntingtin mediates retrograde transport of TrkB receptors. Here again, mutant huntingtin fails to perform its normal functions and, instead, reduces the efficiency of the transport system.

10.8   At the Synapse: Htt and NMDA Receptors

N-Methyl-D-aspartate (NMDA) receptors are the primary CNS receptors for the excitatory neurotransmitter glutamate. To recap, these receptors are dual voltage and neurotransmitter gated. Membrane depolarization relives an Mg^{2+} channel block, thereby enabling subsequent glutamate binding to open the receptor channel and enable Ca^{2+} ions to enter the neuron. Activation of these receptors initiates a host of protective effects in the host neurons through Ca^{2+} mediation of synaptic plasticity, gene transcription, and pro-survival activity (Fig. 10.5). Paradoxically, strong activation of these receptors has deleterious effects though Ca^{2+} driven anti-survival, apoptotic, and gene transcription actions. The term *excitotoxicity* was coined almost a half a century ago by James Olney to describe the toxic effects exerted throughout the brain by excessive calcium entry resulting from glutamatergic hyperactivation of NMDA receptors. The deleterious effects of excessive calcium entry was discussed with regard to both Alzheimer's disease and Parkinson's disease in the preceding chapters.

Recall from Chap. 8 that there are two distinct sets of NMDA receptors—one set at synapses and the other at extrasynaptic sites. The former set acts in a protective, anti-apoptotic capacity, while the latter supports cell death. The differences between the two derive from their activation of opposing cellular signaling pathways and alternate gene expression programs. Both of these regulatory routes are influenced by mHtt. Synaptic NMDA receptors activate TRiC, which by promoting mHtt aggregation prevents the more dangerous monomeric and small oligomer forms

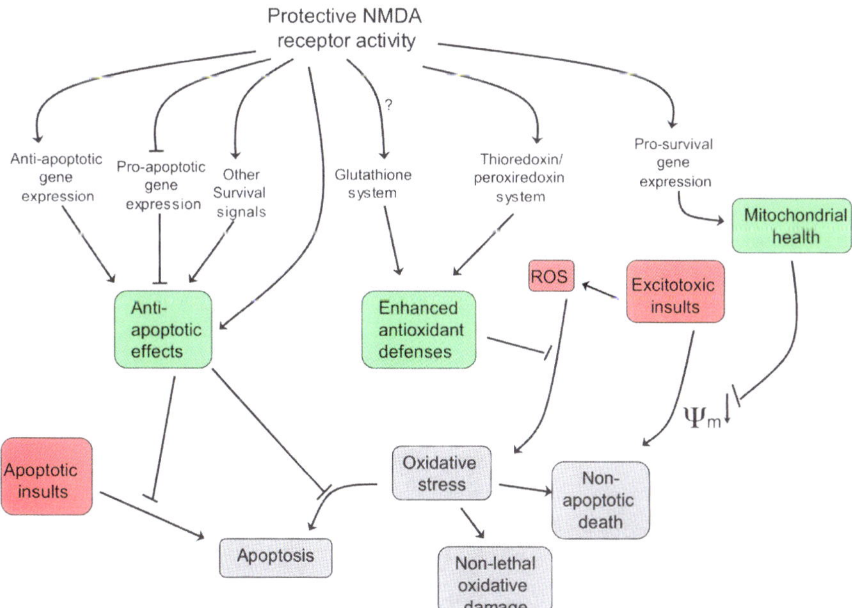

Fig. 10.5 Health-promoting responses to NMDA receptor activation (from Hardingham *Nat. Rev. Neurol.* 11: 682 © 2010 Reprinted by permission from Macmillan Publishers Ltd)

from inhibiting CRE/CREB/PGC-1α gene expression (discussed further in the next section). The extrasynaptic NMDA receptors, in contrast, act through a different agent, the small GTPase known as Ras homolog enriched in striatum (Rhes). The Rhes GTPase does the opposite. It deaggregates mHtt clusters, thereby releasing mHtt monomers and oligomers to inhibit the expression of the protective genes.

10.9   Impairment of Mitochondrial Biogenesis by mHtt

The importance of maintaining mitochondrial health was stressed in the earlier discussions of Parkinson's disease. This theme, in fact, has become a central one in discussions of neurodegeneration, and Huntington's disease is no exception. To see why this is so, consider for a moment brain energetics. The brain accounts for just 2 % of the body's mass yet it consumes 20 % of the oxygen supplied to the resting body. Most of this energy is used for synaptic transmission. It is needed primarily to operate the ion pumps (chiefly Na^+/K^+-ATPases and Ca^{2+}-ATPases) that restore Na^+, K^+, and Ca^{2+} ion concentrations to their resting values and repolarize the membranes following action potential generation and neurotransmitter release at the presynaptic terminal, and NMDA/other ion channel openings on the postsynaptic side. The numbers tell the story. For example, during action potential generation approximately 1×10^9 Na^+ ions enter the neuron requiring the hydrolysis of roughly 4×10^8 ATP molecules to pump them out. For a mean action potential rate of 4 Hz, a total of 3×10^9 ATP molecules per neuron per second need to be supplied by the neuronal metabolic machinery, mostly through oxidative phosphorylation.

To meet this requirement mitochondria are transported and positioned in active presynaptic and postsynaptic terminals. Mitochondrial localization at active synapses is orchestrated by the transient Ca^{2+} influx associated with NMDA receptor activity and voltage-gated calcium channel (VGCC) openings. These signals are sensed by Miro, a mitochondrial GTPase situated in the OMM. Miro possesses a pair of EF hands, modular calcium-binding helix-loop-helix motifs. The arrival of Ca^{2+} ions at Miro is a stop signal and once received Miro working together with another protein, Milton, arrest movement at active axonal and dendritic terminals, thereby enabling the mitochondria to supply energy and maintain calcium homeostasis.

PGC-1α (peroxisome proliferator-activated receptor γ co-activator 1) is a master regulator of mitochondrial gene expression. It accomplishes this by functioning as a platform for the recruitment of transcription factors and cofactors. This protein is highly modular enabling it to bind transcription activators, repressors and RNA. In this capacity, it coordinates the induction of nuclear regulatory factors 1 and 2 (NRF1 and NRF2). Recall that complexes I, III, IV and V of the electron transport chain are encoded by nuclear and mitochondrial genes. In particular, the complex IV (cytochrome c oxidase, COX) holoenzyme is encoded by ten nuclear genes and three mitochondrial genes. NRF1 and NRF2 control the expression of the ten nuclear COX subunits genes, the TFAM (mitochondrial transcription factor) gene, and the transcription factor B (TFB) gene. The latter two sets of factors, in turn, handle the expression of the three mitochondrial COX subunit genes.

These actions are not the only ones carried out by PGC-1α. As noted briefly in Chap. 6, it receives and responds to energy deprivation signals conveyed by AMPK and SIRT1. The former is triggered by elevations in the AMP/ATP ratio and the latter by elevations in the NAD$^+$/NADH ratio. In addition, PGC-1α plays a role in the regulation of ROS levels. It forms transcription activating complexes with CREB, TBP, and associated cofactors TAF4 and TF$_{II}$D at PGC-1α promoters where it helps transcribe genes encoding the ROS handling enzymes superoxide dismutase 1 and 2 (SOD1, 2), and glutathione peroxidase and the adaptive thermogenesis mediator uncoupling protein 1 (UCP-1).

The inhibition of transcription by mHtt extends beyond sequestration. Mutant huntingtin directly inhibits transcription by interfering with the transcription machinery at promoter sites in the nucleus. It targets transcription factors PGC-1α and Sp1 (a regulator of multiple cellular processes), and the transcription coactivator TAF$_{II}$130. In these interactions, the polyQ stretches contribute along with other portions of the mutant protein to alter the gene expression programs normally carried out by these proteins. In addition, mutant huntingtin localizes to mitochondria where it binds to component of the TIM, the translocase responsible for passage of proteins through the IMM and into the matrix. By binding and blocking the TIM, mutant huntingtin prevents protein import, thereby contributing to mitochondrial dysfunction in this additional way.

10.10   The Spinocerebellar Ataxias

10.10.1   Spinocerebellar Ataxias

The autosomal dominant, spinocerebellar ataxias (SCAs) are characterized at the clinical level by unsteady gait, clumsiness, and difficulty in speech. These signs are often accompanied by problems in vision or by mental retardation. Like the other neurological disorders the affected cells die off. In these instances, the main neural populations being impacted reside in the cerebellum and brainstem, with the cerebellar Purkinje cell being the most often targeted cell type. The first of these disorders to be discovered, SCA1, was identified by Huda Zoghbi and Harry Orr in 1993. Since then more than 40 spinocerebellar ataxias have been described. These disorders can be placed according to the nature of the mutation into three groups. These are:

- Protein-coding CAG (polyQ) expanded repeat SCAs;
- Non-protein-coding expanded repeat SCAs, and
- Conventional mutation-associated SCAs (missense, duplications, etc.).

The Purkinje cell, the primary cell of the cerebellum, is responsible for motor coordination. It integrates a large number of excitatory input signals from other regions of the brain along with signals from inhibitory interneurons. It is the sole output unit from the cortical region of the cerebellum. It sends inhibitory GABAergic output signals to neurons in cerebellar nuclei, which, in turn, relay the signals to multiple circuits situated upstream and downstream. The Purkinje cell has a strikingly unusual morphology. In order to integrate a large number of input signals, the Purkinje cell possesses a large dendritic arbor covered with numerous dendritic spines. These receive weak excitatory input from parallel fibers of cerebellar granule neurons and climbing fibers of inferior olive neurons.

10.10.2 Dark Degeneration of Purkinje Cells

Purkinje cells are supported by an extensive and highly expressed calcium handling machinery. This machinery includes an array of receptors and calcium channels, calcium-sensitive protein kinases and phosphatases, and calcium buffers. The importance of this system is provided by studies of SCA2. In this ataxias, arrival of glutamate at AMPA and metabotropic glutamate receptors (mGluRs) results in membrane depolarization. AMPA receptor activation leads to the opening of nearly voltage-gated calcium channels resulting in entry of Ca^{2+} into the cell. mGluR activation stimulates Ca^{2+} release from intracellular stores via the signaling intermediate $InsP_3$ that binds $InsP_3Rs$ in the ER. These combined effects produce a large transient increase in the intracellular calcium concentration. Aberrations in calcium handling has been postulated as playing a major role in *dark degeneration*, a type of excitotoxic cell death of Purkinje cells observed in SCA2 and many other SCAs. This form of cell death exhibits features of both apoptosis and necrosis.

Additional insight into dark degeneration of Purkinje cells is provided by studies of spinocerebellar ataxia 28 (SCA28). This ataxia arises as a consequence of a missense mutation in the gene that encodes AFG_3L_2. This nuclear-encoded mitochondrial protein, together with another protein, paraplegin, forms heterodimeric AFG_3L_2/paraplegin and homodimeric AFG_3L_2 complexes in the inner mitochondrial membrane. These complexes, referred to as the m-AAA proteases, maintain mitochondrial protein quality control. They degrade matrix proteins and also carry out chaperone activities by assisting in the assembly of electron transport chain complexes. Mutations in SPG7, the gene that encodes paraplegin cause a recessive form of the neurological disorder hereditary spastic paraplegia, while mutations in AFG_3L_2 cause SCA28 and another disorder, spastic ataxia neuropathy syndrome. Purkinje cells in which there is a loss of AFG_3L_2 or paraplegin function due to mutations exhibit mitochondrial fragmentation, and loss of proper contacts with the endoplasmic reticulum. Deficits that appear in these diseases include reduced mitochondrial protein synthesis arising from impaired ribosome assembly, axonal transport deficits, loss of mitochondrial protein quality control and maintenance of calcium homeostasis.

10.11   Repeat-Mediated RNA Toxicity

Repeat instabilities resulting in increases in repeat length from generation to generation also occur in noncoding regions such as 5′ UTRs, introns, and 3′ UTRs. Rather than producing toxic protein monomers, oligomers, and aggregates, these expanded repeats generate toxic RNAs. The neurodegenerative diseases generated by these unstable repeats are summarized in Table 10.2. As can be seen, they encompass a variety of disorders. Myotonic dystrophy 1 is the most prominent autosomal dominant muscular dystrophy. Its worldwide frequency is estimated at 1 in 20,000. In this disorder, muscles are unable to relax at will, and undergo weakness, shrinkage, and degeneration. This disorder is also known myotonic muscular dystrophy and as Steinert disease. It is caused by an unstable CTG repeat in the 3′ UTR of the DMPK gene in chromosome 19. A rarer form of this disease, myotonic dystrophy 2 (DM2) involves an unstable CCGT repeat in intron 1 of the ZNF9 gene located on chromosome 3.

Table 10.2 Neurological disorders arising from unstable repeats situated in noncoding regions

Disease	Repeat location	Repeat	Normal repeat length	Disease repeat length
Myotonic dystrophy 1 (DM1)	3′ UTR	CTG	5–37	~50–>2000
Myotonic dystrophy 2 (DM2)	Intron 1	CCTG	<27	75–11,000
FXTAS	5′ UTR	CGG	20–45	55–200
Fragile X syndrome (FXS, FRAXA)	5′ UTR	CGG	6–52	~55–>2000
FRAXE mental retardation	5′end	CGG	6–25	>200
FRA12A mental retardation	5′ UTR	CGG	6–23	?
Friedreich's ataxia (FRDA)	Intron 1	GAA	7–22	>66–>900
Spinocerebellar ataxia 10 (SCA10)	3′ UTR/Intron 9	ATTCT	10–29	280–4500
Spinocerebellar ataxia 12 (SCA12)	5′ UTR/promoter	CAG	<32	51–78
Huntington disease-like 2 (HDL2)	3′ UTR	CTG	6–28	>41

Abbreviations: *FXTAS* fragile X-associated tremor and ataxia syndrome

10.12   Sequestration and Spliceopathies

Studies of the neurological disorders listed in Table 10.2 have led to a simple model of RNA toxicity that seems to account for many of the observed molecular effects. In this picture, referring to DM1 for simplicity, two distinct actions combine to elicit the disease. First, the presence of an expanded nucleotide repeat RNA gives rise to a greatly expanded hairpin structure consisting of a base, stem, and terminal loop that attracts and sequesters numerous proteins in structures called nuclear foci and prevents them from carrying out their cellular tasks. This aspect, sequestration, is the heart of the model, and prominent among the isolated and inactive proteins are RNA splicing factors and transcription factors. In the case of DM1, the splicing factor muscleblind (MBNL) is sequestered.

Secondly, key signaling pathways may become misregulated and inappropriately activated or inactivated. Again, referring to DM1, another key splicing factor, CELF1, which acts in opposition to MBNL is excessively upregulated (as a result of inappropriate hyperphosphorylation by one or more protein kinases). The net result of these two actions (too little MBNL and too much CELF1) is that the wrong spicing isoform is generated resulting in an altered protein conformation and function. From this perspective the disease may be thought of as a spliceopathy. A schematic representation of how sequestration generates a spliceopathy is presented in Fig. 10.6.

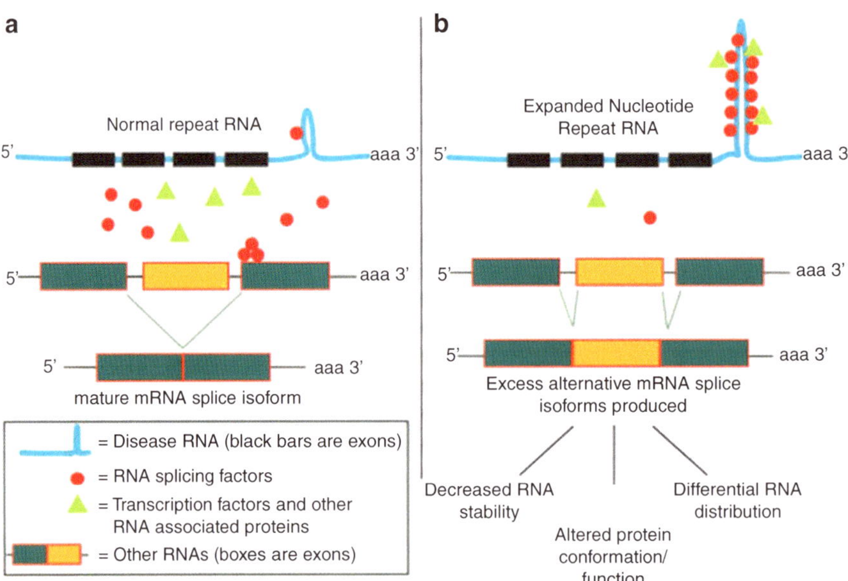

Fig. 10.6 Schematic depiction of the sequestration model of RNA-driven neurological disorders (from Todd *Ann. Neurol.* 67: 291 © 2010 Reprinted by permission from John Wiley and Sons)

10.13 Fragile X Syndrome, R-Loops, and Gene Silencing

During metaphase, chromosomes are highly compacted and can be stained, visualized and studied more easily than at other phases in the cell cycle. When metaphase chromosomes are subjected to replicative stresses brought on, for example, by treatment with agents that partially inhibit DNA synthesis, certain well-defined sites fail to stain and gaps or breaks can be observed. These loci are referred to as *fragile sites*. They are not abnormal occurrences, but instead are found in all individuals and are cell-type specific parts of normal chromosome structure. There are two classes of fragile sites. *Rare fragile sites* are associated with CGG repeat expansions. The most prominent of these, the CGG repeat expansion in the 5′ UTR adjacent to the promoter for the fragile X mental retardation 1 (FMR1) gene located on the X chromosome, gives rise to fragile X syndrome (FXS, FRAXA). Other fragile sites are associated with AT-rich repeats; these are far more frequent in their occurrence and are termed *common fragile sites*. The most prominent of these are encountered in tumors where they become unstable and cause breakdowns in replication fork movement, cell cycle checkpointing, and DNA repair.

Fragile X syndrome causes multiple developmental disabilities; it is the leading cause of mental retardation, and is a major cause of genetically linked autism. It occurs with a frequency of 1 in 4000 for males and 1 in 8000 for females. The FMR1 gene encodes a protein, FMRP, needed for proper synapse development. Individual with intermediate number of repeats, in the 55–200 range, have what is termed a permutation of the FMR1 gene, and exhibit milder forms of the disorder; they have learning disabilities or show signs of autism, or suffer from fragile X-associated tremor and ataxia syndrome (FXTAS).

The FMR1 protein is an mRNA binding protein that shuttles mRNAs to synaptic sites and once there assists in their local translation in response to signals from mGluRs and other agents. Its loss of function leads to synaptic dysfunction and fragile X syndrome. In contrast to the gain-of-toxic function through sequestration model that occurs in the case of myotonic dystrophy, loss-of-function resulting from gene silencing is the central mechanism underlying Fragile X syndrome. This silencing occurs as a consequence of the creation of R-loops.

R-loops are RNA–DNA hybrid structures. They are created when a nascent RNA strand remains attached to its DNA template leaving the non-template DNA strand unattached. These structures can be highly stable and are induced by the expanded CGG repeats (Fig. 10.7). The R-loops impeded transcription in two ways—(1) through their mechanical inhibition of RNA polymerase II, and (2) through their recruitment of enzymes that catalyze the placement of chromatin-condensing histone marks.

Recall that the amino terminals of the core histones extend out from nucleosomes to form histone tails. These tails provide sites for attachment of regulatory groups that influence transcription and are referred to as *histone marks*. The most prominent of these *epigenetic modifications* are methyl groups and acetyl groups that attach to the large number of lysine residues and to lesser extent arginine residues present on histone tails. Attachment of acetyl groups decompresses chromatin. It achieves this by reducing the net positive charge on the tails,

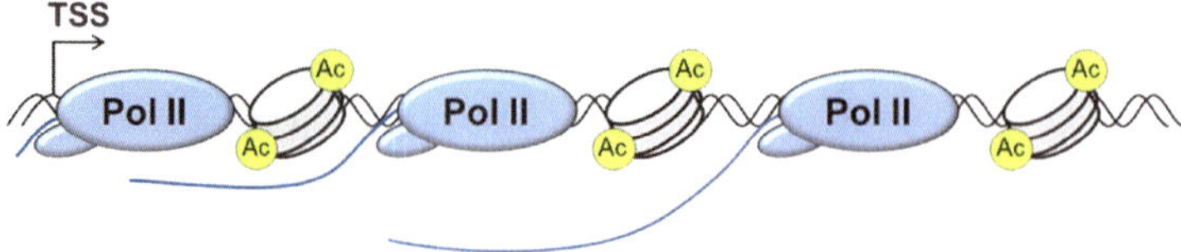

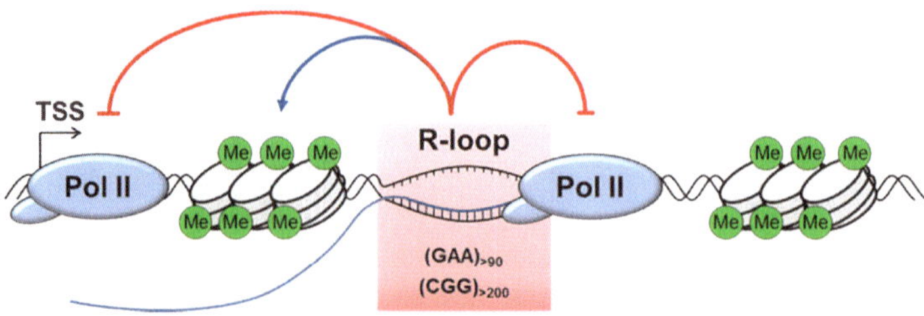

Fig. 10.7 R-loops and gene silencing. Histone marks—*Ac* acetylation, *Me* methylation (from Groh *PLoS Genetics* 10: e1004318 © 2014 reprinted by permission of the authors)

thereby weakening their attraction to the negatively charged DNA strands. Addition of methyl groups can either encourage or repress transcription depending on the specific lysines being targeted and the degree of methylation (mono, di or tri). These posttranslational modifications do not alter the net charge but rather establish sites for attachment of chromatic condensing or decondensing proteins and protein complexes. In the case of fragile X syndrome, chromatin-condensing H3K9me2 (histone H3, lysine 9, di-methylated) marks are added.

10.14   Summary

1. An enormous amount of progress has been made since the first discoveries of the genetic causes of the unstable repeat disorders. It was less than 25 years ago that the gene that encodes the androgen receptor was discovered by La Spada. That event was followed shortly thereafter by the discovery of the gene that causes fragile X syndrome and the gene responsible for Huntington's disease. The discovery of disease-causing genes possessing unstable repeats answered a pervasive mystery—why did the associated disorders such as Huntington's disease and fragile X syndrome increase in severity, and decrease in age of onset, from generation to generation? The answer is that the number of repeats increases in successive generations, and this increase underlies the genetic anticipation phenomenon.

The question of how they do so has also been largely answered. The key observation is that the repeats increase in number because they produce aberrant secondary structures such as slipouts, loops and hairpins. These anomalous structures, built from DNA, DNA–RNA hybrids, or RNA, generate kinetic traps and cause replication forks to slip and stall. As a result the aberrant structures cause additional repeats to be inserted during replication, repair, recombination, and transcription. Once the number of repeats passes a threshold, they alter the resulting DNA and RNA sequences and downstream processing steps sufficiently to produce malformed or inappropriate proteins that induce neurodegenerative diseases.

2. Huntington's disease is caused by an expanded CAG repeat in exon 1 of the HTT gene. Generally speaking, if the number of glutamate repeats in HTT is 35 or less the individual is not susceptible to HD. There is some risk to individuals possessing 36–39 glutamate repeats in HTT. In those situations, the disease will be late-onset and will advance slowly over time. In instances where HTT has 40 or more repeats, the disease onset will occur at a younger age and progress more rapidly, with the severity depending on the number of repeats and other cellular factors.

   Huntington's disease is one of eight (or nine) other polyQ disorders that have been uncovered to date. Although the mutated proteins differ from one another in these disorders, many of them are involved in transcription. That observation led Max Perutz in 1994 to hypothesize that transcription factors and other proteins containing polyQ stretches would bind to each other by means of a polar zipper mechanism. In this binding process, the polyQ chains form β-sheets held together by hydrogen bonds between main-chain and side-chain amides. The seeds grow and eventually form insoluble amorphous aggregates or highly ordered amyloid fibrils, or amalgams of the two.

   One of the most pervasive features of the polyQ disorders is the presence of N-terminal fragments of the mutated proteins. Fragments of huntingtin, the androgen receptor, atrophin-1, ataxin-3, and ataxin-7 have each been found. In the case of mHtt, these fragments are typically generated through proteolytic cleavage of the full-length proteins by caspases and calpains. That the latter proteases are involved is particularly interesting because these proteins are activated by Ca^{2+} signals that play an important role in synaptic function. Lastly, faulty splicing of Htt exon 1 has been found to produce small, highly toxic N-terminal fragments.

   Exon 1 located in the extreme N-terminal encodes a 17 residue Htt^{NT} sequence followed by the expanded polyQ repeats and a 38-residue proline-rich stretch (polyP). Early studies established that expression exon 1 alone was sufficient to recapitulate most of the features of Huntington's disease. Both the Htt^{NT} and polyP influence the folding, aggregation and toxicity of the expanded polyQ repeats. The extent of their influences varies depending on polyQ tract length, modifications to the sequences, and cellular context. In general, the presence of the Htt^{NT} and polyP (1) decrease the solubility of polyQ fragments, (2) destabilize non-fibrillar aggregates and intermediates and (3) accelerate fibril formation. By these means the two flanking sequences decrease the toxicity of the polyQs.

In addition to proteolytic cleavage, huntingtin is subject to phosphorylation, acetylation, palmitoylation, ubiquitination, and SUMOylation. These posttranslational modifications couple cellular signaling and regulatory pathways to huntingtin, thereby rendering its subcellular localization, clearance, and processing responsive to cellular needs. Several clusters of modification sites are present in Htt. One of these is located in Htt^{NT}. That short stretch contains phosphorylation sites at serine 13 and serine 16 and competitive ubiquitination—SUMOylation sites at lysine 6, lysine 9, and lysine 15. As is often the case, ubiquitination targets the protein for proteasomal clearance, while SUMOylation stabilizes the protein and reduces aggregation. Concerted actions by these enzymes take place with phosphorylation preparing the adjacent sites for subsequent Ub/SUMO modifications. By these and similar actions the cell exerts control over huntingtin's stability, clearance, subcellular localization, aggregation, fibrillization, and toxicity.

3. As is the case for Alzheimer's and Parkinson's diseases, loss of protein quality control is central to the generation of Huntington's disease and other expanded repeat disorders. In these disorders, there is a convergence of (1) declining protein quality control with age and (2) the continued presence of malformed proteins that expose hydrophobic surfaces and exhibit a propensity to form soluble monomers and oligomers, and large insoluble aggregates. As a consequence there is a buildup of potentially toxic non-conformers. This failure of clearance, made worse by the mutations, is largely responsible for the diseases.

   Alternatively folded mHtt proteins and protein fragments are difficult-to-impossible to fold into functional conformations. As a consequence, the chaperone network, most notably Hsp70 assisted by the small heat shock proteins and Hsp40 co-chaperone family members, directs these proteins and others like them to the ubiquitin–proteasome system (UPS) and into autophagic-lysosomal pathway for degradation and clearance. The chaperones also promote the formation of the large aggregates and inclusions where the misfolded proteins are the least likely to cause cellular damage. As the imbalance between buildup and clearance of unwanted proteins worsens, the clearance system is overwhelmed by material; other proteins that require clearance are not handled in a timely fashion, and a gradual cascading failure of protein quality control ensues.

4. One of the most striking features of the huntingtin protein is its possession of multiple tandem-repeated structures, each composed of ~47 amino acid residues, known as HEAT repeats. These are rod-like helical structures with a known role in vesicular transport. Htt possesses several dozen of these; these are organized into three clusters, two in its N-terminal domain and one in its C-terminal region. Another feature of huntingtin that provides a clue as to its normal function is its association with lipid membranes. The N-terminal α-helical segment anchors the Htt fragments to lipid membranes and by that means facilitates the formation of close associations between fragments. These aggregation-promoting activities are enhanced by the presence of acidic phospholipids in the membranes.

Close lipid–protein associations are not limited to huntingtin. Rather, the enumeration of disease-causing proteins that associate with lipid membranes forms a veritable who's who list of disease-causing agents. That list includes the prion protein, Aβ peptide, α-synuclein, and huntingtin as well as transthyretin (familial polyneuropathy and systemic amyloidosis), lysozyme (systemic amyloidosis), and islet amyloid polypeptide (type II diabetes).

Mutant huntingtin negatively impacts several cellular functions crucial for synaptic operation:

- In its normal wt form huntingtin functions as a scaffolding protein that facilitates the attachment of cargo to the microtubule-based kinesin and myosin motor proteins. In this role, it mediates the transport of organelles and signaling proteins such as BDNF and its receptor TrkB to and from synaptic terminals. Mutant huntingtin, in contrast, interferes with the transport machinery, thereby causing deficits that can lead to synaptic failure.
- There is a careful balance between synaptic and extrasynaptic NMDA signaling. The former promote cellular homeostasis and survival, while the latter directs the neuron towards an apoptotic fate. Mutant huntingtin interferes with this balance shifting it in the direction of cell death.
- Synaptic transmission is an energy intensive process. In order to meet these requirements mitochondria are positioned in active synaptic terminals and supply ATP through oxidative phosphorylation. Mutant huntingtin interferes with mitochondrial biogenesis. It does so by binding to transcription factors and cofactors at promoter sites for PGC-1α, a master regulator of mitochondrial gene expression and biogenesis. As result energy production is reduced and calcium homeostasis is disturbed.

5. Expanded repeats occurring in non-protein-coding DNA generate a variety of neurological disorders including monotonic dystrophy 1 and 2 and fragile X syndrome. In the case of monotonic dystrophy 1 and 2, expanded nucleotide repeat RNA forms hairpin-like structures leading to splicing errors. This happens because the protuberances recruit and sequester splicing and transcription factors in nuclear foci thus preventing them from carrying out their cellular tasks. These sequestration actions, along with inappropriate activation of other splicing factors, generate alternative splicing errors resulting in altered protein conformations and functions.

Fragile X syndrome is caused by an unstable CGG repeat located in the 5′ UTR adjacent to the FMR1 gene promoter resulting in epigenetic gene silencing. In this case, aberrant DNA–RNA hybrids called R-loops form and these structures carry out two actions. First, they impede RNA polymerase II. Secondly, the structures recruit enzymes that append chromatin-condensing H3K9me2 marks to the histone tails and these posttranslational modifications silence gene transcription.

Bibliography

DiFiglia, M., Sapp, E., Chase, K. O., Davies, S. W., Bates, G. P., Vonsattel, J. P., et al. (1997). Aggregation of huntingtin in neuronal intranuclear inclusions and dystrophic neurites in brain. *Science, 277*, 1990–1993.

Fu, Y. H., Kuhl, D. P. A., Pizzuti, A., Pieretti, M., Sutcliffe, J. S., Richards, S., et al. (1991). Variation of the CGG repeat at the fragile X site results in genetic instability: Resolution of the Sherman paradox. *Cell, 67*, 1047–1058.

La Spada, A. R., Wilson, E. M., Lubahn, D. B., Harding, A. E., & Fischbeck, K. H. (1991). Androgen receptor gene mutations in X-linked spinal and bulbar muscular atrophy. *Nature, 362*, 77–79.

Verkerk, A. J., Pieretti, M., Sutcliffe, J. S., Fu, Y. H., Kuhl, D. P. A., Pizzuti, A., et al. (1991). Identification of a gene (FMR-1) containing a CGG repeat coincident with a breakpoint cluster region exhibiting length variation in fragile X syndrome. *Cell, 65*, 905–914.

Unusual Secondary Structures

Sandou, F., Finkbeiner, S., Devys, D., & Greenberg, M. E. (1998). Huntingtin acts in the nucleus to induce apoptosis but death does not correlate with the formation of intranuclear inclusions. *Cell, 95*, 55–66.

Unstable Repeats

Kovtun, I. V., & McMurray, C. T. (2008). Features of trinucleotide repeat instability in vivo. *Cell Research, 18*, 198–213.

López Castel, A., Cleary, J. D., & Pearson, C. E. (2010). Repeat instability as the basis for human diseases and as a potential target for therapy. *Nature Reviews. Molecular Cell Biology, 11*, 165–170.

Mirkin, S. M. (2007). Expandable DNA repeats and human disease. *Nature, 447*, 932–940.

Perutz, M. F., Johnson, T., Suzuki, M., & Finch, J. T. (1994). Glutamine repeats as polar zippers: Their possible role in inherited neurodegenerative diseases. *Proceedings of the National Academy of Sciences of the United States of America, 91*, 5355–5358.

Structure of the Huntingtin Protein

Abedini, A., & Raleigh, D. P. (2009). A role for helical intermediates in amyloid formation by natively unfolded polypeptides? *Physical Biology, 6*, 015005.

Bhattacharyya, A., Thakur, A. K., Chellgren, V. M., Thiagarajan, G., Williams, A. D., Chellgren, B. W., et al. (2006). Oligoproline effects on polyglutamine conformation and aggregation. *Journal of Molecular Biology, 355*, 524–535.

Darnell, G., Orgel, J. P. R. O., Pahl, R., & Meredith, S. C. (2007). Flanking polyproline sequences inhibit β-sheet structure in polyglutamine segments by inducing PPII-like helix structure. *Journal of Molecular Biology, 374*, 688–704.

Duennwald, M. L., Jagadish, S., Muchowski, P. J., & Lindquist, S. (2006). Flanking sequences profoundly alter polyglutamine toxicity in yeast. *Proceedings of the National Academy of Sciences of the United States of America, 1033*, 11045–11050.

Cleavage and Aberrant Splicing

Gafni, J., Hermel, E., Young, J. E., Wellington, C. L., Hayden, M. R., & Ellerby, L. M. (2004). Inhibition of calpain cleavage of Huntingtin reduces toxicity: accumulation of calpain/caspase fragments in the nucleus. *The Journal of Biological Chemistry, 279*, 20211–20220.

Graham, R. K., Deng, Y., Slow, E. J., Haigh, B., Bissada, N., Lu, G., et al. (2006). Cleavage at the caspase-6 site is required for neuronal dysfunction and degeneration due to mutant Huntingtin. *Cell, 125*, 1179–1191.

Sathasivam, K., Neueder, A., Gipson, T. A., Landles, C., Benjamin, A. C., Bondulich, M. K., et al. (2013). Aberrant splicing of HTT generates the pathogenic exon 1 protein in Huntington disease. *Proceedings of the National Academy of Sciences of the United States of America, 110*, 2366–2370.

Wellington, C. L., Ellerby, L. M., Hackam, A. S., Margolis, R. L., Trifiro, M. A., Singaraja, R., et al. (1998). Caspase cleavage of gene products associated with triplet expansion disorders generates truncated fragments containing the polyglutamine tract. *The Journal of Biological Chemistry, 273*, 9158–9167.

Wellington, C. L., Ellerby, L. M., Gutekunst, C. A., Rogers, D., Warby, S., Graham, R. K., et al. (2002). Caspase cleavage of mutant Huntingtin precedes neurodegeneration in Huntington's disease. *The Journal of Neuroscience, 22*, 7862–7872.

Posttranslational Modifications

Gu, X., Greiner, E. R., Mishra, R., Kodali, R., Osmand, A., Finkbeiner, S., et al. (2009). Serines 13 and 16 are critical determinants of full-length human mutant huntingtin induced disease pathogenesis in HD mice. *Neuron, 64*, 828–840.

Jeong, H., Then, F., Melia, T. J., Jr., Mazzulli, J. R., Cui, L., Savas, J. N., et al. (2009). Acetylation targets mutant huntingtin to autophagosomes for degradation. *Cell, 137*, 60–72.

Steffan, J. S., Agrawal, N., Pallos, J., Rockabrand, E., Trotman, L. C., Slepko, N., et al. (2004). SUMO modification of Huntingtin and Huntington's disease pathology. *Science, 304*, 100–104.

Thompson, L. M., Aiken, C. T., Kaltenbach, L. S., Agrawal, N., Illes, K., Khoshnan, A., et al. (2009). IKK phosphorylates huntingtin and targets it for degradation by the proteasome and lysosome. *The Journal of Cell Biology, 187*, 1083–1099.

HEAT Repeats

Andrade, M. A., & Bork, P. (1995). HEAT repeats in the Huntington's disease protein. *Nature Genetics, 11*, 115–116.

Andrade, M. A., Petosa, C., O'Donoghue, S. I., Müller, C. W., & Bok, P. (2001). Comparison of ARM and HEAT protein repeats. *Journal of Molecular Biology, 309*, 1–18.

Exon 1

Crick, S. L., Ruff, K. M., Garai, K., Frieden, C., & Pappu, R. V. (2013). Unmasking the roles of N- and C-terminal flanking sequences from exon 1 of huntingtin as modulators of polyglutamine aggregation. *Proceedings of the National Academy of Sciences of the United States of America, 110*, 20075–20080.

Kar, K., Jayaraman, M., Sahoo, B., Kodali, R., & Wetzel, R. (2011). Critical nucleus size for disease-related polyglutamine aggregation is repeat length dependent. *Nature Structural & Molecular Biology, 18*, 328–336.

Landles, C., Sathasivam, K., Weiss, A., Woodman, B., Moffitt, H., Finkbeiner, S., et al. (2010). Proteolysis of mutant huntingtin produces an exon 1 fragment that accumulates as an aggregated protein in neuronal nuclei in Huntington disease. *The Journal of Biological Chemistry, 285*, 8808–8823.

Nekooki-Machida, Y., Kurosawa, M., Nukina, N., Ito, K., Oda, T., & Tanaka, M. (2009). Distinct conformations of in vitro and in vivo amyloids of huntingtin-exon1 show different cytotoxicity. *Proceedings of the National Academy of Sciences of the United States of America, 106*, 9679–9684.

Thakur, A. K., Jayaraman, M., Mishra, R., Thakur, M., Chellgren, V. M., Byeon, I. J., et al. (2009). Polyglutamine disruption of the huntingtin exon1 Nterminus triggers a complex aggregation mechanism. *Nature Structural & Molecular Biology, 16*, 380–389.

Htt Misfolding, Oligomerization and Aggregation

Legleiter, J., Mitchell, E., Lotz, G. P., Sapp, E., Ng, C., DiFiglia, M., et al. (2010). Mutant huntingtin fragments form oligomers in a polyglutamine length-dependent manner in vitro and in vivo. *The Journal of Biological Chemistry, 285*, 14777–14790.

Vitalis, A., & Pappu, R. V. (2011). Assessing the contribution of heterogeneous distributions of oligomers to aggregation mechanisms of polyglutamine peptides. *Biophysical Chemistry, 159*, 14–23.

Wacker, J. L., Zareie, M. H., Fong, H., Sarikaya, M., & Munchowski, P. J. (2004). Hsp70 and Hsp40 attenuate formation of spherical and annular polyglutamine oligomers by partitioning monomer. *Nature Structural & Molecular Biology, 11*, 1215–1222.

Waza, M., Adachi, H., Katsuno, M., Minamiyama, M., Sang, C., Tanaka, F., et al. (2005). 17-AGG, an Hsp90 inhibitor, ameliorates polyglutaminemediated motor neuron degeneration. *Nature Medicine, 11*, 1088–1095.

Molecular Chaperones

Behrends, C., Langer, C. A., Boteva, R., Böttcher, U. M., Stemp, M. J., Schaffer, G., et al. (2006). Chaperonin TRiC promotes the assembly of PolyQ expansion proteins in nontoxic oligomers. *Molecular Cell, 23*, 887–897.

Luo, W., Sun, W., Taldone, T., Rodina, A., & Chiosis, G. (2010). Heat shock protein 90 in neurodegenerative diseases. *Mol. Neurodegener. 5*: art. 24.

Månsson, C., Kakkar, V., Monsellier, E., Sourigues, Y., Härmark, J., Kampinga, H. H., et al. (2014). DNAJB6 is a peptide-binding chaperone which can suppress amyloid fibrillization of polyglutamine peptides at substoichiometric molar ratios. *Cell Stress & Chaperones, 19*, 227–239.

Månsson, C., Arosio, P., Hussein, R., Kampinga, H. H., Hashem, R. M., Boelens, W. C., et al. (2014). Interaction of the molecular chaperone DNAJB6 with growing amyloid-beta 42 (Aβ 42) aggregates leads to substoichiometric inhibition of amyloid formation. *The Journal of Biological Chemistry, 289*(45), 31066–31076.

Park, S. H., Kukushkin, Y., Gupta, R., Chen, T., Konagai, A., Hipp, M. S., et al. (2013). PolyQ proteins interfere with nuclear degradation of cytosolic proteins by sequestering the Sis1p chaperone. *Cell, 154*, 134–145.

Tam, S., Spiess, C., Auyeung, W., Joachimiak, L., Chen, B., Poirier, M. A., et al. (2009). The chaperonin TRiC blocks a huntingtin sequence element that promotes the conformational switch to aggregation. *Nature Structural & Molecular Biology, 16,* 1279–1285.

Clearance of mHtt

Arrasate, M., Mitra, S., Schweitzer, E. S., Segal, M. R., & Finkbeiner, S. (2004). Inclusion body formation reduces levels of mutant Huntingtin and the risk of neuronal death. *Nature, 431,* 805–810.

Bennett, E. J., Bence, N. F., Jayakumar, R., & Kopito, R. R. (2005). Global impairment of the ubiquitin-proteasome system by nuclear or cytoplasmic protein aggregates precedes inclusion body formation. *Molecular Cell, 17,* 351–365.

Bjørkøy, G., Lamark, T., Brech, A., Outzen, H., Perander, M., Øvervatn, A., et al. (2005b). p62/SQSTM1 forms protein aggregates degraded by autophagy and has a protective effect on huntingtin-induced cell death. *The Journal of Cell Biology, 171,* 603–614.

Gidalevitz, T., Ben-Zvi, A., Ho, K. H., Brignull, H. R., & Morimoto, R. I. (2006). Progressive disruption of cellular protein folding in models of polyglutamine diseases. *Science, 311,* 1471–1474.

Hipp, M. S., Patel, C. N., Bersuker, K., Riley, B. E., Kaiser, S. E., Shaler, T. A., et al. (2012). Indirect inhibition of 26S proteasome activity in a cellular model of Huntington's disease. *The Journal of Cell Biology, 196,* 573–587.

Holmberg, C. I., Staniszewski, K. E., Mensah, K. N., Matouschek, A., & Morimoto, R. I. (2004). Inefficient degradation of truncated polyglutamine proteins by the proteasome. *The EMBO Journal, 23,* 4307–4318.

Kirkin, V., McEwan, D. G., Novak, I., & Dikic, I. (2009). A role for ubiquitin in selective autophagy. *Molecular Cell, 34,* 259–269.

Martinez-Vicente, M., Talloczy, Z., Wong, E., Tang, G., Koga, H., Kaushik, S., et al. (2010). Cargo recognition failure is responsible for inefficient autophagy in Huntington's disease. *Nature Neuroscience, 13,* 567–576.

Ravikumar, B., Vacher, C., Berger, Z., Davies, J. E., Luo, S., Oroz, L. G., et al. (2004). Inhibition of mTOR induces autophagy and reduces toxicity of polyglutamine expansions in fly and mouse models of Huntington's disease. *Nature Genetics, 36,* 585–595.

Venkatraman, P., Wetzel, R., Tanaka, M., Nukina, N., & Goldberg, A. L. (2004). Eukaryotic proteasomes cannot digest polyglutamine sequences and release them during degradation of polyglutamine-containing proteins. *Molecular Cell, 14,* 95–104,

Wong, Y. C., & Holzbaur, E. L. F. (2014). The regulation of autophagosome dynamics by huntingtin and HAP1 is disrupted by expression of mutant huntingtin, leading to defective cargo degradation. *The Journal of Neuroscience, 34,* 1293–1305.

Huntingtin Membrane Localization and Function

Atwal, R. S., Xia, J., Pinchev, D., Taylor, J., Epand, R. M., & Truant, J. (2007). Huntingtin has a membrane association signal that can modulate huntingtin aggregation, nuclear entry and toxicity. *Human Molecular Genetics, 16,* 2600–2615.

Evangelisti, E., Cecchi, C., Cascella, R., Sgromo, C., Becatti, M., Dobson, C. M., et al. (2012). Membrane lipid composition and its physicochemical properties define cell vulnerability to aberrant protein oligomers. *Journal of Cell Science, 125,* 2416–2427.

Gorbenko, G. F., & Kinnunen, P. K. J. (2006). The role of lipid-protein interactions in amyloid-type protein fibril formation. *Chemistry and Physics of Lipids, 141,* 72–82.

Kegel, K., Kim, M., Sapp, E., McIntyre, C., Castaño, J. G., Aronin, N., et al. (2000). Huntingtin expression stimulates endosomal–lysosomal activity, endosome tubulation, and autophagy. *The Journal of Neuroscience, 20*, 7268–7278.

Rockabrand, E., Slepko, N., Pantalone, A., Nukala, V. N., Kazantsev, A., Marsh, J. L., et al. (2007). The first 17 amino acids of huntingtin modulate its sub-cellular localization and effects on calcium homeostasis. *Human Molecular Genetics, 16*, 61–77.

Axonal Transport of Mitochondria and mHtt

Chang, D. T. W., Honick, A. S., & Reynolds, I. J. (2006). Mitochondrial trafficking to synapses in cultured primary cortical neurons. *The Journal of Neuroscience, 26*, 7035–7045.

Gunawardena, S., Her, L. S., Brusch, R. G., Laymon, R. A., Niesman, I. R., Gordesky-Gold, B., et al. (2003). Disruption of axonal transport by loss of Huntingtin or expression of pathogenic PolyQ proteins in Drosophila. *Neuron, 40*, 25–40.

Nakamura, T., Tu, S., Akhtar, M. W., Sunico, C. R., Okamoto, S. I., & Lipton, S. A. (2013). Aberrant protein s-nitrosylation in neurodegenerative diseases. *Neuron, 78*, 596–614.

Orr, A. L., Li, S., Wang, C. E., Li, H., Wang, J., Rong, J., et al. (2008). N-terminal mutant huntingtin associates with mitochondria and impairs mitochondrial trafficking. *The Journal of Neuroscience, 28*, 2783–2792.

Vesicular Trafficking, BDNF and TrkB

Caviston, J. P., & Holzbauer, E. L. F. (2009). Huntingtin as an essential integrator of intracellular vesicular trafficking. *Trends in Cell Biology, 19*, 147–155.

Gauthier, L. R., Charrin, B. C., Borrell-Pagès, M., Dompierre, J. P., Rangone, H., Cordelières, F. P., et al. (2004). Huntingtin controls neurotrophic support and survival of neurons by enhancing BDNF vesicular transport along microtubules. *Cell, 118*, 127–138.

Liot, G., Zala, D., Pla, P., Mottet, G., Liel, M., & Sandou, F. (2013). Mutant huntingtin alters retrograde transport of TrkB receptors in striatal dendrites. *The Journal of Neuroscience, 33*, 6298–6309.

Yoshii, A., & Constantine-Paton, M. (2010). Postsynaptic BDNF-TrkB signaling in synapse maturation, plasticity and disease. *Developmental Neurobiology, 70*, 304–322.

Zuccato, C., Tartari, M., Crotti, A., Goffredo, D., Valenza, M., Conti, L., et al. (2003). Huntingtin interacts with REST/NRSF to modulate the transcription of NRSE-controlled neuronal genes. *Nature Genetics, 35*, 76–83.

At the Synapse: NMDA Receptors

Milnerwood, A. J., Gladding, C. M., Pouladi, M. A., Kaufman, A. M., Hines, R. M., Boyd, J. D., et al. (2010). Early increase in extrasynaptic NMDA receptor signaling and expression contributes to phenotype onset in Huntington's disease mice. *Neuron, 65*, 178–190.

Okamoto, S. I., Pouladi, M. A., Talantova, M., Yao, D., Xia, P., Ehrnhoefer, D. E., et al. (2009). Balance between synaptic versus extrasynaptic NMDA receptor activity influences inclusions and neurotoxicity of mutant huntingtin. *Nature Medicine, 15*, 1407–1413.

Papouin, T., Ladépêche, L., Ruel, J., Sacchi, S., Labasque, M., Hanini, M., et al. (2012). Synaptic and extrasynaptic NMDA receptors are gated by different endogenous coagonists. *Cell, 150,* 633–646.

Subramaniam, S., Sixt, K. M., Barrow, R., & Snyder, S. H. (2009). Rhes, a striatal specific protein, mediates mutant huntingtin cytotoxicity. *Science, 324,* 1327–1330.

Neuronal Energy Demands and Mitochondrial Trafficking

Attwell, D., & Laughlin, S. B. (2001). An energy budget for signaling in the grey matter of the brain. *Journal of Cerebral Blood Flow and Metabolism, 21,* 1133–1145.

Harris, J. J., Jolivet, R., & Attwell, D. (2012). Synaptic energy use and supply. *Neuron, 75,* 762–777.

Sheng, Z. H. (2014). Mitochondrial trafficking and anchoring in neurons: New insights and implications. *The Journal of Cell Biology, 204,* 1087–1098.

Tl, S. (2013). Mitochondrial trafficking in neurons. *Cold Spring Harbor Perspectives in Biology, 5,* a011304.

Wong-Riley, M. T. T. (1989). Cytochrome oxidase: An endogenous metabolic marker for neuronal activity. *Trends in Neurosciences, 12,* 94–101.

Mutant Huntingtin Alters Mitochondrial Transcription

Cui, L. B., Jeong, H., Borovecki, F., Parkhurst, C. N., Tanese, N., & Krainc, D. (2006). Transcriptional repression of PGC-1α by mutant Huntingtin leads to mitochondrial dysfunction and neurodegeneration. *Cell, 127,* 59–69.

Dunah, A. W., Jeong, H., Griffin, A., Kim, Y. M., Standaert, D. G., Hersch, S. M., et al. (2002). Sp1 and TAFII130 transcriptional activity disrupted in early Huntington's disease. *Science, 296,* 2238–2243.

Nucifora, F. C., Jr., Sasaki, M., Peters, M. F., Huang, H., Cooper, J. K., Yamada, M., et al. (2001). Interference by huntingtin and atrophin-1 with CBP-mediated transcription leading to cellular toxicity. *Science, 291,* 2423–2428.

Scarpulla, R. C. (2011). Metabolic control of mitochondrial biogenesis through the PGC-1 family regulatory network. *Biochimica et Biophysica Acta, 1813,* 1269–1278.

Yano, H., Baranov, S. V., Baranova, O. V., Kim, J., Pan, Y., Yablonska, S., et al. (2014). Inhibition of mitochondrial protein import by mutant huntingtin. *Nature Neuroscience, 17,* 822–831.

Spinocerebellar Ataxias

Casari, G., De Fusco, M., Ciarmatori, S., Zeviani, M., Mora, M., Fernandez, P., et al. (1998). Spastic paraplegia and OXPHOS impairment caused by mutations in paraplegin, a nuclear-encoded mitochondrial metalloprotease. *Cell, 93,* 973–983.

Di Bella, D., Lazzaro, F., Brusco, A., Plumari, M., Battaglia, G., Pastore, A., et al. (2010). Mutations in the mitochondrial protease gene AFG3L2 cause dominant hereditary ataxia SCA28. *Nature Genetics, 42,* 313–321.

Kasumu, A., & Bezprozvanny, I. (2012). Deranged calcium signaling in Purkinje cells and pathogenesis of spinocerebellar ataxia 2 (SCA2) and other ataxias. *Cerebellum, 11,* 630–639.

Maltecca, F., Magnoni, R., Cerri, F., Cox, G. A., Quattrini, A., & Casari, G. (2009). Haploinsufficiency of AFG3L2, the gene responsible for spinocerebellar ataxia type 28, causes mitochondrial-mediated Purkinje cell dark degeneration. *The Journal of Neuroscience, 29,* 9244–9254.

Rugarli, E. I., & Langer, T. (2012). Mitochondrial quality control: A matter of life and death for neurons. *The EMBO Journal, 31,* 1336–1349.

Sequestration and Spliceopathies

Du, H., Cline, M. S., Osborne, R. J., Tuttle, D. L., Clark, T. A., Donohue, J. P., et al. (2010). Aberrant alternative splicing and extracellular matrix gene expression in mouse models of myotonic dystrophy. *Nature Structural & Molecular Biology, 17,* 187–193.

Jiang, H., Mankodi, A., Swanson, M. S., Moxley, R. T., & Thornton, C. A. (2004). Myotonic dystrophy type 1 is associated with nuclear foci of mutant RNA, sequestration of muscleblind proteins and deregulated alternative splicing in neurons. *Human Molecular Genetics, 13,* 3079–3088.

Kalsotra, A., Xiao, X., Ward, A. J., Castle, J. C., Johnson, J. M., Burge, C. B., et al. (2008). A postnatal switch of CELF and MBNL proteins reprograms alternative splicing in the developing heart. *Proceedings of the National Academy of Sciences of the United States of America, 105,* 20333–20338.

Miller, J. W., Urbinati, C. R., Teng-umnuay, P., Stenberg, M. G., Byrne, B. J., Thornton, C. A., et al. (2000). Recruitment of human muscleblind proteins to (CUG) expansions associated with myotonic dystrophy. *The EMBO Journal, 19,* 4439–4448.

Timchenko, L. T., Miller, J. W., Timchenko, N. A., DeVore, D. R., Datar, K. V., Lin, L., et al. (1996). Identification of a (CUG) triplet repeat RNAbinding protein and its expression in myotonic dystrophy. *Nucleic Acids Research, 24,* 4407–4414.

Fragile X Syndrome and R-Loops

Bassell, G. J., & Warren, S. T. (2008). Fragile X syndrome: Loss of local mRNA regulation alters synaptic development and function. *Neuron, 60,* 201–214.

Bear, M. F., Huber, K. M., & Warren, S. T. (2004). The mGluR theory of fragile X mental retardation. *Trends in Neurosciences, 27,* 370–377.

Colak, D., Zaninovic, N., Cohen, M. S., Rosenwaks, Z., Yang, W. Y., Gerhardt, J., et al. (2014). Promoter-bound trinucleotide repeat mRNA drives epigenetic silencing in fragile X syndrome. *Science, 343,* 1002–1005.

Darnell, J. C., Van Driesche, S. J., Zhang, C., Hung, K. Y. S., Mele, A., Fraser, C. E., et al. (2011). FMRP stalls ribosomal translocation on mRNAs linked to synaptic function and autism. *Cell, 146,* 247–261.

Groh, M., Lufino, M. M. P., Wade-Martins, R., & Gromak, N. (2014). R-loops associated with triplet repeat expansions promote gene silencing in Friedrich ataxia and fragile X syndrome. *PLoS Genetics, 10,* e1004318.

Chapter 11
Amyotrophic Lateral Sclerosis and Frontotemporal Lobar Degeneration

Frontotemporal lobar degeneration (FTLD) is the most common cause of dementia in individuals under the age of 60 years, and overall, follows Alzheimer's disease and dementia with Lewy bodies in its frequency of occurrence in the industrialized world. This term encompass a number of syndromes that affect neurons in the frontal and temporal lobes of the brain. These disorders are characterized by deficits in behavior, mood, emotion, or language. The most common form of the disease is frontotemporal dementia (FTD). This form includes three subtypes—behavioral variant FTD (bvFTD), semantic dementia (SD), also known as temporal variant FTD, which affects the ability to assign meaning to words, and progressive nonfluent aphasia (PNFA) that affects the ability to speak. In some instances of FTD, motor neuron disease (MND) is present, as well. In those situations, the disorder is known as FTD-MND. Two other neurological disorders, progressive supranuclear palsy (PSP) and corticobasal syndrome (CBS), complete the enumeration of FTLD disorders. These disorders have a prevalence of 15 per 100,000 with incidences ranging from 2 to 9 per 100,000 per year depending on the age of the individual.

Although it is less well known to the general public than Alzheimer's disease, FTLD has been known and studied for just as long. Arnold Pick (1851–1924) described the disease in 1892, noting the presence of protein tangles, which appeared as large bodies in the cells he had examined. Those deposits along with extracellular plaques and abnormal neurites were studied by Alois Alzheimer, Oskar Fischer (1876–1942), Frederic Lewy and Pick over the next 20 years. The deposits described by Pick are known as Pick bodies. They, like the tangles discussed earlier in the chapter on Alzheimer's disease, are composed mainly of the microtubule protein tau. Since that time an expanded list of disorders characterized by abnormal deposits of tau have been identified, and are known as the tauopathies.

Amyotrophic lateral sclerosis (ALS) is referred to as Lou Gehrig's disease in the USA and as Charcot's disease elsewhere. It is an adult-onset, fatal disorder producing paralysis and primarily affecting adults 45–60 years old. Its progression, typically 3–4 years, is far more rapid than Parkinson's disease, which has a

© Springer International Publishing Switzerland 2015

M. Beckerman, *Fundamentals of Neurodegeneration and Protein Misfolding Disorders*,
Biological and Medical Physics, Biomedical Engineering,
DOI 10.1007/978-3-319-22117-5_11

20 year progression. Its prevalence in the USA is roughly 30,000 individuals with an annual incidence of about 6000. Most cases are sporadic (sALS), but approximately 10 % are inherited (fALS). A landmark finding was the discovery in 1993 by Rosen that missense mutations in the Cu/Zn superoxide dismutase gene, SOD1, were associated with fALS. Shortly thereafter, in 1997, Bruijn reported that ubiquitin-positive inclusions containing ALS-linked mutated SOD1 gene products appear first in astrocytes and then in motor neurons and these inclusions escalate as the disease progresses.

Superoxide dismutase catalyzes the conversion of the dangerous superoxide free radical to far less-dangerous hydrogen peroxide. This enzyme had gained prominence following its initial discovery by McCord and Fridovich in 1969 when its existence was taken as evidence in support of the free radical (oxidative damage) theory of aging. This theory has as its central tenet that free radicals, small molecules containing unpaired electrons that render them highly reactive, accumulate over time in cells and attack macromolecules, thereby placing a limit on an organism's life span. The groundwork for this theory was laid by several discoveries. The first of these was the groundbreaking realization in 1954 by the Argentinian biochemist Rebeca Gerschman (1903–1986) that potentially damaging free radicals are present in cells. She noted that free radicals produced as by-products of oxidative metabolism caused damage to cells in a manner similar to the actions of ionizing radiation such as X-rays. In the same year, Commoner, Townsend, and Pake using an electron-spin resonance spectrometer, which can detect the presence of unpaired electrons, found that reactive oxygen species (ROS) were present in biological materials.

The discovery that free radicals can damage lipids, DNA, and proteins led Denham Harman to advance the free radical theory of aging in a highly influential paper in 1956. The basic idea is that free radicals are continually generated by the mitochondrial electron transport chain (ETC) as a byproduct during cellular respiration. These accumulate over time and induce cellular damage that is responsible for aging. The most prominent of the free radicals is superoxide ($O_2^{\cdot}$). The primary sources of superoxide production from the viewpoint of aging theory are ETC complexes I and III. Small amounts of superoxide are released into the mitochondrial matrix from Complex I and into both the matrix and intermembrane space from Complex III. Under healthy cellular conditions, when the ETC is operating normally and is not excessively stressed, production of superoxide is low and is readily converted to a far less dangerous molecule, hydrogen peroxide (H_2O_2), by resident SODs. That changes over time and under stressed conditions. Considerable effort was expended, once SOD1's involvement in ALS emerged, in exploring this line of reasoning with results that were disappointing and perplexing.

This situation began to evolve in the 2006–2009 time period when a series of discoveries radically changed the conceptual landscape of ALS and FTLD. The discovery in 2006–2007 that the tar-binding protein 43 (TDP-43), a 43-kDa RNA processing protein, was a main component of the ubiquitinated intracellular inclusions seen in the majority of ALS cases began the shift in understanding. These inclusions were found in glial cells and neurons, in many instances of sALS, and also in a large number of cases of FTLD. These studies were followed by several others that established unambiguously that mutated forms of TDP-43 could produce ALS. These findings were followed in 2009 with the discovery that another RNA processing protein, fused in sarcoma (FUS), was present in a different set of inclusions, and that mutated forms of that protein too has a causal connection to ALS. As shown in Table 11.1 mutated forms of these two proteins can not only generate ALS but also produce FTLD, thereby linking the two very different disorders.

Table 11.1 Amyotrophic lateral sclerosis and frontotemporal lobar degeneration genetics and pathology

fALS (%)	sALS (%)	FTLD (%)	Protein	Gene (locus)	Function	Inclusion
20	2		Superoxide dismutase	SOD1 (21q22.1)	Detox enzyme	SOD1
5	<1	<1	Tar-binding protein 43	TDP-43 (1p36.2)	RNA processing	TDP-43
5	<1	<1	Fused in sarcoma	FUS (16p11.2)	RNA processing	FUS
<1			Angiogenin	ANG (14q11.2)	RNA processing	TDP-43
		<1	Ataxin 2	ATXN2 (12q24)	RNA processing	TDP-43
<1	<1		Valosin-containing protein	VCP (9p13.3)	Protein quality control	TDP-43
5	<1		Optineurin	OPTN (10p15-p14)	Protein quality control	TDP-43
<1		<1	Ubiquilin 2	UBQLN2 (Xp11.21)	Protein quality control	TDP-43, FUS
40	5	10	C9orf72	C9ORF72 (9p21.3-p13.3)	(unknown)	TDP-43
		10	Progranulin	PGRN (17q21.31)	Protein quality control	TDP 43
<1		<1	Charged multivesicular protein 2B	CHMP2B (3p11.2)	Protein quality control	p62, TDP-43
		10	Microtubule-associated protein tau	MAPT (17q21)	Cytoskeleton	tau

Columns 1–3 list the approximate frequencies that each mutated protein contributes to familial ALS (fALS), sporadic ALS (sALS), and frontotemporal lobar degeneration (FTLD), while the last column lists the type of disease-associated inclusion. These vary in composition and morphology in the cell types being affected. Adapted from Robberecht 2013 *Nat. Rev. Neurosci* 14: 248; Ling 2013 *Neuron* 79: 416

Yet another discovery further linked together ALS and FTLD. In 2011, an unstable GGGGCC hexanucleotide repeat expansion in the noncoding region of the C9ORF72 gene was shown to cause to fALS and FTLD. This mutation was found to be the single most widespread cause of fALS and FTLD. Disease-causing mechanisms arising from unstable repeat expansions were discussed in the last chapter. These same processes come into play in C9ORF72-associated ALS and FTLD along with several new ones.

In examining the different forms of FTLD several nonoverlapping classes of inclusions can be discerned. One grouping consists of tau-positive, ubiquitin-positive inclusions (FTLD-tau), a second as TDP-43-positive, ubiquitin-positive inclusions (FTLD-TDP-43), and a third as FUS-positive, ubiquitin positive inclusions (FTLD-FUS). Fourth, some inclusions are negative for tau, TDP-43, and FUS; instead, they may contain p62 and are labelled as FTLD-UPS. Lastly, in some rare instances no discernable inclusions can be detected. That group is designated as FTLD-ni. These molecular/histological characterizations underlie the six clinical syndromes named in the first paragraph of the chapter.

In this chapter, the four main genetic constellations underlying ALS and FTLD—SOD1, TDP-43 and FUS, C9ORF72, and tau—are explored starting with SOD1 and then turning to TDP-43 and FUS, and from there to the others. Dynamic, membrane-less organelles such as stress granules, P-bodies, and nucleoli are important loci of action in ALS and FTLD. Their formation and role in these disorders is scrutinized along with a continuation of the discussion of RNA-mismanagement from the previous chapter. Unstable repeat expansions were discussed in the last chapter with Fig. 10.1 depicting the types of problems these repeats cause. That discussion picks up again in an examination of the disease-causing GGGGCC repeat expansion in C9ORF72, and its involvement of non-ATG translation and G-quadruplexes. The chapter concludes with a look into the tauopathy landscape in which a variety of disorders have misfolded and aggregated tau as a common link. Significantly, that family of disorders includes chronic traumatic encephalopathy (CTE) brought on by repeated concussive and sub-concussive injuries to the head suffered by military personnel and athletes involved in violent sports.

11.1 Mutations in the SOD1 Gene Cause Amyotrophic Lateral Sclerosis

Dismutases are enzymes that catalyze the dismutation of their substrates, that is, they simultaneously generate oxidized and reduced forms of a chemical species. The superoxide dismutase (SOD) family consists of three enzymes—cytoplasmic Cu/Zn-SOD (SOD1), mitochondrial Mn-SOD (SOD2), and extracellular EC-SOD (SOD3). These are regarded as the most important antioxidants in the body; they, most notably, convert superoxide anions to molecular oxygen and hydrogen peroxide. Starting with two superoxide molecules the SODs oxidize the first one and reduce the second, as discovered by McCord and Fridovich in 1969.

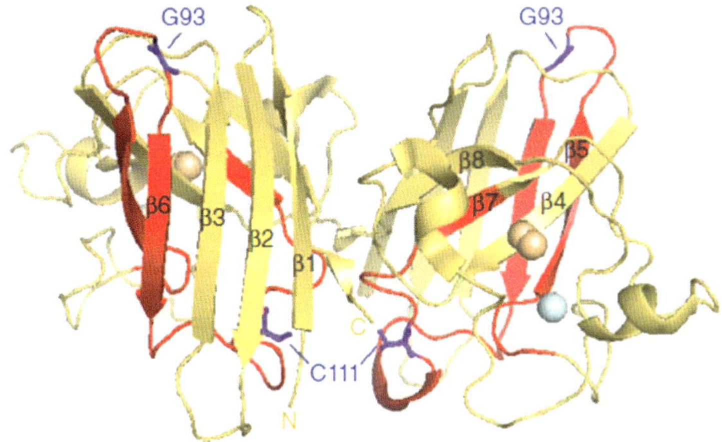

Fig. 11.1 Crystal structure of the SOD1 homodimer. The zinc and copper ions are shown in *cyan* and *orange*, respectively. The G93 mutation site and C111 oxidation site are shown as well. The two events, G93A mutation and C111 oxidation, generate the same misfolded structure (from Bosco *Nat. Neurosci.* 13: 1396 © 2010 Reprinted by permission from Macmillan Publishers Ltd)

Later, in 1993, Rosen found that mutations in SOD1 can produce amyotrophic lateral sclerosis. More than 150 mutations in SOD1 have been uncovered to date. The mutations are situated throughout the protein and produce a variety of structural alterations that result in (1) loss of stability, (2) misfolding, and (3) aggregation. Unlike Aβ and α-synuclein the mutations do not, in general, give rise to amyloids; instead, they tend to produce amorphous aggregates. The most common mutation is the Ala4Val (A4V) mutation. It accounts for 50 % of all the fALS cases in the USA. Like most other SOD1 missense mutations it is inherited in an autosomal dominant manner.

SOD1 contains eight antiparallel β strands, binds two metal atoms, and functions as a dimer (Fig. 11.1). Depending upon the specific mutation, one of several different kinds of functional defects can be produced. They do not, however, produce a loss-of-enzymatic, antioxidant function. Attempts to demonstrate that this was the dominant mode of failure produced negative results as noted in the introductory remarks. Instead, the mutations are widely regarded as primarily generating a gain-of-toxic-function, the specific nature of which has not yet been determined, conclusively, from the underlying loss of stability, misfolding, and loss of solubility/aggregation.

Some of the mutations disrupt the folding of the apoSOD1 monomers. These mutations destabilize the caps that form at the extrema of the folded protein and protect the β-sheet structures from undesirable side reactions. As a result the proteins become "sticky." This is similar to the pathology seen elsewhere where the protective role of edge strands becomes compromised resulting in aggregation.

Other mutations prevent monomers from acquiring their metal cofactors; these mutations have a greater destabilizing effect than mutations that leave this capability alone. Monomers lacking their Cu and Zn ions are more prone to unfold and refold into aggregation-prone conformations. Disulfide bonds are another major factor. Incorrect disulfide bonds prevent dimerization and, instead, enable multimeric aggregation to occur via destabilized monomeric folding intermediates.

There is evidence that oxidative damage can lead to misfolding and aggregation. The central observation is that fully metalized, properly disulfide linked SOD1 is extraordinarily stable. In order to dispose of the enzyme by proteasomes, it first has to be destabilized. This is accomplished through a negative feedback loop in which a buildup in hydrogen peroxide leads to a loss of metal ions as a preliminary step, but the resulting metal-free monomers are sensitive to oxidative misfolding and aggregation. The most significant aspect of this mode of failure is that it ties together SOD1-induced fALS and the far more prevalent sALS in a common pathology, and provides an explanation for the presence of SOD1 inclusions in both forms of the disease.

Lastly, the disease causing mutations in SOD1 produce varying effects on its stability. What these mutations have in common is a tendency to alter SOD1's folding kinetics. Small single domain proteins such as SOD1 should fold fairly rapidly into their native state. In contrast to this expectation, mutant SOD1 proteins take orders of magnitude longer than expected to find their native state. The energy landscapes for these mutants are rugged and the folding pathways are populated by kinetic traps. The formation of a variety of folding intermediates and apo monomers provides ample opportunity for the sampling of aggregation-prone conformers.

11.2 Gain-of-Toxic Function by Mutant SOD1 (mSOD1)

ALS is a disease of upper and lower motor neurons (MNs). It is widely believed (but by no means unanimously) that it begins with the destruction of the postsynaptic apparatus, followed by a dying back of axons at the neuromuscular junction (NMJ) leading to a loss first of lower motor neurons and then of upper ones. This idea is supported by a growing body of evidence typically obtained using the G93A SOD1 transgenic mouse model of ALS. Within this framework, two distinct sets of actions seem to be needed. First, the deterioration of the synapse itself, perhaps due to a single toxic event or small set of events, or alternatively, arising through multiple accumulating events that pile up and initiate the destruction of the NMJ. Secondly, inflammation occurs involving neighboring glial cells activated by extracellular mSOD1. The sustained release of inflammatory molecules by the glia, in turn, potentiates the damage to the neurons and NMJ.

Inflammation plays an important effector role in mutant SOD1-generated ALS. This conclusion is supported by several observations. First, mutant forms of SOD1 are secreted from neurons in secretory vesicles; they are detected in CSF, and are responsive to immunization protocols. Secondly, intracellular SOD1-bearing inclusions are not limited to neurons, but instead appear in nearby glia; these cells have become damaged at some point by mutant SOD1. Astrocytes exposed to mutant SOD1 are toxic to neurons. Paralleling the situation encountered in other neurological disorders, extracellular mutant SOD1 hyperactivates microglia and astrocytes, which, in turn, secrete a variety of toxic inflammatory mediators. Expanding on this notion, astrocytes taken from sporadic ALS sufferers are shown to be toxic to motor neurons. These findings all place astrocytes and microglia downstream of mutant SOD1, in the role of key effectors.

Mutant SOD1-ALS arises primarily through gain-of-toxic-function mechanisms rather than through loss-of-normal-function means. A variety of ways that this can occur have been uncovered through studies of transgenic mice expressing one or more of the predominant mutations with some variations associated with the specific mutation(s) being introduced. Arguably the best established of these disease modes are glutamate excitotoxicity and mitochondrial dysfunction. These two are familiar points of failure, having been encountered in Alzheimer's disease and Parkinson's disease and Huntington's disease.

A seminal observation on the importance of glutamate handling, the first of these widely encountered modes, was made by Rothstein in 1996 who noted that proper uptake of glutamate by astrocyte-resident transporters was critical—it prevents excessive glutamate stimulation of neurons. That these transporters are negatively impacted by mutant SOD1 was established a few years later. More recently, it was found that the presynaptic terminals of the tripartite synapse regulate this activity. Breakdowns in signaling from neuron to astrocyte trigger reductions in transcription and expression of astrocyte-resident glutamate transporters such as GLT1/EAAT2 (excitatory amino acid transporter 2).

One of the earliest signs of a neurological disorder is the appearance of mitochondrial and mitochondrial network abnormalities. The assertion that mSOD1 produces a gain-of-toxic function is based on two broad sets of findings. First, there is the preservation of normal SOD1 dismutase activity in spite of the presence of disease-causing mutations. Secondly, at the same time, mSOD1 localizes to mitochondria that develop a number of striking abnormalities. These include the appearance of numerous membrane-containing vacuoles and a deformed morphology along with the emergence of electron transport chain defects and reduced Ca^{2+} buffering capabilities, and, at the network level, impaired mitochondrial fusion and transport.

There is also evidence that SOD1 misfolding can spread from cell to cell in a prion-like manner. Two papers published in 2011 showed that (1) templated conversion can occur, that is, aggregates composed of misfolded SOD1 can serve as seeds that template the conversion of normal SOD1 to the misfolded form, and (2) propagation can occur, i.e., once seeding has taken place, removal of the introduced seeds does not prevent further conversions, thereby demonstrating that new templates have formed.

11.3   Mutations in TDP-43 and FUS Genes Cause Amyotrophic Lateral Sclerosis and Frontotemporal Lobar Degeneration

The proper maintenance of protein quality control (PQC) requires the participation in one way or another of a substantial fraction of the cellular proteins. Some participants ensure that proteins fold properly into their native conformations, and that they are expressed at the proper levels so that their concentrations are neither too low nor too high. Others transport newly synthesized proteins and mRNAs to locales in the cell where they are needed, and recycle and remove unwanted and damaged proteins and organelles. One of the most dramatic set of findings in the last few years has been discoveries that aberrations in mRNA processing can produce ALS and FTLD.

Mutations in the genes encoding the mRNA processing proteins TDP-43 and FUS give rise to ALS and FTLD. These disorders are characterized by the presence of inclusions bearing one or the other of these proteins, but not SOD1. As a result, these disorders are regarded as having an etiology distinct from that of SOD1-ALS. TDP-43 is a 414-amino-residue-long protein widely expressed throughout the body. It contains two RNA recognition motifs, RRM1 and RRM2, in its central region and a C-terminal glycine-rich domain that mediates protein–protein interactions (Fig. 11.2). As noted in the introduction to the chapter, the involvement of TDP-43

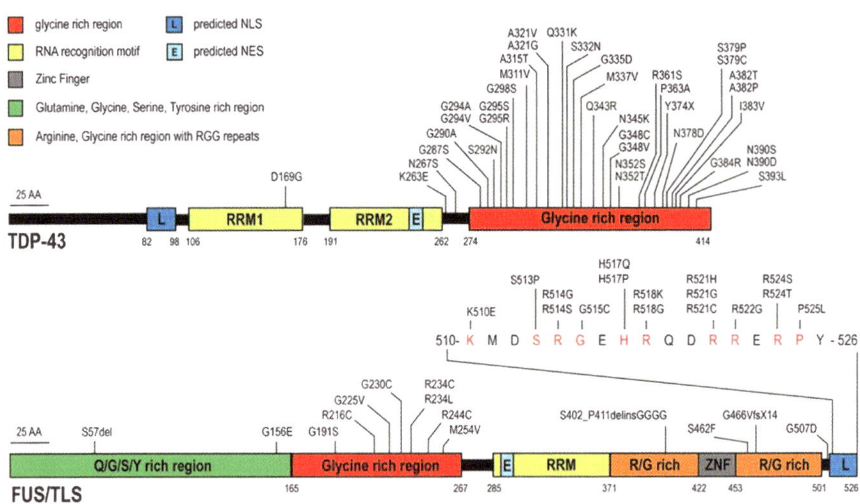

Fig. 11.2 Structure of TDP-43 and FUS along with their disease-causing mutations. These are concentrated in the glycine-rich regions of the two proteins along with a cluster in the C-terminal-most 17 amino acid residues of FUS [see text for details] (from Lagier-Tourenne *Hum. Mol. Genet.* 19, Rev. Iss 1: R46 © 2010 Reprinted by permission from Oxford University Press)

in neurodegeneration was discovered in 2006. Those finding were followed in 2008 by the discovery of a cluster of disease-causing mutations, all but one in the C-terminal region and the lone exception in RRM1 (Fig. 11.2).

Soon after TDP-43's involvement of ALS was uncovered another ALS-causing protein with a structure similar to TDP-43 was found. Discovered in 2009 by two groups, the fused in sarcoma/translocated in liposarcoma (*FUS/TLS*) gene encodes a 526-amino-acid-residue protein FUS that, like TDP-43, can aggregate in ALS-associated inclusions. As shown in Fig. 11.2 FUS contains an N-terminal region enriched in glutamine, glycine, serine and tyrosine (QGSY) residues and a C-terminal zinc-finger motif. Situated in between are a glycine-rich region, an RRM, and an arginine–glycine (RG)-rich region. Most of the disease-causing mutations are concentrated in two regions—the glycine-rich regions, and the extreme C-terminal segment in which there are five targeted arginine residues.

11.4 TDP-43 Localization, Aggregation, and Prion-Like Spread

TDP-43 is found in the nucleus and cytoplasm. It shuttles between the two compartments but under normal conditions is localized primarily in the nucleus. Disease-causing mutations to TDP-43, or other perturbations that disturb its predominant nuclear localization, lead to the formation of cytoplasmic aggregates and produce characteristic disease phenotypes. Consistent with this finding, inclusions are generated when the TDP-43 cytoplasmic concentration is artificially elevated.

TDP-43-bearing inclusions are the preeminent hallmark of frontotemporal lobar dementia and also familial and sporadic forms of amyotrophic lateral sclerosis (Table 11.1). They are also formed in a substantial fraction of Alzheimer's disease cases. Recapitulating a widely encountered phenomenon, the C-terminal region of TDP-43 is often proteolytically cleaved, hyperphosphorylated, and ubiquitinated, operations that enhance its tendency to form insoluble aggregates. Again, as was the case for the other neurodegenerative diseases, the exact role of the inclusions in the disease progression is unclear, but there is a feeling that these deposits are part of a protective response to mislocalization, misfolding, and excessive quantities of the protein.

TDP-43 deposition, like tau pathology, α-synuclein deposition, and SOD1-aggregation spreads from region to region in the brain in a stereotypic manner, giving rise to a progressively worsening sequence of disease stages. It is surmised that aggregates of TDP-43 spread out along the brain's network of anatomical connections, perhaps via exosomes or tunneling nanotubes. This places TDP-43 in an expanding group of proteins that possess some prion-like properties but perhaps lack the latter's ability to convey infectivity between individuals.

11.5   TDP-43 and FUS Are RNA-Processing Proteins

Unlike the situation for SOD1, cellular damage caused by TDP-43 primarily involves a loss-of-normal function. It turns out that three families of proteins have prominent roles in RNA processing and, when malformed, in neurodegeneration. These are, as follows:

- The *heterogeneous nuclear ribonucleoprotein hnRNPs family of RNA-binding proteins* which includes TDP-43. These proteins carry out a variety of tasks in RNA metabolism upon binding to nascent RNAP II transcripts. These include alternative splicing and regulation of translation
- *SR (serine–arginine) repeat-containing RNA-binding proteins.* These proteins usually possess two domains—an RNA-recognition motif (RRM) in their N-terminal and a stretch of SR dipeptides (hence their name) that mediates protein–protein interactions in their C-terminal region.
- The *FET (TET) family of proteins* is less well known than the other two families but does include FUS/TLS as a member. Proteins in this family contain motifs for binding nucleic acids, and participate in, and perhaps coordinate, RNA Poly II mediated transcription and pre-mRNA splicing.

11.5.1   Transcription

TDP-43 and FUS participate in several stages of RNA processing. First, TDP-43 and FUS bind ss and ds DNA, RNA, and proteins. As shown in Fig. 11.3 TDP-43 preferentially binds ss TG-rich sequences in promoters, while FUS has been found to associate with the general transcription machinery as well as with certain gene-specific transcription factors. Transcriptional actions carried out at these sites by TPD-43 and FUS at these sites are primarily inhibitory (Fig. 11.3b–d).

11.5.2   Splicing and miRNA Processing

Secondly, TDP-43 and FUS bind to thousands of mRNAs in their 3′ UTR and contributes to RNA splicing (Fig. 11.3e–g). TDP-43 has a strong preference for GU-rich intronic motifs, especially those occurring in long pre-mRNAs, that is, pre-RNA molecules possessing exceptionally long introns. It binds to noncoding RNAs, introns, and 3′ UTRs of mRNAs, thereby supporting multiple roles. FUS targets a set of mRNAs that have little overlap with those selected by TDP-43. It has some preference for GUGGU motifs but, in general, recognizes multiple RNA-binding motifs resulting in a sawtooth binding pattern and preferentially binds towards the 5′ ends of long introns. Thirdly, TDP-43 and FUS participate in microRNA processing through their membership in the Drosha complex (Fig. 11.5h).

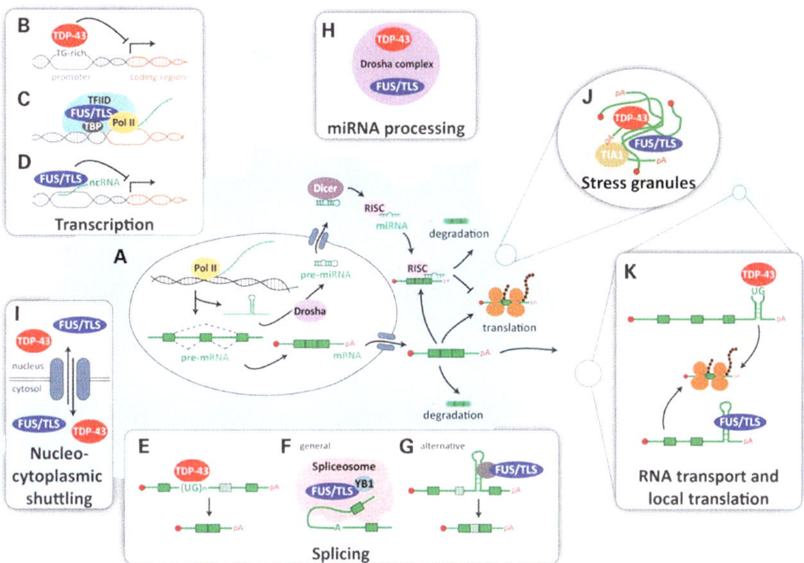

Fig. 11.3 Roles of TDP-43 and FUS in RNA processing. (**a**) Steps in RNA processing; (**b**) TDP-43 binds ss TG-rich promoter sequences and blocks transcription; (**c**) FUS associates with the general transcription machinery; (**d**) in response to DNA damage, FUS is recruited by sense and antisense noncoding RNAs (ncRNAs) and represses transcription; (**e, f, g**) participation of TDP-43 and FUS in pre-mRNA splicing; (**h**) Possible role of TDP-43 and FUS in micro-RNA (miRNA) processing; (**i**) TDP-43 and FUS shuttle between the nucleus and the cytoplasm; (**j**) TDP-43 and FUS are found in stress granules, as will be discussed in the text; (**k**) TDP-43 and FUS are also found in transport granules used to ferry mRNAs to remote sites for local translation (from Lagier-Tourenne *Hum. Mol. Genet.* 19, Rev. Iss 1: R46 © 2010 Reprinted by permission from Oxford University Press)

11.5.3   Autoregulation

Expression levels of proteins essential for RNA processing are tightly controlled. One common method of control is through establishment of negative feedback loops. Proteins such as TDP-43 bind to alternatively spliced sites in the 3′ UTR of their own mRNAs resulting in the appearance of a premature STOP codon. That event triggers nonsense-mediated RNA decay (NMD), a quality control operation whereby transcripts containing premature STOP codons are eliminated to prevent the generation of incomplete proteins. When placed within the abovementioned negative feedback loop, the expression levels are maintained at a constant value. Mutations in TPD-43 that disturb this feedback loop can alter the TDP-43 expression levels.

11.6   TDP-43 and FUS Are Found in Stress Granules, Processing Bodies, and mRNA Transport Granules

11.6.1   Stress Granules and P-Bodies

Cells respond to elevated stress conditions by reallocating resources to operations needed for survival and eventual recovery. In order to conserve energy, translation of mRNAs is halted. The translation-halted mRNAs and their associated RNA binding proteins are then collected and formed into highly dynamic structures called *stress granules*. These dynamic compartments lack a lipid membrane, and can be rapidly assembled and disassembled according to need. A second kind of dynamic, membrane-less aggregate known as a *processing body*, or *P-body*, may also be formed. These are aggregates of translational-stalled mRNAs plus the machinery for halting translation and for degradation.

Both of these compartments contain, in addition to RNAs and RNA-processing proteins, signaling proteins that mediate the response to the stress signals sent from metabolic and other cellular signaling pathways. In addition, the assembly of the stress granules is mediated by TIA-1 (T-cell intracellular antigen 1) proteins. These are RNA-binding proteins that route stalled initiation complexes to nascent stress granules and utilize their prion-like domains to promote their formation. TDP-43 and FUS shuttle between the nucleus and cytoplasm and are found within stress granules when they form. In those locales, they partner with mRNAs and other RNA processing proteins.

11.6.2   mRNA Transport and Local Translation

TDP-43 and FUS also contribute to mRNA transport and local translation. A third type of transiently formed mRNA granule called an *mRNA transport granule* is used by neurons to deliver mRNAs to distal sites such as dendrites where they undergo local protein synthesis (Fig. 11.3k). These granules inhibit protein synthesis while undergoing microtubule-based anterograde transport. TDP-43 and FUS associate with proteins involved in mRNA transport. Some mutations in TDP-43 impair this bidirectional movement of mRNA-bearing granule, perhaps providing an explanation for the observed morphological abnormalities in distal axons sometimes seen in ALS-affected neurons.

11.6.3   Angiogenin and Ataxin-2

TDP-43 and FUS are not the only causal agents of ALS and/or FTLD found to have a role in stress granule dynamics. For example, angiogenin (Ang) promotes their assembly. It is a small 14 kDa ribonuclease that is secreted under conditions of stress.

It is taken up by angiogenin receptors on neighboring cells where it stimulates angiogenesis, proliferation, and pro-survival responses in target epithelial cells, cancer cells, and motor neurons, respectively. A major way that this protein promotes angiogenesis is through its translocation to the nucleus where it promotes rRNA transcription, an action that is permissive for angiogenesis. Since 2006, 17 missense mutations in this protein have been uncovered. In stressed motor neurons, Ang helps reprogram global translation to ensure neuronal survival. It does so by cleaving tNRAs to generate tRNA halves, stress-induced fragments termed tiRNAs. The resulting 5′-tiRNAs subsequently inhibit protein synthesis and promote the assembly of stress granules where the translation stalled mRNAs are collected.

Mutations in the *ATXN2* gene, which encodes the ataxin-2 protein, give rise to two distinct neurological disorders—spinocerebellar ataxia 2 (SCA2) and ALS/FTLD. As noted in the last chapter SCA2 is caused by expanded CAG repeats in the protein-coding region of the *ATXN2* gene. Normally, this gene contains from 15 to 24 CAG repeats. However the gene is unstable and the repeat number can increase from generation to generation. When the number of repeats is 34 or greater, SCA2, which affects cerebellar Purkinje cells, ensues. Intermediate numbers of CAG repeats, from 28 to 33, primarily affects a different set of neurons and produces a different disorder—ALS/FTLD. Several other polyQ-disorder-associated genes have been examined for this propensity, but all failed the test. The implication from this finding is that ability to produce ALS/FTLD is specific to the folding and functions of this particular protein and not to some general property polyQ tracts.

Intermediate repeat-length mutated forms of ataxin-2 associate in an RNA-dependent manner with both wt and mutated TDP-43, the latter more strongly than the former. It also forms associations with FUS and, most importantly, interferes with stress granule and P-body dynamics. One possible result of this interference is to cause these proteins to relocate elsewhere. This protein appears to function as an accelerant, i.e., it enhances the disease-causing propensity of other mutated proteins co-expressed along with it. It not only enhances the harmful effects of mutated TDP-43 and FUS but also that of C9orf72 (soon to be discussed), increases sensitivity to cellular stresses, and alters the mix of ALS and FTLD clinical features being exhibited.

11.7   Phase Transitions Are Mediated by Depletion Attraction, Multivalency, and Low Complexity

11.7.1   Phase Transition Enable the Formation of Liquid Droplets

Recall from elementary chemistry and physics that nonbiological materials exist in one of three phases—gaseous, liquid, or solid. In the gas phase, the constituents are free to move and they undergo little or no interactions with one another. In a liquid phase, weak transient forces are felt that somewhat limit movement, while in a solid

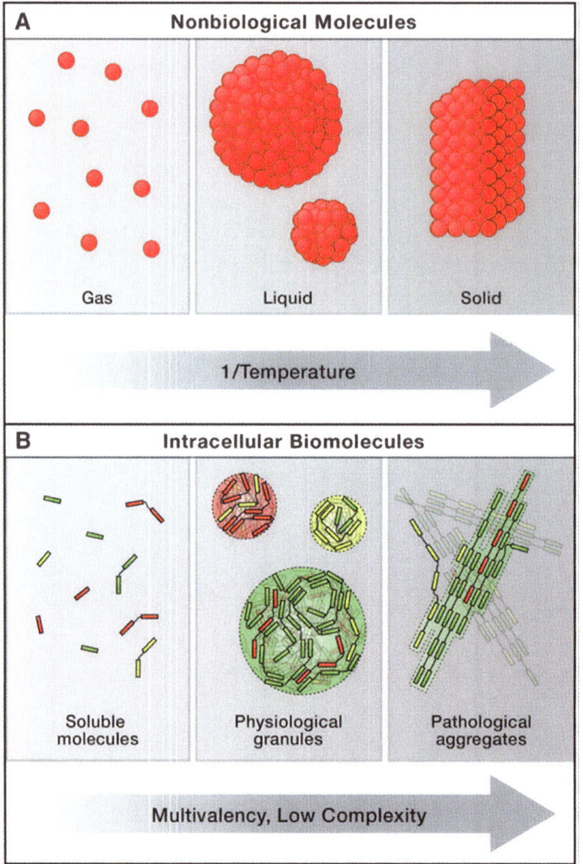

Fig. 11.4 Phase transitions by non-biological molecules and intracellular biomolecules. (**a**) Phase transitions in non-biological materials driven by temperature, and (**b**) phase transitions in cells driven by multivalency and low complexity (from Weber *Cell* 149: 1188 © 2012 Reprinted by permission from Elsevier)

phase the interactions are stable and movements are small. Parameters such as temperature govern the transitions between phases. Something similar can be observed in biological materials. Recall from Chap. 8 that lipid membranes exist in one or more phases depending on the amount of cholesterol contained therein giving rise to specialized microdomains such as lipid rafts and caveolae. A similar partitioning of the cytoplasm (Fig. 11.4) may well occur, and which helps makes order out of the crowded intracellular environment. The self-assembly of proteins land RNA molecules into non-membrane bound compartments such as stress granules and P-bodies is increasingly understood as arising from a phase transitions between a gas-like phase consisting of diffuse soluble molecules and a liquid phase in which the soluble molecules have condensed into a liquid droplet.

That the cytoplasm and its macromolecular constituents might be organized as multi-phase mixtures of liquid droplets ("granules") that self-organize or condense out of a gas-phase diffuse background ("continuous substance") was first posited by Edmund Beecher Wilson (1856–1939), the father of American cell biology. That idea languished for a century, but has recently seen a revival with the recognition that at the mesoscale (100 nm to 1 μm [bacteria] or 10 μm [eukaryotes]) level the cytoplasm and its macromolecular constituents undergo liquid phase separations. This process generates a variety of structures where proteins, lipids, and nucleic acids that need to work together can be rapidly brought into close proximity and then kept there for as long as necessary. In particular, membranes-less dynamic compartments are widely utilized to facilitate protein–RNA interactions. Examples of these structures include cytoplasmic stress granules and P-bodies along with transport granules, Cajal bodies, nucleoli, nuclear speckles, and germ-line P-granules.

11.7.2 Depletion Attraction, Multivalency and Low Complexity Underlie Liquid Droplet Phase Transitions

The ability to form membrane-less compartments containing the correct constituents is made possible by a confluence of physical and chemical properties. First, the structures so formed exhibit surface tension and fluidity; they also fuse together and drip in the presence of shear. These are properties of liquid droplets. Their formation is promoted by macromolecular crowding, which was discussed earlier in Chap. 2. The formal name given to the tendency, or force, arising from macromolecular crowding is *depletion attraction*. This attraction is an entropic one arising from the exclusion of small bodies from the spaces in between large macromolecules. A mathematic formula for the force so generated using a hard-sphere model was presented by Asakura and Oosawa in 1954, and largely confirmed in experiments carried out using optical tweezers. In the case of proteins and RNA, two other properties of the protein constituents—their multivalency and low complexity—further drive the liquid droplet phase transition.

In a *multivalent interaction*, multiple contacts of the same kind are established between a protein and its binding partner. The advantage of this type of non-covalent interaction is that it enables multiple weak and rather nonspecific interactions (i.e., depletion attraction, hydrogen bonds, and van der Waals forces) to produce strong specific bonds, which can be easily broken and the partners separated when necessary.

The other key property needed for liquid droplet formation is *low complexity*, that is, the presence of amino acid sequences that have little diversity in their composition. Repeats are, of course, a prime example of amino acid stretches with little (or no) diversity. (They are also natively disordered.) For example, FUS contains 27 repeats of the tripeptide (G/S)Y(G/S) sequence. In examining the composition of stress granules, it emerges that there is a practical purpose for the repeat sequences.

The proteins containing low complexity (LC) domains found in the stress granules can undergo a phase transition to a hydrogel state in which they form polymerized fibers. Although they are amyloid-like these fibers can readily assemble and disassemble, and accommodate heterotypic polymerization. However, when disease-causing mutations are present these same properties can promote the formation of pathological aggregates (Fig. 11.4b).

11.7.3   Prions Are Low Complexity Proteins

Prions are a good example of low complexity proteins. Prions and prion strains were introduced and discussed in Chap. 7, and again in the subsequent chapters with regard to their remarkable ability to spread misfolding and aggregation from cell to cell. The chief biochemical characteristic of the underlying prion domain was not the details of its primary sequence but simply its overall amino acid composition. Prion-like domains (PrLDs) are low-complexity sequences composed of glycine (G) and the uncharged polar amino acids asparagine (N), glutamine (Q), tyrosine (Y), and serine (S). The structures formed by these amino acids (1) can switch rapidly between noninfectious and infectious conformations, the latter giving rise to different prion strains, and (2) can template the conversion of noninfectious to infectious forms in original or new host cells. These domains are found throughout the human genome. It is estimated that 1 % of all the genes contain domains of this type. These domains are even more common among RNA-binding proteins where 20 % of these proteins contain PrLDs, and that grouping includes both TDP-43 and FUS.

Prion-like domains serve a useful function in their host RNA-processing proteins. They are needed for alternative splicing activities; they stabilize binding to RNAs, and promote optimal RNA hydrogen bond pairing (annealing). They may also mediate the phase transition that underlies the assembly of stress granules, driving the assembly of TDP-43 and FUS into oligomers and linear polymers within the hydrogels. However, when mutated, they tend to drive self-complementary beta-strand assembly.

11.8   Stress Granules Are Cleared by Autophagy

11.8.1   VCP and p62

Several of the entries in Table 11.1 have roles in protein quality control and, more specifically, in autophagic clearance of unwanted and damaged proteins and protein aggregates. One of the listed proteins whose mutated forms lead to ALS is valosin-containing protein (VCP). Mutated forms of this protein also causes inclusion-body myopathy, Paget disease, and Frontotemporal dementia (IBMPFD)/multisystem proteinopathy (MSP). Recall from Chap. 6 that VCP is a versatile AAA-ATPase involved in maintaining protein quality control. It (1) unwinds DNA, (2) extracts proteins from membranes by unfolding and disaggregating them, (3) chaperones proteins to proteasomes, and (4) assists in autophagy. Mutated forms of VCP associate with TDP-43, hnRNPA1, hnRNPA2B1 and other stress-granule-associated proteins. In the presence of VCP mutations, autophagic clearance of stress granules as well as the clearance of damaged mitochondria is impaired.

Another protein involved in autophagy and whose mutated forms are associated with increased susceptibility to ALS is the ubiquitin-binding protein p62/SQSTM1. This scaffold protein plays a role in both proteasomal degradation and through its interactions with LC3 in autophagic pathways (see Fig. 6.8). Inclusion bodies containing p62 are encountered in FTLD, Pick's disease, and several other neurological disorders. Mutant SOD1 associates with p62, and mutations in *SQSTM1*, the gene encoding p62, are found in ALS and FTLD. The p62 protein itself is degraded by autophagy. Impairment through mutations in p62 or in other autophagy-essential proteins results in an excessive accumulation of proteins needing clearance, and a shift in the cellular balance towards stress responsive/cell death pathways.

11.8.2 Ubiquilin 2 and Optineurin

Two other proteins linked to ALS and FTLD and whose mutated forms are tied to defective proteasomal and autophagic-lysosomal pathways are ubiquilin 2 encoded by the UBQLN2 gene and optineurin encoded by the OPTN gene. The ubiquilin 2 protein possesses a ubiquitin-binding motif and a heat-shock chaperone binding domain that stabilizes proteasome substrates. ALS-associated mutations in this protein promote aggregation, formation of cytoplasmic inclusions, and localization to vesicles containing optineurin and p62. Optineurin contains a ubiquitin-binding domain, a motif for interacting with LC3, and a coiled-coil domain that recognizes protein aggregates. It interacts with p62 and LC3 to form an autophagy receptor complex and induces the formation of autophagosomes. It has been found to play a key role in the autophagic removal of ubiquitin-coated *Salmonella enterica* and in the mitophagic removal of parkin-labelled damaged mitochondria.

11.9 Mutations in the *C9ORF72* Gene Are the Leading Cause ALS and FTLD

11.9.1 The C9ORF72 Gene Contains a Disease-Causing Unstable GGGGCC Hexanucleotide Repeat

It had been known for some time that an important locus for autosomal dominant ALS and FTLD resided on the short arm of chromosome 9. Finally, in 2011, two groups reported that the responsible gene, located at 9p21, was *C9ORF72*. In this gene, a

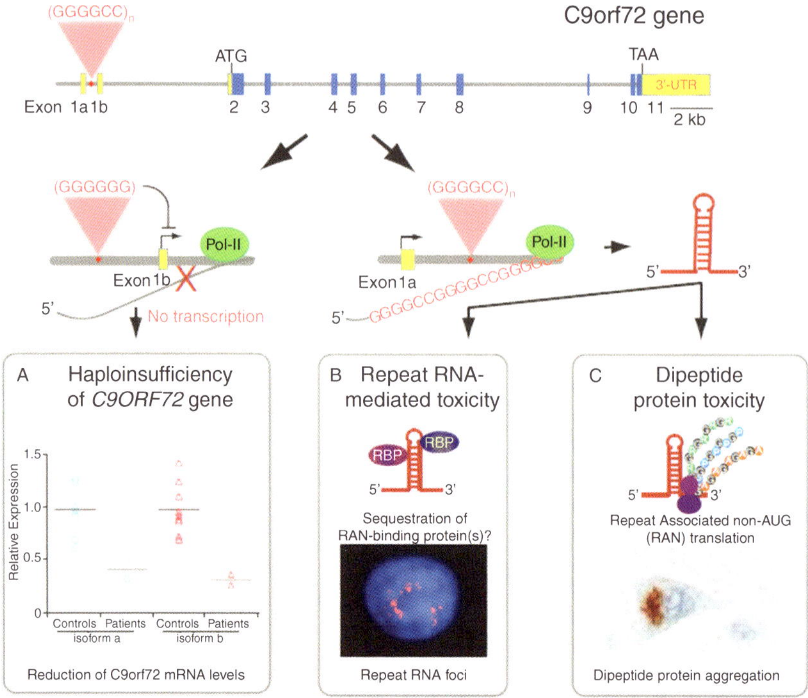

Fig. 11.5 Possible mechanisms of toxicity from GGGGCC repeat expansions contained within the C9ORF72 gene [see text for details] (from Ling *Neuron* 79: 416 © 2013 Reprinted by permission from Elsevier)

GGGGCC hexanucleotide expansion in the noncoding region between exons 1a and 1b (Fig. 11.5) causes ALS and FTLD. This defect is the single most common cause of familial instances of ALS. As shown in Table 11.1 it accounts for roughly 40 % of all fALS cases and also accounts for about 10 % of sALS and FTLD disease situations. Normally, there are from 2 to 23 GGGGCC hexanucleotide repeats, but in the disease-causing *C9ORF72* genes the number of repeats has expanded to the neighborhood of 700–1000.

Unstable repeat expansions in either protein coding or noncoding regions were discussed at length in the last chapter. Continuing that discussion, recall that fragile X syndrome and monotonic dystrophy 1 (MD1) were among the diseases brought on by unstable repeat expansions in non-protein-coding regions. As discussed in Chap. 10, aberrant repeat-generated DNA, RNA, and DNA–RNA hybrid structures such as hairpin and R-loops interfere with transcription, translation, and replication. The hairpins in MD1 sequester essential proteins, while the R-loops in fragile X syndrome silence genes.

Those same classes of defects—loss-of-protein function in the case of gene silencing and gain-of-toxic-RNA function in the case of protein-sequestering R-loops—occur in cases of GGGGCC repeat expansions in *C9ORF72*. That commonality is depicted in Fig. 11.5. The first panel of this figure (Fig. 11.5a) emphasizes the loss of the allele containing the excessively expanded repeat. (In haploinsufficiency, the mutant allele is dominant over the wt allele.) The result is reduced mRNA levels and loss-of-protein function. The second panel (Fig. 11.5b) depicts the formation of RNA foci about hairpin structures. The third panel (Fig. 11.5c) is new and to discuss it two additional sets of recent discoveries need to be introduced. The first of these is the emergence of a phenomenon first observed a century ago, in 1910, and now known to be caused by G-quadruplexes, and the second is the discovery of an unconventional form of translation that bypasses the requirement for an ATG [AUG] (Met) start codon.

11.9.2   GGGGCC Repeats Generate G-Quadruplexes

In 1910, at the same time that Harriette Chick was exploring the heat coagulation of egg albumen and hemoglobin, the Scandinavian medical biochemist Ivar Bang (1869–1918) was studying guanylic acid. He had just published a paper noting its puzzling tendency at high concentrations to form a gel. The mystery of how this could happen lasted for over 50 years, but was finally explained in 1962 when Gellert, Lipsett, and Davies examined the X-ray diffraction pattern generated by the material. They found that sets of four guanines self-assembled into planar squares with open centers, and these structures then stacked one on top of the other to produce the exceptionally stable guanine gels observed by Bang.

Interestingly, these stable structures are formed through Hoogsteen-type hydrogen bonding, an alternative type of hydrogen bonding first described by Karst Hoogsteen in 1963. In this type of bonding, A•T (or G•C) base pairs form with a different geometry from that of Watson–Crick base pairing. Additional stability is provided by the presence of a cation in the central cavity of each planar structure, thereby reducing the electrostatic repulsion of the nearby oxygen atoms. That these structures were more than laboratory curiosities became apparent with the discovery that they cap telomeres. Recall that telomeric ends of eukaryotic chromosomes are capped by repeating guanines interspersed with short A/T-rich sequences. These stretches form an overhang that protects the ends of chromosomes from attack by DNA repair proteins. It emerges that these overhangs form G-quadruplexes.

Interest in these alternative DNA and RNA structures increased further with realization that they are functionally relevant for transcriptional regulation and DNA replication as well as genomic stability. These structures are found in many regions where repetitive sequences are present including upstream promoters and, like the hairpins previously discussed, serve as barriers to replication fork progression.

They are, in fact, an alternative to the hairpins. These two types of structures can interconvert into each other, and into other conformations, according to environmental conditions such as pH, metal ions, or small-molecule metabolites.

11.10   Repeat-Associated Non-ATG (RAN) Translation Causes Neurodegeneration

11.10.1   RAN Translation Generates Toxic Proteins and RNAs

Repeat-associated non-ATG (RAN) translation, the second of the two new features, was first described in 2011 by Zu and coworkers in Laura Ranum's group. They had been studying the CAG•CTG repeat expansion in the *ATXN8OS/ATXN8* genes associated with spinocerebellar ataxia 8 and also the CTG repeat expansion in the *DMPK* gene that causes myotonic dystrophy 1. Both sets of microsatellite expansions occur in noncoding regions of those proteins. What they found was that these microsatellite expansions did not follow the normal rules for translation. First, translation of these sequences was being initiated without the presence of a start ATG [AUG] codon, and secondly, the result of translating the non-protein-coding repeats in this way was a protein, specifically, a number of different homopolymeric peptides. Thus, one needs to consider an expanded repertoire of potentially toxic RNA and protein species and mechanisms. Among the unstable repeats subsequently found to undergo RAN translation were the expanded CGG repeats in the *FMR1* gene responsible for fragile X tremor ataxia syndrome, and the expanded GGGGCC repeats in the C9ORF72 gene that produce ALS and FTLD.

In more detail, each set of malformed transcripts generates a specific collection of RAN translation proteins:

- SCA8—From expanded CAG repeats in the *ATXN8* gene: utilization of all three reading frames (CAG, AGC, and GCA) produces polyglutamine, polyserine, and polyalanine homopolymeric peptides. Similarly, translation from the antisense (opposite strand) CUG, UGC, and GCU reading frames generates polyleucine, polycysteine, and polyalanine, respectively.
- FXTAS—From expanded CGG repeats in the *FMR1* gene: The primary toxic product is a protein containing a long polyglycine stretch in its N-terminal designated by the discoverers as FMRpolyG. That RAN translation product is normally degraded but long repeats resist that process leading to their accumulation in intracellular inclusions, and resulting in cytotoxicity.
- ALS/FTD—From expanded GGGGCC repeats in the *C9ORF72* gene: utilizing all reading frames, six RAN translationdipeptide polymers can be produced. Three of these can be generated from transcription carried out in the sense direction—poly glycine–alanine (GA)$_N$, poly glycine–proline (GP)$_N$, and poly glycine–arginine (GR)$_N$, and another three RAN translation products can be produced from transcription performed in the antisense direction—poly proline–alanine (PA)$_N$, poly proline–glycine (PG)$_N$, and poly proline–arginine (PR)$_N$.

11.10.2   Prematurely Terminated Transcription and RAN Translation Products Disrupt the Normal Operation of the Nucleolus

The nucleolus is the largest structure in the cell nucleus. It is the site of ribosome transcription, processing, and subunit assembly. This membrane-less sub-organelle disassembles when cells divide and reassembles once that process is completed. It reforms around clusters of RNA genes termed nucleolus organizing regions (NORs). Like the other membrane-less compartments, the nucleolus is dynamic and able to alter its protein composition. For example, SR proteins may enter the nucleolus, undergo phosphorylation, and then move out to the nuclear speckles, another dynamic nuclear compartment, where pre-mRNA splicing occurs. The nucleolus participates in the cellular response to stress and helps orchestrate it. In response to energy-related stresses, it can halt energy-consuming replication and gene transcription by sequestering key elements of those processes. And, if situations arise where RNA biogenesis is disrupted so that the cell is no longer viable, it may signal cell death.

Yet another source of toxic species driven by expanded GGGGCC repeats in C9ORF72 can be brought on by the G-quadruplexes (and R-loops). Prematurely terminated transcription may occur in which abortive, less-than-full-length and potentially poisonous transcripts are produced (Fig. 11.6). There is increasing evidence that this process occurs resulting in entities that migrate into the nucleolus. Once there they bind to essential RNA processing proteins and cause their mislocalization. These actions generate nucleolar stress, increase the sensitive of the cells to

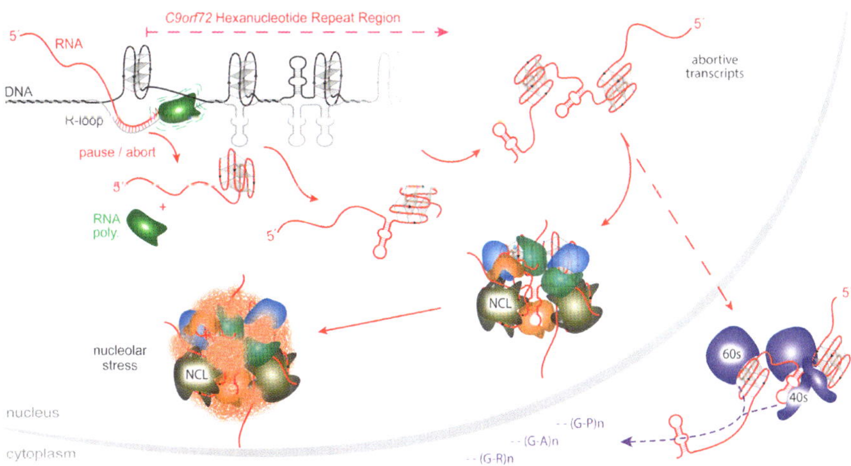

Fig. 11.6 Disruption of translation by R-loops and G-quadruplexes generates prematurely terminated translation products causing nucleolar stress (from Haeusler *Nature* 507: 195 © 2014 Reprinted by permission from Macmillan Publishers Ltd)

age-related and other stresses, and increase their likelihood of undergoing cell death. These entities may be present along with potentially toxic poly-dipeptides and act either individually or in concert with haploinsufficiency and RNA foci to harm cells.

The spectrum of FTLD/ALS disorders arising from the GGGGCC repeat expansions in the noncoding region of the C9ORF72 gene is referred to as c9FTD/ALS. In those instances where RAN translation products are present in neuronal inclusions, those products are termed c9RANTs. Several of the poly dipeptides have been singled out as being particularly dangerous and harm not only the nucleolus but also cause ER stress. For example, it has been found that the c9RANT poly(GA)$_{50}$ protein forms ubiquitin-positive, p62-positive cytoplasmic and nuclear inclusions. These peptides are aggregation prone and generate filamentous structures. They cause ER stress, inhibit the proteasome and stimulate caspase 3 activities. This is in addition to other low-complexity dipeptides that bind hydrogel components, and when introduced into cells migrate to the nucleus, bind nucleoli, disrupt RNA biogenesis, and prove to be toxic to their hosts.

11.11   Mutations in Progranulin Cause Frontotemporal Lobar Degeneration

Progranulin is a 593 amino acid protein containing 7.5 tandem repeats of a 12-cysteinyl granulin motif. These repeats are processed into 6 kDa *granulins* that function as pro-survival growth factors. The peptides are secreted into the extracellular spaces where they are taken up by neighboring cells. In 2006, loss-of-function mutations in the *PGRN* gene that encode progranulin were shown to be a cause of tau-negative FTLD. This protein is secreted by activated microglia, and is taken up by sortilin cell surface receptors on nearly neurons. It is then removed from the cell surface and transported to lysosomes. Its function in lysosomes is unknown, but it may be one involving autophagic clearance.

A variety of FTLD-causing mutations in the *PGRN* gene have been identified. As noted in Table 11.1 these FTLD instances are accompanied by TDP-43-positive inclusions. The primary mechanism underlying these disorders involves a loss-of-protein-function. The disease-causing *PGRN* mutations most often result in loss of their mRNA through NMD resulting in haploinsufficiency of the protein.

Another component of this disease axis is transmembrane protein 106 B (TMEM106B), a protein encoded by a gene at chromosome 7p21. TMEM106B is a genetic modifier of disease risk. The T185 single nucleotide polymorphism (SNP) results in its overexpression and produces an increased risk for developing mutated-progranulin-associated FTLD. In contrast, the S185 SNP reduces TMEM106B expression levels and acts in a protective manner. TMEM106B is found in lysosomes, and there is evidence for poor acidification of lysosomes and perturbations in progranulin function resulting from its overexpression. TMEM106B also acts as a genetic modifier of C9ORF72-associated FTLD where it has been found to increase the chance for survival, as well.

11.12   Mutations in CHMP2B Causes Failed Autophagic Clearance in ALS and FTD3

Frontotemporal dementia linked to chromosome 3 (FTD3) is a rare form of early-onset, autosomal dominant FTLD. It primarily affects neurons in the frontal lobe, and is characterized by ubiquitin-positive, p62-bearing inclusions. This form of FTLD has one known genetic cause—mutations in the gene encoding CHMP2B (charged multivesicular body protein 2B). This protein is a component of the ESCRT-III (endosomal sorting complex required for transport III) complex. Recall from the earlier discussion of extracellular prion spread in Chap. 7 that multivesicular bodies (MVBs) transport receptors and other cell membrane components to lysosomal compartments for degradation and to the Golgi apparatus for recycling back to the cell membrane. They are also needed for autophagy, with defects in this complex associated with a failure of autophagosome/lysosome fusion.

In 2005 and 2006, mutated forms of the *CHMP2B* gene were found to give rise to FTD3 and ALS, respectively. These disease-causing mutations result in C-terminal-truncated forms of the protein. These aberrant proteins disturb endosomal processing by the ESCRT-III complex; they lead to failed autophagy of cytoplasmic ubiquitinated proteins, and to the accumulation of autophagosomes.

11.13   Mutations in the MAPT Gene Are Responsible for Neurological Disorders Collectively Referred to as the Tauopathies

The last entry in Table 11.1 is the microtubule-associated protein tau encoded by the *MAPT* gene. This protein had been introduced earlier in Chap. 8 that dealt with Alzheimer's disease. Recall that the second major observation made by Alzheimer was the presence within diseased neurons of intracellular tangles composed of misfolded tau protein. Tau is natively unfolded and utilizes the addition and subtraction of negative charges supplied through phosphorylation to attach and detach from the microtubules. The tau proteins involved in AD are hyperphosphorylated, modified, and misfolded. In addition, their N- and/or C-terminals are frequently truncated resulting in fragments that are highly aggregation-prone.

In the introductory discussion of Chap. 8, tau was viewed as a major downstream effector of Aβ pathology. However, tau by itself in its mutated, alternatively processed, modified, and altered forms is a potent agent of neurodegeneration. It generates a variety of neurological disorders known as the *tauopathies*. The most prominent of these are:

- Corticobasal degeneration (CBD)
- Progressive supranuclear palsy (PSP)
- Pick's disease (PiD)

- Argyrophilic grain disease (AGD)
- Frontotemporal dementia and parkinsonism linked to chromosome 17 (FTDP-17)
- Chronic traumatic encephalopathy (CTE)

This list includes Pick's disease first described in 1892 by Pick and for whom the disease is named. It was studied further in the early 1900s by Fischer and then by Alzheimer. CBD and PSP were described clinically in the 1960s and those findings were followed in the 1980s and 1990s by the identification of tau-bearing pathological lesions in CBD and PSP. The connection between FTLD-17 and tau was clearly established in the 1994–1998 time period with the linking of the disease to mutations in *MAPT* gene located at chromosome 17q21–23. In contrast to Alzheimer's disease, in the tauopathies, intracellular tangles are found in both glial cells and neurons. The syndromes collected under the rubric "tauopathies" are heterogeneous in their clinical manifestations and affect several regions of the brain. They are characterized by dementia and/or motor defects, and by neuronal loss, gliosis, and, depending on the specific cell type, plaques, inclusions, or neurofibrillary tangles composed of one of several types of tau filaments.

The leading player in these disorders, the microtubule-associated protein tau, was discovered and named in 1975 by Marc Kirschner. Its unusual biophysical/biochemical properties and its normal roles during development and in adult life were uncovered over the next 15 years. Those studies were accompanied by the beginning explorations of tau's pathological properties in disease situations. The current understanding of tau structure is summarized in Fig. 11.7. This figure shows the full-length 441 amino acid protein with its two major domains. The N-terminal domain does not bind microtubules. Instead, it projects outward from the microtubules and for that reason is given the appellation "projection domain". The C-terminal domain contains a number of repeats responsible for microtubule binding and assembly. For that reason the C-terminal domain is called the "assembly domain".

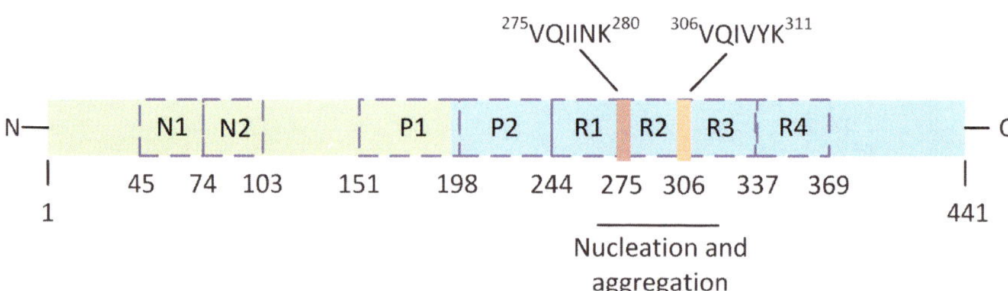

Fig. 11.7 Structure of the full-length 441 amino acid residue Tau protein. Tau consists of an N-terminal projection domain (shown in *green*) and a C-terminal microtubule assembly domain (depicted in *blue*) containing the repeat domains, R1–R4. The C-terminal region located at the beginning of R2 and extending to the beginning of R3 has a propensity for β-sheet structure, and is associated with nucleation and aggregation. P1 and P2 denote proline-rich stretches; N1 and N2 are the regions encoded by exons 2 and 3

Tau is an archetypic example of an intrinsically disordered and natively unfolded protein. These proteins were discussed in Sect. 4.10 where their main features were enumerated. These were: (1) lack of secondary structure and, in its place, an extended open configuration, and (2) an absence of bulky hydrophobic residues and, instead, a large complement of polar and charged residues. These properties endow the protein with flexibility, versatility and environmental responsiveness. Tau is strongly hydrophilic. It possesses 80 threonine or serine residues, 56 negatively charged residues, 58 positively charged residues, and eight aromatic residues. Its overall charge distribution is also highly asymmetric. In addition, the protein contains a number of proline-rich sequences in its central region that are subject to phosphorylation by proline-directed kinases.

The detailed structure of the tau protein depends in large measure on its posttranscriptional processing. In particular, it depends on the alternative splicing of exons 2, 3, and 10 of its pre-mRNA. Exon 10 contains one of four microtubule-binding domains, and absence or presence of that exon produces three-repeat (3R) or four-repeat (4R) tau proteins, respectively. Corticobasal degeneration and progressive supranuclear palsy are primarily 4R tau disorders, while Pick's disease is a 3R tauopathy, and an imbalance between tau isoforms containing 3 and 4 repeats leads to FTDP-17.

While the stages in the development of a tauopathy are not yet known to any degree of certainty, progress has been made. The protein is highly soluble and it must undergo substantive alterations to become aggregation-prone. Hyperphosphorylation certainly plays a role as been stated many times, but by itself is not sufficient to induce aggregation. It can, however, produce detachment from microtubules and promote soluble oligomer formation. Cleavage into fragments helps the formation of aggregates. The development of β-sheet secondary structure that seeds aggregation is clearly important. One possible inducer of this conformation transition is the presence of polyanionic cofactors. Candidates for these actions include heparin, RNAs, and anionic lipid membranes. Another factor that comes into play in the tauopathies is tau aberrant acetylation. That posttranslational modification disrupts tau's ability to bind microtubules, inhibits its clearance, and promotes aggregation.

11.14 Tau Functions and Malfunctions

Tauopathies arise from misfolded, aggregated, overexpressed, and mislocalized tau. Aggregation is promoted by increased β-sheet content, but the large insoluble hyperphosphorylated and ubiquitinated aggregates and inclusions, the tangles, are not thought to be the causative agents of the disorders. Instead, the smaller soluble oligomeric species are believed to be the toxic agents. Early events that presage the emergence of a tauopathy include abnormal synaptic behavior accompanied by memory losses and activation of microglial responses. The development of these abnormalities coincides with the appearance of tau dimeric and soluble higher-order oligomeric species at synaptic sites. These events occur before tangles can be discerned.

Tau is a microtubule-associated protein. Microtubules are dynamic entities that help drive morphological changes that occur in response to synaptic activity in the postsynaptic sites. Among its activities are those that regulate dendritic spine changes central to synaptic plasticity, learning, and memory formation. These activities are coordinated with actions by the actin cytoskeleton, and an emerging view is that tau may mediate some of these plastic changes. This role places tau in the dendritic compartment in addition to its long-established positioning in axons. In disease situations, toxic oligomers collect at these sites.

In Alzheimer's disease, $A\beta_{1-42}$ peptides help drive aberrant, toxic behavior by tau oligomers that involve mislocalization and phosphorylation by the tyrosine kinase Fyn. The corresponding trigger(s) in the other tau-centric disorders await discovery. These are likely to involve one or more of the following: mutation-induced misfolding, mislocalization, cross-seeding and actions by cofactors (e.g., polyanionic species such as α-synuclein), improper phosphorylation and other posttranslational modifications (most notably, acetylation), proteolytic cleavage, overexpression, and failures in clearance through the UPS or autophagy.

Although it is long thought as residing exclusively within cells, tau like α-synuclein has recently been found in the extracellular spaces. Significantly, synaptic activity stimulates its release. Once in the extracellular spaces tau oligomers can diffuse to and be taken up by neighboring neurons and glial cells. Recent efforts to investigate this phenomenon have utilized oligomer-specific antibodies that target the extracellular deposits of oligomeric tau. These explorations show that synaptic function can be recovered by clearing the extracellular deposits.

11.15   Chronic Traumatic Encephalopathy Is a Tauopathy

Chronic traumatic encephalopathy is a progressive neurodegenerative disorder caused by mild repetitive traumatic brain injury (TBI), in which acceleration-deceleration forces are imparted to the brain in each episode. This disorder first came to public attention in 1927 when Osnato and Gilberti published the results of their study of the effects of concussion on the brain, and then a year later in 1928 Harrison Stanford Martland (1883–1954) introduced the term "punch drunk syndrome" to describe the effects of TBI on professional boxers. That lead to the appellation "dementia pugilistica" by Millspaugh in 1937 a term that has persisted until recently when it has been replaced by the more general term *chronic traumatic encephalopathy* first introduced by Macdonald Critchley (1900–1997) in 1949 and again in 1957. That term encompasses repetitive TBI occurring in violent sports such as football, wrestling, rugby, and ice hockey as well as boxing, and repetitive TBI suffered by military personnel.

Repetitive mild and severe TBI both recapitulate many of the theme and processes discussed previously and lead to dementia over time (Fig. 11.8). Significantly, both forms of TBI trigger the buildup of protein aggregates. In the case of repetitive mild TBI, the route involves the development of a tauopathy leading to the onset of CTE. More severe instances of TBI are a well-established risk factor for Alzheimer's

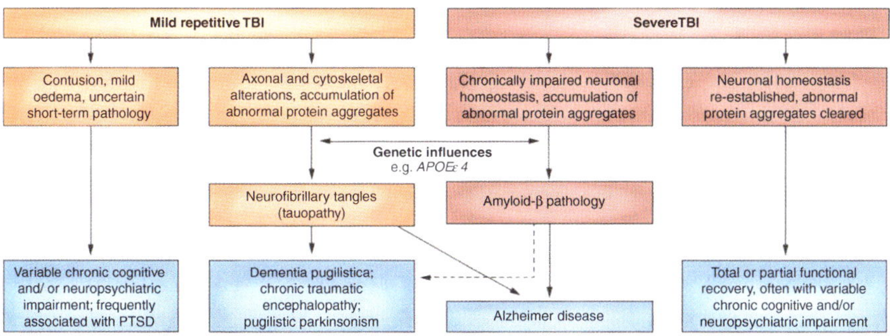

Fig. 11.8 Principal stages in mild repetitive and severe TBI (from DeKosky *Nat. Rev. Neurol.* 9: 192 © 2013 Reprinted by permission from Macmillan Publishers Ltd)

disease and that route involves a more profound loss of proteostasis. Alternatively, there may be a complete restoration of protein homeostasis and recovery from a severe injury. Differences and similarities between these two forms of TBI are summarized in Fig. 11.8.

In more detail, chronic traumatic encephalopathy is a tauopathy with its own set of distinctive features. It is characterized clinically by irritability, impulsivity, aggression, depression and by short-term memory loss and suicidality. These signs usually begin 8–10 years after the initiating episodes of repetitive mild TBI. Over time dementia, gait and speech defects, and Parkinsonism develop. In CTE, neurofibrillary tangles begin to accumulate in the superficial layers (II and III). The distribution is patchy and eventually spreads throughout the cortex. In about half the cases of CTE, amyloid β plaques are present. However, CTE is distinct from AD; in outbreaks of the latter, the NFTs populate deeper layers V and VI and have a different appearance. In addition, TDP-43 inclusions are present in some cases and Lewy bodies in others. Individuals suffering from CTE may develop frontotemporal lobar dementia or Lewy body disease, and there is evidence for a later emergence of Alzheimer's disease (Fig. 11.8).

11.16   Prion-Like Spread of Tau Aggregates Has Been Observed

Tau exhibits considerable conformational variability. There are six isoforms; some are 3R isoforms and others are 4R types. Each of these isoforms can exist in one of several conformational states and give rise to a variety of filament structures. One of the enduring mysteries to the tauopathies has been the identity of the mechanisms underlying the specificity of each to a certain yet different region of the brain. That the biophysical properties of the tau filaments vary from tauopathy to tauopathy may provide an important clue and starting point in the search for an answer.

Tau aggregates possess prion-like properties. Like Aβ and α-synuclein disease-associated tau aggregates can spread from one site to another in the brain. For example, in AD affected brains, they can spread from an initial site to both neighboring and synaptically connected regions. These disease-associated tau aggregates (oligomers and fibrils) are self-propagating; they are capable of seeding the conformational conversion of normal to misfolded tau in recipient cells and, like prions, exhibit stain effects encoded through their differing conformational states.

The mechanism(s) making possible the spread of tau seeds from one cell to another is(are) unknown. Two possibilities, discussed earlier with regard to prions and α-synuclein, are exosomes and tunneling nanotubes. Under normal cellular conditions, considerable quantities of tau can be found in the extracellular spaces (as do many of the others central to neurodegeneration), and as just discussed neuronal activity may regulate its release.

One way that tau aggregates are subsequently taken up by recipient cells is through *macropinocytosis*. This is a non-selective form of endocytosis initiated from cell membrane ruffles. It is widely encountered in the immune system where it is used to capture antigens. In neurodegeneration, not only tau but also prions and SOD1 aggregates are taken up by lipid raft associated macropinocytosis. In the case of tau, it is facilitated by lysine-rich motifs that readily bind heparan sulfate proteoglycans on the cell surface. The involvement of HSPGs acting as a receptor for macropinocytosis is not limited to tau but rather extends to α-synuclein, as well.

11.17   Summary

Amyotrophic lateral sclerosis and frontotemporal lobar degeneration encompasses a broad spectrum of motor and cognitive disorders with each individual sufferer experiencing some combination of the two. The disorders are heterogeneous in their symptoms and depend on the individual's genetic makeup. As was the case for AD and PD, most instances of ALS are sporadic but some are inherited, and most but not all are conveyed in an autosomal dominant fashion. The disorders, like other forms of neurodegeneration, are characterized at the cellular level by the presence of intracellular inclusions. Some inclusions contain the antioxidant protein SOD1. Others contain the RNA-processing proteins TDP-43 or FUS, or they contain the microtubule-associated protein tau, or they contain the autophagic mediator p62.

Considerable progress has been made in identifying possible underlying causes of fALS. Several dozen proteins whose mutated forms are associated with fALS have been discovered to date. Some of these are well accepted as causal entities, while others remain under discussion. The best established of these were listed in Table 11.1. Many of the entries cause both ALS and FTLD and for that reason the two diseases can be through of as two distinct manifestations of a single multisystem pathology affecting several populations of neurons and glia. The primary focus of attention in the field has been on SOD1, TDP-43, FUS, C9orf72, and tau with accompanying studies of the others implicated in this spectrum of neurological disorders.

1. Mutations in the superoxide dismutase gene SOD1 were the first genetically linked cause of ALS and FTLD to be identified. The SOD1 protein resides primarily in the cytosol where it catalyzes the conversion of superoxide to molecular oxygen and hydrogen peroxide. Although the idea that loss of its enzymatic function due to mutations caused neurodegeneration made good sense, that idea failed on several fronts to find experimental support. Instead, it appears that SOD1's disease-causing capabilities must derive through a gain-of-toxic function.

   The two best established of these disease modes are glutamate excitotoxicity and mitochondrial dysfunction. Misfolded and aggregated SOD1 is not only found within neurons but also in the extracellular spaces where it can enter and activate astrocytes. In response, the astrocytes initiate a potent inflammatory response that potentiates damage to the neurons and curtails glutamate uptake. In neurons, misfolded and aggregated SOD1 collects in mitochondria where it causes toxic morphological and functional abnormalities.

2. Failure to maintain protein quality control underlies neurodegeneration. This term is a broad one, and includes activities ranging from cellular logistics and autophagy to RNA-processing. Malfunctions in transport were discussed in earlier chapters. This activity is especially important to neurons given their logistical challenges. Transport failures negatively impact: (1) autophagic clearance of misfolded and aggregated proteins; (2) maintenance of mitochondrial health, calcium homeostasis and localization of mitochondria at sites where they are needed, (3) delivery of mRNAs to distant sites for local translation, and (4) receptor/ligand internalization, cycling and turnover of proteins at synapses.

   RNA-processing has emerged in the last few years as a key activity that can lead to neurodegeneration when it goes awry. The involvement of RNA-processing comes about in several interconnected ways. First, it becomes important when neurons transcribe genes containing unstable repeat expansions in protein-coding or non-protein-coding regions, and then processes the transcripts. Secondly, it comes into play when cellular stresses arise due to various causes including not only metabolic stresses but also aging-related decreases in protein quality control. These activities lead to halting of translation and the formation of dynamic membrane-less sub-organelles such as stress granules and P-bodies to handle the necessary logistics.

3. TDP-43 and FUS are RNA-processing proteins whose mutated forms give rise to ALS and FTLD. These proteins bind DNA, RNA, and proteins, and are involved in RNA-processing activities ranging from transcription to splicing to transport. Both proteins can inhibit transcription; TDP-43 preferentially binds single-stranded TG-rich DNA sequences in promoters, while FUS associates with the general transcription machinery. These two proteins bind thousands of mRNAs in their 3′ UTR and mediate RNA splicing. In addition, these proteins shuttle between the nucleus and cytoplasm, and when translation is halted they migrate together with other RNA-processing proteins into stress granules. They are also found in transport granules, another dynamic membrane-less structure that delivers mRNAs to synaptic sites for local translation. Lastly, these proteins are prominent components of the intracellular inclusions that accompany the most widespread forms of ALS and FTLD.

TDP-43 is a member of the hnRNP family, while FUS is a TET family member. These proteins have two properties important for the formation of dynamic structures such as stress granule, namely, low complexity and multivalency. Together with depletion attraction these properties facilitate phase transitions from independent soluble phases to liquid droplet phases in which proteins that have to work together for a certain amount of time are brought into close proximity to one another. Those three properties underlie the formation of dynamic sub-organelles such as stress granules and nucleoli, and by that means organize the crowded cytoplasm and nucleoplasm.

4. Unstable GGGGCC repeat expansions in C9ORF72 are the leading cause of fALS and FTLD. Unstable repeat expansions were a main topic of discussion in the last chapter. When present in protein-coding regions they cause Huntington's disease and other polyQ disorders. They also cause a variety of neurological disorders when present in non-protein-coding regions. The unstable hexanucleotide repeat expansion in the C9ORF72 gene is situated in the intronic region between exons 1a and 1b.

Like the other repeat expansions the GGGGCC expansion generates aberrant structures in DNA, RNA, and DNA–RNA hybrids. These include G-quadruplex generated by Hoogsteen-type hydrogen bonding, an alternative to Watson–Crick base pairing. The G-quadruplexes and R-loops cause replication fork stalling and other difficulties resulting in malfunction and generation of toxic proteins and RNAs. In the case of C9ORF72, they are responsible for:

- **Haploinsufficiency**—Loss of function resulting from reduced C9orf72 mRNA expression;
- **Toxic RNA (nuclear foci)**—Repeat mediated gain-of-function toxicity from sequestration of RNA binding and processing proteins in nuclear foci;
- **Toxic peptides (RAN translationproducts)**—Repeat mediated generation of toxic dipeptides from expanded GGGGCC repeats;
- **Aborted transcripts**—R-loops and G-quadruplexes cause transcription to be prematurely terminated resulting in formation of aberrant proteins.

5. Mutations in the MAPT gene that encodes the microtubule-associated tau cause neurodegenerative diseases collectively referred to the tauopathies. These disorders are characterized by the presence of one of several different types of tau-bearing filamentous aggregates in neurons and glia in the affected regions of the brain. Chronic traumatic encephalopathy is a tauopathy that has attracted considerable public attention in the last few years due to its development from repetitive mild traumatic brain injuries such as those suffered by athletes in violent sports and soldiers in war zones. This disorder is characterized by the presence of tau-positive and TDP-43-positive intracellular inclusions. Over time the disorder can lead to Parkinsonism and to Alzheimer's disease.

Tau is a natively unstructured low-complexity protein with an open structure enabling it to serve as rails for the microtubules. Like other natively unstructured and intrinsically disordered proteins it is highly flexible; it is sensitive to its environment, and populates an ensemble of states and substates. The mutated tau proteins involved in the tauopathies are misfolded, hyperphosphorylated, and their N- and/or C-terminals are frequently truncated resulting in fragments that are highly aggregation-prone. A key step in becoming aggregation-competent is the occurrence of a conformational change in the C-terminal microtubule assembly domain to a beta-sheet-rich structure. In its disease-promoting conformations, tau aggregates into variety of oligomeric and fibrillar forms, and propagates like prions from cell to cell and from one region of the brain to another.

Prion-like domains are archetypic low complexity domains. They are primarily composed of glycine, asparagine, glutamine, tyrosine and serine. The structures formed by these amino acids can (1) switch rapidly between non-transmissible and transmissible conformations, the latter giving rise to different prion strains, and (2) template the conversion of non-disease-causing to disease-promoting forms in their current or new host cells.

As is the case in all good science the dramatic discoveries of the last few years have given rise to new and perhaps better sets of questions awaiting answer. One of these is how broad is phenomenon of templated and seeded conversion and stabilization of alternative protein forms? Are tau and α-synuclein and TDP-43 and the few others found to-date the only ones or are there a host of others awaiting discovery? Another question is, given that all cells carry out RNA processing, why do the unstable GGGGCC repeats produce just ALS and FTLD (or do they)? These are just two of a great number of questions that may be asked.

Bibliography

Bruijn, L. I., Becher, M. W., Lee, M. K., Anderson, K. L., Jenkins, N. A., Copeland, N. G., et al. (1997). ALS-linked SOD1 mutant G85R mediates damage to astrocytes and promotes rapidly progressive disease with SOD1-containing inclusions. *Neuron, 18*, 327–338.

Gerschman, R., Gilbert, D. L., Nye, S. W., Dwyer, P., & Fenn, W. O. (1954). Oxygen poisoning and x-irradiation. A mechanism in common. *Science, 119*, 623–626.

Harman, D. (1956). Aging: A theory based on free radical and radiation chemistry. *Journal of Gerontology, 11*, 298–300.

McCord, J. M., & Fridovich, I. (1969). Superoxide dismutase: An enzymic function for erythrocuprein (hemocuprein). *The Journal of Biological Chemistry, 244*, 6049–6055.

Rosen, D. R., Siddique, T., Patterson, D., Figlewicz, D. A., Sapp, P., Hentati, A., et al. (1993). Mutations in Cu/Zn superoxide dismutase gene are associated with familial amyotrophic lateral sclerosis. *Nature, 362*, 59–62.

SOD1 Mutations in ALS

Bosco, D. A., Morfini, G., Karabacak, N. M., Song, Y., Gros-Louis, F., Pasinelli, P., et al. (2010). Wild-type and mutant SOD1 share an aberrant conformation and a common pathogenic pathway in ALS. *Nature Neuroscience, 13*, 1396–1403.

Bruijn, L. I., Houseweart, M. K., Kato, S., Anderson, K. L., Anderson, S. D., Ohama, E., et al. (1998). Aggregation and motor neuron toxicity of an ALS-linked SOD1 mutant independent from wild-type SOD1. *Science, 281*, 1851–1854.

Deng, H. X., Hentati, A., Tainer, J. A., Iqbal, Z., Cayabyab, A., Hung, W. Y., et al. (1993). Amyotrophic lateral sclerosis and structural defects in Cu, Zn superoxide dismutase. *Science, 261*, 1047–1051.

Perry, J. J. P., Shin, D. S., Getzoff, E. D., & Tainer, J. A. (2010). The structural biochemistry of the superoxide dismutases. *Biochimica et Biophysica Acta, 1804*, 245–262.

Rakhit, R., Crow, J. P., Lepock, J. R., Kondejewski, L. H., Cashman, N. R., & Chakrabartty, A. (2004). Monomeric Cu, Zn-superoxide dismutase is a common misfolding intermediate in the oxidative models of sporadic and familial amyotrophic lateral sclerosis. *The Journal of Biological Chemistry, 279*, 15499–15504.

mSOD1 Causes ALS

Bisland, L. G., Sahai, E., Kelly, G., Golding, M., Greensmith, L., & Schiavo, G. (2010). Deficits in axonal transport precede ALS symptoms *in vivo. Proceedings of the National Academy of Sciences of the United States of America, 107*, 20523–20528.

Boillée, S., Vande Velde, C., & Cleveland, D. W. (2006). ALS: A disease of motor neurons and their nonneuronal neighbors. *Neuron, 52*, 39–59.

Dadon-Nachum, M., Melamed, E., & Offen, D. (2011). The "dying back" phenomenon of motor neurons in ALS. *Journal of Molecular Neuroscience, 43*, 470–477.

Damiano, M., Starkov, A. A., Petri, S., Kipiani, K., Kiaei, M., Mattiazzi, M., et al. (2006). Neural mitochondrial Ca capacity impairment precedes the onset of motor symptoms in G93A Cu/Zn-superoxide dismutase mutant mice. *Journal of Neurochemistry, 96*, 1349–1361.

Deng, H. X., Shi, Y., Furukawa, Y., Zhai, H., Fu, R., Liu, E., et al. (2006). Conversion to the amyotrophic lateral sclerosis phenotype is associated with intermolecular linked insoluble aggregates of SOD1 in mitochondria. *Proceedings of the National Academy of Sciences of the United States of America, 103*, 7142–7147.

Fischer, L. R., Culver, D. G., Tennant, P., Davis, A. A., Wang, M., Castellano-Sanchez, A., et al. (2004). Amyotrophic lateral sclerosis is a distal axonopathy: Evidence in mice and man. *Experimental Neurology, 185*, 232–240.

Grad, L. I., Guest, W. C., Yanai, A., Pokrishevsky, E., O'Neill, M. A., Gibbs, E., et al. (2011). Intermolecular transmission of superoxide dismutase 1 misfolding in living cells. *Proceedings of the National Academy of Sciences of the United States of America, 108*, 16398–16403.

Haidet-Phillips, A. M., Hester, M. E., Miranda, C. J., Meyer, K., Braun, L., Frakes, A., et al. (2011). Astrocytes from familial and sporadic ALS patients are toxic to motor neurons. *Nature Biotechnology, 29*, 824–828.

Howland, D. S., Liu, J., She, Y., Goad, B., Maragakis, N. J., Kim, B., et al. (2002). Focal loss of the glutamate transporter EAAT2 in a transgenic rat model of SOD1 mutant-mediated amyotrophic lateral sclerosis (ALS). *Proceedings of the National Academy of Sciences of the United States of America, 99*, 1604–1609.

Kong, J., & Xu, Z. (1998). Massive mitochondrial degeneration in motor neurons triggers the onset of amyotrophic lateral sclerosis in mice expressing a mutant SOD1. *The Journal of Neuroscience, 18*, 3241–3250.

Magrané, J., Sahawneh, M. A., Przedborski, S., Estévez, Á. G., & Manfredi, G. (2012). Mitochondrial dynamics and bioenergetics dysfunction is associated with synaptic alterations in mutant SOD1 motor neurons. *The Journal of Neuroscience, 32*, 229–242.

Meyer, K., Ferraiuolo, L., Miranda, C. J., Likhite, S., McElroy, S., Renusch, S., et al. (2014). Direct conversion of patient fibroblasts demonstrates non-cell autonomous toxicity of astrocytes to motor neurons in familial and sporadic ALS. *Proceedings of the National Academy of Sciences of the United States of America, 111*, 829–832.

Münch, C., O'Brien, J., & Bertolotti, A. (2011). Prion-like propagation of mutant superoxide dismutase 1 misfolding in neuronal cells. *Proceedings of the National Academy of Sciences of the United States of America, 108*, 3548–3553.

Rothstein, J. D., Dykes-Hoberg, M., Pardo, C. A., Bristol, L. A., Jin, L., Kunci, R. W., et al. (1996). Knockout of glutamate transporters reveals a major role for astroglial transport in excitotoxicity and clearance of glutamate. *Neuron, 16*, 675–686.

Urushitani, M., Ezzi, S. A., & Julien, J. P. (2007). Therapeutic effects of immunization with mutant superoxide dismutase in mice models of amyotrophic lateral sclerosis. *Proceedings of the National Academy of Sciences of the United States of America, 104*, 2495–2500.

Wong, P. C., Pardo, C. A., Borchelt, D. R., Lee, M. K., Copeland, N. G., Jenkins, N. A., et al. (1995). An adverse property of a familial ALS-linked SOD1 mutation causes motor neuron disease characterized by vacuolar degeneration of mitochondria. *Neuron, 14*, 1105–1116.

Yamanaka, K., Chun, S. J., Boilee, S., Fujimori-Tonou, N., Yamashita, H., Gutmann, D. H., et al. (2008). Astrocytes as determinants of disease progression in inherited amyotrophic lateral sclerosis. *Nature Neuroscience, 11*, 251–253.

Yang, Y., Gozen, O., Watkins, A., Lorenzini, I., Lepore, A., Gao, Y., et al. (2009). Presynaptic regulation of astroglial excitatory neurotransmitter transporter GLT1. *Neuron, 61*, 880–894.

TDP-43 and FUS Mutations in ALS and FTLD

Arai, T., Hasegawa, M., Akiyama, H., Ikeda, K., Nonaka, T., Mori, H., et al. (2006). TDP-43 is a component of ubiquitin-positive tau-negative inclusions in frontotemporal lobar degeneration and amyotrophic lateral sclerosis. *Biochemical and Biophysical Research Communications, 351*, 602–611.

Barmada, S. J., Skibinski, G., Korb, E., Rao, E. J., Wu, J. Y., & Finkbeiner, S. (2010). Cytoplasmic mislocalization of TDP-43 is toxic to neurons and enhanced by a mutation associated with familial amyotrophic lateral sclerosis. *The Journal of Neuroscience, 30*, 639–649.

Kwiatkowski, T. J., Jr., Bosco, D. A., LeClerc, A. L., Tamrazian, E., Vanderburg, C. R., Russ, C., et al. (2009). Mutations in the FUS/TLS gene on chromosome 16 cause familial amyotrophic lateral sclerosis. *Science, 323*, 1205–1208.

Neumann, M., Sampathu, D. M., Kwong, L. K., Truax, A. C., Micsenyi, M. C., Chou, T. T., et al. (2006b). Ubiquitinated TDP-43 in frontotemporal lobar degeneration and amyotrophic lateral sclerosis. *Science, 314*, 130–133.

Sreedharan, J., Blair, I. P., Tripathi, V. B., Hu, X., Vance, C., Rogelj, B., et al. (2008). TDP-43 mutations in familial and sporadic amyotrophic lateral sclerosis. *Science, 319*, 1668–1672.

Vance, C., Rojelj, B., Hortobágyi, T., De Vos, K. J., Nishimura, A. L., Sreedharan, J., et al. (2009). Mutations in FUS, and RNA processing protein, cause familial amyotrophic lateral sclerosis type 6. *Science, 323*, 1208–1211.

TDP-43 Localization, Aggregation and Prion-Like Spread

Alami, N. H., Smith, R. B., Carrasco, M. A., Williams, L. A., Winborn, C. S., Han, S. S. W., et al. (2014). Axonal transport of TDP-43 mRNA granules is impaired by ALS-causing mutations. *Neuron, 81*, 536–543.

Brettschneider, J., Del Tredici, K., Irwin, D. J., Grossman, M., Robinson, J. L., Toledo, J. B., et al. (2014). Sequential distribution of pTDP-43 pathology in behavioral variant frontotemporal dementia (bvFTD). *Acta Neuropathologica, 127*, 423–439.

Furukawa, Y., Kaneko, K., Watanabe, S., Yamanaka, K., & Nukina, N. (2011). A seeding reaction recapitulates intracellular formation of sarkosyl-insoluble transactivation response element (TAR) DNA-binding protein-43 inclusions. *The Journal of Biological Chemistry, 286*, 18664–18672.

Hasegawa, M., Arai, T., Nonaka, T., Kametani, F., Yoshida, M., Hashizume, Y., et al. (2008). Phosphorylated TDP-43 in frontotemporal lobar degeneration and ALS. *Annals of Neurology, 64*, 60–70.

Igaz, L. M., Kwong, L. K., Lee, E. B., Chen-Plotkin, A., Swanson, E., Unger, T., et al. (2011). Dysregulation of the ALS-associated gene TDP-43 leads to neuronal death and degeneration in mice. *The Journal of Clinical Investigation, 121*, 726–738.

Josephs, K. A., Murray, M. E., Whitwell, J. L., Parisi, J. E., Petrucelli, L., Jack, C. R., Jr., et al. (2014). Staging TDP-43 pathology in Alzheimer's disease. *Acta Neuropathologica, 127*, 441–450.

Nonaka, T., Masude-Suzukake, M., Arai, T., Hasegawa, Y., Akatsu, H., Obi, T., et al. (2013). Prion-like properties of pathological TDP-43 aggregates from diseased brains. *Cell Reports, 4*, 124–134.

Winton, M. J., Igaz, L. M., Wong, M. M., Kwong, L. K., Trojanowski, J. Q., & Lee, V. M. Y. (2008). Disturbance of nuclear and cytoplasmic TAR DNA-binding protein (TDP-43) induces disease-like redistribution, sequestration, and aggregate formation. *The Journal of Biological Chemistry, 283*, 13302–13309.

RNA-Processing by TDP-43 and FUS

Ayala, Y. M., De Conti, L., Avendaño-Vázquez, S. E., Dhir, A., Romano, M., D'Ambrogio, A., et al. (2011). TDP-43 regulates its mRNA levels through a negative feedback loop. *The EMBO Journal, 30*, 277–288.

Buratti, E., Dörk, T., Zuccato, E., Pagani, F., Romano, M., & Baralle, F. E. (2001). Nuclear factor TDP-43 and SR proteins promote in vitro and in vivo CFTR exon 9 skipping. *The EMBO Journal, 20*, 1774–1784.

Han, S. P., Tang, Y. H., & Smith, R. (2010). Functional diversity of the hnRNPs: Past, present and perspectives. *The Biochemical Journal, 430*, 379–392.

Lagier-Tourenne, C., Polymenidou, M., Hutt, K. R., Vu, A. Q., Baughn, M., Huelga, S. C., et al. (2012). Divergent roles of ALS-linked proteins FUS/TLS and TDP-43 intersect in processing long pre-mRNAs. *Nature Neuroscience, 15*, 1488–1497.

Liu-Yesucevitz, L., Bassell, G. J., Gitler, A. D., Hart, A. C., Klann, E., Richter, J. D., et al. (2011). Local RNA translation at the synapse and in disease. *The Journal of Neuroscience, 31*, 16086–16093.

Long, J. C., & Caceres, J. F. (2009). The SR protein family of splicing factors: Master regulators of gene expression. *The Biochemical Journal, 417*, 15–27.

Polymenidou, M., Lagier-Tourenne, C., Hutt, K. R., Huelga, S. C., Moran, J., Liang, T. Y., et al. (2011). Long pre-mRNA depletion and RNA missplicing contribute to neuronal vulnerability from loss of TDP-43. *Nature Neuroscience, 14*, 459–468.

Tan, A. Y., & Manley, J. L. (2009). The TET family of proteins: Functions and roles in disease. *Journal of Molecular Cell Biology, 1*, 82–92.

Tollervey, J. R., Curk, T., Rogeli, B., Briese, M., Cereda, M., Kayikci, M., et al. (2011). Characterizing the RNA targets and position-dependent splicing regulation by TDP-43. *Nature Neuroscience, 14*, 452–458.

Stress Granules and Processing Bodies

Buchan, J. R., & Parker, R. (2009). Eukaryotic stress granules: The ins and outs of translation. *Molecular Cell, 36*, 932–941.

Li, Y. R., King, O. D., Shorter, J., & Gitler, A. D. (2013). Stress granules as crucibles of ALS pathogenesis. *The Journal of Cell Biology, 201*, 361–372.

Parker, R., & Sheth, U. (2007). P bodies and the control of mRNA translation and degradation. *Molecular Cell, 25*, 635–646.

Angiogenin and Ataxin-2

Elden, A. C., Kim, H. J., Hart, M. P., Chen-Plotkin, A., Johnson, B. S., Fang, X., et al. (2010). Ataxin-2 intermediate-length polyglutamine expansions are associated with increased risk for ALS. *Nature, 466*, 1069–1075.

Greenway, M. J., Anderson, P. M., Russ, C., Ennis, S., Cashman, S., Donaghy, C., et al. (2006). ANG mutations segregate with familial and 'sporadic' amyotrophic lateral sclerosis. *Nature Genetics, 38*, 411–413.

Ivanov, P., Emara, M. M., Villen, J., Gygi, S. P., & Anderson, P. (2011). Angiogenin-induced tRNA fragments inhibit translation initiation. *Molecular Cell, 43*, 613–623.

Ross, O. A., Rutherford, N. J., Baker, M., Soto-Ortolaza, A. I., Carrasquillo, M. M., DeJesus-Hernandez, M., et al. (2011). Ataxin-2 repeat-length variation and neurodegeneration. *Human Molecular Genetics, 20*, 3207–3212.

Van Damme, J. H., Veldink, J. H., van Blitterswijk, M., Corveleyn, A., van Vught, P. W., Thijs, V., et al. (2011). Expanded ATXN2 CAG repeat size in ALS identifies genetic overlap between ALS and SCA2. *Neurology, 76*, 2066–2072.

Yamasaki, S., Ivanov, P., Hu, G. F., & Anderson, P. (2009). Angiogenin cleaves tRNA and promotes stress-induced translational repression. *The Journal of Cell Biology, 185*, 35–42.

Phase Transitions from Multivalency and Low Complexity/Prion-Like Motifs

Asakura, S., & Oosawa, F. (1954). On interaction between two bodies immersed in a solution of macromolecules. *The Journal of Chemical Physics, 22*, 1255–1256.

Brangwynne, C. P., Eckmann, C. R., Courson, D. S., Rybarska, A., Hoege, C., Gharakhani, J., et al. (2009). Germline P granules are liquid droplets that localize by controlled dissolution/condensation. *Science, 324*, 1729–1732.

Couthouis, J., Hart, M. P., Shorter, J., DeJesus-Hernandez, M., Erion, R., Oristano, R., et al. (2011). A yeast functional screen predicts new candidate ALS disease genes. *Proceedings of the National Academy of Sciences of the United States of America, 108*, 20881–20890.

Crocker, J. C., Matteo, J. A., Dinsmore, A. D., & Yodh, A. G. (1999). Entropic attraction and repulsion in binary colloids probed with a line optical tweezer. *Physical Review Letters, 82*, 4352–4355.

Gilks, N., Kedersha, N., Ayodele, M., Shen, L., Stoecklin, G., Dember, L. M., et al. (2004). Stress granule assembly is mediated by prion-like aggregation of TIA-1. *Molecular Biology of the Cell, 15*, 5383–5398.

Goldschmidt, L., Teng, P. K., Riek, R., & Eisenberg, D. (2010). Identifying the amylome, proteins capable of forming amyloid-like fibrils. *Proceedings of the National Academy of Sciences of the United States of America, 107*, 3487–3492.

Han, T. W., Kato, M., Xie, S., Wu, L. C., Mirzaei, H., Pei, J., et al. (2012). Cell-free formation of RNA granules: Bound RNAs identify features and components of cellular assemblies. *Cell, 149*, 768–778.

Kato, M., Han, T. W., Xie, S., Shi, K., Du, X., Wu, L. C., et al. (2012). Cell-free formation of RNA granules: Low complexity sequence domains form dynamic fibers within hydrogels. *Cell, 149*, 753–767.

King, O. D., Gitler, A. D., & Shorter, J. (2012). The tip of the iceberg: RNA-binding proteins with prion-like domains in neurodegenerative disease. *Brain Research, 1462*, 61–80.

Li, P., Banjade, S., Cheng, H. C., Kim, S., Chen, B., Guo, L., et al. (2012). Phase transitions in the assembly of multivalent signaling proteins. *Nature, 483*, 336–340.

Wilson, E. B. (1899). The structure of protoplasm. *Science, 10*, 33–45.

Stress Granule Clearance by Autophagy

Buchan, J. R., Kolaitis, R. M., Taylor, J. P., & Parker, R. (2013). Eukaryotic stress granules are cleared by granulophagy and Cdc48/VCP function. *Cell, 153*, 1461–1474.

Deng, H. X., Chen, W., Hong, S. T., Boycott, K. M., Gorrie, G. H., Siddique, N., et al. (2011). Mutations in UBQLN2 cause dominant X-linked juvenile and adult onset ALS and ALS/dementia. *Nature, 477*, 211–215.

Johnson, J. O., Mandrioli, J., Benatar, M., Abramzon, Y., Van Deerlin, V. M., Trojanowski, J. O., et al. (2010b). Exome sequencing reveals VCP mutations as a cause of familial ALS. *Neuron, 68*, 857–864.

Komatsu, M., & Ichimura, Y. (2010). Physiological significance of selective degradation of p62 by autophagy. *FEBS Letters, 584*, 1374–1378.

Maruyama, H., Morino, H., Ito, H., Izumi, Y., Kato, H., Watanabe, Y., et al. (2010). Mutations in optineurin in amyotrophic lateral sclerosis. *Nature, 465*, 223–226.

Wild, P., Farhan, H., McEwan, D. G., Wagner, S., Rogov, V. V., Brady, N. R., et al. (2011). Phosphorylation of the autophagy receptor optineurin restricts Salmonella growth. *Science, 333*, 228–233.

Mutations in C9ORF72 Cause ALS and FTLD

DeJesus-Hernandez, M., Mackenzie, I. R., Boeve, B. F., Boxer, A. L., Baker, M., Rutherford, N. J., et al. (2011b). Expanded GGGGCC hexanucleotide repeat in noncoding region of C9ORF72 causes chromosome 9p-linked FTD and ALS. *Neuron, 72*, 245–256.

Donnelly, C. J., Zhang, P. W., Pham, J. T., Haeusler, A. R., Mistry, N. A., Vidensky, S., et al. (2013a). RNA toxicity from ALS/FTD C9ORF72 is mitigated by antisense intervention. *Neuron, 80*, 415–428.

Fratta, P., Mizielinska, S., Nicoll, A. J., Zioh, M., Fisher, E. M. C., Parkinson, G., et al. (2012). C9orf72 hexanucleotide repeat associated with amyotrophic lateral sclerosis and frontotemporal dementia form RNA G-quadruplexes. *Scientific Reports, 2*, 1016.

Renton, A. E., Majounie, E., Waite, A., Simón-Sánchez, J., Rollinson, S., Gibbs, J. R., et al. (2011b). A hexanucleotide repeat expansion in C9ORF72 is the cause of chromosome 9p21-linked ALS-FTD. *Neuron, 72*, 257–268.

G-Quadruplexes

Bang, I. (1910). Untersuchungen über die Guanylsäure. *Biochemische Zeitschrift, 26*, 293–311.

Gellert, M., Lipsett, M. N., & Davies, D. R. (1962). Helix formation by guanylic acid. *Proceedings of the National Academy of Sciences of the United States of America, 48*, 2013–2018.

Hoogsteen, K. (1963). The crystal and molecular structure of a hydrogen-bonded complex between 1-methylthymine and 9-methyladenine. *Acta Crystallographica, 16*, 907–916.

Lam, E. N. I., Beraldi, D., Tannahil, D., & Balasubramanian, S. (2013). Quadruplex structures are stable and detectable in human genomic DNA. *Nature Communications, 4*, 1796.

Nikolova, E. N., Kim, E., Wise, A. A., O'Brien, P. J., Andricioaei, I., & Al-Hashimi, H. M. (2011). Transient Hoogsteen base pairs in canonical duplex DNA. *Nature, 470*, 498–502.

Sundquist, W. I., & Klug, A. (1989). Telomeric DNA dimerizes by formation of guanine tetrads between hairpin loops. *Nature, 342*, 825–829.

Williamson, J. R., Raghuraman, M. K., & Cech, T. R. (1989). Monovalent cation-induced structure of telomeric DNA: The G-quartet model. *Cell, 59*, 871–880.

C9orf72 and Repeat-Associated Non-ATG (RAN) Translation

Ash, P. E. A., Bieniek, K. F., Gendron, T. F., Caulfield, T., Lin, W. L., DeJesus-Hernandez, M., et al. (2013). Unconventional translation of C9ORF72 GGGGCC expansion generates insoluble polypeptides specific to c9FTD/ALS. *Neuron, 77*, 639–646.

Boulon, S., Westman, B. J., Hutten, S., Boisvert, F. M., & Lamond, A. I. (2010). The nucleolus under stress. *Molecular Cell, 40*, 216–227.

Donnelly, C. J., Zhang, P. W., Pham, J. T., Haeusler, A. R., Mistry, N. A., Vidensky, S., et al. (2013b). RNA toxicity from ALS/FTD C9ORF72 expansion in mitigated by antisense intervention. *Neuron, 80*, 415–428.

Haeusler, A. R., Donnelly, C. J., Periz, G., Simko, E. A. J., Shaw, P. G., Kim, M. S., et al. (2014). C9orf72 nucleotide repeat structures initiate molecular cascades of disease. *Nature, 507*, 195–200.

Kwon, I., Xiang, S., Kato, M., Wu, L., Theodoropoulos, P., Wang, T., et al. (2014). Poly-dipeptides encoded by the C9orf72 repeats bind nucleoli, impeded RNA biogenesis and kill cells. *Science, 345*, 1139–1145.

Mizielinska, S., Grönke, S., Niccoli, T., Ridler, C. E., Clayton, E. L., Devoy, A., et al. (2014). C9orf72 repeat expansions cause neurodegeneration in Drosophila through arginine-rich proteins. *Science, 345*, 1192–1194.

Mori, K., Weng, S. M., Arzberger, T., May, S., Rentzsch, K., Kremmer, E., et al. (2013). The C9ofr72 GGGGCC repeat is translated into aggregating dipeptide-repeat proteins in FTD/ALS. *Science, 339*, 1335–1338.

Todd, P. K., Oh, S. Y., Krans, A., He, F., Selier, C., Frazer, M., et al. (2013). CGG repeat associated translation mediates neurodegeneration in fragile X tremor ataxia syndrome. *Neuron, 78*, 440–455.

Zu, T., Gibbens, B., Doty, N. S., Gomes-Pereira, M., Huguet, A., Stone, M. D., et al. (2011). Non-ATG-initiated translation directed by microsatellite expansions. *Proceedings of the National Academy of Sciences of the United States of America, 108*, 260–265.

Progranulin, TMEM106B and C9ORF72

Baker, M., Mackenzie, I. R., Pickering-Brown, S. M., Gass, J., Rademakers, R., Lindholm, C., et al. (2006). Mutations in progranulin cause taunegative frontotemporal dementia linked to chromosome 17. *Nature, 442*, 916–919.

Cruts, M., Gijselinck, I., van der Zee, J., Engelborghs, S., Wils, H., Pirici, D., et al. (2006). Null mutations in progranulin cause ubiquitin-positive frontotemporal dementia linked to chromosome 17q21. *Nature, 442*, 920–924.

Hu, F., Padukkavidana, T., Vaegter, C. B., Brady, O. A., Zheng, Y., Mackenzie, I. R., et al. (2010). Sortilin-mediated endocytosis determines levels of the frontotemporal dementia protein, progranulin. *Neuron, 68*, 654–667.

Van Blitterswijk, M., Mullen, B., Nicholson, A. M., Bieniek, K. F., Heckman, M. G., Baker, M. C., et al. (2014). TMEM106B protects C9ORF72 expansion carriers against frontotemporal dementia. *Acta Neuropathologica, 127*, 397–406.

Van Deerlin, V. M., Sleiman, P. M. A., Martinez-Lage, M., Chen-Plotkin, A., Wang, L. S., Graff-Radford, N. R., et al. (2010). Common variants at 7p21 are associated with frontotemporal lobar degeneration with TDP-43 inclusions. *Nature Genetics, 42*, 234–239.

CHMP2B and Autophagic Clearance

Cox, L. E., Ferraiuolo, L., Goodall, E. F., Heath, P. R., Higginbottom, A., Mortiboys, H., et al. (2010). Mutations in CHMP2B in lower motor neuron predominant amyotrophic lateral sclerosis (ALS). *PLoS ONE, 5*, e9872.

Filimonenko, M., Stuffers, S., Raiborg, C., Yamamoto, A., Malerød, L., Fisher, E. M. C., et al. (2007). Functional multivesicular bodies are required for autophagic clearance of protein aggregates associated with neurodegenerative disease. *The Journal of Cell Biology, 179*, 485–500.

Lee, J. A., Belgneux, A., Ahmad, S. T., Young, S. G., & Gao, F. B. (2007). ESCRT-III dysfunction causes autophagosome accumulation and degeneration. *Current Biology, 17*, 1561–1567.

Parkinson, N., Ince, P. G., Smith, M. O., Highley, R., Skibinski, G., Andersen, P. M., et al. (2006). ALS phenotypes with mutations in CHMP2B (charged multivesicular body protein 2B). *Neurology, 67*, 1074–1077.

Skibinski, G., Parkinson, N. J., Brown, J. M., Chakrabarti, L., Lloyd, S. L., Hummerich, H., et al. (2005). Mutations in the endosomal ESCRTIIIcomplex subunit CHMP2B in frontotemporal dementia. *Nature Genetics, 37*, 806–808.

Tau and the Tauopathies

Buée, L., Bussière, T., Buée-Scherrer, V., Delacourte, A., & Hof, P. R. (2000). Tau protein isoforms, phosphorylation and role in neurodegenerative disorders. *Brain Research Reviews, 33*, 95–130.

Cleveland, D. W., Hwo, S. Y., & Kirschner, M. W. (1977). Physical and chemical properties of purified tau factor and the role of tau in microtubule assembly. *Journal of Molecular Biology, 116*, 227–247.

Goedert, M., Spillantini, M. G., Cairns, N. J., & Crowther, R. A. (1992). Tau proteins of Alzheimer's paired filaments: Abnormal phosphorylation of all six brain isoforms. *Neuron, 8*, 159–168.

Goedert, M., Spillantini, M. G., Jakes, R., Rutherford, D., & Crowther, R. A. (1989). Multiple isoforms of human microtubule-associated protein tau: Sequence and localization in neurofibrillary tangles of Alzheimer's disease. *Neuron, 3*, 519–526.

Kidd, M. (1963). Paired helical filaments in electron microscopy of Alzheimer's disease. *Nature, 197*, 192–193.

Mukrasch, M. D., Bibow, S., Korukottu, J., Jeganathan, S., Biernat, J., Griesinger, C., et al. (2009b). Structural polymorphism of 441-residue tau at single residue resolution. *PLoS Biology, 7*, e1000034.

Rebeiz, J. J., Kolodny, E. H., & Richardson, E. P., Jr. (1967). Corticodentatonigral degeneration with neuronal achromasia: A progressive disorder of late adult life. *Transactions of the American Neurological Association, 92*, 23–26.

Steele, J. C., Richardson, J. C., & Olzewski, J. (1964). Progressive supranuclear palsy: A heterogeneous degeneration involving brain stem, basal ganglia and cerebellum with vertical gaze and pseudobulbar palsy, nuchal dystonia and dementia. *Archives of Neurology, 10*, 333–359.

Weingarten, M. D., Lockwood, A. H., Hwo, S. Y., & Kirschner, M. W. (1975). A protein factor essential for microtubule assembly. *Proceedings of the National Academy of Sciences of the United States of America, 72*, 1858–1862.

Normal and Abnormal Function of Tau and Tau Oligomers

Berger, Z., Roder, H., Hanna, A., Carlson, A., Rangachari, V., Yue, M., et al. (2007). Accumulation of pathological tau species and memory loss in a conditional model of tauopathy. *The Journal of Neuroscience, 27*, 3650–3662.

Bhaskar, K., Konerth, M., Kokiko-Cochran, O. N., Cardona, A., Ransohoff, R. M., & Lamb, B. T. (2010). Regulation of tau pathology by the microglial fractalkine receptor. *Neuron, 68*, 19–31.

Castillo-Carranza, D. L., Sengupta, U., Guerrero-Muñoz, M. J., Lasagna-Reeves, C. A., Gerson, J. E., Singh, G., et al. (2014). Passive immunization with tau oligomer monoclonal antibody reverses tauopathy phenotypes without affecting hyperphosphorylated neurofibrillary tangles. *The Journal of Neuroscience, 34*, 4260–4272.

Cohen, T. J., Guo, J. L., Hurtado, D. E., Kwong, L. K., Mills, I. P., Trojanowski, J. Q. et al. (2011). The acetylation of tau inhibits its function and promotes pathological aggregation. *Nature Communications 2*, art. 252.

Giasson, B. I., Forman, M. S., Higuchi, M., Golbe, L. I., Graves, C. L., Kotzbauer, P. T., et al. (2003). Initiation and synergistic fibrillization of tau and alpha-synuclein. *Science, 300*, 636–640.

Guo, J. L., Covell, D. J., Daniels, J. P., Iba, M., Stieber, A., Zhang, B., et al. (2013). Distinct α-synuclein strains differentially promote tau inclusions in neurons. *Cell, 154*, 103–117.

Jaworsky, J., Kapitein, L. C., Gouveia, S. M., Dortland, B. R., Wulf, P. S., Grigoriev, I., et al. (2009). Dynamic microtubules regulate dendritic spine morphology and synaptic plasticity. *Neuron, 61*, 85–100.

Lasagna-Reeves, C. A., Castillo-Carranza, D. L., Sengupta, U., Clos, A. L., Jackson, G. R., & Kayed, R. (2011). Tau oligomers impair memory and induce synaptic and mitochondrial dysfunction in wild-type mice. *Molecular Neurodegeneration 6*: art. 39.

Ma, Q. L., Zuo, X., Yang, F., Ubeda, O. J., Gant, D. J., Alaverdyan, M., et al. (2012). Curcumin suppresses soluble tau dimers and corrects molecular chaperone, synaptic, and behavioral deficits in aged human tau transgenic mice. *The Journal of Biological Chemistry, 288*, 4056–4065.

Min, S. W., Cho, S. H., Zhou, Y., Schroeder, S., Haroutunian, V., Seeley, W. W., et al. (2010). Acetylation of tau inhibits its degradation and contributes to tauopathy. *Neuron, 67*, 953–956.

Pooler, A. M., Phillips, E. C., Lau, D. H., Noble, W., & Hanger, D. P. (2013). Physiological release of endogenous tau is stimulated by neuronal activity. *EMBO Reports, 14*, 389–394.

Sydow, A., Van der Jeugd, A., Zheng, F., Ahmed, T., Balschun, D., Petrova, O., et al. (2011). Tau-induced deficits in synaptic plasticity, learning, and memory are reversible in transgenic mice after switching off the toxic tau mutant. *The Journal of Neuroscience, 31*, 2511–2525.

Yamada, K., Holth, J. K., Liao, F., Stewart, F. R., Mahan, T. E., Jiang, H., et al. (2014). Neuronal activity regulates extracellular tau in vivo. *The Journal of Experimental Medicine, 211*, 387–393.

Yoshiyama, Y., Higuchi, M., Zhang, B., Huang, S. M., Iwata, N., Saido, T. C., et al. (2007). Synaptic loss and microglial activation precede tangles in a P301S tauopathy mouse model. *Neuron, 53*, 337–351.

Chronic Traumatic Encephalopathy

Blennow, K., Hardy, J., & Zetterberg, H. (2012). The neuropathology and neurobiology of traumatic brain injury. *Neuron, 76*, 886–899.

Critchley, M. (1957). Medical aspects of boxing, particularly from neurological standpoint. *British Medical Journal, 1*, 357–362.

DeKosky, S. T., Blennow, K., Ikonomovic, M. D., & Gandy, S. (2013). Acute and chronic traumatic encephalopathies: Pathogenesis and biomarkers. *Nature Reviews. Neurology, 9*, 192–200.

Johnson, V. E., Stewart, W., & Smith, D. H. (2013). Axonal pathology in traumatic brain injury. *Experimental Neurology, 246*, 35–43.

Martland, H. S. (1928). Punch Drunk. *Journal of the American Medical Association, 91*, 1103–1107.

McKee, A. C., Cantu, R. C., Nowinski, C. J., Hedley-Whyte, E. T., Gavett, B. E., Budson, A. E., et al. (2009). Chronic traumatic encephalopathy in athletes: Progressive tauopathy following repetitive head injury. *Journal of Neuropathology and Experimental Neurology, 68*, 709–735.

McKee, A. C., Stein, T. D., Nowinski, C. J., Stern, R. A., Daneshvar, D. H., Alvarez, V. E., et al. (2013). The spectrum of disease in chronic traumatic encephalopathy. *Brain, 136*, 43–64.

Millspaugh, J. A. (1937). Dementia pugilistica. *US Naval Medical Bulletin, 35*, 297–303.

Prion-Like Spread by Tau Fibrils

Clavaguera, F., Bolmont, T., Crowther, R. A., Abramowski, D., Frank, S., Probst, A., et al. (2009). Transmission and spreading of tauopathy in transgenic mouse brain. *Nature Cell Biology, 11*, 909–913.

De Calignon, A., Polydoro, M., Suárez-Calvet, M., William, C., Adamowicz, D. H., Kopeikina, K. J., et al. (2012). Propagation of tau pathology in a model of early Alzheimer's disease. *Neuron, 73*, 685–697.

Frost, B., Jacks, R. L., & Diamond, M. I. (2009). Propagation of tau misfolding from the outside to the inside of a cell. *The Journal of Biological Chemistry, 284*, 12845–12852.

Guo, J. L., & Lee, V. M. Y. (2011). Seeding of normal tau by pathogenic tau conformers drives pathogenesis of Alzheimer's-like tangles. *The Journal of Biological Chemistry, 286*, 15317–15331.

Holmes, B. B., DeVos, S. L., Kfouri, N., Li, M., Jacks, R., Yanamandra, K., et al. (2013). Heparan sulfate proteoglycans mediate internalization and propagation of specific proteopathic seeds. *Proceedings of the National Academy of Sciences of the United States of America, 110*, E3138–E3147.

Iba, M., Guo, J. L., McBride, J. D., Zhang, B., Trojanowski, J. Q., & Lee, V. M. Y. (2013). Synthetic tau fibrils mediate transmission of neurofibrillary tangles in a transgenic mouse model of Alzheimer's-like tauopathy. *The Journal of Neuroscience, 33*, 1024–1037.

Kfouri, N., Holmes, B. B., Jiang, H., Holtzman, D. M., & Diamond, M. I. (2012). Trans-cellular spread of tau aggregation by fibrillary species. *The Journal of Biological Chemistry, 287*, 19440–19451.

Lasagna-Reeves, C. A., Castillo-Carranza, D. L., Sengupta, U., Guerrero-Munoz, M. J., Kiritoshi, T., Neugebauer, V., et al. (2012). Alzheimer brainderived tau oligomers propagate pathology from endogenous tau. *Scientific Reports, 2*, 700.

Sanders, D. W., Kaufman, S. K., DeVos, S. L., Sharma, A. M., Mirbaha, H., Li, A., et al. (2014). Distinct tau prion strains propagate in cells and mice and define different tauopathies. *Neuron, 82*, 1271–1288.

Index

© Springer International Publishing Switzerland 2015

375

M. Beckerman, *Fundamentals of Neurodegeneration and Protein Misfolding Disorders*,
Biological and Medical Physics, Biomedical Engineering,
DOI 10.1007/978-3-319-22117-5